cpo science

Foundations of
**Physical
Science**

Tom Hsu, Ph.D.

Properties of matter

FIBER OPTICS

H_2

wind energy

simple machines

ACCELERATION

Third Edition

MOMENTUM

FRICTION

Author

Thomas C. Hsu, Ph.D., Applied Plasma Physics, Massachusetts Institute of Technology

Co-Authors

Scott Eddleman, M.Ed., Harvard University; **Mary Beth Abel**, M.S., Biological Sciences, University of Rhode Island; **Patsy Eldridge**, M.Ed., Tufts University; **Stacy Kissel**, M.Ed., Physics Education, Boston College

Editorial Team

Lynda Pennell – Senior Editor
Polly Crisman – Graphics Manager/Illustrator
Bruce Holloway – Senior Designer/Illustrator
Jesse Van Valkenburgh – Designer/Illustrator
Susan Gioia – Administrator
Sara Desharnais – Electronic Production Specialist

Contributing Writers and Editors

Laine Ives, Erik Benton, Laura Preston, Laurette Viteritti, Leslie Sheen, Shannon Donovan, Alyson Mazza, Christine Golden

Equipment Design and Materials

Thomas Narro – Senior Vice President
Danielle Dzurik – Industrial Designer
Kathryn Gavin – Purchasing and Quality Control Manager
Matthew Connor – Senior Quality Analyst

Assessment

Mary Ann Erickson, David H. Bliss

Reviewers

Mark Baker - Diamond, OH
Dr. Nicholas Benfaremo - South Portland, ME
Nancy Baker Cazan - SAMM Center, Massillon, OH
Ann Cleary - Medina, OH
Jean A. Cyders - Canton, OH
Deirde L. Davenport, William Miller - Beaumont, TX
Dr. Gregorio Garcia - Brownsville, Texas
Cort Gillen - Cypress, TX
Alan P. Gnospelius - San Antonio, Texas
Lisa Q, Gothard - East Canton, OH
Liz Gregory - Austin, TX
James Max Hollon - Indiana
William C. Huckeba - Irving, TX
Chrystal Brooke Johnson - Irving, TX
Daniel Klein - Navarre, OH
Kathleen Kuhn - Uniontown, OH
Jay Kurima - Fort Worth, Texas
Ed Laubacher - Uniontown, OH
Matt Leatherberry - Minerva, OH
John C. Lineweaver - Abilene, Texas

Anita K. Marshall - Cypress, TX
Maria Estela Martinez - Irving, TX
Chandra R. Maxey - Irving, TX
Alyson Mazza - Salem, NH
Thomas McArthur - North Canton, OH
Michael Mihalik - Uniontown, OH
William Miller - Ft Worth, TX
Kathryn Schommer Neuenschwander - Indiana
Stacey L. Nunley - Detroit, Michigan
Joel C. Palmer Ed. D. - Mesquite, Texas
Neil Parrot - Massillon, OH
Jeff Pickle - North Canton, OH
Emily O. Price - Waxahachie, Texas
Julie Randolph - Beaumont, TX
Steve Remenaric - Massillon, OH
Jay Don Steele - Floresville, TX
Ginger Torregrossa - Montgomery, TX
Timothy Totten - Navarre, OH
Melissa Vela - Brookline, NH
Michael Vela - Concord, MA

Elizabeth Volt - Cleveland, Ohio
David Warner - Massillon, OH
Dan Williamson - Irving, TX
Kathleen Mills - Rosharon, TX
Jaclyn L. Ziders - Uniontown, OH

Special Thanks
David Pinsent
J. Michael Williamson
John Kwasnoski
Emily Dagon
Jen Wallace
Heinz Rudolf
Dr. Edith Widder
Mr. Garth Fletcher
Brenda Gillian
Robin Hurst
Media Relations, Woods Hole Oceanographic Institution
Liz Goehring, Ridage 2000 and National Science Foundation
Brookline High School Science Students
Brookline High School, Massachusetts

Foundations of Physical Science – Third Edition

Copyright © 2018 CPO Science

ISBN: 978-1-62571-845-7
Part Number: 1576056

Printing 3—July 2018

Printed by Webcrafters Inc., Madison, WI

CPO Science
80 Northwest Boulevard, Nashua, New Hampshire 03063
(800) 932-5227
www.cposcience.com

TABLE OF CONTENTS

Unit 7 Electricity and Magnetism

Unit 8 Waves, Sound, and Light

Throughout your textbook you will see this icon, which stands for STEM (Science, Technology, Engineering, and Math). STEM is an integrated way to approach learning that connects academic disciplines to real-world situations.

Unit 1

Science and Measurement

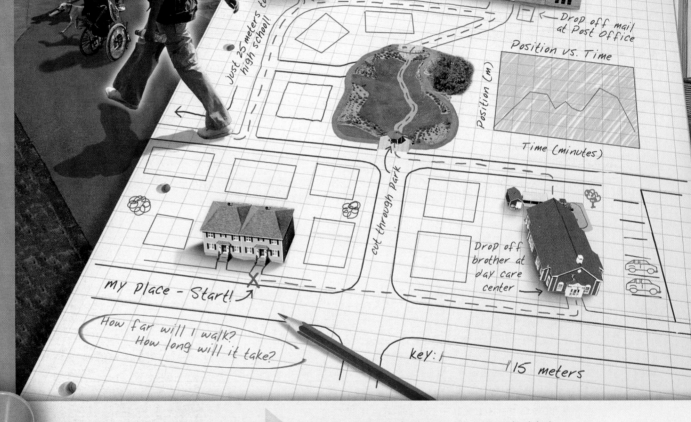

Pick up a bottle of water at store

CornerStore

just 25 meters to high school!

Drop off mail at Post Office

Position vs. Time

Position (m)

Time (minutes)

cut through park

Drop off brother at day care center →

my place - Start! ↗

How far will I walk? How long will it take?

key: ├──┤ 15 meters

Try this at home

Does a 1-cup dry measuring cup hold the same amount as a 1-cup liquid measuring cup? Fill a 1-cup measuring cup that is meant to be used for dry cooking ingredients with water. Pour the water into a plastic cup and mark the water level, then discard the water (use it to water a plant!). Now measure out 1 cup of water in a liquid measuring cup. Pour this into the plastic cup and mark the level. How do the amounts compare? Why should scientists use a standardized system of measurement?

2

Measurement

Measurement is to physical science what power tools are to a house builder; what clues are to a detective; what musical notes are to a musician. Scientists measure dimensions, distances, temperature, mass, force, electrical current, and the list could go on for pages. Scientists want to discover the natural laws of the universe. Measurements give us information about the world around us—reliable facts that form the basis of scientific theories that explain how the world works. In this chapter, you will make measurements and learn how to convert from one unit of measurement to another. You will also learn how to decide if one measurement is significantly different from another, or if it is essentially the same. These skills will be used many times throughout this physical science course as you collect data to learn how things work.

CHAPTER 1 INVESTIGATIONS

1A: Measurement
Are you able to use scientific tools to make accurate measurements?

1B: Conversion Chains
How can you use unit canceling to solve conversion problems?

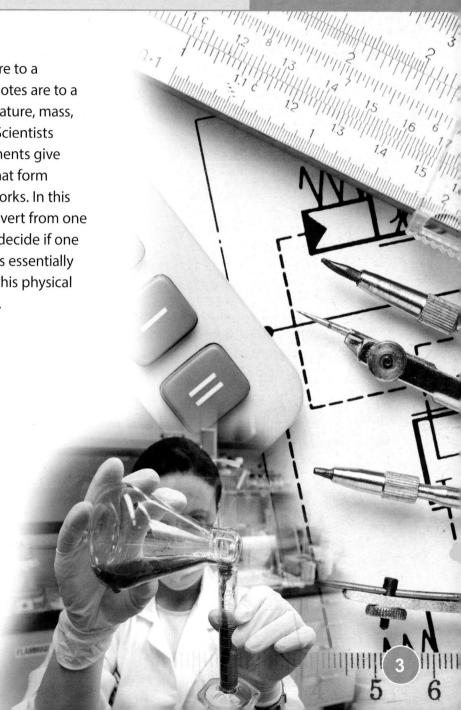

1.1 **Measurements**

A rocket must reach a speed of over 40,000 kilometers per hour to break free from Earth's gravity and get into space. The rocket has to travel very fast—it covers about 7 miles each *second*! How do kilometers and miles compare? Which is a longer distance: one kilometer, or one mile? The answer is one mile, but why do we talk about distance in two different units? Kilometers and miles are two common ways to describe distance, but scientists prefer to use kilometers. Read on to find out why.

Measurement and units

Measurements When studying physical science, you will make many measurements. Distance, time, mass, volume, weight, and temperature are just some of the quantities you will measure. A **measurement** is a determination of the amount of something. A measurement has two parts: a *number value* and a *unit* (Figure 1.1). For example, 2 meters (2 m) is a measurement because it has a number value, 2, and a unit, meters.

Units A **unit** is a standard amount that everyone agrees on. Without units, the numbers in a measurement don't make any sense. For example, if you asked someone to "walk 22," she would not know how far to go. Do you want her to walk 22 meters, 22 miles, or 22 centimeters (the height of this textbook)? If you say "walk 22 meters," then you have given her enough information because the unit "meters" tells her how to understand the quantity "22." An important rule of science is to *always include the correct units with number values.*

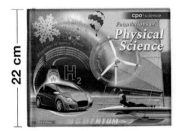

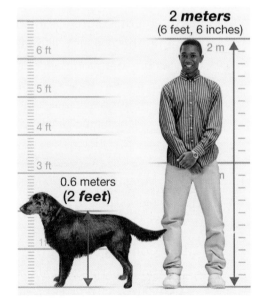

Figure 1.1: *A measurement includes a number value and a unit. Two* meters *is much taller than 2* feet!

Two common measurement systems

English System of measurement The **English System** is used for everyday measurements in the United States. Miles, yards, feet, inches, pounds, pints, quarts, gallons, cups, and teaspoons are all English System units. However, only one or two countries other than the United States still use this old system of measurement.

Measuring with SI units During the 1800s, a new system of measurement—the Metric System—was developed in France and was quickly adopted by other European and South American countries. The goal of this system was for all units of measurement to be related, and for the units to form a base-10, or decimal, system. In 1960, the Metric System was revised and simplified, and a new name was adopted—International System of Units, or **SI** for short. The acronym SI comes from the French name *Le Système International d'Unités.* Today, the United States is the only industrialized nation that has not switched completely to SI.

Scientists use SI Almost all fields of science worldwide use SI units because they are so much easier to work with. In the English System, there are 12 inches in a foot, 3 feet in a yard, and 5,280 feet in a mile. These are not easy numbers to remember. In the metric system, there are 10 millimeters in a centimeter, 100 centimeters in a meter, and 1,000 meters in a kilometer. Factors of 10 are easier to remember and work with mathematically than 12, 3, and 5,280 (Figure 1.2).

United States uses both systems Did you know that you use both English and SI units in your daily life? In many other countries, people use SI units for all measurements. Do you think the United States will ever use SI units for all measurements?

Table 1.1: Everyday SI measurements in the United States

Measurement	Unit	Symbol	Usage
length	millimeter	mm	nails and screws, tools, pencil lead
length	meter	m	track and field sports, Olympic swimming pools
volume	liter	L	1- and 2-liter soda bottles
mass	milligram	mg	medication, nutrition labels
power	kilowatt	kW	electricity

VOCABULARY

English System – measurement system used for everyday measurements in the United States

SI – International System of Units used by most countries for everyday measurement and used by the scientific community worldwide

Prefix	Meaning	Number
giga (G)	1 billion	1,000,000,000
mega (M)	1 million	1,000,000
kilo (k)	1 thousand	1,000
centi (c)	one-hundredth	0.01
milli (m)	one-thousandth	0.001
micro (μ)	one-millionth	0.000001

Figure 1.2: *SI prefixes.*

The International System of Units (SI)

Units allow people to communicate amounts. To make sure their measurements are accurate, scientists use a set of standard units that have been agreed upon around the world. Table 1.2 shows the units in the International System of Units, known as SI.

Table 1.2: Common SI units

Measurement approximations	Unit	Value
Length width of pinky finger = 1 cm	**meter (m)** kilometer (km) decimeter (dm) centimeter (cm) millimeter (mm) micrometer (mm) nanometer (nm)	1 km = 1,000 m 1 dm = 0.1 m 1 cm = 0.01 m 1 mm = 0.001 m 1 mm = 0.000001 m 1 nm = 0.000000001 m
Volume	**cubic meter (m³)** cubic centimeter (cm³) liter (L) milliliter (mL)	1 cm³ = 0.000001 m³ 1 L = 0.001 m³ 1 mL = 0.000001 m³
Mass 1 large paper clip = 1 gram	**kilogram (kg)** gram (g) milligram (mg)	1 g = 0.001 kg 1 mg = 0.000001 kg
Temperature 21°C = room temperature	Kelvin (K) Celsius (°C)	0°C = 273 K 100°C = 373 K

STUDY SKILLS

Learn to think SI

How long is a centimeter? How heavy is 1 gram? How much is a milliliter? The easy way to "think SI" is to remember some simple measurements. Take a look at the pictures in the table at the left, and see if you can remember them.

1. 1 cm is about the width of your little finger.
2. 1 mL is about the same volume as 10 drops of water.
3. 1 g is about the mass of one large paper clip.
4. 21 degrees Celsius is a comfortable room temperature.

Learning to think SI is like learning a new language; the more practice you have, the easier it is to understand.

Digital data storage and SI prefixes

A byte is a unit of computer data storage. When you add SI prefixes to any unit, you change the size of the unit as you can see in the chart below and in Figure 1.3. It's difficult to imagine a quantity as large as one quadrillion! One quadrillion bytes equals 1,000 trillion—that's a petabyte.

SI prefixes for decimal multiples

Number	Factor	Name	Symbol
1 000 000 000 000 000	10^{15}	peta	P
1 000 000 000 000	10^{12}	tera	T
1 000 000 000	10^{9}	giga	G
1 000 000	10^{6}	mega	M
1 000	10^{3}	kilo	k
100	10^{2}	hecto	h
10	10^{1}	deca	da
0.1	10^{-1}	deci	d
0.01	10^{-2}	centi	c
0.001	10^{-3}	milli	m
0.000 001	10^{-6}	micro	μ
0.000 000 001	10^{-9}	nano	n
0.000 000 000 001	10^{-12}	pico	p

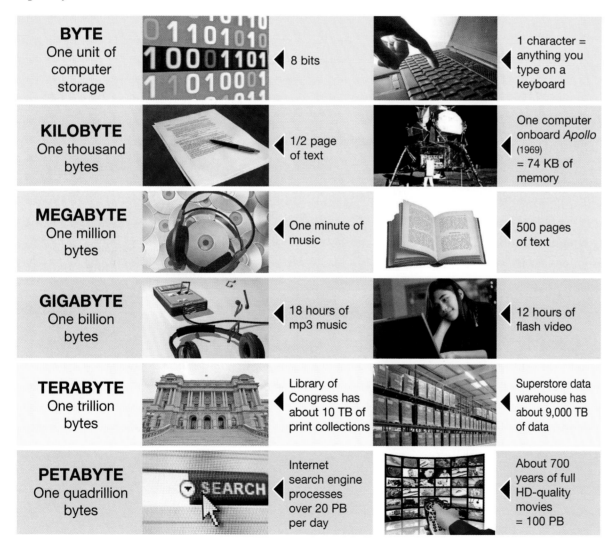

BYTE — One unit of computer storage — 8 bits — 1 character = anything you type on a keyboard

KILOBYTE — One thousand bytes — 1/2 page of text — One computer onboard *Apollo* (1969) = 74 KB of memory

MEGABYTE — One million bytes — One minute of music — 500 pages of text

GIGABYTE — One billion bytes — 18 hours of mp3 music — 12 hours of flash video

TERABYTE — One trillion bytes — Library of Congress has about 10 TB of print collections — Superstore data warehouse has about 9,000 TB of data

PETABYTE — One quadrillion bytes — SEARCH — Internet search engine processes over 20 PB per day — About 700 years of full HD-quality movies = 100 PB

Figure 1.3: *Use these prefixes on any SI unit to change its size. A nanometer is one billion times smaller than a meter!*

Section 1.1 *Review*

1. Explain, using your own example, why you must always give a unit when reporting a measurement.

2. Draw two columns on your paper. Label the first column *SI* and the second column *English System*. Sort this list and write the units in the correct column: inch, centimeter, yard, teaspoon, milliliter, bushel, gallon, liter, mile, gram, quart, pint, kilometer, pound.

3. Explain two reasons why *SI* is easier to use than the English System.

4. An external computer flash drive can store 1 gigabyte of digital data. How many bytes is this?

5. Which is larger: a megawatt or a kilowatt? How many times larger is it?

6. Put these units in order from smallest to largest: decimeter, meter, kilometer, millimeter, centimeter, nanometer, micrometer.

7. Your friend asks you for a glass of water and you bring her 5 milliliters of water. Is this more or less than what she was probably expecting? Explain your reasoning.

8. The length of a sheet of U.S. standard (letter-size) paper is closest to:
   a. 8 centimeters
   b. 11 centimeters
   c. 29 centimeters
   d. 300 centimeters

9. A nickel weighs about:
   a. 0.1 gram
   b. 5 grams
   c. 50 grams
   d. 100 grams

10. Why do you suppose the United States still uses the English System for everyday measurements, while almost every other country uses SI? Give several possible reasons.

CHALLENGE

Everyday English and SI units

How many different ways are English and SI units used to measure everyday things in the United States? Speed is measured in miles per hour (mph). Is that an English or SI unit? Is gasoline sold in English or SI units? What is that unit? Here is a list of things that are commonly measured. Make a chart that shows what unit is most commonly used to measure each thing in the United States, and show whether that unit belongs to the English System or SI. You may be surprised at how much we use *both* systems!

- gasoline
- road map distances
- aspirin/pain reliever tablets
- mechanical pencil lead
- ski length
- milk
- large soda bottles
- electricity
- time
- body weight

1.2 **Time and Distance**

Measurement is a key skill and concept in physical science. In this section, you will learn about measuring two fundamental properties of the universe: time and distance.

Time

Time in science We often want to know how things change over time. For example, a car rolls down a hill over time. A hot cup of coffee cools down over time. The laws of physical science tell us how things change over time.

What time is it? Time is used two ways (Figure 1.4). One way is to identify a particular moment in the past or in the future. For example, saying your 18th birthday party will be on January 1, 2020, at 2:00 p.m. identifies a particular moment in the future for your party to start. This is the way "time" is usually used in everyday conversation.

How much time? The second way is to describe a *quantity* of time. The question "How much time?" is asking for a quantity of time. A quantity of time is also called a *time interval*. Any calculation involving time that you do in physical science will always use time intervals, *not* time of day.

Time in seconds Many problems in science use time in seconds. For calculations, you may need to convert hours and minutes into seconds. For example, the timer (below) shows 2 hours, 30 minutes, and 45 seconds.

Hours   Minutes   Seconds

$2 : 30 : 45$

How many total seconds does this time interval represent? There are 60 seconds in a minute, so multiply 30 minutes by 60 to get 1,800 seconds. There are 3,600 seconds in an hour, so multiply 2 hours by 3,600 to get 7,200 seconds. Add up all the seconds to get your answer: $45 + 1{,}800 + 7{,}200 = 9{,}045$ seconds.

What time is it?

September
S M T W T F S
1 2
3 4 5 6 7 8 9
10 11 12 13 14 15 16
17 18 19 20 21 22 23
25 26 27 28 29 30

11:52 a.m.
September 6
2017

How much time?

2 hours,
22 minutes, 42 seconds

Figure 1.4: *There are two different ways to understand time.*

CHALLENGE

What is your reaction time?

Sit at a table and rest your arm on the table, with your hand hanging off the edge. Have a friend dangle a metric ruler just above your thumb and index finger. When your friend drops the ruler, catch it quickly between your thumb and finger. Record the centimeter mark where you caught the ruler. Approximate reaction times are: 0.10 seconds for 5 cm, 0.14 s for 10 cm, 0.18 s for 15 cm, 0.20 s for 20 cm, 0.23 s for 25 cm, and 0.25 s for 30 cm.

Do several trials and discuss.

Distance

What is distance? **Distance** is the amount of space between two points (Figure 1.5). You can also think of distance as how far apart two objects are. You probably have a good understanding of distance from everyday experiences, like the distance from your house to school, or the distance between your city and the next town. The concept of distance in physics is the same, but the actual distances may be much larger or much smaller than anything you measure in everyday life.

Distance is measured in units of length Distance is measured in units of **length**. The English System uses inches, feet, yards, and miles for length units. One foot equals 12 inches. Do you know how many feet are in a yard? There are three feet in a yard. How many yards are in a mile? There are 1,760 yards in a mile. Did you know those answers? These numbers are not easy to remember. The SI units of length are much easier to use, because they are based on powers of 10, and the prefixes tell you something about the unit value. For example, the prefix *centi-* means one hundredth, so you know that a centimeter is 100 times smaller than a meter. There are 100 centimeters in a meter. The unit "inch" does not tell you anything about how it is related to a foot. There are 12 inches in a foot, but you wouldn't know that from the unit name!

SI distance unit The **meter** is a basic SI distance unit. In 1791, a meter was defined as one ten-millionth of the distance from the North Pole to the equator. Today, a meter is defined more accurately using the speed of light (Figure 1.6). The meter was used as a starting point for developing the other SI units.

Useful prefixes Prefixes are added to the names of basic SI units. Prefixes describe very small or very large measurements. There are many SI unit prefixes, but these three are commonly used with meters to measure distance.

Prefix	Prefix + meter	Compared to 1 meter
kilo-	kilometer	1,000 times bigger
centi-	centimeter	100 times smaller
milli-	millimeter	1,000 times smaller

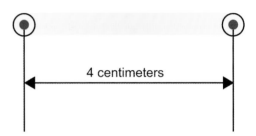
4 centimeters

Figure 1.5: *Distance is the amount of space between two points.*

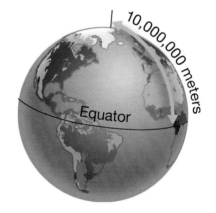

Figure 1.6: *In 1791, a meter was defined as 1/10,000,000 of the distance from Earth's North Pole to the equator. Today, a meter is defined more accurately as the distance that light travels in a fraction of a second.*

The meter stick A meter stick is a good tool to use for measuring ordinary lengths in the laboratory. A meter stick is 1 meter long and is divided into millimeters and centimeters. Figure 1.7 shows a meter stick next to objects of different lengths. Can you see how the meter stick is used to measure the length shown for each object?

Using a centimeter ruler Using a meter stick or a centimeter ruler to make distance or length measurements is easy. Each centimeter is divided into 10 smaller units, called millimeters. Try using the centimeter rulers below to find the measurement of the length each object. Don't peek at the answers!

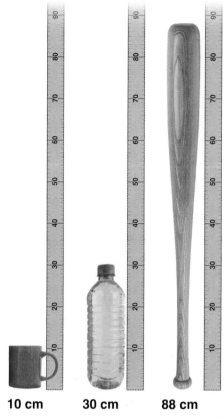

10 cm   30 cm   88 cm

Figure 1.7: *Reading a meter stick.*

SOLVE FIRST LOOK LATER

The measurements are:

bolt: 4.70 cm

pencil: 7.90 cm

pushpin: 2.60 cm

Section 1.2 *Review*

1. What are two different ways to understand time? Explain and give examples.

2. How many minutes are there in 1.5 hours? Don't forget to show your work!

3. Convert 330 minutes to hours, and show your work.

4. Men in the age group of 18–34 years need to be able to run a marathon in 3 hours and 10 minutes to qualify for the Boston Marathon. How many seconds is this? Show your work!

5. Study the table in Figure 1.8 to answer the following questions.
   a. Which movies are longer than 2 hours?
   b. Which (if any) movies are longer than 3 hours?
   c. Convert the running time of *Harry Potter and the Sorcerer's Stone* to hours and minutes.
   d. Does any movie have a running time of less than 1.5 hours? If so, which one(s)?

6. Your teacher says, "There are 100 centimeters in a meter, and this fact is revealed in the unit's name (centimeter). There are 3 feet in one yard, but this fact is not revealed in the unit's name (yard)." Explain what your teacher means by this.

7. Which is larger? Copy each pair of units and circle the one that is the largest quantity for each pair.
   a. 1 centimeter or 1 meter
   b. 1 millimeter or 1 centimeter
   c. 1 kilometer or 1 meter

8. Which is larger? Copy each pair of measurements and circle the length that is the longest for each pair.
   a. 42 mm or 10 cm
   b. 15 mm or 0.15 cm
   c. 10 mm or 2 cm

Movie	Running time (min)
Finding Nemo	101
Star Wars 4: A New Hope	121
Harry Potter and the Sorcerer's Stone	152
E.T. the Extraterrestrial	115
Jaws	124
King Kong (2005)	188
Titanic (1997)	194
Back to the Future	116

Figure 1.8: *Question 5.*

1.3 Converting Units

When describing the length of a ski, you could say that it is 150 centimeters or 1.5 meters. The ski length is the same—the only thing that is different is the measurement unit. Unit conversion is an important skill in the language of measurement.

Why convert?

What does it mean to "convert"? Suppose you empty your coin bank and count out 1,565 pennies. You probably would rather think of that quantity of money in dollars. How do you figure out how many dollars you have? Well, you convert the 1,565 cents to the dollar amount. Because there are 100 pennies in a dollar, you divide 1,565 by 100. This is the same as moving the decimal point two places to the left. 1,565 pennies and 15.65 dollars represent the same amount of money.

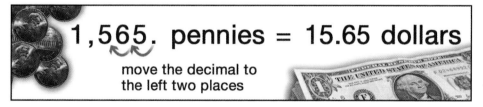

1,565. pennies = 15.65 dollars

move the decimal to the left two places

Converting SI units Converting SI units is just as easy as converting pennies to dollars. Suppose a snail can travel about 65 millimeters in one minute. In 10 minutes, it can go 10 times as far (65 × 10), or 650 mm. It's hard to visualize 650 mm. You know that a meter stick is relatively close in size to a yard stick, which you are familiar with. If you convert millimeters to meters, you might be able to better visualize how far the snail can travel in 10 minutes.

How do you convert millimeters to meters?

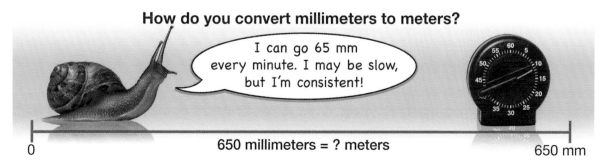

I can go 65 mm every minute. I may be slow, but I'm consistent!

650 millimeters = ? meters

0   650 mm

Solving Problems: Converting SI Units

When you convert from one SI unit to another, you multiply or divide by a series of tens. This conversion tool will help you move the decimal point.

kilo	hecto	deca	meter gram liter	deci	centi	milli
1,000	100	10		0.1	0.01	0.001

Convert 650 millimeters to meters.

1. **Looking for:**   You are asked for the distance in meters.

2. **Given:**   You are given the distance in millimeters.

3. **Relationships:**   There are 1,000 millimeters in 1 meter.

4. **Solution:**   1. Find the millimeter place, and put your pencil on that space.

    2. Move your pencil to the meters place, and count how many spaces you move your pencil, including the last landing space.

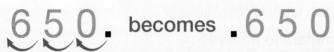

3 spaces to the left

    3. Now move the decimal point in 650 to the left 3 places.

$$650. \text{ becomes } .650$$

$$650 \text{ mm} = 0.650 \text{ m}$$

Your turn...

a. Convert 142 kilometers to meters.

b. Convert 754,000 centimeters to kilometers.

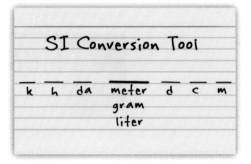

SOLVE FIRST LOOK LATER

a.   142,000 m

b.   7.54 km

Converting between English and SI units

The problem of multiple units It would be nice if everyone always used the same unit for length, like the meter. Unfortunately, many different units of length are used for different things, and in different places. In the United States, you will find inches, feet, and miles used more commonly than centimeters, meters, and kilometers. Sometimes you will need to convert from English to SI units.

Comparing English and SI units Downhill skis come in many different centimeter lengths. If you stand a ski up next to you, the ski should come up as high as your chin. Suppose the distance from your toes to chin is 4.5 feet. What length skis should you buy, in centimeters? To answer the question you need to convert from feet to centimeters. To do the conversion you multiply 4.5 feet by a **conversion factor**. A conversion factor is a ratio that has the value of one. Study the problem-solving steps on the next page to see how to set up a conversion using conversion factors. This method of converting units is called **dimensional analysis**.

SCIENCE FACT

English and SI units

Suppose you are working on your bicycle and the wrench you select is one size too small. The illustration below shows that it is easier to choose the next bigger size if you use SI units.

4.5 feet - - - - - - - - - - - - - - - - - ? centimeters

Wrenches in Inches
(English Units)

3/8"

7/16"

Wrenches in Millimeters
(SI Units)

11

10

Which is the bigger wrench in each pair?

Solving Problems: Converting Units

Convert 4.5 feet to centimeters.

1. **Looking for:**   You are asked for a length in centimeters.

2. **Given:**   You are given the length in feet.

3. **Relationships:**   There are 30.48 cm in one foot (you can look this up in a conversion table).

4. **Solution:**   1. Write down the given measurement and a multiplication symbol.

$$4.5 \text{ feet} \times$$

     2. Create a conversion factor by drawing a fraction bar and copying the given unit (feet) into the bottom of the fraction. Next, put the unit you are looking for in the numerator (cm). Put the number "1" next to the larger unit (foot) and for the smaller unit, write down how many of them equal 1 of the larger unit (30.48 cm).

$$4.5 \text{ feet} \times \left(\frac{30.48 \text{ cm}}{1 \text{ foot}} \right)$$

     3. Cancel like units in the problem setup. This is how you keep track of how well your dimensional analysis setup is working. Your goal is to cancel all units except the one you are solving for (cm).

$$4.5 \text{ feet} \times \left(\frac{30.48 \text{ cm}}{1 \text{ foot}} \right) = 137 \text{ cm}$$

     4. Now you are ready to do the math! This problem setup tells you to multiply 4.5 by 30.48. The answer is 137 cm.

Your turn...

a. Convert 175 yards to meters. (You might need more than one fraction!)

b. Convert 2.50 inches to millimeters. (More than one fraction is needed!)

STUDY SKILLS

Handy conversion factors

Copy these handy conversion factors so you can use them anytime you need to set up a unit-canceling problem like the one on this page. *Note*: You can flip these fractions around as needed; the "1" (larger unit) isn't always in the denominator.

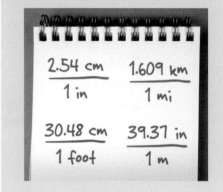

$$\frac{2.54 \text{ cm}}{1 \text{ in}} \qquad \frac{1.609 \text{ km}}{1 \text{ mi}}$$

$$\frac{30.48 \text{ cm}}{1 \text{ foot}} \qquad \frac{39.37 \text{ in}}{1 \text{ m}}$$

SOLVE FIRST LOOK LATER

a. 160 m

b. 63.5 mm

Section 1.3 *Review*

1. What does it mean to "convert" from one unit to another? Give an example.

2. What are the general mathematical operations you use when converting from one SI unit to another?

3. How many meters do you cover in a 10 km (10-K) race?

4. An Olympic swimming pool is 50 meters long. You swim from one end to the other four times.
    a. How many meters do you swim?
    b. How many kilometers do you swim?
    c. How many centimeters do you swim?

5. In the United States, a standard letter-sized piece of paper is 8.5 inches wide by 11 inches long. The international standard for a letter-sized piece of paper is different. The international standard is based on SI units: 21.0 cm wide by 29.7 cm long.
    a. Convert 21.0 cm to inches. Show your dimensional analysis setup.
    b. Convert 29.7 cm to inches. Show your dimensional analysis setup.
    c. State the dimensions, in inches, of the international standard for a letter-sized piece of paper.
    d. Which piece of paper is longer: a U.S. letter-sized piece of paper, or an international letter-sized piece of paper?
    e. Suppose the United States adopted the international standard for letter-sized paper. Explain at least two things that might result from this change.

6. The height of an average adult person is closest to:
    a. 1.0 meter
    b. 1.8 meters
    c. 5.6 meters

JOURNAL

Find out!

What are the official measurements for an Olympic swimming pool? Create a table in your journal with the answers:

- length of pool
- width of pool
- number of lanes
- lane width
- water temperature
- depth

KEYWORDS

Do an Internet search using the keywords "international paper size." Write a report of your findings about the standards for paper sizes. Do all countries use the same size paper for letters? How was the international standard paper size defined? What are some interesting outcomes of having different standard paper sizes in different countries? What surprised you the most about your research? If requested, cite your sources using a required format.

1.4 Working with Measurements

All measurements involve a degree of uncertainty. The object in Figure 1.9 is definitely longer than 2.6 centimeters. But how much longer? Not everyone would agree on the third digit of the measurement. Is it 2.63 cm or 2.65 cm? In this section, you will explore different ways to work with measured quantities where every measurement involves some amount of error or uncertainty. How much is acceptable? Read on to find out.

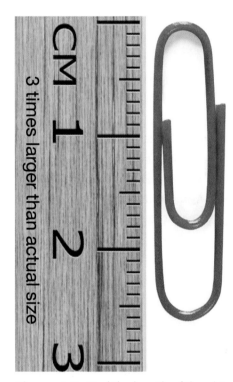

VOCABULARY

significant digits – meaningful digits in a measured quantity

Significant digits

Uncertainty in measurements In the real world, it is *impossible* to make a measurement of the exact true value of anything (except when counting). Using a meter stick, the paper clip is 2.65 cm. To a scientist this number means "between 2.62 and 2.67 cm." The last digit, 5, representing the smallest amount, is always considered to be rounded up or down. **Significant digits** are the meaningful digits in a measured quantity. For the paper clip, the third digit serves to tell someone that the object is about halfway between 2.6 and 2.7 cm long. Therefore, we say there are three useful or significant digits in the length measurement. It is important to be honest when reporting a measurement, so readers know how much resolution it has. We do this by using significant digits.

Using significant digits in math problems What happens when you use measured quantities with different numbers of significant digits in a math problem? A shoe is 38 cm long and you want to convert the length to inches:

$$38 \text{ cm} \times \left(\frac{1 \text{ inch}}{2.54 \text{ cm}} \right) = ?$$

To find the answer, divide 38 by 2.54 and you get 14.960629. This answer has an artificially large number of significant digits (eight!). An answer involving measured quantities should have no more resolution than the starting measurement with the least number of significant digits. The correct answer to this conversion problem is rounded up to 15 inches, because 38 centimeters has two significant digits. Study the next page for more help with using significant digits in math problems.

Figure 1.9: *Find the length of the object in centimeters. How many digits does your answer have?*

3 times larger than actual size

Solving Problems: Significant Digits

What is the area of an 8.5 inch by 11.0 inch piece of paper?

1. **Looking for:**   You are asked for an area.

2. **Given:**   You are given the width, 8.5 inches, and the height, 11.0 inches.

3. **Relationships:** Area = width × length

4. **Solution:**   Area = 8.5 inch × 11.0 inch

   Area = 93.5 square inches

93.5 has three significant digits. The width measurement had only two significant digits, and the length measurement had three significant digits. So how many significant digits should your answer have? That's right, the answer can have no more significant digits than the measurement with the least number. In this case, because the width measurement only had two significant digits, your answer can only have two. You must round 93.5 square inches to 94 square inches. The correct answer is 94 square inches.

Your turn...

a. How many significant digits does each of these numbers have? 40 cm, 4 cm, 4.0 cm, 40. cm, 45 cm, 450 cm, 450. cm

b. Convert 1.10 miles to kilometers and report your answer with the correct number of significant digits. Use the relationship 1 mi = 1.6 km.

STUDY SKILLS

Which digits are significant?

Digits that are *always significant*:

1. Non-zero digits.
2. Zeros between two significant digits.
3. All *final* zeros to the *right* of a decimal point.

Digits that are *never significant*:

1. Leading zeros in a decimal number (0.002 cm has only one significant digit.)
2. Final zeros in a number that does not have a decimal point.

Note: A decimal point is used after a whole number ending in zero to indicate that a final zero *is* significant. Thus, 50. cm has two significant digits, not one.

SOLVE FIRST LOOK LATER

a. 40 cm: **1**, 4 cm: **1**, 4.0 cm: **2**, 40. cm: **2**, 45 cm: **2**, 450 cm: **2**, 450. cm: **3**

b. 1.8 km

Accuracy, precision, and resolution

Accuracy The words *accuracy* and *precision* have special meanings in science that are a little different from how people use these words in daily conversation (Figure 1.10). **Accuracy** is how close a measurement is to the true value. An accurate clock or watch will give a time reading that is the same as or extremely close to the official time from a government standard. An accurate golf putt is one that falls in the hole. A very accurate golf drive would be a hole-in-one!

Precision Precision does not have the same meaning as accuracy. **Precision** describes how close together repeated measurements or events are to one another. Precise clocks throughout a school would all read the same time at any given moment. School clocks can be precise without being accurate. Can you explain how this could be true? If I hit three different golf balls off the same tee, and each one of them goes into the same sand trap, I have good precision but poor accuracy!

Resolution **Resolution** is another important term to understand when you are working with measured quantities. Resolution refers to the smallest interval that can be measured (Figure 1.10). The resolution of a centimeter ruler is 0.5 mm. This is because, if you look closely, you can tell if a measurement falls right on a millimeter mark, or between millimeter marks. The resolution on most classroom clocks is 1 second. Without a second hand, the resolution of a clock would be only 1 minute.

Resolution in images The word *resolution* often appears in connection with digital cameras or high definition (HD) TV. A high resolution image is very sharp and high quality. For example, an HDTV image has 1,980 dots in the horizontal direction. A standard TV image has only 640 dots. A feature that is two dots wide in an HDTV image is just a blur on a standard TV. You can think of resolution as the "sharpness" of a measurement. A measurement with lots of resolution is a very "sharp" measurement. A timer that measures seconds to four decimal places has a resolution of one ten-thousandth of a second. A stopwatch that measures seconds to two decimal places has a lower resolution of one one-hundredth of a second.

accuracy – how close a measurement is to an accepted or true value

precision – how close together or reproducible repeated measurements are

resolution – the smallest interval that can be measured

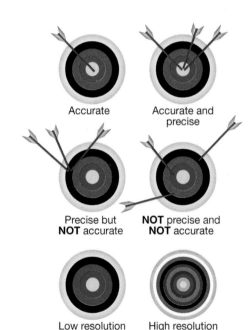

Figure 1.10: *Accuracy, precision, and resolution.*

Comparing measurements

VOCABULARY

significant difference – two results are only significantly different if their difference is much larger than the estimated error

Reproducible measurements Precise measurements are *reproducible*. This means a measurement gives the same result if you or anyone else makes the same measurement again in the same way. This brings up a key question: How can you tell if two results are the same when both contain uncertainty (often called *error*)?

Measurements that are the same are not significantly different In everyday conversation, "same" means two numbers that are exactly the same, like 2.56 and 2.56. When comparing scientific results, however, "same" means "*not significantly different.*" **Significant differences** are differences that are much larger than the estimated uncertainty (or error) in the results. That means two results are "the same" *unless* their difference is greater than the estimated error. *This is important to remember.*

Comparing data sets You will collect lots of data when you do the investigations for this physical science course. Once you collect data, you usually compare measurements to check for differences. For example, suppose two different groups of students use electronic timers to measure the time it takes a cart to pass between two points on a ramp. Each group makes four measurements and takes the average (Figure 1.11). Are the results different, or are they the same? The numbers 0.3352 and 0.3349 *are* different. But is the difference *significant*, or could it just be uncertainty?

Finding estimated error To answer the question, we need to estimate the uncertainty, or error. When we estimate error in a data set, we will *assume the average is the exact value*. Our estimated error will be the average of the differences (use absolute value, drop negative signs) between each measured value and the group average time (see problem-solving steps on the next page).

Looking for significant differences Use the estimated error to decide if two results are the same or if they are significantly different. If the difference in the averages is at least three times larger than the estimated error, you can assume the difference is significant. In Figure 1.11, the difference between the groups' time averages is only 0.0003 seconds. This is *not* three times larger than the estimated error of 0.0002 seconds. We conclude that the two experiments produced "the same" result, meaning that the results are not significantly different.

	Group 1 time (s)	Est. error (+/−)	Group 2 time (s)	Est. error (+/−)
1	0.3356	0.0004	0.3346	0.0003
2	0.3351	0.0001	0.3353	0.0004
3	0.3349	0.0003	0.3349	0.0000
4	0.3352	0.0000	0.3347	0.0002
AVG	0.3352	0.0002	0.3349	0.0002

Figure 1.11: *The groups have different time averages. Are the averages close enough to be called "the same" result? Note: The estimated error is calculated by taking the group average time and subtracting each individual trial time. The estimated error is an absolute value; drop any negative signs.*

Solving Problems: **Comparing Data Sets**

Table 1.3: Comparing data

Trial	Group 1 mass (g)	Est. error (+/−)	Group 2 mass (g)	Est. error (+/−)
1	2.6	**0.0**	2.1	**0.0**
2	2.5	**0.1**	2.2	**0.1**
3	2.8	**0.2**	2.1	**0.0**
AVERAGE	**2.6**	**0.1**	**2.1**	**0.03**

Two groups of students were each given their own small container of candy mints. Each group counted out 15 mints and divided them into 3 piles of 5 mints each. Then they found the mass of each collection of 5 mints. Are the average masses significantly different, or are they the same?

1. **Looking for:**   Significant difference between two data sets.

2. **Given:**   Masses in Table 1.3.

3. **Relationships:**   • estimated trial error = avg mass – trial mass (drop negative signs)

   • If difference between averages is *at least three time the largest estimated error average*, then we will conclude that the results are significantly different.

4. **Solution:**   See bold numbers in Table 1.3 for answers.
   The difference between the averages: (2.6 – 2.1) = 0.5
   The difference of 0.5 is five times greater than the largest estimated error (0.1), so the results are significantly different; the groups probably had different brands of mints!

Your turn...

a. Study the table in Figure 1.12. Are the group averages significantly different? Show your work to prove your answer.

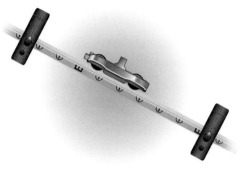

Group 1 time (s)	Group 2 time (s)
0.1776	0.2134
0.1780	0.2130
0.1772	0.2129
0.1777	0.2137

Figure 1.12: *Two groups of students record how long it takes a cart to pass between two points on a ramp. Each group does four time trials. Are their averages significantly different?*

SOLVE FIRST LOOK LATER

Group 1 avg = 0.1776 s

Group 2 avg = 0.2133 s

Difference = 0.0357 s; this is significant. One group had the ramp at a greater angle than the other. Can you tell which one had the higher ramp angle? (Group 1)

Section 1.4 *Review*

1. Which of these measurements has 3 significant digits? (*There may be more than one correct answer choice.*)

   a. 29.3 cm

   b. 290 cm

   c. 0.029 cm

   d. 290. cm

2. Convert 345 cm to inches. Show your dimensional analysis setup and report your answer with the correct number of significant digits. (1 in = 2.54 cm)

3. Four different students measure the length of a toy cart. The manufacturer reports the length to be 10.5 cm. Study the data in Figure 1.13 and answer the questions below.

   a. Are the measurements accurate? Explain why or why not.

   b. Are the measurements precise? Explain why or why not.

   c. Do the measurements all have the same resolution? Explain.

4. What does it mean when two measured quantities are different but not *significantly* different?

5. All measurements contain some error. Why is this a true statement?

6. Suppose you are going to measure the length of a pencil in centimeters. What should you do to get the most accurate measurement? If you give the ruler to three different friends, what should they do to achieve good precision?

7. Refer to Figure 1.14 to answer these questions:

   a. What is the resolution of the stopwatch?

   b. Time measurements from a stopwatch are not very precise. Why not?

Measurement taken by	Length of toy cart (cm)
Jessica	10
Marco	10.5
Julius	10.8
Steve	11

Figure 1.13: *Question 3.*

5 minutes and 32.14 seconds

Figure 1.14: *Question 7.*

Nanotechnology

It's a Small World After All

If you could shrink yourself to the size of a molecule, what do you think you would see? You might think you had just been dropped into an enormous climbing structure made from magnetic toys. You'd see atoms arranged in all sorts of amazing geometric patterns—like rings of benzene, tetrahedron-shaped methane molecules, and the intricate DNA double helix.

Chemists have known for years that, in nature, atoms are arranged in many different geometric patterns, and that each of these patterns forms a substance with its own unique properties.

Nanotechnology is the exciting new branch of science that attempts to arrange individual atoms into specific patterns, forming molecules that can do incredible things. Someday, these molecules may be able to repair damaged cells in the body, fight diseases, form tiny circuits, and create super-strong, extremely lightweight materials.

Nanotechnology is based on a unit of measurement called a *nanometer*, which is one billionth of a meter. Comparing a nanometer with a meter is like comparing a marble with the planet Earth. What registers on the nanometer scale? A double-helix DNA molecule has a width of two nanometers. Bacteria have a width of about 200 nanometers. A sheet of paper has a thickness of 100,000 nanometers.

By manipulating the nanoworld, scientists can influence the visible world. This is because an object's observable properties, such as strength and electrical conductivity, begin on the nano or molecular level.

Nanotech: More common than you think

Nanotechnology is currently used to make many everyday products. Small hair-like molecules, or *nanoparticles*, of zinc oxide are currently used as a water and stain repellent on clothing. These nanoparticles have the ability to pass liquid molecules along from one to the next so water or other liquids cannot seep into the cloth. Nanoparticles are also used in glass to scatter water into small drops to disperse it across a windshield faster. Silver nanoparticles embedded in antimicrobial bandages break up bacterial cell membranes, bind to bacteria's DNA, and interrupt bacteria metabolism.

How carbon atoms are arranged

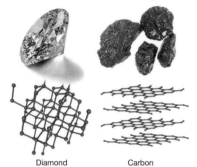

Diamond   Carbon

The wonder nanoparticle: Carbon nanotubes

Why is steel stronger than wood? It has to do with the particular molecules and atoms that make up each material and the bonds between those particles. The bonds between steel particles are stronger than those between wood particles.

Now imagine a substance that is hundreds of times stronger than steel but weighs six times less. What you're imagining actually exists. It's carbon made from carbon nanotubes.

Carbon has two well-known forms, or *allotropes*: diamond and graphite. As you know, diamond is much stronger than graphite; this is due to the arrangement and stronger bonds between the carbon atoms in a diamond allotrope's structure.

The carbon nanotube is a third kind of carbon, with capabilities far beyond the materials we use today. This allotrope is composed of cylinders of carbon atoms a few nanometers in width but thousands in length. Their special structure allows carbon nanotubes (CNTs) to have unique properties. They are excellent conductors of electricity and heat because they can conduct with little resistance, behaving as tiny electrical wires or thermal energy "pipelines." They also allow the carbon material to be very, very strong and lightweight.

Carbon nanotubes are still in development because of the difficulty in processing them for mass production. However, NASA and many companies are researching their use with polymers and epoxies for construction of spacecraft, aircraft, cars, and skyscrapers. CNTs are more commonly used in electrical circuit applications because smaller amounts of them are needed and because of their outstanding electrical properties.

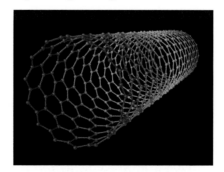

The carbon nanotube allotrope is usually manufactured in laboratories under high heat, high pressure, and controlled conditions.

Nanomedicine: The new frontier

Did you ever wonder how medicines work in the body? A medicine's effectiveness depends in part on its *bioavailability*, or how well the medicine can travel through the blood to the part of the body where it is needed. Due to their size, nanoparticles incorporated into a drug can pass easily through cell membranes, making bioavailability more efficient. The size of nanoparticles is also important for fighting disease because many diseases affect processes within the cell itself, such as the production of proteins that govern immune responses. Diseases like rheumatoid arthritis and multiple sclerosis elevate the production of these proteins, activating the immune responses of the body unnecessarily. Diseases like these can only be impeded by drugs that have the ability to penetrate a cell membrane.

One current application of nanotechnology is in the surgical removal of certain types of tumors. Surgeons can inject cadmium selenide nanoparticles into a tumor to locate its boundaries. When exposed to UV light, these nanoparticles glow and effectively illuminate the tumor so the surgeon can remove it more precisely.

What if biological nanomachines could seek out a broken part of a cell and fix it? What if a bio nanomachine could seek out cancer cells and destroy them? How can a nanomachine mimic nature's ability to heal? These are the cutting-edge questions that nanomedicine scientists are trying to answer. Their research includes the development of molecular machines that have the ability to mimic nature's healing processes by entering cells, sensing dysfunction, and making appropriate modifications by either repairing damaged cells or manufacturing new ones. Scientists predict these cell-repair machines will have the ability to open and close cell membranes and correct a single molecular disorder like DNA damage.

Questions:

1. A nanometer is what fraction of a meter?

2. Name two practical applications of nanotechnology.

3. Research: Use the Internet keyword search: "Teslaphoresis" to learn about a recent discovery in nanotechnology. Briefly explain what Teslaphoresis is and how it could be used.

Chapter 1 Assessment

Vocabulary

Select the correct term to complete each sentence.

accuracy	mass	significant difference
conversion factor	measurement	
dimensional analysis	meter	significant digits
distance	precision	unit
English System	resolution	
length	SI	

Section 1.1

1. A _____ is a standard amount, like a kilometer or a gallon, that is used to communicate different quantities.

2. The _____ is a measurement system used for everyday measurements in the United States.

3. _____ is the international system of units used by scientists worldwide.

4. When someone determines the amount of something using a value and a unit, they are making a(n) _____ .

Section 1.2

5. _____ describes how far it is from one place to any other place.

6. The amount of space between two points is measured in units of _____ .

7. A(n) _____ is a unit of length in SI that equals 100 centimeters.

Section 1.3

8. A ratio that has a value of one and is used when setting up unit conversion problems is called a(n) _____ .

9. A method of using conversion factors and unit canceling to solve a unit conversion problem is called _____ .

Section 1.4

10. Meaningful digits in a measured quantity are known as _____ .

11. _____ refers to the smallest interval that can be measured.

12. When you describe how close a measured quantity is to a true or accepted value, you are describing its _____ .

13. _____ describes how close together repeated measurements are.

14. If the difference between two results is larger than the estimated error, the result is called a(n) _____ .

Concepts

Section 1.1

1. Explain, using examples, how SI and English systems of measurement are both used in daily life in the United States.

2. All SI units use a common set of prefixes. For example, you can have milligrams, milliliters, and millimeters. What does *milli* mean in each case? How are these units similar? How are they different?

Section 1.2

3. What are the two different ways to understand time? Give examples to support your explanation.

4. In the following list of units, which are SI units of length? mm, yd, cm, mi, m, g, mg, lb, oz, km, ml

5. What unit is represented by the smallest intervals printed on a centimeter ruler?

Section 1.3

6. How do you use the SI conversion tool to perform metric conversions? Explain, step-by-step, using your own example.

7. Why can't you use the SI conversion tool to convert from SI to English units?

8. The dimensional analysis method of unit conversion is sometimes called "unit canceling." Explain why this is a good name for the method.

Section 1.4

9. Suppose you are measuring the height of a small child. What will determine the number of significant digits you record?

10. Why do you often have to round off answers to math problems that involve measured quantities?

11. Compare and contrast the terms *accuracy*, *precision*, and *resolution*. What do they have in common? How are they different?

12. How can two experimental results be considered "the same" if the numbers are not exactly the same?

Problems

Section 1.1

1. Which of the following is closest to 2 cm?
   a. the width of your pinky finger
   b. the length of a dollar bill
   c. the length of a small paper clip

2. Rank these units from smallest to largest: micrometer, nanometer, kilometer, centimeter, meter.

Section 1.2

3. Arrange the following intervals of time from shortest to longest: 160 seconds, 2 minutes, 2 minutes 50 seconds.

4. Write 3,800 seconds in hours, minutes, and seconds.

5. Report the length of the object shown below.

6. How many millimeters is represented by 6.7 cm?

Section 1.3

7. Convert 54 grams to kilograms.

8. Convert 26 decimeters to meters.

9. Convert 1,200 meters to millimeters.

10. Convert 525 pounds to kilograms. Show your dimensional analysis setup. (1 kilogram = 2.2 pounds)

11. A runner completes a 4,000.-meter race. How many yards did she run? Show your dimensional analysis setup.

Section 1.4

12. A meter stick with millimeters as its smallest graduation is used to measure a wood block. Which value correctly represents the resolution of the best measurement that can be made?
   a. 20 cm
   b. 20.5 cm
   c. 205.5 mm
   d. 205.53 mm

Wind-Up Toy Travel Times

Group 1 time (s)	Group 2 time (s)
2.56	1.23
2.62	1.29
2.75	1.22
2.65	1.24

13. Two groups of students test the same wind-up toy. Each group conducts four trials to find out how long it takes the toy to travel 1 meter. Study their data in the table above and answer the questions.

   a. Find the average time for each group's four trials.
   b. Estimate the average error for each group.
   c. Which group had the best precision? Explain.
   d. What was the resolution of the stopwatch?
   e. The group averages are quite different. Are they significantly different? Explain, and use a very simple math problem to prove your answer.
   f. One of the groups actually had the wind-up toy travel down a ramp for 1 meter, and the other group had their toy travel across a flat surface. Which group was which? Explain how you know.

Applying Your Knowledge

Section 1.1

1. Do some research to find out what influenced the development of the International System of Units. Where did the system originate? When did other countries decide to adopt the system? Did the United States adopt the system? (You may be surprised at what your research will reveal!)

2. Do you think the United States will ever switch completely to SI? Why or why not?

Section 1.2

3. What is the distance from Earth to the Moon? Is that distance changing? Do some research to find out.

Section 1.3

4. Why do you think it is necessary to know how to convert from English to SI units and vice versa? Give your own example.

Section 1.4

5. Measuring time is necessary for many Olympic events. Choose one Olympic event and write a report on how time is measured and how much resolution is necessary. If requested, cite your sources using a required format. Use technology to present your findings.

6. You are asked to find the area of a room that measures 24.5 meters by 21 meters. How many significant digits should the answer have?

Science Skills

No one knows the importance of including the correct unit with a measurement like NASA (National Aeronautics and Space Administration). On December 11, 1998, NASA launched the Mars Climate Orbiter. This spacecraft was designed to orbit Mars and collect weather data. When the orbiter reached Mars on the 23rd of September in 1999, its guidance system brought it too close to the surface, and the orbiter was destroyed by atmospheric friction. Why did the guidance system malfunction? NASA discovered that the guidance software, instead of using the SI unit of newtons for force data, used the English unit of pounds instead. The orbiter computer was expecting the guidance data to be in newtons. Because of this incorrect unit, the orbiting altitude was off by a factor of 4.45. The low-altitude orbit destroyed the spacecraft. In this chapter, you will measure important properties of matter like mass, weight, volume, and density. Don't forget to report the correct unit when you record science measurement data!

Image courtesy of NASA

CHAPTER 2 INVESTIGATIONS

2A: Mass, Volume, and Indirect Measurement
How can you find the mass of a single rice grain?

2B: Density
How is an object's density related to its volume, mass, and tendency to sink or float?

2.1 Measuring Mass and Volume

How many gallons of gasoline do I need to fill the tank of this car? Do I have enough sugar to make a batch of brownies? Will this suitcase fit in the airplane's overhead compartment? Every day, people need to measure various amounts of matter. In this section, you will review the definitions of mass and volume and how to measure these important properties of matter.

Mass

What is mass?   **Mass** describes the amount of matter in an object. **Matter** is anything that has mass and takes up space. All matter has mass. A car has more mass than a bicycle because the car contains more matter. Steel, plastic, rubber, and glass are different kinds of matter, and a car has a lot more of each kind than a bicycle does (Figure 2.1).

Kilograms   The SI unit for mass is the **kilogram** (kg). A bunch of bananas has a mass of about 1 kilogram. An average adult has a mass of about 55 kilograms. Common masses for vehicles range from a bicycle (about 12 kg) to a motorcycle (about 200 kg) to a car (1,000–2,000 kg). Try it for yourself: Can you think of something with a mass of 1 kilogram? Is the object you are thinking of heavier or lighter than this book?

Grams   The kilogram is too large a unit to be convenient for measuring small masses. One **gram** (g) is one one-thousandth of a kilogram. One large paper clip has a mass of about one gram (1,000 large paper clips = about 1 kilogram). Grams are also convenient for analyzing food. Look at the nutrition label on a food package and you will find the protein, fat, and carbohydrate content listed in grams per serving.

Which has more matter?

Figure 2.1: *A car contains more matter than a bicycle; therefore, it has more mass.*

Measuring mass in the laboratory

Using a mass balance   You will usually use an electronic scale (used as a balance) to measure mass in your lab investigations. Most lab scales display mass in grams. For example, the scale in Figure 2.2 shows the mass of six steel nuts to be 96.2 grams. The resolution of most classroom scales is to the nearest tenth of a gram. A tenth of a gram is a tiny mass, and scales are therefore sensitive (and quite delicate). Never drop things onto a scale! Instead, set things gently on the scale. You can also damage the scale if you press down on the cover plate with your hand. This can be tempting, but don't do it!

Converting grams to kilograms   For some calculations, you might have to convert masses from grams to kilograms. To convert a mass in grams to kilograms, you need to divide by 1,000 since there are 1,000 grams in 1 kilogram. You can see the dimensional analysis setup in Figure 2.2. You can also use your metric conversion tool from the previous chapter.

Masses you will consider in science   Ordinary objects tend to have masses between a few grams and a few hundred kilograms. However, you will encounter a *much* wider range of masses in science. A bacterium has a mass of 0.000000001 kg! That seems small—but then an atom has a mass a thousand billion times smaller than a bacteria. Science also involves large masses, such as planets and stars. A star like our Sun has a mass of 2 million trillion trillion kilograms! It may seem strange, but stars and bacteria are made of the same kinds of matter, and can both be measured in the same unit of kilograms!

Do you say "mass" or "weight"?   Does it seem awkward to say "I'm going to *find the mass of* these objects" instead of "I'm going to *weigh* these objects"? In everyday language, we use the word *weight* but hardly ever use the term *mass*. However, physical science is about the true nature of how the universe works, so it is time for us to learn and understand the difference between mass and weight—and there is a big difference! Read on to find out what the difference is. When you use the triple beam balance or electronic scale in your classroom laboratory, you are finding the *mass* of objects, not the weight.

Electronic scale

Converting from grams to kilograms

$$96.2\ \cancel{g} \times \frac{1\ kg}{1,000\ \cancel{g}} = 0.0962\ kg$$

$$96.2\ g = 0.0962\ kg$$

Figure 2.2: *A scale displays mass in grams. You may need to convert grams to kilograms when doing calculations.*

Mass and weight are different

Mass vs. weight   We tend to use the terms *mass* and *weight* interchangeably, but they are not the same thing. Mass is the amount of matter in an object. **Weight** is a measure of the pulling force of gravity on an object. Your mass is constant throughout the universe, but your weight can change, depending on which planet you are on! Since most of us live and work on Earth, it is easy to forget that mass and weight are different.

Mass is constant, weight can change   Figure 2.3 compares the mass and weight of a bag of flour on Earth and on the Moon. A 2.3-kg bag of flour has the same mass no matter where it is in the universe. The *weight* of the bag of flour, however, is *less* on the Moon. This is because the force of gravity on the Moon is six times less than the force of gravity on Earth.

Mass is fundamental   Although mass and weight are related quantities, always remember the difference. Mass is a fundamental property of an object measured in kilograms (kg). Weight is a force that depends on the pulling force of gravity and is measured in *newtons* (N). Use the Venn diagram below to review similarities and differences between mass and weight.

Comparing mass and weight

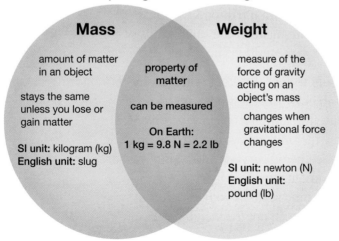

Mass

amount of matter in an object

stays the same unless you lose or gain matter

SI unit: kilogram (kg)
English unit: slug

property of matter

can be measured

On Earth:
1 kg = 9.8 N = 2.2 lb

Weight

measure of the force of gravity acting on an object's mass

changes when gravitational force changes

SI unit: newton (N)
English unit: pound (lb)

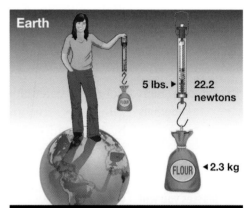

Earth

5 lbs. ▶   22.2 newtons

FLOUR   ◀ 2.3 kg

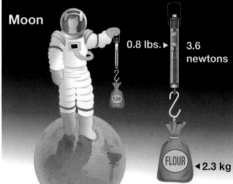

Moon

0.8 lbs. ▶   3.6 newtons

FLOUR   ◀ 2.3 kg

Figure 2.3: *A 2.3-kg bag of flour has the same mass everywhere, but its weight will be less on the Moon, where the force of gravity is less.*

Volume

Volume
Volume is the amount of space an object takes up. The fundamental unit of volume in SI is the cubic meter (m^3). However, a cubic meter is a relatively huge volume for laboratory work, about the size of the inside of a refrigerator. More convenient, smaller units include cubic centimeters (cc or cm^3), liters (L), and milliliters (mL). One cubic centimeter is the volume of a cube measuring 1 cm on each side. One cubic centimeter is the same volume as one milliliter. One liter is 1,000 milliliters (1 L = 1,000 mL).

Measuring the volume of liquids
You can measure the volume of liquids by pouring them into a *graduated cylinder.* A graduated cylinder has markings that show volume in milliliters (mL). To read a graduated cylinder correctly, follow these two rules:

1. Read the mark at eye level.

2. You will notice that the surface of the liquid forms a curve rather than a straight line (Figure 2.4). This curve is called the *meniscus.* Read the volume at the center of the meniscus.

Volume of solids
You have probably already learned to measure the volume of some solid shapes. The volume of a rectangular solid (a shoebox shape), for example, is found by multiplying length times width times height. The volume of a sphere is $4/3pr^3$, with *r* equal to the radius of the sphere.

The displacement method
You can find the volume of an irregular shape using a technique called displacement. To *displace* means to "take the place of" or to "push aside." You can find the volume of an irregularly shaped object by putting it in water and measuring the amount of water displaced.

Making a displacement measurement
Here's how to use the displacement method to find the volume of a house key. Fill a 100-mL graduated cylinder with 50 mL of water (Figure 2.5) and then add the key. The water level will rise, because the key displaces some water. If the level now reads 53.0 mL, you know that the key displaced 3.0 mL of water. The volume of the key is equal to the volume of the water it displaces. The key has a volume of 3.0 milliliters (mL), or 3.0 cubic centimeters (cm^3).

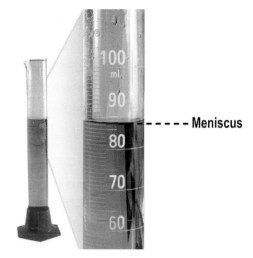

----- Meniscus

Figure 2.4: *The meniscus of water has a concave shape. Read the mark at the center of the meniscus, which is the bottom of the curve.*

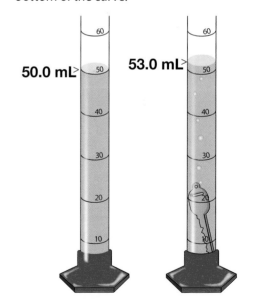

50.0 mL▷   53.0 mL▷

Figure 2.5: *The key displaced 3.0 mL of water.*

Comparing mass and volume

Mass and volume are different
Mass and volume are not the same thing; they are two different properties of matter (Figure 2.6). Mass is the amount of matter in an object, and **volume** is the space that matter takes up. Breakfast cereal, for example, is not sold by volume. As boxes of cereal are shipped from plant to warehouse to store, the contents "settle." Settling means the individual bits of cereal jostle and nest into each other so the same quantity that once filled the bag takes up less space when "packed" more tightly. By the time you buy it, a cereal box may appear to be only three-fourths full! For this reason, cereal is measured in grams, to reflect the true amount of cereal in the box. An equal mass of cereal is placed into each box at the factory. Breakfast cereal is sold by mass, not volume, because mass is a more fundamental property that only changes if the amount of cereal (matter) changes.

Mass vs. volume
Can you think of something that has a relatively large amount of mass, but fits in a small space? How about a construction brick or a cement block? A brick has quite a bit of mass for the amount of space it takes up. A brick is smaller than a loaf of bread, but it has considerably more mass.

Volume vs. mass
Can you think of something that has a relatively small amount of mass, but takes up quite a bit of space? How about a large, fake boulder made of foam? You would easily be able to lift the foam boulder even if it were 1 meter across. On the other had, a real boulder 1 meter in diameter has as much mass as 12 average-sized people!

Mass and volume are related!
We naturally assume that an object's mass is proportional to its size. Big objects do tend to have more mass than small ones. Don't be fooled! Mass and volume are related, but are not the same thing at all. The type of matter an object is made of and how that matter is distributed have an enormous effect on the object's mass. Foam is light and airy, or *low density*, and even a large volume of foam has very little mass. Brick is heavy, or *high density*, and therefore contains a lot more mass per cubic centimeter than foam. You will explore the property of density in the next section.

VOCABULARY

volume – the amount of space taken up by matter

Foam boulder

Cement block

Figure 2.6: *Mass and volume are two different properties of matter.*

Section 2.1 *Review*

1. If object A has more mass than object B, does object A contain more matter? Explain.

2. Copy Table 2.1 and fill in the missing boxes. All answers can be found in Section 2.1. (*Hint*: Don't forget to check the illustrations.)

Table 2.1: Measuring matter

Property of matter	SI unit (abbreviation)	English unit (abbreviation)
Mass	Kilogram (kg)	?
Weight	?	Pound (lb)
Volume (liquids)	?	Fluid ounce (fl oz)
Volume (solids)	?	Cubic inch (in³)

3. *Mass is constant, but weight can change with location.* Explain.

4. On Earth, 1 kg = 9.8 N = 2.2 lbs. On the Moon, 1 kg = 1.6 N = 0.37 lbs. Use these relationships to answer the following questions. Show your work.

   a. What is the weight, in newtons, of a 50.-kg person on Earth?

   b. What is the weight, in newtons, of a 50.-kg person on the Moon?

   c. What is the mass, in kilograms, of a 100.-lb. person on Earth?

   d. What is the mass of the person in question 4c on the Moon?

5. Explain, using numbered steps, how you could find the volume of a small, irregularly shaped rock.

6. Read the following statements. If the *italicized* word is used correctly, answer "correct." If the italicized word is not used correctly, rewrite the statement with the correct word.

   a. Taylor's *weight* is 65 kilograms.

   b. I went on a diet and my *mass* decreased.

   c. I went to the Moon and I lost *mass*, even though I didn't lose any of my matter.

TECHNOLOGY

Scale vs. balance

How does a **scale** work? An electronic scale, like one you might have in your bathroom, measures the gravitational force between an object and Earth. A scale that displays grams or kilograms has actually measured *weight* and calculated the mass from the weight. How does a **balance** work? A balance measures the mass of an object by comparing it with objects whose masses are known (Figure 2.7).

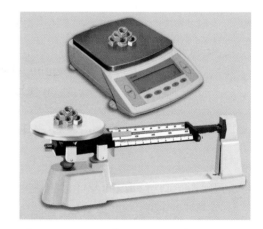

Figure 2.7: *An electronic scale and a triple beam balance.*

2.2 **Determining Density**

Mass and volume are different properties of matter, but they are related. For instance, a solid block of wood and a solid block of steel can have the same volume, but they would *not* have the same mass. The steel block has a lot more mass than the wood block. Because of the mass difference, the wood block floats in water and the steel block sinks. Whether an object floats or sinks is related to the object's density. This section will explain density, a property of all matter.

Density is a property of matter

Density is mass per unit volume   **Density** describes how much mass is in a given volume of a material. Steel has high density; it contains 7.8 grams of mass per cubic centimeter (7.8 g/cm³). Aluminum, as you might predict, has a lower density; a one-centimeter cube has a mass of only 2.7 grams (2.7 g/cm³).

Comparative densities
(20°C at sea level)

Steel
7.8 g/mL

Aluminum
2.7 g/mL

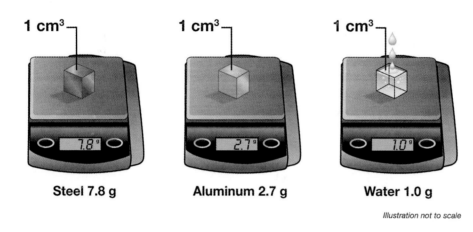

1 cm³ —

1 cm³ —

1 cm³ —

Steel 7.8 g

Aluminum 2.7 g

Water 1.0 g

Illustration not to scale

Water
1.0 g/mL

Air
0.001 g/mL

The density of water and air   Liquids and gases are matter and have density. The density of water is about one gram per cubic centimeter. The density of air is lower, of course—much lower. The air in your classroom has a density of about 0.001 grams per cubic centimeter (0.001 g/cm³) (Figure 2.8).

Figure 2.8: *The density of steel, aluminum, water, and air expressed in grams per milliliter (1 mL = 1 cm³).*

Units of density

Density in units of grams per milliliter
Your laboratory investigations will typically use density in units of grams per milliliter (g/mL). The density of water is 1 gram per milliliter. That means 1 milliliter of water has a mass of 1 gram.

Density in g/cm³ and kg/m³
Some problems use density in units of grams per cubic centimeter (g/cm³). Since 1 milliliter is exactly the same volume as 1 cubic centimeter, the units of g/cm³ and g/mL are actually the same. For measuring large objects, it is easier to use density in units of kilograms per cubic meter (kg/m³). Figure 2.9 gives the densities of some common substances in both units.

Converting units of density
To convert from one unit of density to the other, remember that 1 g/cm³ is equal to 1,000 kg/m³. To go from g/cm³ to kg/m³, you multiply by 1,000. For example, the density of ice is 0.92 g/cm³. This is the same as 920 kg/m³. To go from kg/m³ to g/cm³, you divide by 1,000. For example, the density of aluminum is 2,700 kg/m³. Dividing by 1,000 gives a density of 2.7 g/cm³.

Material	(kg/m³)	(g/cm³)
Platinum	21,500	21.5
Lead	11,300	11.3
Steel	7,800	7.8
Titanium	4,500	4.5
Aluminum	2,700	2.7
Glass	2,700	2.7
Granite	2,600	2.6
Concrete	2,300	2.3
Plastic	2,000	2.0
Rubber	1,200	1.2
Liquid water	1,000	1.0
Ice	920	0.92
Ash (wood)	600	0.67
Pine (wood)	440	0.44
Cork	120	0.12
Air (avg.)	0.9	0.0009

Figure 2.9: *Density of some common materials.*

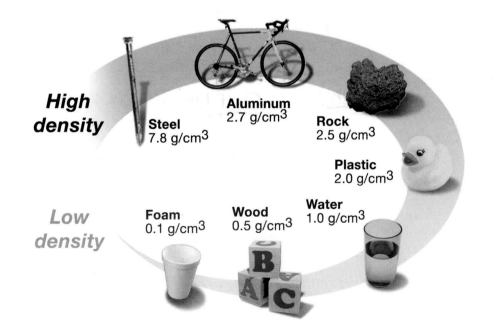

SOLVE IT!

Ipe (pronounced ee-pay) is a Brazilian hardwood that can be used as a durable (and expensive!) construction material for decks, docks, and other outdoor projects. Every cubic foot of ipe weighs 69 pounds. Use dimensional analysis to convert the density of ipe to g/cm³. How does the density compare to other woods and materials on the list above?

Density of solids and liquids

Material density is independent of shape Density is a property of material independent of quantity or shape. For example, a steel nail and a steel cube have different amounts of matter and therefore different masses (Figure 2.10). They also have different volumes, but they have the same density. Dividing mass by volume gives the same density for the nail and the cube because both are made of steel.

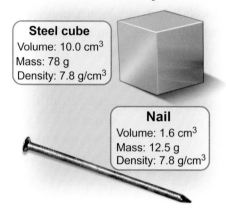

> Steel cube
> Volume: 10.0 cm³
> Mass: 78 g
> Density: 7.8 g/cm³
>
> Nail
> Volume: 1.6 cm³
> Mass: 12.5 g
> Density: 7.8 g/cm³

Density of a material is the same no matter what the size or shape of the material.

Liquids tend to be less dense than solids of the same material The density of a liquid is usually a little *less* than the density of the same material in solid form. Take the example of solder (pronounced sod-der). Solder is a metal alloy used to join metal surfaces.

Figure 2.10: *The density of a steel nail is the same as the density of a solid cube of steel.*

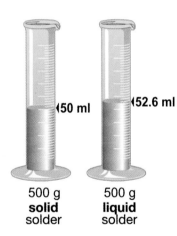

50 ml   52.6 ml

500 g
solid
solder

500 g
liquid
solder

A total of 500 g of solid solder fills a volume of 50 mL. The density of solid solder is 10 g/mL. The same mass of melted (liquid) solder takes up 52.6 mL. Liquid solder has a lower density of 9.5 g/mL. The density of a liquid is lower because the atoms are not packed as tightly as in a solid. Picture a brand-new box of toy blocks. When you open the box, the blocks are tightly packed, like the atoms in a solid. Now imagine dumping the blocks out of the box, and then pouring them back into the original box again. The same number of jumbled blocks take up more space, like the atoms in a liquid (Figure 2.11).

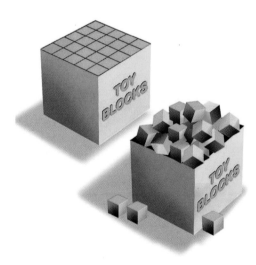

Water is an exception Water is an exception to this rule. The density of solid water, or ice, is *less* than the density of liquid water. When water molecules freeze into ice crystals, they form a pattern that has an unusually large amount of empty space. The water molecules in ice are actually farther apart than they are in liquid water. Because of this, ice floats in liquid water.

Figure 2.11: *The same number (or mass) of blocks arranged in a tight, repeating pattern take up less space than when they are jumbled up.*

Determining density

Finding density   To find the density of a material, you need to know the mass and volume of a sample of the material. You can calculate density using the formula below.

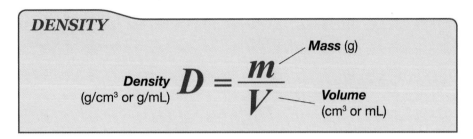

DENSITY

Density (g/cm³ or g/mL) $D = \dfrac{m}{V}$ *Mass* (g) — *Volume* (cm³ or mL)

Aluminum block
Mass = 10.8 grams
Volume = 4 cm³
Density = 2.7 g/cm³

Density gives information about atoms and molecules   Density gives us information about how tightly the atoms or molecules of a particular material are "packed." Diamond is made of carbon atoms and has a density of 3.5 g/cm³ (3,500 kg/m³). The carbon atoms in diamond are relatively tightly packed. Paraffin wax is also made mostly of carbon atoms, but the density is only 0.87 g/cm³ (870 kg/m³). The density of paraffin is low because the carbon atoms are mixed with hydrogen atoms in long molecules that take up a lot of space. The molecules in paraffin are not as tightly packed as the atoms in diamond.

Hollow

Aluminum soda can
Mass = 10.8 grams
Volume = 400 cm³
Average density = 0.027 g/cm³

The average density of a hollow object   Suppose you have a piece of aluminum foil, a length of aluminum wire, and an aluminum brick. At the same temperature and pressure, the aluminum making each of these has the same density. It does not matter whether the aluminum is shaped into a brick, flat sheet, or long wire. The density is 2.7 g/cm³ as long as the object is made of solid aluminum.

If an object is hollow, its average density is less than the density of the material the object is made from. Suppose a small block of aluminum with a mass of 10.8 grams is used to make a soda can (Figure 2.12). Both the solid block of aluminum and the soda can have a mass of 10.8 grams, but the hollow can has a much larger volume. The can has 100 times the volume of the block, so its density is 100 times less.

Figure 2.12: *The aluminum block and can have the same mass but different volumes and densities. The density of the aluminum can is called its average density because it also includes the air inside the can as part of the volume.*

Solving Problems: Calculating Density

To find:	Use:
Density	$D = \dfrac{m}{V}$
Volume	$V = \dfrac{m}{D}$
Mass	$m = D \times V$

A solid wax candle has a volume of 1,700 mL. The candle has a mass of 1.5 kg (1,500 g). What is the density of the candle?

1. **Looking for:**   You are asked for the density.

2. **Given:**   You are given the mass and volume.

3. **Relationships:** Density is mass divided by volume.

4. **Solution:**   Density = 1,500 g ÷ 1,700 mL = 0.88 g/ml

    (answer should have only 2 significant digits)

Your turn...

a. Look at Figure 2.13. A student measures the mass of five steel hex nuts to be 96.2 g. The hex nuts displace 13 mL of water. Calculate the density of the steel in the hex nuts.

b. The density of granite is about 2.60 g/cm³. How much mass would a solid piece of granite have that measures 2.00 cm × 2.00 cm × 3.00 cm?

c. Ice has a density of about 0.920 g/cm³. What is the volume of 100 g of ice?

Figure 2.13: *A student measures the volume and mass of five steel hex nuts.*

SOLVE FIRST LOOK LATER

a. 7.4 g/mL; b. 31.2 g; c. 109 cm³

Section 2.2 *Review*

1. Define density, write the formula (from memory!), and give two different units used to measure density.

2. One cubic centimeter (cm³) is the same volume as one _____ .

3. A material's density is the same, no matter how large or small the sample is, or what its shape is, as long as it is a solid, uniform piece of the material. Explain how this is possible and give an example.

4. The density of balsa wood is about 170 kg/m³. Convert to g/cm³. Why do you think balsa wood, rather than oak or ash, is commonly used for building models? (Use evidence from Figure 2.9 on page 37.)

5. A certain material has a density of 0.2 g/cm³. Is this material better for building a bridge or for making sofa cushions? Explain, using evidence from Figure 2.9 on page 37.

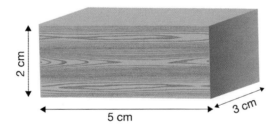

6. The piece of wood shown above has a mass of 20 grams. Calculate its volume and density. Then use Figure 2.9 on page 37 to determine which type of wood it is. What are the two factors that determine a material's density?

7. The density of maple wood is about 755. kg/m³. What is the mass of a solid piece of maple that has a volume of 640. cm³?

SOLVE IT!

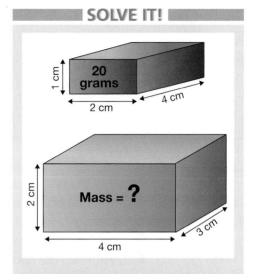

Two toy blocks are made of the same type of material. One has a mass of 20 grams and its dimensions are 2 cm × 4 cm × 1 cm. The second block measures 4 cm × 3 cm × 2 cm. Calculate the mass of the second block.

2.3 **Graphing**

Mass, volume, and density are common properties of matter that we measure. Once we have measured and collected data, it is often necessary to organize it visually, and look for relationships. A **graph** is a visual way to organize data. In this section, we will focus on creating and interpreting scatterplots (*XY* graphs). There are other types of graphs, but scatterplots are the most useful for organizing and presenting physical science data.

◼◼◼◼◼ **VOCABULARY** ◼◼◼◼◼

graph – a visual representation of data

scatterplot (or *XY* graph) – a graph of two variables thought to be related

Types of graphs

Scatterplots, bar graphs, pie graphs, and line graphs

Most graphs are either scatterplots, bar, pie, or line graphs. A **scatterplot** or *XY* **graph** is used to determine if two variables are related. For example, the more hex nuts you have, the more space they take up (Graph A). Scatterplots are commonly used in science, and you will create many of them from the data you collect in your investigations.

A *bar graph* compares groups of information (Graph B). A *pie graph* is a circular graph that shows how a whole is divided up into percentages. (Graph C). A "connect-the-dots" *line graph* is often used to show trends in data over time (Graph D). Strictly speaking, a line graph does not usually show cause and effect. For example, a line graph of a stock price may change over time, but it is not the *time* that causes the change to happen.

A. Scatterplot

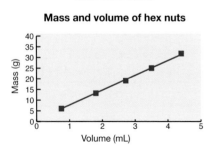

B. Bar graph

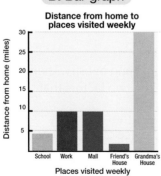

C. Pie graph

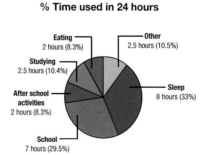

D. Line graph

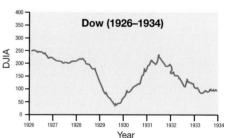

Making a scatterplot or *XY* graph

independent variable – a variable that you believe might influence another variable; the independent variable can also be called the *manipulated variable*

dependent variable – the variable that you believe is influenced by the independent variable; the dependent variable can also be called the *responding variable*

Independent and dependent variables

Scatterplots show how a change in one variable influences another variable. The **independent variable** is the variable you believe might influence another variable. It is often controlled by the experimenter, and is sometimes called the *manipulated variable*. The **dependent variable** is the variable that may be influenced by the independent variable, and can also be called the *responding variable*.

An example

Pressure is measured in units of *atmospheres*. You live at Earth's surface under a pressure of 1 atmosphere. Pressure is critical to safe scuba diving. As a diver goes deeper under water, she has to think about pressure. How does an increase in depth affect the pressure? What sort of graph would best show the relationship between pressure and depth? Figure 2.14 shows depth and pressure data for the ocean.

Step 1: Assign the *x*- and *y*-axes

In this example, depth is the independent or manipulated variable. The diver can choose her depth in the water. The independent variable always goes on the *x*-axis of a graph. The dependent variable always goes on the *y*-axis. In this example, pressure is the dependent variable. Pressure depends on the diver's depth in the water.

Step 2: Make a scale

To create a depth vs. pressure graph, you first make a scale. When talking about a graph, *scale* refers to how each axis is divided up to fit the range of data values. Use the formula below to make a scale for any graph.

$$\text{value per box on graph} = \frac{\text{data range}}{\text{number of boxes on axis}}$$

A quick rule of thumb to use for creating scales is to try counting first by ones, then twos, then fives, then tens. One of these should work most of the time. For example, if the data range for the *x*-axis is 0 to 40 units and the *x*-axis on your graph covers 20 boxes, each box would be worth 2 units.

Depth (m) (*x*-axis)	Pressure (atm) (*y*-axis)
0	1.0
5	1.5
10	2.0
15	2.5
20	3.0
25	3.5
30	4.0
35	4.5
40	5.0

Figure 2.14: *Depth of the ocean and pressure data.*

Step 3: Plot your data   Using the data in Figure 2.14, plot each point by finding the *x*-value and tracing the graph upward until you get to the correct *y*-value. Make a dot for each point. Draw a smooth curve that shows the pattern of the points.

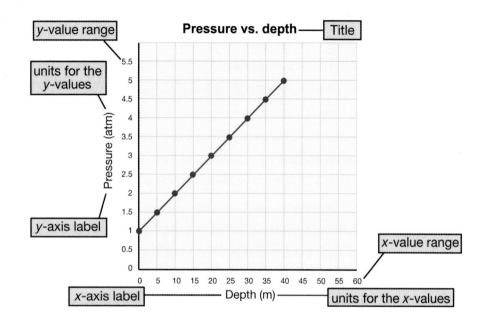

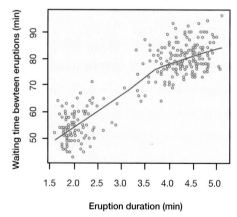

STUDY SKILLS

Key elements of a scatterplot

MIXES TUCS

M: maximize your graph (use all of the graph paper!)

IX: Independent variable on *x*-axis (dependent variable on *y*-axis)

ES: Equally spaced scale increments (start at 0)

T: Title (*y*-variable vs. *x*-variable)

U: Units and labels on both axes

CS: Continuous smooth curve to connect the data points

Step 4: Create a title   Create a title for your graph. Also, be sure to label each axis including units (shown above).

If time is a variable   Like many rules, there are important exceptions. Time is an exception to the rule about which variable goes on which axis. *When time is one of the variables on a graph, it usually goes on the x-axis.* This is true even though you may not think of time as an independent variable.

Using scatterplots in science   When scientists create scatterplots, they are usually working with large amounts of data. Figure 2.15 shows a scatterplot of data for the Old Faithful geyser in Yellowstone National Park, Wyoming. The graph shows there are generally two types of eruptions: short-wait-short-duration, and long-wait-long-duration. This discovery about the geyser activity would be hard to make without the visual aid of the scatterplot!

Figure 2.15: *Waiting time vs. eruption duration for Old Faithful.*

Identifying relationships between variables on a graph

Patterns indicate relationships When there is a relationship between the variables, the graph shows a clear pattern. The speed and distance variables (below left) show a direct relationship. In a **direct relationship**, when one variable increases, so does the other.

When there is no relationship, the graph looks like a collection of dots. No pattern appears. The number of musical groups a student listed in one minute and the last two digits of his or her phone number are examples of two variables that are not related.

VOCABULARY

direct relationship – a relationship in which one variable increases with an increase in another variable

inverse relationship – a relationship in which one variable decreases when another variable increases

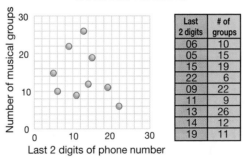

Strong relationship between variables

Distance (cm)	Speed (cm/s)
10	99
20	140
30	171
40	198
50	221
60	242
70	262
80	280
90	297

No relationship between variables

Last 2 digits	# of groups
06	10
05	15
15	19
22	6
09	22
11	9
13	26
14	12
19	11

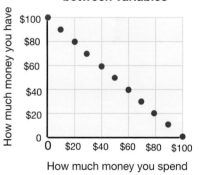

Inverse relationship between variables

Figure 2.16: *Graphs of inverse relationships slope down to the right.*

Inverse relationships Some relationships are inverse. In an **inverse relationship**, when one variable increases, the other decreases. If you graph how much money you have against how much you spend, you see an inverse relationship. The more you have, the less you spend. Graphs of inverse relationships always slope down and to the right (Figure 2.16).

What type of relationship does the depth vs. pressure graph on the previous page show? The depth vs. pressure scatterplot shows a strong direct relationship. That makes sense. The deeper you go, the more water is on top of you, pushing down and creating more pressure.

Reading a graph

Using a graph to make a prediction Suppose you measure the speed of a car at four places on a ramp. Can you figure out the speed at other places without having to actually measure it? As long as the ramp and car are set up the same, the answer is yes! A graph can give you an accurate answer even without doing the experiment. Look at the example below to see how. The students doing the experiment measured and graphed the speed of the car at 20, 40, 60, and 80 cm. They want to know the speed at 50 cm.

1. Start by finding 50 cm on the *x*-axis.

2. Draw a line vertically upward from 50 cm until it hits the curve that fits the points that were measured.

3. Draw a line across horizontally to the *y*-axis.

4. Use the scale on the *y*-axis to read the predicted speed.

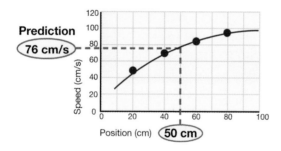

Large graphs are more precise For this example, the graph predicts the speed to be 76 cm/s. You will get the best predictions when the graph is big enough to show precise measurements. That's why you should draw your graphs so they fill as much of the graph paper as possible.

A graph is a form of a model A graph is a simple form of a model. Remember, a model is a relationship that connects two or more variables. Scientists use models to make and test predictions.

SOLVE IT!

A student measures the mass of water collected every five minutes on a rainy day. Design a graph to show the student's data. Estimate how many minutes it took for 20 grams of water to be collected.

Time (min)	Mass (g)
0	0
5	17
10	26
15	38
20	49

Time is the independent variable; therefore, mass is the dependent variable. The mass axis should go from 0 to at least 50 grams. The time axis should go from 0 to at least 20 minutes. The graph shows that 20 grams of rainwater fell in the first 7.5 minutes.

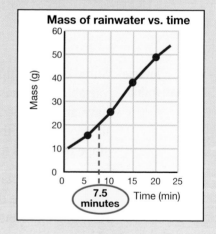

Section 2.3 *Review*

1. Scatterplots, bar graphs, pie graphs, and line graphs all have different purposes. Which type of graph best fits each purpose?

   a. Grouping data for comparison

   b. Comparing parts of a whole

   c. Seeing if two variables are related, such as in cause and effect

2. For each pair of variables, identify which is the independent and which is the dependent variable.

   a. How much gas is left in the gas tank vs. how far the car has traveled

   b. How much money you've spent vs. how much money is in your wallet

   c. How far a toy car traveled vs. how much time went by

3. You have a small tank of water. Suppose you make waves in the tank, and measure their speed in different depths of water. Which is the independent variable, and which is the dependent variable?

4. Make a scatterplot using the data below.

Water depth (cm)	Wave speed (cm/s)
0	0
1	29.8
2	43.3
3	52.1
4	59.2
5	64.4
6	69.3

5. Use the graph of wave speed vs. water depth to answer the following questions.

   a. What happens to wave speed as the depth of the water increases?

   b. What would the average wave speed be at 4.5 cm?

STUDY SKILLS

Four steps to making a graph

Step 1: Determine which variable is independent and which is dependent. The dependent variable (responding variable) goes on the *y*-axis and the independent variable (manipulated variable) goes on the *x*-axis. If *time* is one of the variables, it goes on the *x*-axis.

Step 2: Make a scale for each axis by counting boxes to fit your largest value. Count by multiples of 1, 2, 5, or 10.

Step 3: Plot each point by finding the *x*-value and tracing upward until you get to the corresponding *y*-value.

Step 4: Draw a smooth curve that shows the trend of the points. Do not just connect the dots with straight lines.

2.4 Solving Problems

People use problem-solving skills daily. Doctors collect information about patients to figure out what is causing their pain. Mechanics gather information about a car to figure out how to fix the engine. To solve a problem, you use what you already know to figure out something you want to know. Many physical science problems ask you to calculate something using math formulas.

A four-step technique
The method for solving problems has four steps (Figure 2.17). Follow these steps and you will be able to see a way to the answer most of the time. You will at least make progress toward the answer almost every time. Keep in mind that there is often more than one way to solve a problem. Sometimes you will have to use creativity to find information or to use the information given to figure out a solution.

Solved example problems are provided
Throughout this book, you will find full-page example problems that have been solved for you. Below the solved problem, there is a "Your turn..." section of practice problems. The answers to the practice problems are to the right. Always remember to write out the steps when you are solving problems on your own. If you make a mistake, you will be able to look at your work and figure out where you went wrong. You have already seen several example problem pages in this unit. The example on the next page shows how to follow the problem-solving steps.

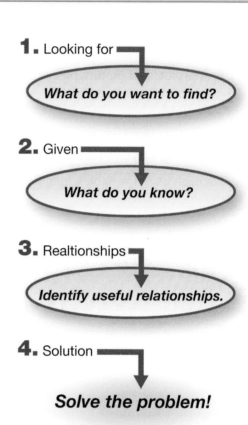

1. Looking for
What do you want to find?

2. Given
What do you know?

3. Realtionships
Identify useful relationships.

4. Solution
Solve the problem!

Figure 2.17: *Follow these steps and you will be able to find the answer to a problem most of the time.*

Step	What to do
1. Looking for:	What is the problem asking for? Figure out exactly what variables or values need to be in the answer.
2. Given:	What information are you given? Sometimes this includes numbers or values. Other times it includes descriptive information to interpret.
3. Relationships:	What relationships exist between what you are asked to find and what you are given? Suppose you are given mass and volume and you are asked to find the density. The relationship to use is $D = m/V$.
4. Solution:	Combine the relationships with what you know to find what you are asked for. Once you complete steps 1 through 3, you will be able to see how to solve most problems.

Solving Problems: Example

A 6-gram marble, placed in a graduated cylinder of water, raises the water from 30 mL to 32 mL. Calculate the marble's volume and density.

1. **Looking for:** You are looking for the marble's volume and density.

2. **Given:** You are given the mass and the results of water displacement.

3. **Relationships:** *water displaced = marble's volume;* $D = \dfrac{m}{V}$

4. **Solution:** *water displaced* = 32 mL – 30 mL = 2 mL = *Volume*

$$D = \frac{6\ g}{2\ mL} = 3\ g\,/\,mL$$

Your turn...

Set up the problem below by writing out steps 1 through 3. This will give you some practice with the problem-solving steps. No need to find the solution unless you want to, of course!

a. Calculate the volume and density of a block that has the dimensions 10 cm × 5 cm × 4 cm and a mass of 400 grams.

SCIENCE FACT

Dense water

The Dead Sea between Israel and Jordan is said to have the saltiest water of all the bodies of water on Earth. This makes the water very dense. Visitors can float effortlessly on the surface of the Dead Sea. In fact, you can't swim underwater in the sea because it is so dense, your body is forced to float. On average, the Dead Sea is nearly nine times saltier than average ocean water. Is the Dead Sea really dead? Almost! Only some bacteria can survive the high mineral content to live in the Dead Sea.

SOLVE FIRST / LOOK LATER

Looking for: *V* and *D*

Given: *m* and block dimensions

Relationships:

$V = L \times W \times H$;

$D = m/V$

Solution:

V = 10 cm × 5 cm × 4 cm

V = 200 cm³

D = 400 g/200 cm³ = 2 g/cm³

How to solve design problems

Different kinds of problems   Consider the following two problems.

- How far do you travel in 2 hours at 60 mph on a straight road?

- Create a container that will protect a raw egg from breaking when dropped 10 meters onto a sidewalk.

"Formula" problems   The first problem has a single answer, 120 miles. You apply what you know (distance = speed × time) to the information you are given and find the answer.

Design problems   The second problem is more challenging (and more fun, too). You have to use what you know to design a solution that solves the problem. Unlike "formula problems," design problems have many correct solutions limited only by your creativity, ingenuity, skill, and patience (Figure 2.18). The important thing is to create something that "does the job" and fits the requirements. In the egg drop problem, typical requirements are that the container must have a mass less than 1 kilogram and cannot be something simply purchased "off the shelf."

Solving design problems   Here are some useful steps to help you solve design problems.

1. Write down everything your solution needs to accomplish.

2. Write down every *constraint* that must also be met. Constraints are limits on cost, weight, time, materials, size, or other things.

3. Think up an idea that *might* work. Talking with others, doing research, and trying things out are all ways to help.

4. Follow the design cycle (shown above).

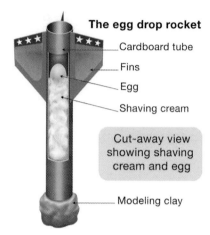

The egg drop rocket
- Cardboard tube
- Fins
- Egg
- Shaving cream

Cut-away view showing shaving cream and egg

- Modeling clay

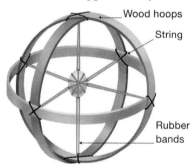

The egg and hoop frame
- Wood hoops
- String
- Rubber bands

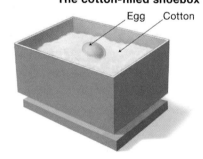

The cotton-filled shoebox
- Egg   Cotton

Figure 2.18: *Some successful solutions to the egg drop problem.*

Section 2.4 *Review*

1. What are two different types of problems you will be asked to solve in this physical science course? Give an example of each.

2. Describe two benefits of following the 4-step method of problem solving.

3. Devise a memory device that could help you remember the 4 problem-solving steps. Describe the memory device.

4. For each of the following, set up the problem using the 4-step method of problem solving, and find the answer. Use the list of relationships at the right to help you with your setups.

   a. Downhill skiing burns about 600 calories per hour. How many calories would you burn if you skied for 3.5 hours?
   b. What is your mass in kilograms if you weigh 120 pounds on Earth?
   c. A car is moving at 65 miles per hour (mph). How many feet can it travel in one second?
   d. The density of Dead Sea surface water is about 1.166 g/mL. How much mass, in grams, would 2 liters of this salty water have?

5. There are four steps to help you solve design problems. What are the four steps?

6. Find a solution to the following design problem (Figure 2.19): *What is the average mass of one grain of rice?* You must use an electronic scale that measures to the nearest tenth of a gram. Also, your answer must be precise to one percent. Use the steps outlined on the previous page for solving design problems. Try out your solution in class, with your teacher's permission and the right materials. Be sure to write down what you are trying to accomplish, what your constraints are, and what your idea is for finding the answer. Then show how you would follow the steps of the design cycle. *Hint*: One grain of rice will not register any mass on the electronic scale. The smallest mass you can measure on the scale with one percent precision is 10 grams $(0.01x = 0.1 \text{ g}; x = 10 \text{ g})$. $\boxed{\text{STEM}}$

Helpful relationships

Use these to help you with question 4.

1 mile = 5,280 feet

1 hour = 3,600 seconds

Cals = Calories/hr × activity hrs

1 kg = 2.2 lb.

1 kg = 9.8 N

$m = D \times V$

1 L = 1,000 mL

Figure 2.19: *Question 6: What is the average mass of one grain of rice?*

STEM Density
and Ocean Currents

Atlantic Ocean
(less salty, less dense)

Mediterranean salt water
(more salty, more dense)

Continental slope

In a remote mountainous region of southeastern Venezuela, water spills over a rocky ledge and tumbles down an amazing 979-meter vertical drop—that's almost a full kilometer! It is the world's highest above-ground waterfall, a spectacular site known as Angel Falls.

Did you know that there are underwater waterfalls in the ocean that are about the same height as Angel Falls? While it may seem strange for water to fall through water, it really happens due to density differences in ocean water coming from different sources.

Saltier water is denser, so when the Mediterranean water enters the less dense Atlantic, it sinks down along the continental slope. At a depth of about 1,000 meters the waterfall becomes an underwater river, which then separates into clockwise-flowing eddies that may continue to spin westward for more than two years. These eddies often merge with others to form giant, salty whirlpools up to 100 kilometers in diameter. Scientists dub these eddies "Meddies" because they originate from water that comes from the Mediterranean Sea.

The Meddies are part of a larger system of ocean currents that plays a critical role in regulating Earth's climate patterns. Currents absorb, store, and release huge amounts of heat and carbon dioxide at different times and places. Complex surface ocean currents and deep ocean currents are driven by a combination of factors. Density is one important factor that influences these ocean current patterns.

Underwater waterfalls

At the Strait of Gibraltar, where the Mediterranean Sea empties into the Atlantic Ocean, there's a giant underwater waterfall. Why? The water flowing out of the Mediterranean Sea is very salty due to low rainfall and the high rate of evaporation there. (When ocean water evaporates, the salt is left behind.)

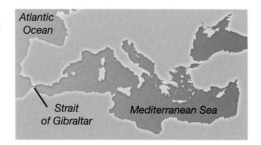

Atlantic
Ocean

Strait
of Gibraltar

Mediterranean Sea

Tracking meddies

Dr. Amy Bower studies these swirling masses of warm, salty ocean currents. She is a senior scientist in the Department of Physical Oceanography at the Woods Hole Oceanographic Institution in Woods Hole, Massachusetts. Dr. Bower and her colleagues use several tools to investigate the location and movement of Meddies. Images from satellites are used to measure sea surface temperature, color, and

height. But to get information about ocean currents far below the surface, Dr. Bower uses a device called a RAFOS float.

The RAFOS float is a glass tube about two meters long containing electronic devices. RAFOS floats can carry multiple sensors for measuring important ocean properties, including pressure, temperature, and dissolved oxygen. A device is attached to make the floats "isopycnal," or density-following, so that they drift in ocean water of a specific density as it rises, falls, and swirls along.

The bottom of the float contains an "acoustic hydrophone," which is basically an underwater sound sensor. The hydrophone picks up sound signals from acoustic beacons (underwater "beepers") that are anchored at specific, known locations. The sound from several beacons is recorded at one time, which makes it possible to pinpoint the float's exact location. At the end of the float's pre programmed mission, it drops its ballast weights, rises to the surface, and beams its data to orbiting satellites.

Dr. Bower receives the information from the satellites and uses it to construct a map of the journey taken by the float. Maps such as these are used to build our overall understanding of how different bodies of water interact to transfer heat, carbon dioxide, and salt from one place to another in the oceans.

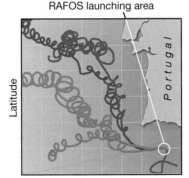

Each colored line represents the path of one float.

Challenges and opportunities

Dr. Bower loves her job. She explains, "Since childhood, I've always had an insatiable curiosity about how the Earth works. I grew up on the coast north of Boston in Rockport, Massachusetts, and spent a lot of my time exploring the rocky beaches there. In high school, I mostly enjoyed physics and math, and so decided to pursue the physical, rather than biological, aspects of oceanography. I was introduced to the field of physical oceanography as an undergraduate in a program called Sea Semester offered by Sea Education Association in Woods Hole. This field of oceanography allows me to explore the oceans from a physical perspective."

When she was in her mid-20s, Dr. Bower was diagnosed with macular degeneration, an incurable eye condition that damages the area of the retina responsible for central vision. While she has some peripheral vision remaining, she is unable to drive, read normal-sized print, or recognize faces. She hasn't let this visual impairment stop her from leading research cruises and doing significant research.

When asked what she finds challenging in her work, Dr. Bower replied, "It's challenging to function at a professional level in a sighted world. The most challenging aspect otherwise is to raise research funding." Scientists frequently have to write grant proposals to organizations that can provide financial support for their research. Effective communication and persuasive writing skills are critical for scientists so that they can continue to do what they love—to use the tools of science to gain greater understanding of the natural world.

Dr. Bower's research has contributed to our understanding of the complex deep ocean current system that profoundly affects Earth's climate. It has also given her an opportunity to travel to many interesting parts of the world. "My favorite part of the job is going to sea on research vessels, an opportunity that few individuals ever have. I like especially research cruises to less explored areas."

Questions:

1. Why is the Mediterranean sea water so salty?

2. Why do oceanographers study Meddies?

3. What skills help scientists obtain research funding?

Photo of Dr. Bower by David Fisichella, Woods Hole Oceanographic Institution

Chapter 2 *Assessment*

Vocabulary

Select the correct term to complete each sentence.

density	graph	mass
dependent variable	independent variable	scatterplot
direct relationship		volume
gram	inverse relationship	weight
	kilogram	

Section 2.1

1. _____ is the amount of space taken up by an object.

2. The amount of matter an object contains is called _____ .

3. The basic SI unit of mass is the _____ .

4. One one-thousandth of a kilogram: _____ .

5. _____ is a measure of the pulling force of gravity.

Section 2.2

6. The mass per unit volume of a material is its _____ .

Section 2.3

7. A(n) _____ is shown by a continuous smooth curve rising from left to right on a scatterplot.

8. The _____ can also be called the manipulated variable.

9. A(n) _____ is also called an XY graph, and it is often used to determine if one variable causes an effect in another variable.

10. The _____ is always plotted on the *y*-axis of a scatterplot.

11. A(n) _____ is a visual representation of data; there are four major types.

12. When one variable decreases as another increases, you have a(n) _____ .

Concepts

Section 2.1

1. Draw a Venn diagram to show the similarities and differences between mass and volume.

2. Describe how to properly find the volume of a liquid in a graduated cylinder. Use the term *meniscus* in your answer.

3. What are two precautions to take when using an electronic scale, so you don't damage the equipment?

4. What is the difference between mass and weight?

5. At the top of Mt. Everest in the Himalayas, the force of gravity is slightly less than it is at sea level. Would your weight be a little greater or a little less on the top of this mountain? Explain.

Section 2.2

6. In general, how do the densities of a material in solid and liquid form compare? Name a common exception to the general rule.

7. A cube of solid steel and a cube of solid aluminum are both covered with a thin plastic coating, making it impossible to identify the cubes based on color. Referring to Figure 2.9 on page 37, tell how you could determine which cube is steel and which is aluminum.

8. Which makes better packing material: a high-density or low-density material? Why?

Section 2.3

9. Why is the scatterplot the most commonly used type of graph in science?

10. A blank graph grid is 20 boxes by 20 boxes. You want to plot a data set on this graph. The range of x-axis values is 0–60. The range of y-axis values is 0–15. Sketch the best scale to use that would maximize the graph size.

11. You wish to make a graph of the height of the Moon above the horizon every 15 minutes between 9:00 p.m. and 3:00 a.m. during one night.
    a. What is the independent variable?
    b. What is the dependent variable?
    c. On which axis should you graph each variable?

Section 2.4

12. Copy the problem-solving steps table below on a piece of paper or in a notebook. Fill in the table as you solve this problem. Platinum is a valuable metal. The density of platinum is 21.4 g/cm^3. Suppose you have a 113-cm^3 disk of pure platinum. What is the mass of the disk?

Step	What to do
1. Looking for:	
2. Given:	
3. Relationships:	
4. Solution:	

13. You have an idea for making a new type of backpack that is more comfortable to wear. Describe how you would use the design cycle to turn your idea into a practical solution. STEM

Problems

Remember: Use problem-solving steps to solve these problems.

Section 2.1

1. Make the following calculations for the weight, in newtons, of a 60-kg person on:
    a. Earth (1 kg = 9.8 N)
    b. Mars (1 kg = 3.7 N)
    c. Find the weight, in pounds, of the person on Earth (1 kg = 2.2 lbs.)
    d. Find the weight, in pounds, of the person on Mars. (*Hint*: Use information from question 1b to help you figure out the correct relationship to use.)

Section 2.2

2. A chunk of paraffin (wax) has a mass of 50.4 grams and a volume of 57.9 cm^3. What is the density of paraffin?

3. A large amount of the gold reserve for the United States is stored in the Fort Knox Bullion Depository vault in Kentucky. Much of it is in the form of bars with the dimensions 7 in × 3 5/8 in × 1 3/4 in. The gold has a density of 19,300 kg/m^3. Calculate the mass of one gold bar (1 in = 2.54 cm). If you picked up this gold bar, would it be more like picking up a can of soda, a gallon of water, or a box of books? Set up this problem using your problem-solving steps and you will avoid a lot of confusion!

Section 2.3

Use the graph below to answer questions 4 through 7. This graph was created by a student who measured the mass and volume of a collection of hex nuts from a hardware package. Each hex nut was made of the same material, and each was the same size and shape.

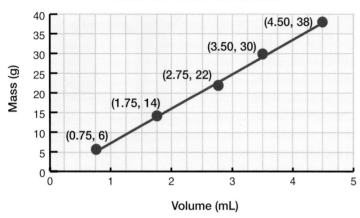

Mass and volume of hex nuts

4. Each data point in the mass/volume graph represents adding an additional hex nut to the group. The first data point shows the mass and volume of one hex nut, the second data point shows the mass and volume of two hex nuts together, and so on up to five hex nuts.

    a. What is the mass and volume of one hex nut?
    b. What is the mass and volume of five hex nuts together?
    c. What do you predict the mass and volume of six hex nuts would be?

5. What type of relationship exists between mass and volume on the mass/volume graph?

6. Make a quick sketch of what you think the scatterplot would look like if you used random hex nuts of different materials and sizes, rather than a collection that is all the same.

7. *Challenge*: Can you use information from this graph to determine what material the hex nuts are made from? You can use information from Figure 2.9 on page 37 to help you. Explain how you arrived at your answer.

Applying Your Knowledge

Section 2.1

1. Matter is anything that has mass and volume. Is air considered to be matter? Describe a simple demonstration you could do to prove your answer.

2. Write a procedure for finding the volume of air spaces in an ordinary kitchen sponge.

Section 2.2

3. Deep ocean currents are caused by differences in ocean water density. What two things can cause density differences in ocean water? How does the density difference actually cause the movement of deep ocean water? Do some research to find the answers to these questions. Be sure to cite your references.

Section 2.3

4. Look through recent newspapers and/or magazines to find at least one example of a scatterplot, bar graph, pie graph, and line graph. Photocopy or cut out the graph examples and create a small poster that illustrates the differences between these types of graphs.

The Scientific Process

In 2014, on Mont Ventoux in France, Guy Martin of the United Kingdom climbed into a gravity race car with no motor. With one push, he raced downhill with nothing else but gravity for energy, and set a world record of 85.6 mph! His fiercely competitive engineering team had created the slipperiest low-friction car they could in four months' time, using high-tech materials and parts. At the speeds he was traveling, the car also had to be safe so safety features were designed such as roll-cage protection, a safety harness, and an energy-absorbing steering column.

How did his car reach such high speeds without using a motor? How did the engineering team design his car to be so fast? Answers to these questions involve experiments and variables. Read on and you will find out how engineers learn to make things better, faster, and more efficient.

CHAPTER 3 INVESTIGATIONS

3A: Measuring Time
How is time measured accurately?

3B: Experiments and Variables
How do you design a valid experiment?

3.1 Inquiry and the Scientific Method

Scientists believe the universe follows a set of rules called **natural laws**. Everything that happens obeys the same natural laws. Unfortunately, the natural laws are not written down, nor are we born knowing them. *The primary goal of science is to discover what the natural laws are.* Over time, we have found that the most reliable way to discover natural laws is through *scientific inquiry.*

What "inquiry" means

Inquiry is learning through questions

How is science like solving a mystery?

Learning by asking questions is called **inquiry** (Figure 3.1). Inquiry resembles a crime investigation with a mystery to solve. Something illegal happened and the detective must figure out who did it. Solving the mystery means accurately describing who did what, when they did it, and how. The problem is that the detective never actually *saw* what happened. The detective must **deduce** what happened in the past from information collected in the present.

Searching for evidence

In the process of inquiry, the detective asks lots of questions related to the mystery. The detective searches for evidence and clues that help answer the questions. Eventually, the detective comes up with a *theory* about what happened. The theory is a description of what must have occurred in the crime, down to the smallest details.

How do you know you have learned the truth?

At first, the detective's theory is only one possible explanation among several of what *might have happened*. The detective must have evidence to back up the theory. To be accepted, a theory must pass three demanding tests. First, it must be supported by enough evidence. Second, there cannot be even a *single* piece of evidence that contradicts the theory. Third, the theory must be unique because if two theories both fit the facts equally well, you cannot tell which is correct. When the detective arrives at a theory that passes all three tests, he believes he has "solved" the mystery by using the process of inquiry.

natural law – a rule that describes an action or set of actions in the universe and that can sometimes be expressed by a mathematical statement

inquiry – a process of learning that starts with asking questions and proceeds by seeking the answers to the questions

deduce – to figure something out from known facts using logical thinking

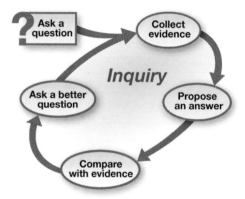

Figure 3.1: *The steps in learning through inquiry.*

Scientific evidence

What counts as scientific evidence? In science, the only way to know if you are right is to test your idea against real evidence. But what types of evidence qualify as *scientific* evidence? Do feelings or opinions count as scientific evidence? Does what other people think qualify as scientific evidence? The answer to both questions is no. Because evidence is so important in science, there are exacting rules defining what counts as scientific evidence.

An example of scientific evidence Scientific evidence may include numbers, tables, graphs, words, pictures, sound recordings, or other information. The important thing is that the evidence accurately describes what happens in the real world (Figure 3.2). Scientific evidence may be collected without doing experiments in a laboratory. For example, Galileo used his telescope to look at the Moon. He recorded what he saw by sketching in his notebook. Galileo's sketches are considered scientific evidence.

When is evidence considered scientific? Scientific evidence must be **objective** and **repeatable**. *Objective* means the evidence should describe *only what actually happened* as exactly as possible. *Repeatable* means that others who look the same way at the same thing will observe the same results. Galileo's sketches describe in detail what he actually saw through the telescope. That means the sketches are *objective*. Others who looked through his telescope saw the same thing. That makes the sketches *repeatable*. Galileo's sketches are good scientific evidence because they are both objective and repeatable. Galileo's sketches helped convince people that the Moon was actually a world like Earth with mountains and valleys. This was not what people believed in Galileo's time.

Communicating scientific evidence with exact definitions It is important that scientific evidence be clear, with no room for misunderstanding. For this reason, scientists define concepts like "force" and "weight" very clearly. Usually, the scientific definition is similar to the way you already use the word, but more exact. For example, your "weight" in science means the force of gravity pulling on the mass of your body.

Examples of scientific evidence

Pictures or sketches that show actual observations

Time (s)	Speed (m/s)	Position (m)
0.0	0.00	0.00
0.2	0.83	0.08
0.4	1.66	0.33
0.6	2.50	0.75
0.8	3.33	1.33
1.0	4.16	2.08

Measurements and data

Figure 3.2: *Some examples of scientific evidence.*

Scientific theories

How theories are related to natural laws
A scientific **theory** is a human attempt to describe a natural law. For example, if you leave a hot cup of coffee on the table, eventually it will cool down. Why? There must be some natural law that explains what causes the coffee to cool. A good place to start looking for the law is by asking what it is about the coffee that makes it hot. Whatever quality that creates "hot" must go away or weaken as the coffee gets cold (Figure 3.3). The question of what causes hot and cold puzzled people for a long time.

The theory of caloric
Before 1843, scientists believed (a theory) that heat was a kind of fluid (like water) that flowed from hotter objects to colder objects. They called this fluid *caloric*. People thought hot objects had more caloric than cold objects. When a hot object touched a cold object, the caloric flowed between them until the temperatures were the same.

Testing the theory
The caloric theory explained what people knew at the time. However, a big problem came up when people learned to measure weight accurately. Suppose caloric really did flow from a hot object to a cold object. That means an object should weigh more when it's hot than it does when it's cold. Experiments showed this was not true. Precise measurements showed that objects have the same weight, whether hot or cold. The caloric theory was soon abandoned because it could not explain this new evidence.

How theories are tested against evidence
Scientists are always testing theories against new experiments and new evidence. One of two things can happen when new evidence is found.

1. The current theory correctly explains the new evidence. This gives us confidence that the current theory is the right one.

   OR

2. The current theory does *not* explain the new evidence. This means there is a new (or improved) theory waiting to be discovered that can explain the new evidence (as well as all the old evidence).

What makes these two cups of coffee different?

Hot coffee
70°C

Cold coffee
21°C

Figure 3.3: *A question that might begin inquiry into what "heat" really is.*

Hypotheses

The hypothesis Based on observations and evidence, a good detective evaluates many different theories for what might have happened. Each different theory is then compared with the evidence. The same is true in science, except that the word *theory* is reserved for a single explanation supported by lots of evidence collected over a long period of time. Instead of *theory*, scientists use the word **hypothesis** to describe a possible explanation for a scientific mystery.

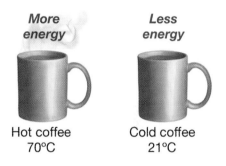

VOCABULARY

hypothesis – a possible explanation that can be tested by comparison with scientific evidence

More energy   Less energy

Hot coffee 70°C   Cold coffee 21°C

Many experiments are done to see if heat might be energy.

Hypothesis
Suppose heat is really a form of energy. Then it would not have any mass.

Theory
Heat is a form of energy.

Figure 3.4: *A hot cup of coffee has more heat energy than a cold cup of coffee. As coffee cools, its heat energy is transferred to the air in the room. As a result, the air is warmed.*

Theories start out as hypotheses Theories in science start out as hypotheses. The old explanation that heat was the fluid caloric was an incorrect hypothesis, one of many leading up to the modern theory of heat. The first hypothesis that heat is a form of energy was made by a German doctor, Julius Mayer, in 1842, and was confirmed by experiments done by James Joule in 1843. Energy has no weight, so Mayer's hypothesis explained why an object's weight remained unchanged whether it was hot or cold. After many experiments, Mayer's hypothesis that heat was a form of energy became the *theory* of heat that we accept today (Figure 3.4).

Hypotheses must be testable to be scientific A scientific hypothesis must be testable. That means it must be possible to collect evidence that proves whether the hypothesis is true or false. This requirement means that *not all hypotheses* can be considered by science. For instance, it has been believed at times that creatures are alive because of an undetectable "life force." This is not a scientific hypothesis because there is no way to test it. If the "life force" is undetectable, that means no evidence can be collected that would prove whether it exists or not. Science restricts itself only to those ideas that may be proved or disproved by actual evidence.

The scientific method

Learning by chance In their early years, children learn about the world by trial and error. Imagine a small child trying to open a jar. She will try what she knows: biting the lid, pulling on it, shaking the jar, dropping it, until—by chance, she twists the lid. It comes off. She puts it back and tries twisting it again— and the lid comes off again. The child learns by trying many different things and then *remembering what works*.

Learning by the scientific method It takes a long time to learn by randomly trying everything. What's worse, you can never be sure you tried *everything*. The **scientific method** is a much more dependable way to learn.

The Scientific Method

1. Scientists observe nature, then develop or revise hypotheses about how things work.

2. The hypotheses are tested against evidence collected from observations and experiments.

3. Any hypothesis that correctly accounts for all of the evidence from the observations and experiments is a potentially correct theory.

4. A theory is continually tested by collecting new and different evidence. Even one single piece of evidence that does not agree with a theory will force scientists to return to step one.

Why the scientific method works The scientific method is the underlying logic of science. It is a careful and cautious way to build an evidence-based understanding of our natural world. Each theory is continually tested against the results of observations and experiments. Such testing leads to continued development and refinement of theories to explain more and more different things. The way people came to understand the solar system is a good example of how new evidence leads to new and better theories (Figure 3.5).

Early civilizations thought Earth was covered by a dome on which the Sun, stars, and planets moved.

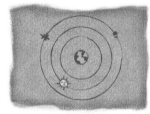

During the Middle Ages, people thought the Sun, stars, and planets circled Earth which sat in the center.

Today we know Earth and the planets orbit around the Sun, and the stars are very far away.

Figure 3.5: *Three different models for Earth and the solar system that were believed at different times in history.*

Section 3.1 *Review*

1. Which of the following is an example of deduction?
   a. Hector calls the weather service to find out if the temperature outside is below freezing.
   b. Caroline looks out the window and concludes that the temperature is below freezing because she sees that the puddles in her neighbor's driveway are frozen.

2. Describe the relationship between a hypothesis, a theory, and a natural law.

3. To be correct, a scientific theory must be everything **except**
   a. supported by every part of a large collection of evidence
   b. considered to be unchangeable even if new scientific evidence disproves it
   c. testable by comparison with scientific evidence
   d. an explanation of something that actually occurs in the natural world or in man-made technology

4. Julie, a third-grade student, believes that the Moon disappears on certain days every month. Explain why the following information is or is not scientific evidence that can be used to evaluate Julie's hypothesis.
   a. Julie sometimes cannot see the Moon all night, even though the sky is clear.
   b. Anne, Julie's older sister, thinks the phases of the Moon are caused by the Moon's position in its orbit around Earth.

5. When describing scientific evidence, what is the meaning of the word *repeatable*?

6. Which of the following is an example of learning through inquiry?
   a. Miguel is told that hot objects, like a cup of coffee, cool off when left on the table in a cooler room.
   b. Erik wonders what happens to hot objects if you remove them from the stove. He puts a thermometer in a pot of boiling water and observes that the water cools off once it's removed from the heat source.

3.2 Experiments and Variables

An **experiment** is a situation specifically set up to investigate something. For example, you could do an experiment in your class to investigate how fast a car moves as it travels down a ramp (Figure 3.6). An experiment is designed around a **system**, or a group of variables that are related in some way.

Experiments

Experiments tell us how variables are related   The goal of any experiment is to understand the relationship between variables. For example, what is the relationship between the speed of the car and the angle of the ramp? To answer the question, you set up the experiment with the ramp attached to different holes in the stand. Each hole puts the ramp at a different angle. You measure the speed of the car at different ramp angles to see how (and if) the speed changes when the angle is changed. A **variable** is a factor that affects how an experiment works.

Changing one variable at a time   In a simple, ideal experiment, *only one variable is changed at a time*. You can assume that any changes you see in other variables were caused by the one variable you changed. If you change more than one variable at a time, it's hard to tell which one caused the changes in the others. You would not learn much from the results of that experiment.

The experimental variable   The variable you change in an experiment is called the **experimental variable**. This is usually the variable that you can freely manipulate. For the experiment with a car on a ramp, the angle of the ramp is the experimental variable. If you were experimenting with different brands of fertilizer on tomato plants, the experimental variable would be the brand of fertilizer.

Control variables   The variables you keep the same are called **control variables**. If you are changing the angle of the ramp, you want to keep the mass of the car the same each time you roll the car. Mass is a control variable. You also want to keep the position of the photogate the same. Photogate position is also a control variable. You will also want to have the same release technique for the car each time it rolls down the ramp. If you want to test different angles, the ramp angle should be the *only* variable you change.

■■■■■ **VOCABULARY** ■■■■■

experiment – a situation specifically set up to investigate relationships between variables

system – a group of variables that are related

experimental variable – the variable you change in an experiment

variable – a factor that affects how an experiment works

control variable – a variable that is kept constant (the same) in an experiment

Hole #	Time (s)
5	0.0286
7	0.0220
9	0.0198
11	0.0160
13	0.0125

Figure 3.6: *A car rolling downhill can be an experiment.*

Experimental techniques

Experiments often have several trials

Many experiments are done over and over with only one variable changed. For example, you might roll a car down a ramp 10 times, each with a different angle. Each time you run the experiment is called a **trial**. To be sure of your results, each trial must be as similar as possible to all the others. The only change should be the one variable you are testing.

trial – each time an experiment is tried

experimental technique – the exact procedure that is followed each time an experiment is repeated

procedure – a collection of all the techniques you use to do an experiment

Experimental technique

Your **experimental technique** is how you actually do the experiment. For example, you might release the car using one finger on top. If this is your technique, you want to do it the same way every time. When you place the photogate on the track, you make sure the gate is always perpendicular to the track. By developing a good technique, you make sure your results accurately show the effects of changing your experimental variable. If your technique is sloppy, you may not be able to determine if your results are due to technique or changing your variable.

Procedures

The **procedure** is a collection of all the techniques you use to do an experiment. Your procedure for testing the ramp angle might have several steps. Good scientists keep careful track of their procedures so they can come back another time and repeat their experiments. Writing the procedures down in a lab notebook is a good way to keep track (Figure 3.7).

Scientific results must always be repeatable

Scientific discoveries and inventions must always be testable by someone other than you. If other people can follow your procedure and get the same results, then most scientists would accept your results as being true. Writing good procedures is the best way to ensure that others can repeat and verify your experiments. This is a good thing to keep in mind when you write your own procedure for an experiment. Write it with enough detail that someone else could follow the procedure and do the experiment exactly the way you did it.

Communicating your results

A *lab report* is a good way to communicate the results of an experiment to others. It should contain your research question, hypothesis, experiment procedures and data, and conclusion. If you give an oral report to your class, colorful charts and graphs are a good way to show your data. This is how scientists present the results of their experiments to other scientists.

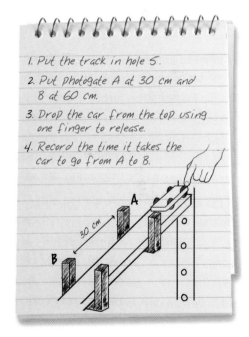

Figure 3.7: *A notebook keeps your observations and procedures from getting lost or being forgotten.*

Experiments: Then and now

Electricity and magnetism in the 1800s

Michael Faraday, a British scientist, made some important discoveries while experimenting with electricity and magnets. This is a great example of how one experiment often leads to another. Faraday's original question was, "How are electricity and magnetism related?"

The original experiment

Faraday designed a controlled experiment with a loop of wire and a magnet. When he moved the magnet through the loop of wire, an electric current was produced in the wire. In previous experiments, he had generated electricity by using homemade batteries, but the magnet experiment was different. Moving a magnet through the wire loop was enough to produce an electric current in the wire, without using the chemical reactions of his homemade batteries. The opposite was also true. When Faraday rotated a wire through a magnetic field, an electric current was produced in the wire (Figure 3.8).

A new experiment based on the old one

NASA (National Aeronautics and Space Administration) has conducted a modern version of Faraday's electromagnetism experiments. In simple terms, Earth is like a giant magnet. Magnetic field lines extend out from Earth into space. What would happen if Faraday's experiment were performed in space? What if you dragged a wire through Earth's magnetic field? Could an electric current be produced in the wire? This became an important mission for the space shuttle in 1996 (Figure 3.9).

The world's most amazing electricity generator

NASA scientists worked with Italian scientists to design equipment for the experiment. They made a special satellite and connected it to the space shuttle with more than 20 kilometers of a special insulated copper cable. As the shuttle orbited Earth, scientists released the tethered satellite and conducted 12 different experiments while dragging the cable through Earth's magnetic field at speeds over 15,000 mph! The satellite was equipped with many instruments to study the electricity generated in the cable. As the cable cut through Earth's magnetic field, 3,500 volts of electricity were produced, and a current of 0.5 amperes was generated. Faraday's experiment worked in outer space! Unfortunately, the tether broke during the experiment and the satellite was lost, but not before scientists gathered enormous amounts of interesting data.

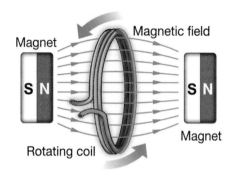

Figure 3.8: *Electric current is created when a coil rotates in a magnetic field.*

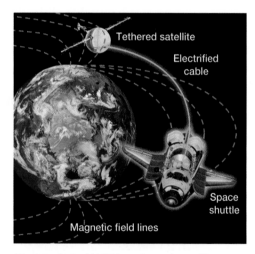

Figure 3.9: *NASA's tethered satellite experiment from shuttle mission STS-75. An electric current was created when the cable was dragged through Earth's magnetic field.*

Section 3.2 *Review*

1. Why is experimentation so important to science?

2. Why is it important, in an ideal, simple experiment, to change only one variable from trial to trial?

3. What is the difference between an experimental variable and a control variable? Give an example to explain your answer.

4. You are planning an experiment to find out which detergent is the best at removing grass stains from cotton fabric. Think about how you might do this experiment and what kinds of variables are involved. Make a list of two variables that would be a part of the experimental system and two variables that would not be a part of this system.

5. Suppose you have three identical drinking cups: one made of plastic, one foam, and one paper. You want to find out which cup will keep your hot cocoa hot for the longest time. **STEM**
   a. Phrase a formal question for this experiment.
   b. What is your hypothesis?
   c. What is the experimental variable?
   d. What are three important control variables?
   e. What type of evidence will you collect to test your hypothesis?
   f. *Challenge*: Conduct your experiment and summarize your findings.

6. Think of an experiment you did in a past science class.
   a. Describe the experiment.
   b. What was the experimental variable?
   c. What were two control variables?
   d. What was the outcome of the experiment?

7. When an open cup of water sits on a table, it will evaporate over time. Do all liquids evaporate at the same rate? Suppose you conduct an experiment to see how quickly water, rubbing alcohol, and nail polish remover evaporate. Describe three important techniques you will have to follow to make sure your experimental procedure is repeatable and objective. **STEM**

▬▬▬ CHALLENGE ▬▬▬

Science and serendipity

Not all discoveries in science are made using the scientific method! In fact, many important new discoveries and inventions happen by trial and error, a lucky experiment, or by accident. The word *serendipity* describes an event in which a valuable discovery is made by accident.

1. Describe a situation in which you made a serendipitous discovery.

2. Think about an object you use everyday. Find out how it was invented. Was this invention the result of serendipity? Why or why not?

▬▬▬ SOLVE IT! ▬▬▬

For question 5, predict what a graph or graphs of the data you collect would look like. Sketch the graph or graphs. The graph(s) should support the hypothesis you made in 5b.

3.3 The Nature of Science and Technology

Science is a way of knowing that is based on evidence, logic, and skepticism. The purpose of scientific study is to learn about our natural world. Technology, such as mobile phones and medical devices, uses scientific knowledge to create devices like mobile phones and medical instruments that meet needs and solve problems. Science and technology are closely related, as you will see in this section.

Ethical traditions

Truthful reporting Scientists all over the world conduct experiments every day, in colleges and universities, in industry, and for government agencies. Truthful reporting is the most important tradition of science. When scientists collect data, organize it, report it, and write about their results and conclusions, they must be unbiased and honest in their communication.

Scientific journals and peer review How do scientists get the word out about their findings? Often, scientists write a report of their experiments and submit it to a *scientific journal*. A scientific journal is a publication, like a magazine, that comes out on a regular basis. There are hundreds of major scientific journals in print. Before a paper is published in a scientific journal, the work is reviewed by peers. Only if the work is approved by independent scientists can the paper be published.

Science news for everyone Scientific journals can be very technical, and cannot be read like an ordinary magazine. Other periodicals such as *Popular Science* or *Scientific American* are also published. Their articles are less technical. They are selected from the thousands of papers that are published each month in scientific journals. These magazines are not scientific journals, but they are a good source of current science research findings. Daily news broadcasts on television, radio, and the Internet also carry headlines about recent scientific discoveries. Sometimes these headlines are unintentionally misleading, because they are just quick summaries of technical research explanations. Keep this in mind when you hear science news "sound bites." Remember, good science is always repeatable, durable, based on evidence, and unbiased.

▮▮▮▮ JOURNAL ▮▮▮▮

Scientific journals

Perhaps you are keeping a journal for your science class. Your science journal can contain notes, thoughts, reflections, scientific data, experimental procedures, tables, graphs, and lab reports.

A *scientific journal* is a specific kind of publication that is different from a science journal you might create in class. A scientific journal is a periodical publication that contains the results and conclusions of many different experiments. All of the papers (you might think of them as articles) submitted to a scientific journal must be reviewed by peers and accepted before they are published.

Have you heard of any of these scientific journals?

- *Nature*
- *Science*
- *Proceedings of the National Academy of Sciences*
- *Journal of the American Medical Association*
- *Journal of the American Chemical Society*
- *Advances in Physics*

Science and technology

Inventions solve problems

You are surrounded by inventions, from the toothbrush you use to clean your teeth to the computer you use to do your school projects (and play games). Where did these inventions come from? Most of them came from a practical application of science knowledge.

What is technology?

Science helps us understand the natural world. **Technology** is the application of science to meet human needs and solve problems. All technology—from the windmill to the supersonic jet—arises from the perception that "There must be a better way to do this!" Although technology is widely different in the details, there are some general principles that apply to all forms of technological design or innovation. People who design technology to solve problems are called **engineers**. Scientists study the natural world to learn the basic principles behind how things work. Engineers use scientific knowledge to create or improve inventions that solve problems.

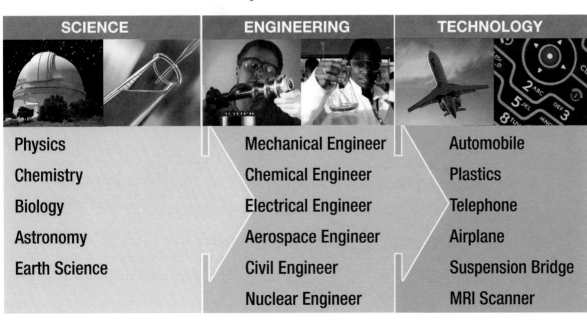

SCIENCE	ENGINEERING	TECHNOLOGY
Physics	Mechanical Engineer	Automobile
Chemistry	Chemical Engineer	Plastics
Biology	Electrical Engineer	Telephone
Astronomy	Aerospace Engineer	Airplane
Earth Science	Civil Engineer	Suspension Bridge
	Nuclear Engineer	MRI Scanner

TECHNOLOGY

GPS

GPS stands for *Global Positioning System*. A GPS receiver can determine its position to within a few meters anywhere on Earth's surface. How does this work? How does the GPS receiver "know" its position?

There are 24 satellites in orbit around Earth that transmit radio signals as part of a global navigation system. At any one time, you can receive signals from anywhere from 6 to 11 of those satellites in the sky, all transmitting their unique codes and locations. A GPS receiver works by comparing the signals from four different GPS satellites.

STEM Engineering

VOCABULARY

prototype – a working model of a design that can be tested to see if it works

engineering cycle – a process used to build and test devices that solve technical problems

A sample engineering problem

Suppose you are given a box of toothpicks and some glue, and are told to build a bridge that will hold a brick without breaking. After doing research, you come up with an idea for how to make the bridge. Your idea is to make the bridge from four structures connected together. Your idea is called a *conceptual design*.

Basic structure (truss)   BRICK   Conceptual design for bridge

The importance of a prototype

You need to test your idea to see if it works. If you could figure out how much force it takes to break *one* structure, you would know if four structures will hold the brick. Your next step is to build a **prototype** and test it. Your prototype should be close enough to the real bridge that what you learn from testing can be applied to the actual bridge.

Testing the prototype

You test the prototype by applying more and more force until it breaks. You learn that your structure (called a truss) breaks at a force of 5 newtons. The brick weighs 25 newtons. Four trusses are not going to be enough. You have two choices now. You can make each truss stronger, by using thread to tie the joints. Or you could use more trusses in your bridge (Figure 3.10). The *evaluation* of test results is a necessary part of any successful design. Testing identifies potential problems in the design in time to correct them.

Changing the design and testing again

If you decide to build a stronger structure, you will need to make another prototype and test it again. Good engineers often build many prototypes and keep testing them until they are successful under a wide range of conditions. The process of design, prototype, test, and evaluate is the **engineering cycle**. The best inventions go through the cycle many times, being improved after each cycle until all the problems are worked out.

Engineering Cycle

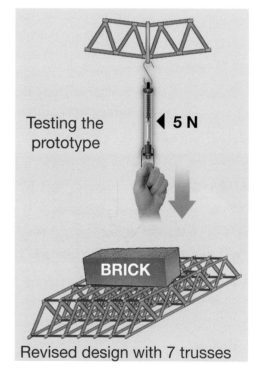

Testing the prototype   ◀ 5 N

BRICK

Revised design with 7 trusses

Figure 3.10: *Testing the prototype tells you if it is strong enough. Testing often leads to an updated design, such as this one that uses more trusses.*

Section 3.3 *Review*

1. Why are scientific journals such as *Nature* and *Journal of the American Medical Association* extremely important to scientific progress?

2. Suppose a scientist conducts a series of experiments and the results are so amazing, she wants to share them with other experts in the field. Why would it be risky for her to make the results public on her own Web site before publishing them in a scientific journal?

3. Identify whether each item is an example of science or technology.
   a. digital music player
   b. Newton's laws of motion
   c. atomic theory
   d. windmill
   e. Universal law of gravitation
   f. GPS
   g. biochemistry
   h. maglev train

4. Discuss what might happen if an automobile manufacturer began making and selling cars based on a prototype that went through only one engineering design cycle.

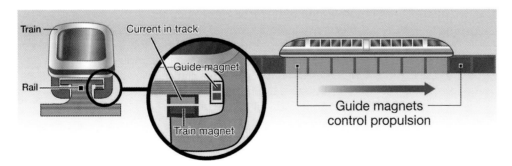

A maglev train track has electromagnets in it that both lift the train and pull it forward. See the technology box in the sidebar.

TECHNOLOGY

Magnetic levitation

In the previous section, you read about Michael Faraday's experiments with electromagnetism. A powerful new technology based on Faraday's experiments is currently in development. Magnetically levitated, or maglev, train technology uses electromagnetic force to lift a train a few inches above its track (see figure below). By "floating" the train on a powerful magnetic field, the friction between wheel and rail is eliminated. Maglev trains can reach high speeds using less power than an ordinary train. In 2015, a Japanese seven-car maglev train carrying 49 passengers reached a record speed of 375 mph! Maglev trains are now being developed and tested in Germany and the United States. Many engineers believe maglev technology will become the standard for mass transit systems over the next 100 years. Perhaps someday you will commute to work on a maglev train!

Ethics
in Medical Research

Dr. Louis Cantilena enjoys being close to the cutting edge of science. As director of the Division of Clinical Pharmacology and Medical Toxicology at the Uniformed Services University of Health Science (USUHS), Cantilena leads a team of researchers who look for and test new drugs as well as antidotes for poisons. Their medical research studies the effects of these potential treatments on human subjects. In his

work as a professor, doctor, and researcher, he sees firsthand the enormous benefits of scientific advancement through research.

Cantilena's research unit is funded largely by federal institutions such as the Food and Drug Administration (FDA), the National Institutes of Health (NIH), and the U.S. Army. The research unit is responsible for using taxpayers' dollars wisely and ethically and for maintaining the trust of the public. Cantilena knows that ethics plays a crucial role in any scientist's work.

What does ethics mean?

Ethics can be defined as "standards of conduct that enable a person to determine what behavior is right or wrong in a given situation." Because medical research involves the use of human subjects and its results can be used to either harm or help the human population, the application of ethics is vital.

The vast majority of scientists are very careful with both the process of research and the reporting of results. They realize that accuracy and honesty are required to serve the purposes of discovery and advancement in science. Peer review, publication in scientific journals, and informal exchanges of ideas are the most common ways that scientists share and critique one another's work. Naturally, some errors do occur in research, but usually they are not intentional or unethical. However, when an individual scientist decides to deceive others about the nature of his or her research, it is called ethical misconduct. Some examples of ethical misconduct include making up results, changing data, or not giving the proper credit for someone else's ideas.

Truthfulness and trust

Cantilena explains that scientists are not helping anyone by not telling the truth about research results. While it may be tempting to advance one's career, obtain more research funds, or gain fame for oneself, the aims of science are not served by unethical behavior.

According to Cantilena, the main effect of unethical practices in medical research is the loss of trust. A scientist who commits fraud risks losing the trust of his or her colleagues, those funding the research project, the subjects or patients involved, and the general public. This type of incident can also result in the loss of an entire set of data. For example, when the FDA finds fraud in a clinical trial, all the results from the research unit where the fraud originated will be discounted.

Even when research leads to undesired results, it is imperative to tell the truth about those results. In fact, negative results can lead to important discoveries. At one time, Cantilena's team was asked to test a drug that was being developed for the treatment of cocaine abuse. Their research found the drug to be dangerous to the heart and development was stopped. While it can be considered unfortunate not to have developed such a treatment, future patients were spared potential heart problems and taxpayers' money was saved.

Problems and progress

The findings of medical research may not always be popular, but they can save lives. In the 1990s, Cantilena's team found a problem with a widely prescribed antihistamine, Seldane. In 1990, Seldane ranked fifth in new prescriptions dispensed in the United States. It was used to treat allergy symptoms without inducing drowsiness. Cantilena was asked to consult on a case of a patient taking Seldane who had an unusual heart rhythm associated with unexplained fainting. Doctors suspected that the interaction between Seldane and another drug might be the cause of the symptoms. Cantilena's research team began a study on Seldane, funded by the FDA, to learn more about this interaction. Their research confirmed a connection between the drug interaction and potentially fatal changes in heart rhythms. As a result, the FDA took Seldane off the market in 1997. Cantilena's discovery was very important, and not only regarding Seldane; it led to similar studies on heart rhythm problems due to other drug interactions.

Of the patients waiting at the outpatient department of Apac Hospital in Northern Uganda, the majority are mothers of children under five years old with malaria.

Cantilena's work also benefits the global community. His team just finished an important study for the U.S. Army on a drug for the treatment of malaria. Malaria is a disease that kills worldwide, claiming many children as its victims. One problem in treating malaria is that there is only one intravenous drug that can be used to treat severe cases. Unfortunately, this particular drug can damage the patient's heart. A safer treatment is needed.

The U.S. Army isolated an active ingredient in a Chinese herb and made it into an intravenous drug called artesunate. They asked Cantilena's team to test it on volunteers. Their research found artesunate to be safe and well tolerated. It is now being tested on malaria-infected patients in Africa, and they hope it will be approved soon by the FDA. Imagine the worldwide implications if this drug is approved and made available!

Cantilena's team is guided not only by their internal understanding of ethics but by the many external oversights that are in place in the scientific community. For example, Cantilena's research unit is regularly subjected to audits of their data. In addition, the Institutional Review Board from their university also monitors how the researchers explain the risks of a clinical trial to their volunteers before they sign an agreement to participate. Cantilena says that while these sorts of oversights may sometimes feel cumbersome, they can be used as a resource to improve the research process and to do self-assessment.

While his work requires personal and professional standards of an exacting nature, Cantilena is more than willing to maintain his ethical responsibilities as a medical researcher. It is work that he enjoys. After all, as he puts it, he gets to "see science as it is being discovered."

Questions:

1. Define ethics.

2. Name three types of ethical misconduct in scientific research.

3. You are part of a research group studying the effectiveness of a new drug. Your team found that the drug doesn't work any better than a sugar pill and your team hasn't found any undesirable side effects. A team member wants to alter the results so that the drug appears to be effective. That way, your research will continue to receive funding. What would you say to your group? Why?

Dr. Cantilena photo courtesy of Leslie A.H. Sheen.
Apac Hospital photo by Toshihiro Horii, Department of Molecular Protozoology, Research Institute for Microbial Diseases, University of Osaka, Osaka, Japan.
Licensing: http://creativecommons.org/licenses/by/2.5/

Chapter 3 *Assessment*

Vocabulary

Select the correct term to complete each sentence.

control variable	experimental variable	prototype
deduce	hypothesis	repeatable
engineer	inquiry	scientific method
engineering cycle	natural laws	technology
experiment	objective	theory
experimental technique	procedure	trial

Section 3.1

1. To _____ is to figure something out from known facts using logical evidence.

2. A scientific explanation supported by lots of evidence collected over a long period of time is a(n) _____ .

3. Scientific evidence that is _____ can be seen by others if they repeat the same experiment.

4. Learning by asking questions and seeking the answers is called _____ .

5. A(n) _____ is a possible scientific explanation that can be tested by comparison with scientific evidence.

6. _____ evidence documents only what actually happened in an experiment as exactly as possible.

7. Scientists believe the universe follows a set of "rules" known as _____ .

8. The _____ is a process of learning that begins with a hypothesis and proceeds to collect evidence to confirm or disprove the hypothesis.

Section 3.2

9. When you run an experiment multiple times, you conduct several _____ (s).

10. The thing you are testing (changing) in an experiment is the _____ .

11. Something you keep the same from trial to trial in an experiment is called the _____ .

12. A step-by-step account of all that you do when conducting a particular experiment is called the _____ .

13. The way you release a car on a ramp while conducting an experiment is an example of _____ .

14. A(n) _____ is a situation specifically set up to investigate relationships between variables.

Section 3.3

15. A(n) _____ is a working model of a design that can be tested to see if it works.

16. A process used to build devices that solve technical problems is the _____ .

17. _____ is the application of science to meet human needs and solve problems.

18. A professional who uses scientific knowledge to create or improve technology is a(n) _____ .

Concepts

Section 3.1

1. Explain the difference between a theory and a hypothesis.

2. For each example, write whether it could be considered scientific evidence (**S**) or not (**N**).

   a. _____ an artist's watercolor painting of an oak leaf

   b. _____ the time for a car to drive once around a track

   c. _____ the number of each different color of candy-coated chocolate in a bag of candy

3. Indicate which of the following hypotheses are testable and scientific (**S**) and which are not (**N**).

   a. _____ Your brain produces undetectable energy waves.

   b. _____ Life forms do not exist in other galaxies.

   c. _____ Earth completes 1 rotation every 24 hours.

Section 3.2

4. Explain the difference between experimental variables and control variables.

5. What is the difference between experimental technique and procedure? Give an example to support your explanation.

Section 3.3

6. Science and technology are related, but they are not the same. What is the difference?

7. Scientist and engineer are two different career options. How does their work differ?

Problems

Section 3.1

1. Suppose you turn on your digital music player and it doesn't work. Describe how you could use the scientific method to figure out what's wrong. **STEM**

Section 3.2

2. Monique wants to see what happens when she drops a marble from different heights into a baking tray that has a thick layer of very soft modeling dough pressed inside. She predicts that the closer the marble is to the dough when she drops it, the deeper the marble's indentation will be.

   a. What is Monique's hypothesis?

   b. What is the experimental variable?

   c. What are two control variables?

   d. What evidence will be collected?

   e. Write a step-by-step procedure for the experiment.

   f. Do you think the data Monique collects will confirm or disprove her hypothesis? Explain your reasoning. **STEM**

Section 3.3

3. You have an idea for making a homemade shoe that will allow you to walk on open egg cartons without crushing any eggs. How could you use the engineering cycle to design it? **STEM**

Applying Your Knowledge

Write a caption for this illustration and include one or more vocabulary terms from the chapter that best describe the illustration. Then make a detailed and labeled sketch that shows what this progression of design might look like in the year 3000. Write a paragraph justifying one of your design features.

1896

1920

1950

2020?

Unit 2

Motion and Force

Lift

Airfoil

momentum

velocity

Speed

Time

acceleration

centripetal force

gravity

Try this at home

Find a toy car, a piece of cardboard, and three to five books that are nearly the same thickness. Make a ramp for the car by tilting the cardboard against one of the books. Set this up on a flat surface. Let the car roll down the ramp and see how far it goes once it leaves the ramp. Now, put a second book on top of the first to make the ramp steeper. See how far the car goes once it leaves the ramp. Using a ruler or measuring tape, record the distance the car travels. Continue stacking the books to make the ramp steeper. Does the car go farther each time? Why or why not?

Motion

How long can you stand perfectly still? Ten seconds? A minute? Even if you stand still, things inside your body are moving, like your heart and lungs. And even fast asleep, your body is not really at rest with respect to the universe! The 24-hour rotation of Earth is carrying you around at several hundred miles per hour. Every 365 days, Earth completes a 584-million-mile orbit around the Sun. To make this trip, Earth (with you on its surface) is rushing through space at the astounding speed of 67,000 miles per hour! To understand nature, we need to think about motion. How do we describe going from here to there? Whether it is a toy car rolling along a track or Earth rushing through space, the ideas in this chapter apply to all motion. Position, speed, and acceleration are basic ideas we need in order to understand the physical world.

67,000 mph

CHAPTER 4 INVESTIGATIONS

4A: Speed
Can you predict the speed of the car as it moves down the track?

4B: Acceleration
What is acceleration?

4.1 Position, Speed, and Velocity

Where are you right now? How fast are you moving? To answer these questions precisely, you need to use the concepts of position, speed, and velocity. These ideas apply to ordinary objects, such as cars, bicycles, and people. They also apply to microscopic objects the size of atoms and to enormous objects like planets and stars. Let's begin our discussion of motion with the concept of position.

The position variable

Position as a variable   You may do an experiment in your class that uses a car on a track. How do you tell someone exactly where the car is at any given moment? The answer is by measuring its **position**. Position is a variable. The position of the car describes where the car is relative to the track. In the diagram below, the position of the car is 50 centimeters (cm). That means the center of the car is at the 50 cm mark on the track.

Where is the car if it moves 20 cm to the right?

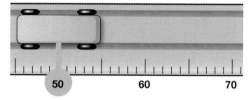

Figure 4.1: *If the car moves 20 cm to the right, its position will be 70 cm.*

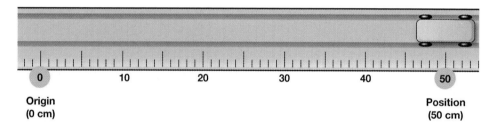

Position and distance   Position and distance are similar but not the same. Both use units of length. However, position is given relative to an **origin**. The origin is the place where position equals 0 (near the left end of the track above). Here's an example of the difference between position and distance. Assume the track is 1 meter long. Suppose the car moves a *distance* of 20 cm away from the 50 cm mark. Where is it now? You know a distance (20 cm) but you still don't know where the car is. It could have moved 20 cm to the right or 20 cm to the left. Saying the car is at a *position* of 70 cm tells you where the car *is*. A position is a unique location relative to an origin (Figure 4.1).

Speed

Speed is a motion variable

The variable speed describes how quickly something moves. To calculate the **speed** of a moving object, you divide the distance it moves by the time it takes to move. For example, if you drive 120 miles (the distance) and it takes you 2 hours (the time), your speed is 60 miles per hour (60 mph = 120 miles ÷ 2 hours). The lowercase letter v is used to represent speed.

SPEED

$$\text{Speed (cm/s)} \quad v = \frac{d}{t} \quad \begin{array}{l} \textbf{Distance (cm)} \\[1.2em] \textbf{Time (s)} \end{array}$$

Units for speed

The units for speed are distance units over time units. If distance is in kilometers and time in hours, then speed is in kilometers per hour (km/h). Other metric units for speed are cm per second (cm/s) and meters per second (m/s). Your family's car probably shows speed in miles per hour (mph). Table 4.1 shows different units commonly used for speed.

Table 4.1: Common units for speed

Distance	Time	Speed	Abbreviation
meters	seconds	meters per second	m/s
kilometers	hours	kilometers per hour	km/h
centimeters	seconds	centimeters per second	cm/s
miles	hours	miles per hour	mph

Average speed and instantaneous speed

When you divide the total distance of a trip by the time taken, you get the **average speed**. Figure 4.2 shows an average speed of 100 km/h. But think about actually driving though Chicago. On a real trip, your car will slow down and speed up. Sometimes your speed will be higher than 100 km/h, and sometimes lower (even 0 km/h!). The speedometer shows you the car's **instantaneous speed**. The instantaneous speed is the *actual* speed an object has at any moment.

VOCABULARY

speed – describes how quickly an object moves, calculated by dividing the distance traveled by the time it takes

average speed – the total distance divided by the total time for a trip

instantaneous speed – the actual speed of a moving object at any moment

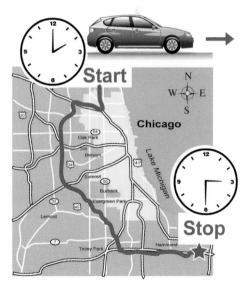

$$\frac{150 \text{ kilometers}}{1.5 \text{ hours}} = 100 \text{ kilometers (km/h)}$$

Figure 4.2: *A driving trip with an average speed of 100 km/h.*

Solving Problems: Speed

How far will you go if you drive for 2 hours at a speed of 100 km/h?

1. **Looking for:**	You are asked for a distance.	
2. **Given:**	You are given the speed and the time.	
3. **Relationships:**	distance = speed × time	
4. **Solution:**	distance = (100 km/h) × (2 h) = 200 km	

Your turn...

a. You travel at an average speed of 20 km/h in a straight line to get to your grandmother's house. It takes you 3 hours to get to her house. How far away is her house from where you started?

b. What is the speed of a snake that moves 20 meters in 5 seconds?

c. A train is moving at a speed of 50 km/h. How many hours will it take the train to travel 600 kilometers?

SCIENCE FACT

300,000,000 m/s

The speed limit of the universe

The fastest speed in the universe is the speed of light. Light moves at 300 million meters per second (3×10^8 m/s). If you could make light travel in a circle, it would go around the Earth 7.5 times in one second! Scientists believe the speed of light is the ultimate speed limit in the universe.

SOLVE FIRST LOOK LATER

a. Your grandmother's house is 60 km away from where you started.

b. The snake's speed is 4 m/s.

c. It will take the train 12 hours to travel 600 kilometers.

Vectors and velocity

Telling "in front" from "behind" How can you tell the difference between one meter in front of you and one meter behind you? The variable of *distance* is not the answer. The distance between two points can only be positive (or zero). You can't have a negative distance. For example, the distances between the ants in Figure 4.3 are either positive or zero. Likewise, one meter in front of you and one meter behind you both have the same distance: 1 meter.

Using positive and negative numbers The answer is to use *position* and allow positive and negative numbers. In the diagram below, positive numbers describe positions to the right (or in front) of the origin. Negative numbers are to the left of (or behind) the origin.

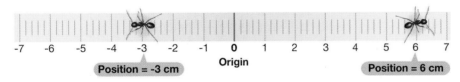

Position can be positive or negative.

Position = -3 cm   Origin   Position = 6 cm

Vectors Position is an example of a kind of variable called a *vector*. A **vector** is a variable that tells you a direction as well as an amount. Positive and negative numbers are enough information for a variable when the only directions are forward and backward. When up–down and right–left are also possible directions, vectors get more complicated.

Velocity Like position, motion can go right, left, forward, or backward. We use the term **velocity** to mean speed with direction. Velocity is positive when moving to the right or forward. Velocity is negative when moving to the left or backward (Figure 4.4).

The difference between velocity and speed Velocity is a vector, speed is not. In regular conversation, you might use the two words to mean the same thing. In science, they are related but different. Speed can have only a positive value (or zero) that tells you how far you move per unit of time (like meters per second). Velocity is speed *and* direction. If the motion is in a straight line, the direction can be shown with a positive or negative sign. The sign tells whether you are going forward or backward, and the quantity (speed) tells you how quickly you are moving.

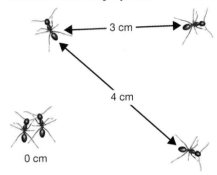

Distance is always positive or zero.

3 cm

4 cm

0 cm

Figure 4.3: *Distance is always a positive value or zero.*

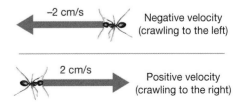

Velocity is speed and direction.

–2 cm/s — Negative velocity (crawling to the left)

2 cm/s — Positive velocity (crawling to the right)

Figure 4.4: *Velocity can be a positive or a negative value.*

Keeping track of where you are

A robot uses vectors *Sojourner* is a small robot sent to explore Mars (Figure 4.5). As it moved, *Sojourner* needed to keep track of its position. How did *Sojourner* know where it was? It used its velocity vector and a clock to calculate every move it made.

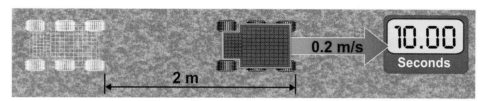

Figure 4.5: *Sojourner is a robot explorer that landed on Mars in 1997 (NASA/JPL).*

Use two variables to find the third one Any formula that involves speed can also be used for velocity. For example, you move 2 meters if your *speed* is 0.2 m/s and you keep going for 10 seconds. But did you move forward or backward? You move –2 meters (backward) if you move with a *velocity* of –0.2 m/s for 10 seconds. Using the formula with velocity gives you the *position* instead of *distance*.

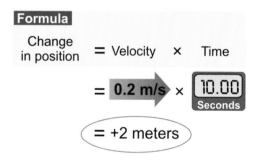

Formula

Change in position $=$ Velocity $\times$ Time

$=$ 0.2 m/s $\times$ 10.00 Seconds

$=$ +2 meters

Figure 4.6: *The change in position or distance is the velocity multiplied by the time.*

Word formulas		Equation
speed = distance ÷ time	velocity = distance ÷ time	$v = \dfrac{d}{t}$
distance = speed × time	distance = velocity × time	$d = vt$
time = distance ÷ speed	time = distance ÷ velocity	$t = \dfrac{d}{v}$

Forward and backward movement Suppose *Sojourner* moves forward at 0.2 m/s for 10 seconds. Its velocity is +0.2 m/s. In 10 seconds, its position changes by +2 meters.

Now, suppose *Sojourner* goes backward at 0.2 m/s for 4 seconds. This time, the velocity is –0.2 m/s. The change in position is –0.8 meters. *A change in position is velocity × time* (Figure 4.6).

Adding up a series of movements The computer in *Sojourner* adds up +2 m and –0.8 m to get +1.2 m. After these two moves, *Sojourner*'s position is 1.2 meters in front of where it was. *Sojourner* knows where it is by keeping track of each move it makes. It adds up each change in position using positive and negative numbers to come up with a final position (Figure 4.7).

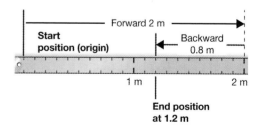

Figure 4.7: *Each change in position is added up using positive and negative numbers.*

Maps and coordinates

Two dimensions If *Sojourner* was crawling on a straight board, it would have only two choices for direction. Positive is forward and negative is backward. Out on the surface of Mars, *Sojourner* has more choices. It can turn and go sideways! The possible directions include north, east, south, west, and anything in between. A flat surface is an example of *two dimensions*. We say *two* because it takes two number lines to describe every point (Figure 4.8).

North, south, east, and west One way to describe two dimensions is to use north–south as one number line, or **axis**. Positive positions are north of the origin. Negative positions are south of the origin. The other axis goes east–west. Positive positions on this axis are east of the origin. Negative positions are west of the origin.

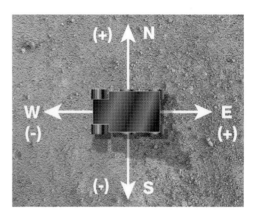

Figure 4.8: *A flat surface has two perpendicular dimensions: north–south and east–west. Each dimension has positive and negative directions.*

Coordinates describe position

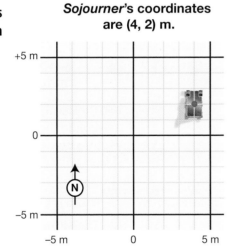

Sojourner's coordinates are (4, 2) m.

Sojourner's exact position can be described with two numbers. These numbers are called **coordinates**. The graph at the left shows *Sojourner* at the coordinates of (4, 2) m. The first number (or coordinate) gives the position on the east–west axis. *Sojourner* is 4 m east of the origin. The second number gives the position on the north–south axis. *Sojourner* is 2 m north of the origin.

Maps A graph using north–south and east–west axes can accurately show where *Sojourner* is. The graph can also show any path *Sojourner* takes, curved or straight. This kind of graph is called a *map*. Many street maps use letters on the north–south axis and numbers for the east–west axis. For example, the coordinates F-4 identify the square that is in row F, column 4 of the map shown in Figure 4.9.

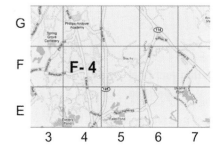

Figure 4.9: *Street maps often use letters and numbers for coordinates.*

Vectors on a map

A trip with a turn   Suppose you run east for 10 seconds at a speed of 2 m/s. Then you turn and run south at the same speed for 10 more seconds (Figure 4.10). Where are you compared to where you started? To get the answer, you figure out your east–west changes and your north–south changes separately.

Figure each direction separately   Your first movement has a velocity vector of +2 m/s on the east–west axis. After 10 seconds your change in position is +20 meters (east). There are no more east–west changes because your second movement is north–south only. Your second movement has a velocity vector of –2 m/s north–south. In 10 seconds you moved –20 meters. The negative sign means you moved south.

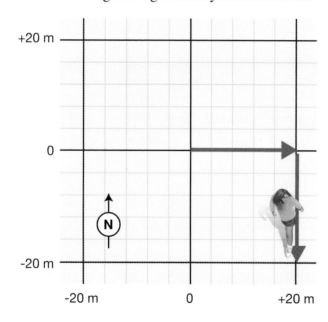

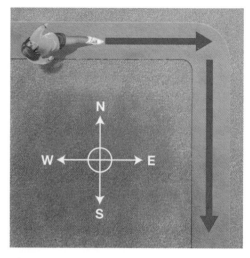

Figure 4.10: *A running trip with a turn.*

Captain Vector's hidden treasure

Use these velocity vectors to determine the location of Captain Vector's hidden pirate treasure. Your starting place is (0, 0).

1. Walk at a velocity of 1 m/s south for 10 seconds.

2. Then jog at a velocity of 3 m/s east for 5 seconds.

3. Run at a velocity of 5 m/s north for 2 seconds.

4. Then walk backward south at a velocity of 0.5 m/s for 2 seconds.

Where is the treasure relative to your starting place?

Figuring your final position   Now add up any east–west changes to get your final east–west position. Do the same for your north–south position. Your new position is (+20 m, –20 m).

 **Solving Problems: Velocity**

A train travels at 100 km/h heading east to reach a town in 4 hours. The train then reverses and heads west at 50 km/h for 4 hours. What is the train's position now?

1. **Looking for:** You are asked for position.

2. **Given:** You are given two velocity vectors and the times for each.

3. **Relationships:** change in position = velocity × time

4. **Solution:** The first change in position is (+100 km/h) × (4 h) = +400 km.
The second change in position is (–50 km/h) × (4 h) = –200 km.
The final position is (+400 km) + (–200 km) = +200 km. The train is 200 km east of where it started.

Your turn...

a. You are walking around your town. First you walk north from your starting position and walk for 2 hours at 1 km/h. Then you walk west for 1 hour at 1 km/h. Finally, you walk south for 1 hour at 2 km/h. What is your new position relative to your starting place?

b. A ship needs to sail to an island that is 1,000 km south of where the ship starts. If the captain sails south at a steady velocity of 30 km/h for 30 hours, will the ship make it?

SOLVE FIRST LOOK LATER

a. Your new position is 1 kilometer west of where you started.

b. No, because 30 km/h × 30 h = 900 km. The island is still 100 km away.

Section 4.1 *Review*

1. What is the difference between distance and position?

2. From an origin you walk 3 meters east, 7 meters west, and then 6 meters east. Where are you now relative to the origin?

3. What is your average speed if you walk 2 kilometers in 20 minutes?

4. Give an example where instantaneous speed is different from average speed.

5. A weather report says winds blow at 5 km/h from the northeast. Is this description of the wind a speed or velocity? Explain your answer.

6. What velocity vector will move you 200 miles east in 4 hours traveling at a constant speed?

7. Give an example of a situation in which you would describe an object's position in:
   a. one dimension
   b. two dimensions
   c. three dimensions

8. A movie theater is 4 kilometers east and 2 kilometers south of your house.
   a. Give the coordinates of the movie theater. Your house is the origin.
   b. After leaving the movie theater, you drive 5 kilometers west and 3 kilometers north to a restaurant. What are the coordinates of the restaurant? Use your house as the origin.

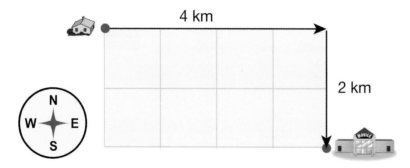

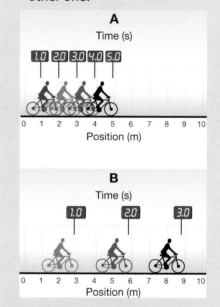

CHALLENGE

Look at the graphic below and answer the following questions.

1. How fast is each cyclist going in units of meters per second*?

2. Which cyclist is going faster? How much faster is this cyclist going compared to the other one?

*The word *per* means "for every" or "for each." Saying "5 kilometers per hour" is the same as saying "5 kilometers for each hour." You can also think of *per* as meaning "divided by." The quantity before the word *per* is divided by the quantity after it.

4.2 Graphs of Motion

Consider the phrase "a picture is worth a thousand words." A graph is a special kind of picture that can quickly give meaning to a lot of data (numbers). You can easily spot relationships on a graph. It is much more difficult to see these same relationships in columns of numbers. Compare the table of numbers to the graph in Figure 4.11 and see if you agree!

■■■■ VOCABULARY ■■■■

constant speed – speed that stays the same and does not change

The position vs. time graph

Recording data Imagine you are helping a runner who is training for a track meet. She wants to know if she is running at a **constant speed**. Constant speed means the speed stays the same. You mark the track every 50 meters. Then you measure her time at each mark as she runs. The data for your experiment is shown in Figure 4.11. This is position vs. time data because it tells you the runner's position at different points in time. She is at 50 meters after 10 seconds, 100 meters after 20 seconds, and so on.

Position and time data for a runner

Time (s)	Position (m)
0	0
10	50
20	100
30	150

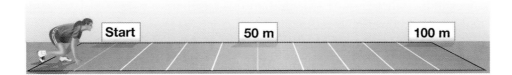

Graphing data To graph the data, you put position on the vertical (*y*) axis and time on the horizontal (*x*) axis. Each row of the data table makes one point on the graph. Notice the graph goes over to the right 10 seconds and up 50 meters between each point. This makes the points fall exactly in a straight line. The straight line tells you the runner moves the same distance during each equal time period. *An object moving at a constant speed always creates a straight line on a position vs. time graph.*

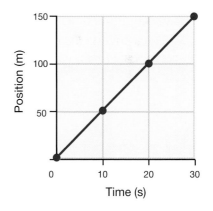

Runner's position vs. time

Figure 4.11: *A data table and a position vs. time graph for a runner.*

Calculating speed The data shows that the runner took 10 seconds to run each 50-meter segment. Because the time and distance were the same for each segment, you know her speed was the same for each segment. You can use the formula $v = d/t$ to calculate the speed. Dividing 50 meters by 10 seconds tells you her constant speed was 5 meters per second.

Graphs show relationships between variables

Relationships between variables   Think about rolling a toy car down a ramp. You theorize that steeper angles on the ramp will make the car go faster. How do you find out if your theory is correct? You need to know the relationship between the variables *angle* and *speed*.

Patterns on a graph show relationships   A good way to show a relationship between two variables is to use a graph. A graph shows one variable on the vertical (or *y*) axis and the second variable on the horizontal (or *x*) axis. Each axis is marked with the range of values the variable has. In Figure 4.12, the *x*-axis (angle) has values between 0 and 60 degrees. The *y*-axis (time) has average speed values between 0 and 300 cm/s. You can tell there is a relationship because all the points on the graph follow the same curve that slopes up and to the right. The curve tells you instantly that the average speed increases as the ramp gets steeper.

Recognizing a relationship from a graph   The relationship between variables may be strong or weak, or there may be no relationship at all. In a strong relationship, large changes in one variable make similarly large changes in the other variable, like in Figure 4.12. In a weak relationship, large changes in one variable cause only small changes in the other. The graph on the right (below) shows a weak relationship. When there is no relationship, the graph looks like scattered dots (below left). The dots do not make an obvious pattern (a line or curve).

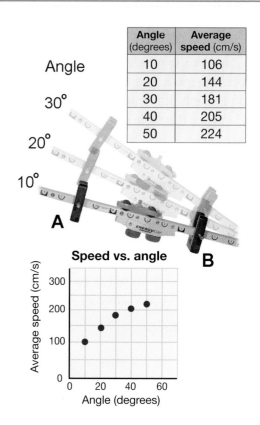

Angle (degrees)	Average speed (cm/s)
10	106
20	144
30	181
40	205
50	224

Figure 4.12: *This graph shows that the average speed between A and B increases as the angle of the track increases.*

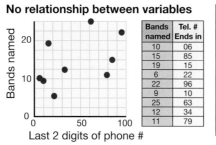

Bands named	Tel. # Ends in
10	06
15	85
19	15
6	22
22	96
9	10
25	63
12	34
11	79

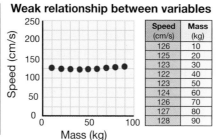

Speed (cm/s)	Mass (kg)
126	10
125	20
123	30
122	40
123	50
124	60
126	70
127	80
128	90

Slope

Comparing speeds You can use position vs. time graphs to quickly compare speeds. Figure 4.13 shows a position vs. time graph for two people running along a jogging path. Both runners start at the beginning of the path (the origin) at the same time. Runner A (blue) takes 100 seconds to run 600 meters. Runner B (red) takes 150 seconds to go the same distance. Runner A's speed is 6 m/s (600 ÷ 100) and runner B's speed is 4 m/s (600 ÷ 150). Notice that the line for runner A is *steeper* than the line for runner B. A steeper line on a position vs. time graph means a faster speed.

A steeper line on a position vs. time graph means a faster speed.

Calculating slope The "steepness" of a line is called its slope. The **slope** is the ratio of the *rise* (vertical change) divided by *run* (horizontal change). The diagram below shows how to calculate the slope of a line. Visualize a triangle with the slope as the hypotenuse. The rise is the height of the triangle. The run is the length along the base. Here, the x-axis is time and the y-axis is position. The slope of the graph is therefore the distance traveled divided by the time it takes, or the speed. The units are the units for the rise (meters) divided by the units for the run (seconds), meters per second, or m/s.

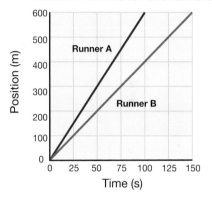

Figure 4.13: *A position vs. time graph for two runners.*

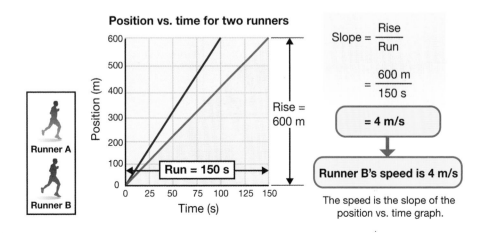

Speed vs. time graphs

Constant speed on a speed vs. time graph The speed vs. time graph has speed on the *y*-axis and time on the *x*-axis. The bottom graph in Figure 4.14 shows the speed vs. time for the runners. The top graph shows the position vs. time. Can you see the relationship between the two graphs? The blue runner has a speed of 5 m/s. The speed vs. time graph shows a horizontal line at 5 m/s for the entire time. On a speed vs. time graph, constant speed is shown with a straight horizontal line. At any point in time between 0 and 60 seconds, the line tells you the speed is 5 m/s.

Another example The red runner's line on the position vs. time graph has a less steep slope. That means her speed is slower. You can see this immediately on the speed vs. time graph. The red runner shows a line at 4 m/s for the whole time.

Calculating distance A speed vs. time graph can also be used to find the *distance* the object has traveled. Remember, distance is equal to speed multiplied by time. Suppose we draw a rectangle on the speed vs. time graph between the *x*-axis and the line showing the speed. The area of the rectangle (shown below) is equal to its length times its height. On the graph, the length is equal to the time and the height is equal to the speed. Therefore, the area of the graph is the speed multiplied by the time. This is the distance the runner traveled.

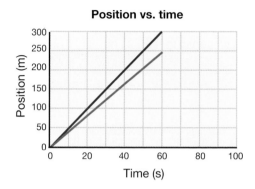

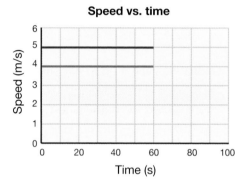

Figure 4.14: *The position vs. time graph (top) shows the exact same motion as the speed vs. time graph (bottom).*

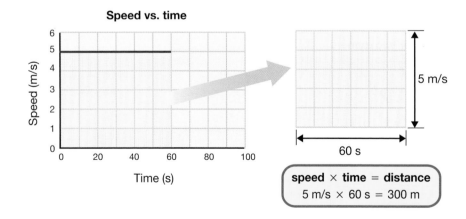

Section 4.2 *Review*

Position vs. time for two runners

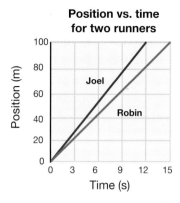

1. On a graph of position vs. time, what do the *x*-values represent? What do the *y*-values represent?

2. Explain why time is an independent variable and position is a dependent variable in a position vs. time graph.

3. What does the slope of the line on a position vs. time graph tell you about an object's speed?

4. The graph in Figure 4.15 shows the position and time for two runners in a race. Who has the faster speed, Robin or Joel? Explain how to answer this question without doing calculations.

5. Calculate the speed of each runner from the graph in Figure 4.15.

Figure 4.15: *Questions 4, 5, and 6.*

6. The runners in Figure 4.15 are racing. Predict which runner will get to the finish line of the race first.

7. Maria walks at a constant speed of 2 m/s for 8 seconds.
   a. Draw a speed vs. time graph for Maria's motion.
   b. How far does she walk?

8. Which of the three graphs below corresponds to the position vs. time graph in Figure 4.16?

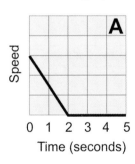 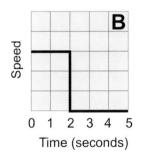 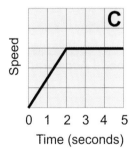

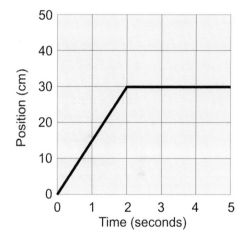

Figure 4.16: *Questions 8 and 9.*

9. Between which times is the speed zero for the motion shown on the position vs. time graph in Figure 4.16?

4.3 Acceleration

Constant speed is easy to understand. However, almost nothing moves with constant speed for long. When a driver steps on the gas pedal, the speed of the car increases. When the driver brakes, the speed decreases. Even while using cruise control, the speed goes up and down as the car's engine adjusts for hills. Acceleration, an important concept in physics, is how we describe changes in speed or velocity.

VOCABULARY

acceleration – the rate at which velocity changes

An example of acceleration

Definition of acceleration What happens if you coast down a long hill on a bicycle? At the top of the hill, you move slowly. As you go down the hill, you move faster and faster—you accelerate. **Acceleration** is the rate at which your speed (or velocity) changes. If your speed increases by 1 meter per second (m/s) each second, then your acceleration is 1 m/s per second.

Time (s)	Speed (m/s)
1	1
2	2
3	3
4	4
5	5

Time (s)	Speed (m/s)
1	2
2	4
3	6
4	8
5	10

Acceleration = 1 m/s each second | Acceleration = 2 m/s each second

Acceleration can change Your acceleration depends on the steepness of the hill. If the hill is a gradual incline, you have a small acceleration, such as 1 m/s each second. If the hill is steeper, your acceleration is greater, perhaps 2 m/s per second.

Acceleration on a speed vs. time graph Acceleration is easy to spot on a speed vs. time graph. If the speed changes over time, then there is acceleration. Acceleration causes the line to slope up on a speed vs. time graph (Figure 4.17). The graph on the top shows constant speed. There is zero acceleration at constant speed because the speed does not change.

Constant speed (acceleration = 0)

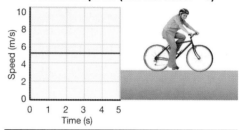

Acceleration of 1 m/s each second

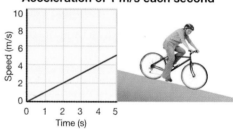

Acceleration of 2 m/s each second

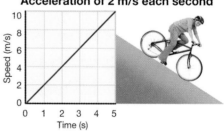

Figure 4.17: *Speed vs. time graphs showing constant speed (top) and acceleration (middle and bottom).*

Speed and acceleration

The difference between speed and acceleration

Speed and acceleration are not the same thing. You can be moving (non-zero speed) and have no acceleration (think *cruise control*). You can also be accelerating and not moving! But if the brakes are applied and the car slows down, it is accelerating because the speed is now changing (faster to slower).

Example: Acceleration in cars

Acceleration describes how quickly speed changes. More precisely, acceleration is the change in velocity divided by the change in time. For example, suppose a powerful sports car changes its speed from 0 to 60 mph in 5 seconds. In English units, the acceleration is 60 mph ÷ 5 seconds = 12 mph/second. In SI units, 60 mph is about the same as 100 km/h. The acceleration is 100 km/h ÷ 5 seconds, or 20 km/h/s (Figure 4.18). A formula you can use to calculate acceleration is shown below.

$$\text{ACCELERATION}$$

$$\underset{\text{Acceleration (m/s}^2)}{a} = \frac{\overset{\text{Change in velocity (m/s)}}{v_{finish} - v_{start}}}{\underset{\text{Time (s)}}{t}}$$

Acceleration in metric units

To calculate acceleration, you divide the change in velocity by the amount of time it takes for the change to happen. If the change in speed is in kilometers per hour, and the time is in seconds, then the acceleration is in km/h/s or *kilometers per hour per second*. An acceleration of 20 km/h/s means that the speed increases by 20 km/h *every second*.

What does "units of seconds squared" mean?

The time units for acceleration are often written as seconds squared or s^2. For example, acceleration might be 50 meters per second squared or 50 m/s^2. The steps in Figure 4.19 show how to simplify the fraction m/s/s to get m/s^2. Saying *seconds squared* is just a math-shorthand way of talking. It is better to think about acceleration in units of speed change per second (that is, meters per second *per second*).

Sports car

Speed goes from 0 to 100 km/h in 5 seconds

$$\text{Acceleration} = \frac{\text{Change in speed}}{\text{Time}}$$

$$= \frac{60 \text{ mph}}{5 \text{ seconds}}$$

$$= \frac{100 \text{ km/h} - 0 \text{ km/h}}{5 \text{ s}}$$

$$= 20 \text{ km/h per second}$$

$$= 20 \text{ km/h/s}$$

Figure 4.18: *The acceleration of a sports car.*

Plug in values:

$$\frac{50 \text{ m/s}}{s}$$

Clear the compound fraction:

$$\frac{50 \text{ m}}{s} \times \frac{1}{s} = \frac{50 \text{ m}}{s \times s}$$

Find units:

$$50 \frac{\text{m}}{s^2}$$

Figure 4.19: *How do we get m/s^2?*

Solving Problems: Acceleration

A sailboat moves at 1 m/s. A strong wind increases its speed to 4 m/s in 3 seconds (Figure 4.20). Calculate the acceleration.

1. **Looking for:**	You are asked for the acceleration in m/s².	
2. **Given:**	You are given the initial speed in m/s (v_1), final speed in m/s (v_2), and the time in seconds.	
3. **Relationships:**	Use the formula for acceleration: $a = \dfrac{v_2 - v_1}{t}$	
4. **Solution:**	$a = \dfrac{4 \text{ m/s} - 1 \text{ m/s}}{3 \text{ s}} = \dfrac{3 \text{ m/s}}{3 \text{ s}} = 1 \text{ m/s}^2$	

Your turn...

a. Calculate the acceleration of an airplane that starts at rest and reaches a speed of 45 m/s in 9 seconds.

b. Calculate the acceleration of a car that slows from 50 m/s to 30 m/s in 10 seconds.

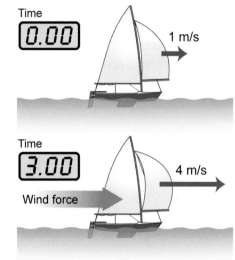

What is the acceleration?

Figure 4.20: *An acceleration example.*

SOLVE FIRST LOOK LATER

a. 5 m/s²

b. -2 m/s2

Acceleration on motion graphs

Acceleration on a speed vs. time graph
A speed vs. time graph is useful for showing how the speed of a moving object changes over time. Think about a car moving on a straight road. If the line on the graph is horizontal, then the car is moving at a constant speed (top of Figure 4.21). The upward slope in the middle graph shows increasing speed. The downward slope of the bottom graph tells you the speed is decreasing. The word *acceleration* is used for any change in speed, up or down.

Positive and negative acceleration
Acceleration can be positive or negative. Positive acceleration in one direction adds more speed each second. Things get faster. Negative acceleration in one direction subtracts some speed each second. Things get slower. People sometimes use the word *deceleration* to describe slowing down.

Acceleration on a position vs. time graph
The position vs. time graph is a *curve* when there is acceleration. Think about a car that is accelerating (speeding up). Its speed increases each second. That means it covers more distance each second. The position vs. time graph gets steeper each second. The opposite happens when a car is slowing down. The speed decreases so the car covers less distance each second. The position vs. time graph gets shallower with time, becoming flat when the car is stopped.

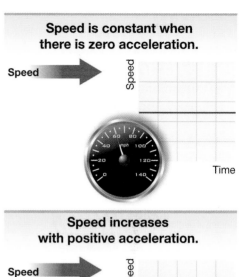
Speed is constant when there is zero acceleration.

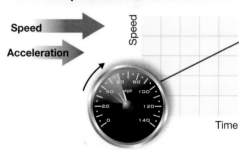
Speed increases with positive acceleration.

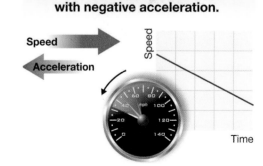
Speed decreases with negative acceleration.

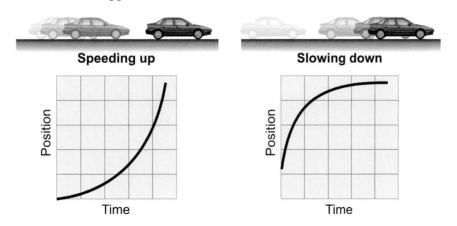
Speeding up

Slowing down

Figure 4.21: *Three examples of motion showing constant speed (top) and acceleration (middle, bottom).*

Free fall

The definition of free fall An object is in **free fall** if it is accelerating due to the force of gravity and no other forces are acting on it. A dropped ball is almost in free fall from the instant it leaves your hand until it reaches the ground. The "almost" is because there is a little bit of air friction that *does* make an additional force on the ball. A ball thrown upward is also in free fall after it leaves your hand. Even going up, the ball is in free fall because gravity is the only significant force acting on it.

The acceleration of gravity

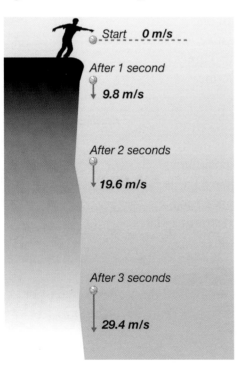

Start   *0 m/s*

After 1 second

9.8 m/s

After 2 seconds

19.6 m/s

After 3 seconds

29.4 m/s

If air friction is ignored, objects in free fall on Earth accelerate downward, increasing their speed by 9.8 m/s every second. The value 9.8 m/s² is called the **acceleration due to gravity**. The small letter *g* is used to represent its value. When you see the lowercase letter *g* in a physics question, you can substitute the value 9.8 m/s².

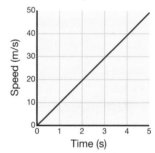

Free fall speed vs. time

Time (s)	Speed (m/s)
0	0
1	9.8
2	19.6
3	29.4
4	39.2
5	49.0

Figure 4.22: *A dropped ball increases its speed by 9.8 m/s each second, so its constant acceleration is 9.8 m/s².*

Constant acceleration The speed vs. time graph in Figure 4.22 is for a ball in free fall. Because the graph is a straight line, the speed increases by the same amount each second. This means the ball has a *constant acceleration*. Make sure you do not confuse constant speed with constant acceleration! Constant acceleration means an object's *speed* changes by the same amount each second.

Acceleration and direction

A change in direction is acceleration   If an object's acceleration is *zero*, the object can only move at a constant speed *in a straight line* (or be stopped). A car driving around a curve at a constant speed is accelerating (in the "physics sense") because its direction is changing (Figure 4.23). Acceleration occurs whenever there is a change in speed, direction, or both.

What "change in direction" means   What do we mean by "change in direction"? Consider a car traveling east. Its velocity is drawn as an arrow pointing east. Now suppose the car turns southward a little. Its velocity vector has a new direction.

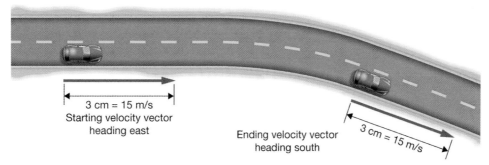

3 cm = 15 m/s
Starting velocity vector
heading east

Ending velocity vector
heading south

3 cm = 15 m/s

Drawing vectors   When drawing velocity arrows, the length represents the speed. A 2 cm arrow stands for 10 m/s (22 mph). A 4 cm arrow is 20 m/s, and so on. At this *scale*, each centimeter stands for 5 m/s. You can now find the change in velocity by measuring the length of the arrow that goes from the old velocity vector to the new one.

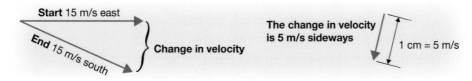

Start 15 m/s east

End 15 m/s south

Change in velocity

The change in velocity is 5 m/s sideways

1 cm = 5 m/s

Turns are caused by sideways accelerations   The small red arrow in the graphic above represents the difference in velocity before and after the turn. The change vector is 1 centimeter long, which equals 5 m/s. Notice the speed is the same before and after the turn! However, the change in direction is a *sideways* change of velocity. This change is caused by a *sideways acceleration*.

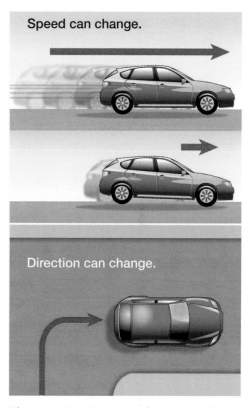

Speed can change.

Direction can change.

Figure 4.23: *A car can change its velocity by speeding up, slowing down, or turning. The car is accelerating in each of these cases.*

Curved motion

Acceleration and curved motion
Like velocity, acceleration has direction and is a vector. Curved motion is caused by sideways accelerations. Sideways accelerations cause velocity to change direction, which results in turning. Turns create curved motion.

projectile – an object moving through space and affected only by gravity

An example of curved motion
As an example of curved motion, imagine a soccer ball kicked into the air. The ball starts with a velocity vector at an upward angle (Figure 4.24). The acceleration of gravity changes the direction of the velocity vector more toward the ground during each second the ball is in the air. Therefore, gravity accelerates the ball downward as it moves through the air. Near the end of the motion, the direction of the ball's velocity vector is angled down toward the ground. The path of the ball makes a bowl-shaped curve called a *parabola*.

Projectile motion - - - - - - - - -
Velocity vector ⟶

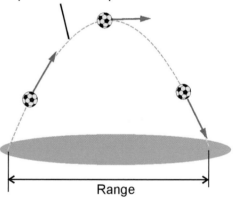

The shape of the ball's path is called a parabola.

Range

Projectiles
A soccer ball is an example of a **projectile**. A projectile is an object moving under the influence of only gravity. The action of gravity is to constantly turn the velocity vector more and more downward. Flying objects such as airplanes and birds are *not* projectiles, because they are affected by forces generated from their own power.

Figure 4.24: *A soccer ball in the air is a projectile. The path of the ball is a bowl-shaped curve called a parabola.*

Circular motion

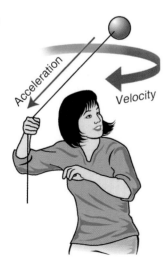

Acceleration

Velocity

Circular motion is another type of curved motion. An object in circular motion has a velocity vector that constantly changes direction. Imagine whirling a ball around your head on a string. You have to pull the string to keep the ball moving in a circle. Your pull accelerates the ball toward you. That acceleration is what bends the ball's velocity into a circle with you at the center. Circular motion always has an acceleration that points toward the center of the circle. In fact, the direction of the acceleration changes constantly so it *always* stays pointed toward the center of the circle.

Section 4.3 *Review*

1. Nearly all physics problems will use the unit m/s^2 for acceleration. Explain why the seconds are squared. Why isn't the unit given as m/s, as it is for speed?

2. Suppose you are moving left (negative) with a velocity of –10 m/s. What happens to your speed if you have a *negative* acceleration? Do you speed up or slow down?

3. A rabbit starts from a resting position and moves at 6 m/s after 3 seconds. What is the acceleration of the rabbit? (Figure 4.25)

4. You are running a race and you speed up from 3 m/s to 5 m/s in 4 seconds.
   a. What is your change in speed?
   b. What is your acceleration?

5. Does a car accelerate when it goes around a corner at a constant speed? Explain your answer.

6. A sailboat increases its speed from 1 m/s to 4 m/s in 3 seconds. What will the speed of the sailboat be at 6 seconds if the acceleration stays the same? (Figure 4.26)

7. The graph at the right shows the speed of a person riding a bicycle through a city. Which point (A, B, or C) on the graph is a place where the bicycle has speed but no acceleration? How do you know?

8. What happens to the speed of an object that is dropped in free fall?

9. A ball is in free fall after being dropped. What will the speed of the ball be after 2 seconds of free fall?

10. What happens when velocity and acceleration are not in the same direction? What kind of motion occurs?

11. The Earth moves in a nearly perfect circle around the Sun. Assume the speed stays constant. Is the Earth accelerating or not?

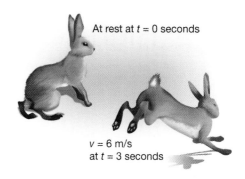

At rest at t = 0 seconds

v = 6 m/s
at t = 3 seconds

Figure 4.25: *Question 3.*

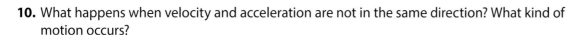

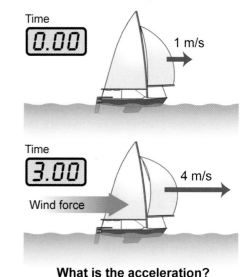

Time
0.00
1 m/s

Time
3.00
4 m/s
Wind force

What is the acceleration?

Figure 4.26: *Question 6.*

STEM High-Tech
Animal Trackers

April 22, 2007—A young harp seal was found stranded on a beach in Virginia's Chincoteague National Wildlife Refuge. He appeared slightly thin with some superficial injuries. Park rangers, optimistic that he would heal on his own, placed him under observation. Unfortunately, park visitors didn't heed requests to keep a respectful distance from the seal.

For the seal's and the public's safety, he was captured and sent to the Virginia Aquarium Stranding Center. Veterinarians treated him with antibiotics, and soon he was consuming 10 pounds of herring a day. In less than a month, the seal grew from 35 to 66 pounds.

During that time, a 13-year-old girl asked her birthday party guests to bring donations to the Aquarium's Stranding Response Program instead of gifts. With the money she collected, the aquarium purchased a satellite tag to track the seal's movements.

On May 19, 2007, the tag was attached and the healthy seal was released back to the ocean.

What is a satellite tag?

A satellite tag is a palm-sized, salt-water-resistant data collector with an antenna attached. It is glued to the fur of a seal's upper back, where it remains until the seal molts and the tag falls off.

With the satellite tag attached to his back, the seal moves toward the ocean.

The tag records information including the time, date, dive depth, dive duration, and amount of time at the surface over the last six hours. When the seal surfaces, the tag transmits this data to satellites orbiting Earth. Sometimes there are no satellites overhead when the animal surfaces, so data isn't received every day.

When data is received, instruments on the satellite record the location of the tag and relay the data to processing computers back on Earth. Organizations such as WhaleNet (Internet keyword search: "whalenet") make this information available online, where it is used by marine scientists, government and conservation organizations, and students.

The seal's journey: Position, time, and speed

WhaleNet's Satellite Tagging Observation Program (STOP) provided the following information about the seal's journey.

Date	Time (GMT)	Time elapsed since previous point (h:min)	Latitude	Longitude	Distance traveled from previous point
05/19/07	10:06	0	36.850 N	76.283 W	0 km (This is the release location–First Landing State Park, Virginia.)
05/30/07	04:45	258:39	42.195 N	65.554 W	1,096 km
06/03/07	07:27	98:42	44.317 N	63.137 W	307 km
06/05/07	19:20	59:53	45.294 N	60.812 W	214 km
06/11/07	03:11	127:51	45.749 N	59.440 W	119 km
06/16/07	20:16	137:05	47.669 N	58.009 W	240 km
06/19/07	08:11	59:55	46.594 N	56.125 W	186 km
06/25/07	13:17	149:06	48.523 N	51.069 W	437 km
06/28/07	06:25	65:08	50.412 N	51.192 W	210 km
07/03/07	08:46	122:21	54.127 N	54.070 W	458 km
07/05/07	00:40	39:54	54.889 N	55.558 W	128 km
07/09/07	19:08	114:28	56.665 N	59.970 W	340 km

This information can be used to determine the seal's average speed on each leg of his journey. To calculate his average speed on the first leg:

1. Convert elapsed time from h:min to hours.

   258 hours 39 minutes = 258 39/60 hours = 258.65 hours

2. Plug the values into the speed formula: speed = distance / time.

   Speed = 1,096 km / 258.65 h = 4.237 km/h

New insights, improved coexistence

Knowing the seal's average speed at various points on his journey can help us gain insight into his behavior. For example, between June 5 and June 11, his average speed slowed significantly. During that time, he remained in a small area just off the coast of Cape Breton Island. The satellite data suggests that this area may be a "critical habitat" for the harp seal. What was he doing there? Resting? Feeding? Finding answers to these questions can help us make better decisions about how and when we humans use this coastal region.

J. Michael Williamson, WhaleNet's founder and director, explains, "Similar data from tagging right whales has led to changes in shipping lanes around the whale's feeding areas and slowed shipping traffic through areas where whale calves are born. Satellite tagging research studies have led to many new laws and guidelines governing human activities around endangered species."

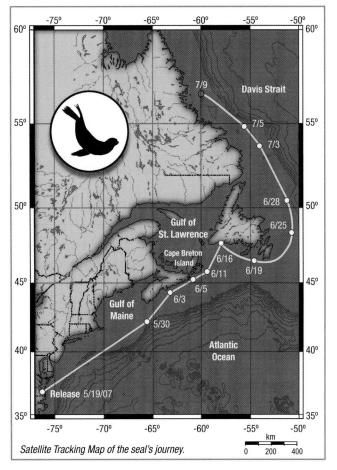

Satellite Tracking Map of the seal's journey.

Sea ice formed late and broke up early for seven of the eleven years between 1996 and 2007. Satellite tagging data helps us monitor how animals respond to these changing conditions. Some seals travel farther north. Others have tried to adapt to new habitats—for example, seals have given birth on land instead of ice. There the pups face new predators like foxes, wolves, and domestic and wild dogs—animals that don't hunt on ice.

Marine scientists share information about seal population activity with government agencies that monitor seal hunting and fishing industries. If the seal population declines, new regulations could be enacted to restrict hunts and/ or protect the seal's food sources and critical habitat areas, while areas with abundant resources can be opened to the fishing industry. The more we learn about how animals interact with their environments, the better decisions we can make about how we as humans use the oceans.

What's nice about sea ice?

Satellite tagging data can help us understand more about how animals adapt to changes in their environment. For example, marine scientists are paying careful attention to how far up the Davis Strait harp seals travel. Harp seals stop their northward journey when they run into sea ice, rather than swimming under it, since they need to breathe air like we do.

Harp seals rest, mate, molt, and grow new coats on the sea ice. They also give birth and nurse their pups on the ice. If the ice breaks up before the pups are weaned, the pups may drown or be crushed between large chunks of ice.

Questions:

1. What was the seal's average speed between June 5 and June 11, 2007?

2. Name two ways satellite tagging can help humans make better decisions about how we use the oceans.

3. **Research:** Using an Internet keyword search for "WhaleNet," find out what marine animal species are currently tagged. Use the website resources to create your own map of one animal's journey. Compare your animal's top speed to the harp seal's. What questions do you have about your animal's travels?

Chapter 4 *Assessment*

Vocabulary

Select the correct term to complete each sentence.

acceleration	acceleration due to gravity	average speed
axis	constant speed	coordinates
dependent variable	free fall	graph
independent variable	origin	position
projectile	slope	speed
vector	velocity	

Section 4.1

1. Speed with direction is called _____ .

2. A variable that is described using both a number and a direction is called a _____ .

3. The _____ is the place where position equals zero.

4. The _____ of an object is given relative to an origin.

5. The formula for _____ is distance divided by time.

6. _____ is speed that does not change over time and _____ is the total distance divided by the total time of a trip.

7. The _____ of the origin of a graph are (0, 0).

8. The x-_____ is horizontal on a graph.

Section 4.2

9. A mathematical diagram using two axes to represent the relationship between variables is a(n) _____ .

10. The _____ of a line is the ratio of rise to run.

11. The variable usually represented on the x-axis of a graph is the _____ .

12. The variable usually represented on the y-axis of a graph is the _____ .

Section 4.3

13. The rate at which velocity changes is defined as _____ .

14. An object moving in a curved path and affected only by gravity is called a(n) _____ .

15. An object accelerating under only the force of gravity is said to be in _____ .

16. An object in free fall will accelerate toward Earth at 9.8 m/s², the _____ .

Concepts

Section 4.1

1. What is the speed of an object that is standing still?

2. Name three common units for measuring speed.

3. Write the form of the speed equation that you would see in each of the following scenarios. Let v = speed, t = time, and d = distance.
   a. You know distance and speed and want to find the time.
   b. You know time and distance and want to find the speed.
   c. You know speed and time and want to find the distance.

4. How are the variables speed and velocity different? How are they similar?

5. Are the following directions usually considered positive or negative? Write + for positive or – for negative.

a. _____ up
b. _____ down
c. _____ left
d. _____ right
e. _____ north
f. _____ south
g. _____ east
h. _____ west

6. If you are given *x-y* axes coordinates of (4, 9), which axis is represented by the number 9?

Section 4.2

7. You do an experiment to find out how much light is needed to make house plants grow taller. The two variables in this experiment are the amount of light and the height of the plants. Which variable is the dependent variable and which is the independent variable? Explain your answer.

8. Look at the graph and answer the following questions.

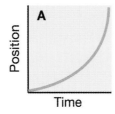

Position vs. time for two runners

a. What is the speed of runner B at 100 seconds?

b. For how many seconds has runner A been running at the 300-meter position?

c. Make a sketch of this graph in your notebook. Add a line to the graph that represents a third runner who has a speed that is slower than the speeds of runner A and B. This new line should begin at the origin of the graph.

9. Which of the graphs below shows an object that is stopped?

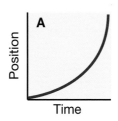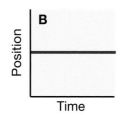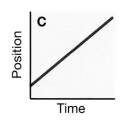

10. Which of the graphs above shows an object moving at a constant speed?

Section 4.3

11. How would it be possible for an object to be traveling with constant speed and still be accelerating?

12. Can an object have a speed of zero while it has an acceleration that is not zero? Explain.

13. Which of these graphs show acceleration occurring?

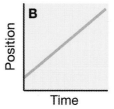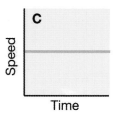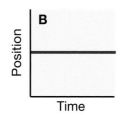

Problems

Section 4.1

1. Your starting place on a track is 30 centimeters. What is your new position if you move 10 centimeters to the left of this position?

2. A high-speed train travels at 300 km/h. How long (in hours) would it take the train to travel 1,500 km at this speed?

3. Lance Armstrong's teammate, George Hincapie, averaged a speed of 33.6 km/h in the 15th stage of the Tour de France, which took 4.00 hours. How far (in kilometers) did he travel in the race?

4. It takes Brooke 10 minutes to run 1 mile. What is her speed in miles per minute?

5. You are traveling on the interstate highway at a speed of 65 mph. What is your speed in km/h? The conversion factor is 1.0 mph = 1.6 km/h.

6. Use the speed equation to complete the following chart.

Distance (m)	Speed (m/s)	Time (s)
	10	6
45	5	
100		2

7. A pelican flies at a speed of 52 km/h for 0.25 hours. How many miles does the pelican travel? The conversion factor is 1.6 km/h = 1.0 mph.

8. A snail crawls 300 cm in 1 hour. Calculate the snail's speed in each of the following units.
   a. centimeters per hour (cm/h)
   b. centimeters per minute (cm/min)
   c. meters per hour (m/h)

9. If it takes 500 seconds for the light from the Sun to reach Earth, what is the distance to the Sun in meters? (The speed of light is 300,000,000 meters/second.)

10. Look at the graph below and give the coordinates for each point.

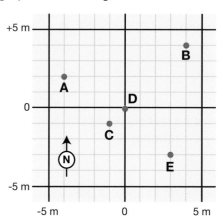

11. A train travels 50 km/h south for 2 hours. Then the train travels north at 75 km/h for 5 hours. Where is the train now, relative to its starting position?

12. You want to arrive at your friend's house by 5 p.m. Her house is 240 kilometers away. If your average speed will be 80 km/h on the trip, when do you need to leave your house in order to get to her house in time?

13. Starting from school, you bicycle 2 km north, then 6 km east, then 2 km south.
   a. How far did you cycle?
   b. What is your final position compared to your school?
   c. How far and in what direction must you travel to return to school?

14. If you walk 8 blocks north and then 3 blocks south from your home, what is your position compared to your home? What distance did you walk?

15. You use an x-y plane to represent your position. Starting at (+150 m, –50 m), you walk 20 meters west and 30 meters north. What are your new coordinates?

16. A bird flies from its nest going north for 2 hours at a speed of 20 km/h and then goes west for 3 hours at 15 km/h. What are the distance coordinates for the bird relative to its nest?

Section 4.2

17. Draw the position vs. time graph for a person walking at a constant speed of 1 m/s for 10 seconds. On the same axes, draw the graph for a person running at a constant speed of 4 m/s.

18. Calculate the speed represented by each position vs. time graph below.

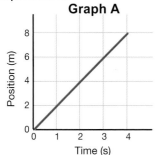

Graph A

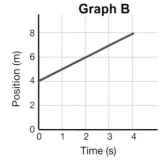

Graph B

19. Draw the speed vs. time graph that shows the same motion as each position vs. time graph above.

Section 4.3

20. A loaded garbage truck has low acceleration. It takes 10 seconds to go from 0 km/h to 100 km/h. What is the acceleration of the garbage truck? How much slower is the acceleration of the garbage truck compared to the acceleration of the sports car in Figure 4.18 on page 93?

Garbage truck
Speed goes from 0 to 100 km/h in 10 seconds

21. When a ball is first dropped off a cliff in free fall, it has an acceleration of 9.8 m/s². What is its acceleration as it gets closer to the ground? Assume no air friction.

22. Why is the position vs. time graph for an object in free fall a curve?

23. Draw a speed vs. time graph for an object accelerating from rest at 2 m/s².

24. Draw a speed vs. time graph for a car that starts at rest and steadily accelerates until it is moving at 40 m/s after 20 seconds. Then answer the following questions.

   a. What is the car's acceleration?

   b. What distance did the car travel during the 20 seconds?

25. Draw a speed vs. time graph for each of the following situations.

   a. A person walks along a trail at a constant speed.

   b. A ball is rolling up a hill and gradually slows down.

   c. A car starts out at rest at a red light and gradually speeds up.

Applying Your Knowledge

Section 4.1

1. If you take a one-hour drive at an average speed of 65 mph, is it possible for another car with an average speed of 55 mph to pass you? Explain your answer.

2. Make up your own problem! You want to end up 3 meters south of a starting point. Write a 5-step velocity vector problem that will get you to this point. You must travel in at least three directions before you get to your end point.

3. Answer the following questions.

a. A herd of wild animals moves in the following directions from a starting point in search of water: 10 km north, 3 km east, 7 km west, 20 km south, and 4 km east. Where does the herd end up relative to its starting point?

b. A watering hole is 2 km west and 2 km south of the starting point. Does the herd make it to the watering hole? If not, write down the directions the herd would need to follow to get to the watering hole from their end position.

Section 4.2

4. Oliver is warming up for a track meet. First he walks 1 m/s for 100 seconds. Then he runs at 3 m/s for 200 seconds. His shoe comes untied, so he stops for 20 seconds to tie it. Finally he runs at 4 m/s for 200 seconds.

a. Draw a position vs. time graph of Oliver's motion. (*Hint*: Use the table below to calculate Oliver's position during each part of his warm up.)

b. Draw a speed vs. time graph of Oliver's motion.

c. What is the total distance that Oliver travels?

d. What is Oliver's average speed during his 520-second warm up?

Speed (m/s)	×	Time (s)	=	Distance (m)

Section 4.3

5. Look at the graph below and make up a story involving motion that would create a graph shaped like the one below.

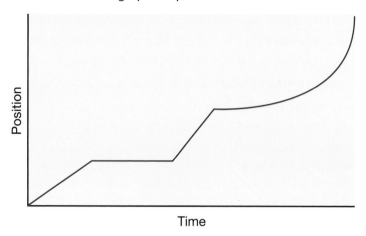

6. Now draw a speed vs. time graph that shows the same motion as the position vs. time graph above.

Forces

Every year, there are competitions that require strength and a knowledge of force. Athletes compete in events with names like the Giant Log Lift, the Pillars of Hercules, the Atlas Stones, and the Plane Pull. As you might imagine, moving a giant log or a plane requires a tremendous amount of force. How can these athletes achieve these amazing feats? There is a good chance that during their training, they thought about how best to apply force so that they could lift a giant log or pull a plane or lift a 160-kilogram Atlas Stone.

Forces are created and applied every time something moves. Forces such as weight are even present when things are not moving. Your body uses forces even when your heart is beating or when you are walking upstairs. And force is necessary when you want to pick up or move something that is very heavy. Understanding forces is fundamental to understanding how tasks are best accomplished in nature and by people. Read this chapter to learn more about how forces are created, measured, described, and used in daily life.

CHAPTER 5 INVESTIGATIONS

5A: What Is a Newton?
What is force and how is it measured?

5B: Friction
How does friction affect motion?

5.1 Forces

Force is a very important concept in physics and in everyday life. In this chapter, you will learn where forces come from, how they are measured, and how they are added and subtracted.

The cause of forces

What are forces? A **force** is a push or pull. Technically, force is the *action* that has the ability to change motion. You need force to start an object moving. You also need force to change an object's motion if it is already moving. Forces can increase or decrease the speed of a moving object. Forces can also change the direction in which an object is moving.

How are forces created? Forces are created in many ways. For example, your muscles create force when you swing a tennis racket. On a windy day, the movement of air can create forces. Earth's gravity creates a force called *weight* that pulls on everything around you. Each of these actions creates forces and through those forces, each can change an object's motion.

Some causes of forces

| Muscles | Moving matter (like wind) | Massive objects (like planets) |

The four elementary forces All of the forces we know of in the universe come from four elementary forces. Figure 5.1 describes the four elementary forces. If you study physics or chemistry, you may need to know about the strong or weak force. These forces are only important inside the atom and in certain kinds of radioactivity. However, the electromagnetic force and gravity are important in almost all areas of human life and technology.

The four elementary forces

Strong nuclear force
Electromagnetic force
Weak force
Gravity

Strong nuclear force
This force holds the nucleus of an atom together. This force is very strong but only reaches a very short distance.

Electromagnetic force
This force acts between positive and negative charges. This force holds atoms together in molecules.

Weak force
This force causes some kinds of radioactivity.

Gravity
This force causes all masses to attract each other. Your weight comes from the mass of Earth attracting the mass of your body.

Figure 5.1: *All forces in the universe come from only four elementary forces.*

Units of force

Pounds   Imagine mailing a package at the post office. How does the postal clerk know how much you should pay? You are charged a certain amount for every pound of *weight*. The **pound** (lb) is a unit of force commonly used in the United States. When you measure weight in pounds on a scale, you are measuring the force of gravity acting on an object (Figure 5.2). For smaller amounts, pounds are divided into ounces (oz). There are 16 ounces in 1 pound.

The origin of the pound   The pound is based on the Roman unit *libra*, which means "balance." This is why the abbreviation for pound is "lb." The word *pound* comes from the Latin *pondus*, meaning "weight." The definition of a pound has varied over time and from country to country.

Newtons   Although we use pounds all the time in our everyday life, scientists prefer to measure forces in *newtons*. The **newton** (N) is a metric unit of force. The newton is defined by how much a force can change the motion of an object. A force of 1 newton is the exact amount of force needed to cause a mass of 1 kilogram to speed up (or slow down) by 1 m/s each second (Figure 5.2). We call the SI unit of force the *newton* because force is defined by Newton's laws. The newton is a useful way to measure force because it connects force directly to its effect on motion.

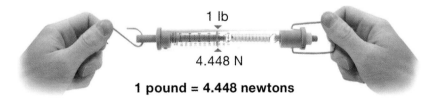

1 lb

4.448 N

1 pound = 4.448 newtons

Unit conversions   The newton is a smaller unit of force than the pound. One pound of force equals 4.448 newtons. That makes the newton a little less than a quarter of a pound. This is about the weight of a stick of butter. As another example, a 100-pound person weighs 444.8 newtons. In SI units, the mass of a 100-pound person (on Earth) is about 45 kilograms. If you do the math (444.8 ÷ 45), you will find that 1-kg of mass has a weight of 9.8 newtons of force.

Pound
One pound (lb) is about the weight of 0.454 kg of mass.

0.454 kg

1 lb

Gravity

Newton
One newton (N) is the force it takes to change the speed of a 1-kg mass by 1 m/s in 1 second.

Time (s)

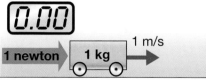

0.00

1 newton   1 kg   1 m/s

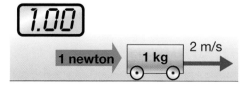

1.00

1 newton   1 kg   2 m/s

Figure 5.2: *The definitions of the newton and pound.*

The force vector

Force is a vector   The direction of a force makes a big difference in what the force does. That means force is a *vector*, like velocity or position. To predict the effect of a force, you need to know both its *strength* and its *direction*. Strength is usually measured in newtons. Direction may be given in words, such as 5 newtons *down,* or in diagrams. Arrows are often used to show the direction of forces in diagrams (Figure 5.3).

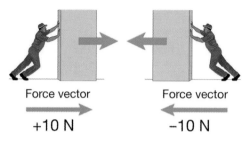

Force vector
+10 N   Force vector
−10 N

Figure 5.3: *Positive and negative numbers are used to indicate the direction of force vectors.*

Using positive and negative numbers   Forces may be assigned positive and negative values to tell their directions. For example, suppose a person pushes with a force of 10 newtons to the right (Figure 5.3). The force vector is +10 N. A person pushing with the same force to the left would create a force vector of −10 N. The negative sign indicates that the −10 N force is in the opposite direction from the +10 N force. We usually choose positive values to represent forces directed up, to the right, to the north, or to the east.

Drawing a force vector   It is sometimes helpful to show both the strength and direction of a force vector as an arrow on a graph. The length of the arrow represents the strength of the force. The arrow points in the direction of the force. The *x*- and *y*-axes show the strength of the force in the *x* and *y* directions.

Scale   When drawing a force vector to show its strength, you must choose a scale. For example, suppose you want to draw a force of 5 N pointing straight up (*y*-direction). You might use a scale of 1 cm = 1 N. At this scale, the force vector is a 5-cm-long arrow pointing up, along the *y*-axis on your graph (Figure 5.4). A 5 N horizontal force would be drawn along the *x*-axis with a 5-cm-long arrow pointing to the right.

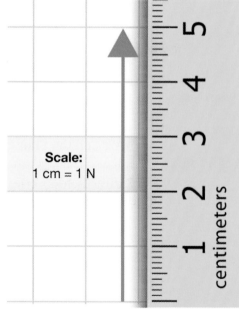

Scale:
1 cm = 1 N

centimeters

Figure 5.4: *You must use a scale when drawing a vector.*

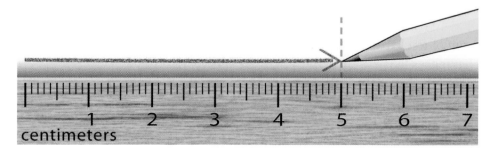

How forces act

Contact forces There are two ways that objects can affect each other through forces. One way is the result of direct contact. The force between two people pulling on a rope is a good example of a force that occurs through direct contact (Figure 5.5). A contact force is transmitted by matter directly touching other matter. The wind acting to slow a parachute is also a contact force because air is matter. The force comes from air contacting the parachute. In the next section, you will learn about *friction*, another contact force.

Forces that act through space Now think about Earth and the Moon. If Earth were to disappear, the Moon would sail off into space by itself. The Moon doesn't fly off because a force exists between Earth and the Moon. That force is called *gravity*. Gravity provides the force that keeps Earth and the Moon together in orbit. But exactly how does "gravity" get from Earth to the Moon? Space is empty of matter, so the force cannot be a contact force.

Some examples The force of gravity between Earth and the Moon appears to be what people once called "action-at-a-distance." The force between two magnets is another force that acts at a distance; so is the force that causes electricity. Table 5.1 summarizes the two kinds of forces.

Tensional force
Air resistance
Applied force
Gravitational force

Figure 5.5: *Contact forces and a force that acts through a force field.*

Table 5.1: Types of forces

Contact forces	"At-a-distance" forces
Friction	Gravity
Normal force	Electricity
Tension, air resistance, spring	Magnetism

The force field Today, we know that a true "action-at-a-distance" force is impossible. The force of gravity actually acts in two steps. First, the mass of Earth creates a *gravitational field* that fills the space around Earth with potential energy. Second, the gravitational field of Earth creates the force on the Moon. The gravitational force is carried from Earth to the Moon by a *force field*. In fact, if Earth were to vanish instantly, the Moon would be affected by Earth's gravity for a few seconds! This is because the force field "flows" between Earth and the Moon quickly, *but not instantly.*

STUDY SKILLS

Using force to define forces

Pick a term that is listed in Table 5.1 but is not described on this page— friction, normal force, or spring force. Find out what the term means. You might ask someone who is knowledgeable or complete research and find the answer on your own.

Contact forces from ropes and springs

Two ways contact forces occur   Ropes and springs are often used to make and apply forces. Ropes are used to transfer forces or change their direction. Springs are used to make and control forces.

Tension

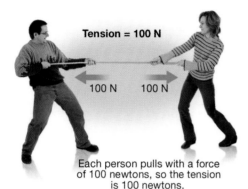

Tension = 100 N

100 N   100 N

Each person pulls with a force of 100 newtons, so the tension is 100 newtons.

The pulling force carried by a rope is called **tension**. *Tension always acts along the direction of the rope.* A rope carrying a tension force is stretched tight and pulls with equal strength on either end. For example, two people in the diagram at left are each pulling on the rope with a force of 100 newtons. The tension in the rope is 100 newtons. Ropes do *not* carry pushing forces. This is obvious if you have ever tried pushing a rope!

Springs

Two of the many kinds of springs are extension springs and compression springs. Extension springs are designed to be stretched.

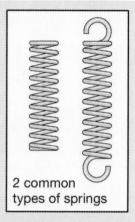

2 common types of springs

Spring forces

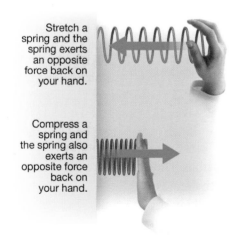

Stretch a spring and the spring exerts an opposite force back on your hand.

Compress a spring and the spring also exerts an opposite force back on your hand.

Springs are used to make or control forces. A spring creates a force when you stretch it or squeeze it away from its resting shape. The force from a spring always acts to return the spring to its resting shape. If you stretch a spring (**extension**), the spring acts to make itself shorter, pulling back on your hand. If you squeeze a spring (**compression**), the spring tries to get longer again and pushes back on your hand.

They often have loops on either end. Compression springs are designed to be squeezed. They are usually flat on both ends. Can you find both types of springs in your classroom?

1. What is the spring used for?

2. What would happen to the force if the spring broke?

Spring forces vary in strength   The force created by a spring is proportional to the ratio of the extended or compressed length divided by the original (resting) length. If you stretch a spring twice as much, it makes a force that is twice as strong.

Gravity

Gravitational force depends on mass The force of gravity on an object is called *weight*. At Earth's surface, gravity exerts a force of 9.8 N on every kilogram of mass. Therefore, on Earth, the weight of any object is its mass multiplied by 9.8 N/kg. For example, a 1-kilogram mass has a weight of 9.8 N, a 2-kilogram mass has a weight of 19.6 N, and so on. Because weight is a force, it is measured in units of force such as newtons and pounds.

Earth

98 newtons

10 kg

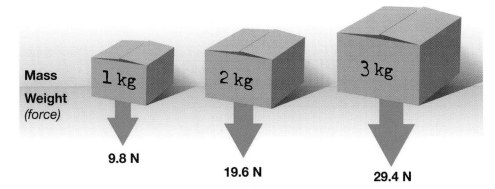

Mass

Weight *(force)*

1 kg   2 kg   3 kg

9.8 N   19.6 N   29.4 N

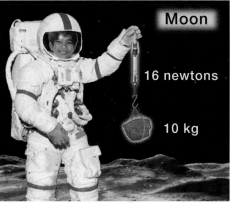

Moon

16 newtons

10 kg

Weight vs. mass In Chapter 2, you learned that weight and mass are not the same. Mass is a fundamental property of matter measured in kilograms (kg). Weight is a *force* measured in *newtons* (N). Weight depends on mass *and* gravity. For example, how much you weigh depends on your mass and the strength of gravity at your location. It is easy to confuse mass and weight because they seem similar. Heavy objects (more weight) have lots of mass, and light objects (less weight) have little mass. But always remember the difference when doing physics.

Figure 5.6: *A 10-kilogram rock weighs 98 newtons on Earth but only 16 newtons on the Moon.*

Weight is a force that depends on mass and gravity.

Weight is less on the Moon A 10-kilogram rock has the same mass no matter where it is in the universe. The rock's *weight*, however, depends on where it is located. On Earth, the rock weighs 98 newtons. But on the Moon, it weighs only 16 newtons (Figure 5.6). The rock weighs six times less on the Moon because *gravity* is six times less on the Moon.

Calculating weight

The weight formula The weight formula lets you calculate the weight of an object if you know the object's mass and the strength of gravity at the object's location. Three forms of the weight formula are given in Table 5.2. Use the appropriate form to find weight, mass, or the strength of gravity if you know any two of the three values.

WEIGHT

$$W = mg$$

Weight (N) W

Strength of gravity (N/kg)

Mass (kg)

"g" the symbol for gravity The strength of gravity at Earth's surface is so important to our everyday life that we give it a special symbol, the lowercase letter "g." When you see a "g" in a formula, you can usually substitute the value $g = 9.8$ N/kg. Of course, that assumes the formula is being applied at the surface of Earth! Elsewhere in the universe, "g" has different values. You should recognize that the value of 9.8 N/kg is the same as g (9.8 m/s^2) but with different units. This is no coincidence. According to Newton's second law, a force of 9.8 newtons acting on 1 kilogram produces an acceleration of 9.8 m/s^2. For this reason, the value of g can also be used as 9.8 N/kg. Which units you choose depends on whether you want to calculate acceleration or the weight force. Both units are actually identical: 9.8 N/kg = 9.8 m/s^2.

Table 5.2: Different forms of the weight formula

Use . . .	if you want to find . . .	and you know . . .
$W = mg$	weight (W)	mass (m) and strength of gravity (g)
$m = W/g$	mass (m)	weight (W) and strength of gravity (g)
$g = W/m$	strength of gravity (g)	weight (W) and mass (m)

Different ways to show "divided by"

Below are three different ways to show the equation "mass equals weight divided by gravity." Notice the different ways to show "divided by." You should familiarize yourself with all three versions.

Mass = weight divided by gravity

$m = W/g$

$m = \dfrac{W}{g}$

$m = W \div g$

Some notes about drawing force vectors

1. Force vectors should always be drawn in the direction of the force they represent.

2. Force vectors should be drawn to scale if possible, with length proportional to strength.

3. A force on a surface can be shown as pointing toward the surface or away from it. What matters is that the direction is clear so you know what the net force is in a certain direction.

Solving Problems: Weight and Mass

Calculate the weight of a 60-kilogram person (in newtons) on Earth and on Mars ($g = 3.7$ N/kg on Mars) (Figure 5.7).

1. **Looking for:**   You are asked for a person's weight on Earth and Mars.

2. **Given:**   You are given the person's mass and the value of g on Mars.

3. **Relationships:** $W = mg$

4. **Solution:**   For the person on Earth:
$W = mg$
$W = (60 \text{ kg})(9.8 \text{ N/kg}) = 588$ newtons

For the person on Mars:
$W = mg$
$W = (60 \text{ kg})(3.7 \text{ N/kg}) = 222$ newtons

Notice that while the masses are the same, the weight is much less on Mars.

Your turn...

a. Calculate the mass of a car that weighs 19,600 N on Earth.

b. A 70-kg person travels to a planet where he weighs 1,750 N. What is the value of g on that planet?

Figure 5.7: *How does the weight of a person on Earth compare to the weight of the same person on Mars?*

SOLVE FIRST LOOK LATER

a. 2,000 kg

b. 25 N/kg

Section 5.1 *Review*

1. Name three situations in which force is created. Describe the cause of the force in each situation.

2. Which of the following are units of force?

   a. kilograms and pounds
   b. newtons and pounds
   c. kilograms and newtons

3. Which is greater: a force of 10 N or a force of 5 lbs?

4. A rope is used to apply a force to a box. Which drawing shows the force vector drawn correctly?

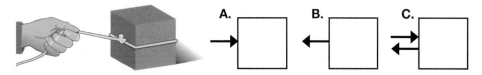

5. What is the difference between contact forces and forces that act through a force field?

6. A spring is stretched as shown. Which drawing shows the force exerted *by the spring*? (Hint: *Not* the force *on* the spring.)

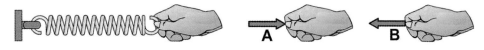

7. If the strength of gravity is 9.8 newtons per kilogram, that means:

   a. each newton of force equals 9.8 pounds
   b. each pound of force equals 9.8 newtons
   c. each newton of mass weighs 9.8 kilograms
   d. each kilogram of mass weighs 9.8 newtons

8. An astronaut in a spacesuit has a mass of 100 kilograms. What is the weight of this astronaut on the surface of the Moon where the strength of gravity is approximately 1/6 that of Earth?

9. What is the weight (in newtons) of a bowling ball that has a mass of 3 kilograms?

5.2 **Friction**

Friction is a force that resists motion. Friction is found everywhere in our world. You feel the effects of friction when you swim, ride in a car, walk, and even when you sit in a chair. Friction can act when an object is moving or when it is at rest. Many kinds of friction exist. Figure 5.8 shows some common examples.

Some causes of friction

The cause of friction Imagine looking through a microscope at two smooth surfaces touching each other. You would see tiny hills and valleys on both sides. As surfaces slide (or try to slide) across each other, the hills and valleys grind against each other, and this is a cause of friction. The tiny hills may change shape or wear away. If you rub sandpaper on a piece of wood, friction affects the wood's surface and makes it either smoother (hills wear away) or rougher (hills change shape).

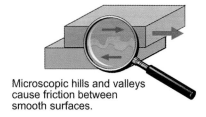

Microscopic hills and valleys cause friction between smooth surfaces.

Two surfaces are involved Friction depends on *both* of the surfaces that are in contact. The force of friction on a rubber hockey puck is very small when it is sliding on ice. But the same hockey puck sliding on a piece of sandpaper experiences a large friction force. When the hockey puck slides on ice, a thin layer of water between the rubber and the ice allows the puck to slide easily. Water and other liquids, such as oil, can greatly reduce the friction between surfaces.

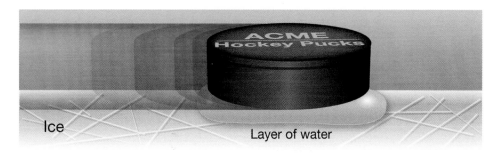

ACME Hockey Pucks

Ice

Layer of water

friction – a force that resists motion

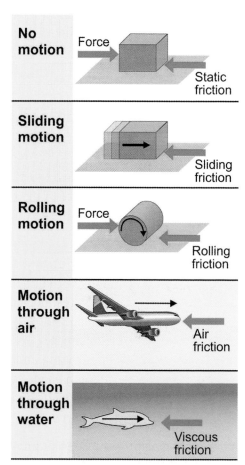

No motion — Force — Static friction

Sliding motion — Sliding friction

Rolling motion — Force — Rolling friction

Motion through air — Air friction

Motion through water — Viscous friction

Figure 5.8: *There are many types of friction.*

Identifying friction forces

Direction of the friction force Friction is a force, measured in newtons just like any other force. You draw the force of friction with a force vector. To figure out the direction of the vector, always remember that *friction resists motion between surfaces.* The force of friction acting *on* a surface always points opposite the direction of the motion *of that surface.* Imagine pushing a heavy box across the floor (Figure 5.9). If the box is moving to the right, then friction acts to the left against the surface of the box touching the floor. If the box were moving to the left instead, the force of friction would point to the right. This is what we mean when we say friction resists motion.

Sliding friction **Sliding friction** is a force that resists dry sliding motion between any two surfaces. If you push a box across the floor toward the right, sliding friction acts toward the left, slowing down the motion of the box. The friction force acts between the floor and the bottom surface of the box. Let's say you stop pushing the box, but it keeps moving. Sliding friction continues to work and eventually slows the box to a stop.

Static friction **Static friction** keeps an object that is standing still (at rest) from starting to move. Imagine trying to push a heavy box with a small force. The box stays at rest because the static friction force acts against your force and cancels it out. As you increase the strength of your push, the static friction also increases. Eventually your force becomes strong enough to overcome static friction and the box starts to move (Figure 5.9). The force of static friction balances your force up to a limit. The limit of the static friction force depends on the types of surfaces and the weight of the object you are pushing.

Comparing sliding and static friction How does sliding friction compare with static friction? If you have ever tried to move a heavy sofa or refrigerator, you probably know the answer. *It is harder to get something moving than it is to keep it moving.* This is because static friction is almost always greater than sliding friction at slow speeds.

VOCABULARY

sliding friction – the friction force that resists the motion of an object moving across a surface

static friction – the friction force that resists the motion between two surfaces that are not moving

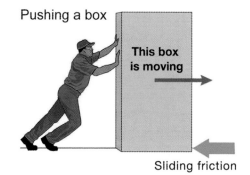

Pushing a box

This box is moving

Sliding friction

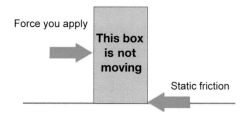

Force you apply

This box is not moving

Static friction

Figure 5.9: *The direction of the force of friction is opposite the direction the box is pushed.*

A model for friction

Different amounts of friction   The amount of friction generated when a box is pushed across a smooth floor is very different from when it is pushed across a carpet. This is because friction depends on materials, roughness, how clean the surfaces are, and other factors. Even the friction between two identical surfaces changes as the surfaces are polished by sliding motion. No single formula can accurately describe all types of friction.

An example   An easy experiment to measure friction is to pull a piece of paper across a table with a force scale. The paper slides smoothly, and the scale measures almost no force. Now put a brick on the piece of paper (Figure 5.10). Friction increases and you must pull with a greater force to move the paper.

Friction depends on the force between surfaces   Why does the brick have an effect on friction? The two surfaces in contact are still the paper and the tabletop. The brick causes the paper to press harder into the table's surface. The tiny hills and valleys in the paper and in the tabletop are pressed together with a much greater force, so the friction increases. The same is true of most dry sliding friction. Increasing the force that pushes surfaces together increases the amount of friction.

It takes very little force to slide paper across a table.

Adding a brick on top of the paper greatly increases the friction force.

Figure 5.10: *Friction increases greatly when a brick is placed on the paper.*

The greater the force squeezing two surfaces together, the greater the friction force.

Why sliding friction increases with weight   The friction force between two smooth, hard surfaces is approximately proportional to the force squeezing the surfaces against each other. Consider sliding a heavy box across a floor. The force between the bottom of the box and the floor is the weight of the box. Therefore, the force of friction is proportional to the weight of the box. If the weight doubles, the force of friction also doubles.

Other kinds of friction act differently   This rule is NOT true if one or both surfaces are wet, or if they are soft. Rubber is soft compared to pavement. The friction between rubber and pavement also depends on how much rubber is contacting the road. Wide tires have more friction (traction) than narrow tires.

Reducing the force of friction

All surfaces experience some friction

Unless a force is constantly applied, friction will slow all motion to a stop eventually. For example, bicycles have low friction, but even the best bicycle slows down as you coast on a level road. It is impossible to completely eliminate friction. However, many clever inventions have been devised to reduce friction. You use them every day.

Lubricants reduce friction in machines

Putting a liquid such as oil between two sliding surfaces keeps them from touching each other. The tiny hills and valleys don't become locked together, so there is less friction. The liquid also keeps the surfaces from wearing away as quickly. You add oil to a car's engine so that the moving parts slide or turn with less friction. Even water can be used to reduce friction between objects if they are not too hot.

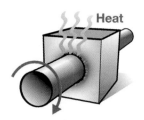

Sliding friction
• Larger forces
• More heat

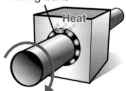

Rolling friction
• Smaller forces
• Less heat

Figure 5.11: *The friction between a shaft (the long pole in the picture) and the inner surface of the hole produces a lot of heat. Friction can be reduced by placing ball bearings between the shaft and the hole surface.*

Ball bearings

A ball bearing you might find in a machine

Ball bearings reduce friction in rotating motion (Figure 5.11). Ball bearings change sliding motion into rolling motion, which has much less friction. For example, a metal shaft rotating in a hole rubs and generates a lot of friction. Ball bearings that go between the shaft and the inside surface of the hole allow the shaft to spin more easily. The shaft rolls on the bearings instead of rubbing against the walls of the hole. Well-oiled bearings rotate easily and greatly reduce friction.

Magnetic levitation

Another method of decreasing friction is to separate the two surfaces with a cushion of air. A hovercraft floats on a cushion of air created by a large fan. Magnetic forces can also be used to separate surfaces. A magnetically levitated (or maglev) train uses magnets that run on electricity to float on the track once the train is moving (Figure 5.12). There is no contact between train and track, so there is far less friction than with a standard train on tracks. The ride is smoother, so maglev trains can move at very fast speeds. Maglev trains are not widely used yet because they are much more expensive to build than regular trains. They may become more popular in the future.

Figure 5.12: *With a maglev train, there is no contact between the moving train and the rail—and thus there is little friction.*

Using friction

Friction is useful for brakes and tires

What part of a bicycle brake is designed to increase friction?

There are many occasions when friction is very useful. For example, the brakes on a bicycle create friction between the brake pads and the rim of the wheel. Friction makes the bicycle slow down or stop. Friction is also needed to make a bicycle move. Without friction, the bicycle's tires would not grip the road.

Tires designed for bad weather

Friction is also important to anyone driving a car. Tires are specially designed to maintain friction on pavement in rain or snow. Tire treads have grooves that allow space for water to be channeled away where the tire touches the road (Figure 5.13). This allows good contact between the rubber and the road surface. Special groove patterns along with tiny slits are used on snow tires to increase traction in snow. These grooves and slits keep snow from getting packed into the treads.

Nails

Friction keeps nails in place (Figure 5.14). When a nail is hammered into wood, the wood pushes against the nail on all sides. The force of the wood against the nail surface creates a lot of friction. Each hit of the hammer pushes the nail deeper into the wood. The deeper the nail goes, the more surface there is for friction to grab onto.

Cleated shoes

Cleats

Shoes are designed to increase the friction between your foot and the ground. Many athletes, including football and soccer players, wear shoes with cleats. Cleats are like teeth on the bottom of the shoe that dig into the ground. Players wearing cleats can apply much greater force against the ground to help them move and to keep them from slipping.

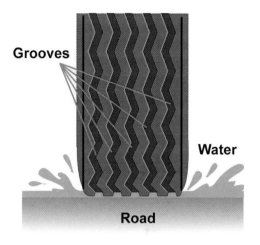

Grooves

Water

Road

Figure 5.13: *Grooved tire treads allow space for water to be channeled away from the road–tire contact point, allowing for more friction in wet conditions.*

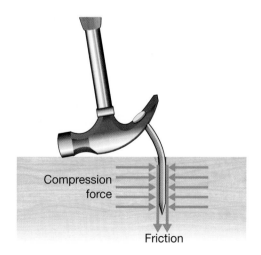

Compression force

Friction

Figure 5.14: *Friction is what makes nails hard to pull out, and what gives nails the strength to hold things together.*

Friction and energy

Friction changes energy of motion into heat

Why does rubbing your hands together make them warmer?

Earlier we learned that energy moves through the action of forces. Energy also changes into different forms. Friction changes energy of motion into heat energy. You may have noticed that rubbing your hands together quickly can make them warmer. You are feeling the effect of friction changing energy of motion into heat.

Heat in machines

Friction is always present in any machine with moving parts. In small machines, the forces are low and the amount of heat produced by friction may be small. A sewing machine is an example of a small machine. Larger machines have more problems with heat. In many machines, oil is pumped around moving parts. The oil does two important things. First, oil reduces friction so less heat is generated. Second, the oil absorbs the heat and carries it away from the moving parts. Without the flow of cooling oil, moving parts in an engine would heat up too much and melt.

Friction causes wear

Which rocks have been worn by friction and which have not?

Another way friction changes energy is by wearing away moving parts. You have probably noticed that objects that slide against each other often get rounded or smoothed. Each time two moving surfaces touch each other, tiny bits of material are broken off by friction. Breaking off bits of material uses energy. Sharp corners and edges are rounded off, and flat surfaces may be scratched or even polished smooth and shiny. This is why water flowing over stones in a stream causes the stones to be rounded and smooth.

TECHNOLOGY

Heat and machines

Every machine releases heat from friction. The faster the parts move, and the larger the forces inside the machine, the more heat is released. Electronic machines, like computers, are no exception, even though they may have no moving parts! Electricity moving through wires also creates friction.

If a machine gets too hot, parts can melt and the machine may stop working. Because of this, many machines have special systems, parts, and designs to get rid of unwanted heat energy.

Here are three machines you probably see every day. How is excess heat removed from each one?

Computer
Vacuum cleaner
Car engine

Section 5.2 *Review*

1. It is a common practice to put oil in a car and to change the car's oil once in a while. Why do cars need oil?

2. Which TWO of the following statements are true?
   a. Sliding friction is typically greater than static friction at slow speeds.
   b. Static friction is typically greater than sliding friction at slow speeds.
   c. Sliding friction occurs at rest and static friction occurs in motion.
   d. Static friction occurs at rest and sliding friction occurs in motion.

3. If the force squeezing two surfaces together is decreased, the force of dry sliding friction between the two surfaces will most likely:
   a. increase
   b. decrease
   c. stay about the same

4. Name three devices or inventions that are designed to **decrease** friction.

5. Name three devices or inventions that are designed to **increase** friction.

6. True or false? A well-oiled machine has no friction. Explain your answer.

7. A box is sliding across the floor from left to right. Which diagram correctly shows the force of friction acting on the box?

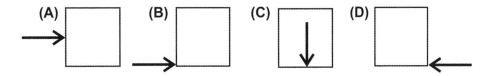

8. True or false? Friction makes energy vanish. Explain your answer.

9. True or false? Electronic machines with no moving parts experience friction and get hot because electricity is moving through the wires.

10. Water is useful for reducing friction between objects. However, you also learned that (due to friction) moving water causes rough rocks to become smooth. How can this be?

JOURNAL

You can count on friction!

Friction is a part of your daily life.

Write a paragraph telling how the events of your day would not have been possible without friction.

Then imagine the world suddenly had much more friction than normal. Write a paragraph telling how your day would have been affected.

CHALLENGE

Design a new shoe!

If it weren't for friction, it would be hard to walk! We need to be able to place our feet on a hard surface and push off from it to move forward.

Invent a new shoe that would be suitable for an environment of your choice. For example, you might want to design a shoe for mountain climbing or for walking on the Moon!

Make a sketch of your shoe and write an explanation about the research you did to develop the best design.

5.3 **Forces and Equilibrium**

We almost never feel only one force. For example, while walking, friction and weight are two forces that both act on you at the same time. As you might expect, it is the total of *all* forces acting on your body that determines how you move. This section is about how forces can be added and subtracted.

■■■■ **VOCABULARY** ■■■■

net force – the sum of all forces acting on an object

balanced forces – combined forces that result in a zero net force on an object

Adding forces

An example   The sum of all the forces on an object is called the **net force**. The word *net* means total. *Net force* also means that the direction of each force is considered when multiple forces are added. Consider a flying airplane (Figure 5.15). Four forces act on the plane: weight, drag (air friction), the thrust of the engines, and the lift force caused by the flow of air over the wings. For a plane to fly at a constant speed on a level path, the forces must all balance. **Balanced forces** result in a net force of zero.

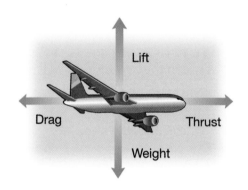

Figure 5.15: *Four forces act on a plane as it flies.*

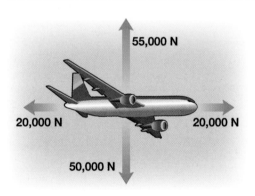

A pilot must always be aware of these four forces and know how to change them in order to speed up, slow down, lift off, and land. For example, to speed up, there must be a net force in the forward direction. The thrust must be greater than the drag. To climb, there must be an upward net force. The lift force must be greater than the weight.

Adding x-y components   To calculate the net force on an object, you must add the forces in each direction separately. Remember to define positive and negative directions for both the *x*-direction and the *y*-direction. In the diagram above, +*x* is to the right and +*y* is up. The net force in the *x*-direction is zero because the +20,000 N and −20,000 N add up to zero. The net force in the *y*-direction is +5,000 N (+55,000 N − 50,000 N). The plane climbs because there is a positive (upward) net force.

Equilibrium

Net force can be zero or not zero

When many forces act on the same object, either:

The net force is zero, or

The net force is NOT zero.

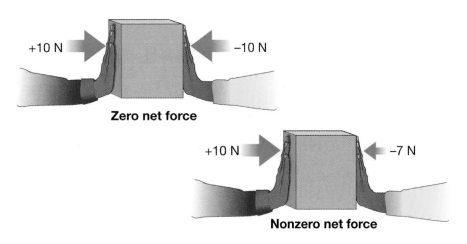

Zero net force

Nonzero net force

equilibrium – the state in which the net force on an object is zero

When the net force is zero . . .

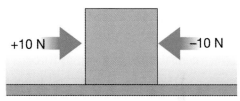

An object at rest will stay at rest.

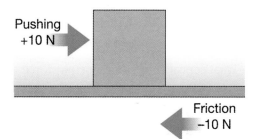

An object in motion will stay in motion.

Figure 5.16: *Objects are in equilibrium when the net force is zero.*

Definition of equilibrium

When the net force on an object is zero, we say the object is in **equilibrium**. Equilibrium does NOT mean there are no forces! Equilibrium means all forces cancel each other out, leaving zero net force. For example, when the net force is zero, an object at rest will stay at rest. Interestingly, an object can be in motion at constant speed and still be in equilibrium. This happens when a pushing force and a friction force are equal but opposite in direction so the object does not speed up or slow down (Figure 5.16).

Using equilibrium to find unknown forces

The idea of equilibrium is often used in reverse. Instead of thinking "an object in equilibrium stays at rest," we think "an object at rest must be in equilibrium." If an object is at rest, *the net force on it must be zero*. This fact often allows us to find the strength and direction of forces that must be there even if we don't know what they are.

When net force is not zero

If the net force is NOT zero, then the motion of an object will change. An object at rest will start moving. An object that is moving may change its velocity. In other words, unbalanced forces cause *acceleration*.

Normal forces

Definition of normal force
Imagine a book sitting on a table (Figure 5.17). Gravity pulls the book downward with the force of the book's weight. The book is at rest, so the net force must be zero. But what force balances the weight? The table exerts an upward force on the book called the **normal force**. The word *normal* here has a different meaning from what you might expect. In mathematics, *normal* means "perpendicular." The force that the table exerts is perpendicular to the table's surface. The normal force is also sometimes called the *support force*.

When normal force is created
A normal force is created whenever an object is in contact with a surface. The normal force has *equal strength* to the force pressing the object into the surface, which is often the object's weight. The normal force has *opposite direction* to the force pressing the object into the surface. For example, the weight of a book presses down on the table's surface. The normal force is equal in strength to the book's weight but acts upward on the book, in the opposite direction from the weight.

What normal force acts on
The normal force acts on the object pressing into the surface. That means, in this example, the normal force *acts on the book*. The normal force is created by the book *acting on the table*.

Strength of the normal force
What happens to the normal force if you put a brick on top of the book? The brick makes the book press harder into the table. The book does not move, so the normal force must be the same strength as the total weight of the book and the brick (Figure 5.18). The normal force acting on the book increases to keep the book in balance.

How the normal force is created
How does a table "know" how much normal force to supply? The answer is that normal force is very similar to the force exerted by a spring. When a book sits on a table, it squeezes the atoms in the table together by a tiny amount. The atoms resist this squeezing and try to return the table to its natural thickness. The greater the table is compressed, the larger the normal force it creates. The matter in the table acts like a bunch of very stiff springs. You don't see the table compress because the amount of compression is very small.

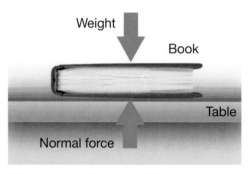

Figure 5.17: *The normal force and the weight are equal in strength and opposite in direction.*

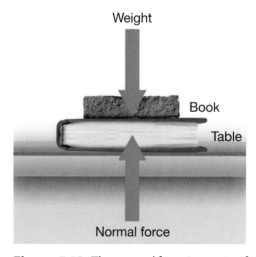

Figure 5.18: *The normal force is greater if a brick is placed on the book.*

The free-body diagram

Forces on a free-body diagram How can you keep track of many forces with different directions? The answer is to draw a **free-body diagram**. A free-body diagram contains only a single object, like a book or a table. All connections or supports are taken away and replaced by the forces they exert on the object. An accurate free-body diagram includes *every* force acting on an object, including weight, friction, and normal forces.

An example As an example of a free-body diagram, consider a stack of books weighing 30 newtons resting on a table that weighs 200 newtons. The books are on one corner of the table so that their entire weight is supported by one table leg. Figure 5.19 shows a free-body diagram of the forces acting on the table.

Finding the forces Because the table is in equilibrium, the net force on it must be zero. The weight of the books acts on the table making a 30 N force. The weight of the table acts on the floor. At every point where the table touches the floor (each leg), a normal force is created. The correct free-body diagram shows six forces. The normal force at each of three legs is one-quarter the weight of the table (50 newtons). The leg beneath the book also supports the weight of the book (50 N + 30 N = 80 N).

The purpose of a free-body diagram By separating an object from its physical connections, a free-body diagram helps you identify all forces and where they act. A normal force is usually present at any point where an object is in contact with another object or surface. Forces due to weight may be assumed to act directly on an object, often at its center.

Positive and negative forces There are two ways to handle positive and negative directions in a free-body diagram. One way is to make all upward forces positive and all downward forces negative. The second way is to draw all the forces in the direction you believe they act on the object. When you solve the problem, if you have chosen correctly, all the values for each force are positive. If one comes out negative, it means the force points in the opposite direction from what you guessed.

Real-life situation

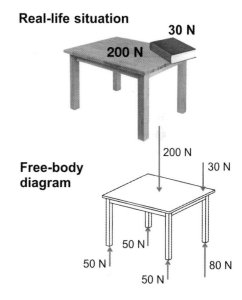

Free-body diagram

Equilibrium (net force = 0)

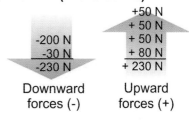

Downward forces (-)   Upward forces (+)

Figure 5.19: *A free-body diagram showing the forces acting on a table that has a stack of books resting on one corner.*

Solving Problems: **Equilibrium**

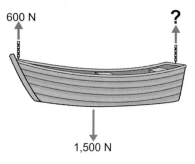

600 N   ?

1,500 N

Figure 5.20: *What is the force exerted by the other chain that is supporting the boat?*

Two chains are used to support a small boat weighing 1,500 newtons. One chain has a tension of 600 newtons (Figure 5.20). What is the force exerted by the other chain?

1. **Looking for:**	You are asked for an unknown tension in a chain.	
2. **Given:**	You are given the boat's weight in newtons and the tension in one chain in newtons.	
3. **Relationships:**	The net force on the boat is zero.	
4. **Solution:**	Draw a free-body diagram. The force of the two chains must balance the boat's weight. $600 \text{ N} + F_{chain2} = 1{,}500 \text{ N}$ $F_{chain2} = 900 \text{ N}$	

Your turn...

a. A person with a weight of 400 N is sitting motionless on a swing (Figure 5.21). For the swing to be in equilibrium, what is the tension force in each rope holding up the swing?

b. A heavy box weighing 1,000 N sits on the floor. You press down on the box with a force of 450 N. What is the normal force on the box?

c. A cat weighing 40 N stands on a chair. If the normal force on each of the cat's back paws is 12 N, what is the normal force on each front paw? (You can assume the force is the same on each front paw.)

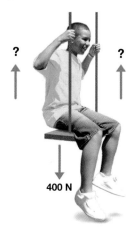

?   ?

400 N

Figure 5.21: *What is the tension force in each rope holding up the swing?*

SOLVE FIRST LOOK LATER

a. The upward force from both ropes must be 400 N, so the force in each rope is 200 N.

b. 1,450 N

c. 8 N

Section 5.3 *Review*

1. What is the relationship between net force and balanced forces?

2. Make two free-body diagrams. The first diagram should show a net force of zero on an object, and the other diagram should show a net force that is not zero.

3. If an object is accelerating, can the net force acting on it ever be zero? Explain your answer.

4. If you push down on a table with a force of 5 newtons, what is the normal force pushing back on you?

5. The diagram in Figure 5.22 shows three forces acting on a pencil. What is the net force acting on the pencil?

6. If an object is in equilibrium,
    a. the net force on the object is zero.
    b. the object has zero total mass.
    c. no forces are acting on the object.
    d. only normal forces are acting on the object.

7. A train is climbing a gradual hill. The weight of the train creates a downhill force of 150,000 newtons. Friction creates an additional force of 25,000 newtons acting in the same direction (downhill) (Figure 5.23). How much force does the train's engine need to create so the train is in equilibrium (going uphill at constant speed)?

8. Draw a free-body diagram of your own body sitting on a chair. Include all forces acting on your body.

9. If a force has a negative value, such as –100 N, that means the force
    a. is less than 100 N in strength.
    b. acts in the opposite direction from a +100 N force.
    c. is a normal force.

10. A child weighing 200 newtons is sitting in the center of a swing. The swing is supported evenly by two ropes, one on each side. What is the tension force in one of the ropes?

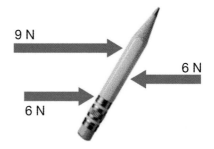

9 N

6 N

6 N

Figure 5.22: *Question 5.*

Weight = 150,000 N

Friction = 25,000 N

Figure 5.23: *Question 7.*

Parabolic Flights

*Have you ever seen video footage of astronauts floating in a spacecraft? The term **weightlessness** is used to describe this experience, but did you know that you can achieve this feeling even while Earth's gravity is pulling on you?*

Weightlessness

Every object with mass exerts gravitational force (also called weight) on every other object. As objects move farther apart, this force weakens. To be truly weightless, you would have to go to a location infinitely far away from every planet, star, moon, and other piece of mass in the universe. However, you feel weightless when the force of gravity is the only force on you, and no other force is acting to balance your weight.

For example, suppose you weigh 500 newtons. You are standing on the ground. Gravity pulls you down with a force of 500 newtons, and the floor pushes you up with a force of 500 newtons. You are aware of your weight because you feel the ground pushing up on your feet.

Now imagine you are in an elevator when the cable snaps. The elevator and your body experience free fall. You do not feel the force of the floor pushing up on your feet because the elevator is falling at the same rate as your body. Gravity is still pulling on you, but you feel weightless!

Astronauts in orbit around Earth feel weightless for the same reason. An orbiting spacecraft has a horizontal speed, but as it moves sideways, gravity causes it to fall around the Earth. Astronauts float because the spacecraft is falling as fast as they are.

Parabolic flights

Before going on missions, astronauts must practice working in an environment in which they feel weightless. This training is done on airplanes that fly in a path called a *parabola*. A parabola is the curved path an object follows when it is launched from the ground. If you kick a soccer ball up at an angle, its path is a parabola.

When flying in a parabola, the pilot maneuvers the plane so its path matches the path the passengers would follow if they were launched at an angle into the air at the speed of the plane. The passengers float around and feel weightless because the plane does not exert any forces on them. Each parabola lasts approximately 30 seconds, and a plane makes up to 50 parabolas during a flight.

NASA conducts parabolic flights to train astronauts.

NASA has been conducting parabolic flights since the 1950s to train astronauts. Scientists and college students have also gone on parabolic flights to perform a wide variety of chemistry, biology, and physics experiments. They have studied how weightlessness affects muscles, bones, blood circulation, digestion, and respiration. This research helps NASA learn how long missions may affect astronauts.

ZERO-G

Since 2004, thrill seekers have been able to take a ride on a parabolic flight simply for fun. Non-astronauts can have the experience of floating, spinning, and flying through the air with the Zero Gravity Corporation (ZERO-G). The company uses a specially modified Boeing 727-200 plane called the G-FORCE-ONE to take people on parabolic flights.

Now, non-astronauts can have the experience of floating, spinning, and flying through the air.

Passengers attend a training session so they know what to expect in flight. The plane is obviously not a typical jet. The rear third of the plane contains 35 seats. The front two-thirds is an open area called the Floating Zone. It is 90 feet long, with a padded floor and walls. During take-off, passengers remain in their seats.

Once the plane reaches an elevation of approximately 25,000 feet, passengers move to the Floating Zone and lie on the floor. ZERO-G flights contain three types of parabolas: Martian gravity (1/3 Earth gravity), Lunar gravity (1/6 Earth gravity), and zero gravity. The Martian and Lunar parabolas are not as steep as the zero gravity parabolas, so the floor of the plane provides some support force to the passengers. The effect is a feeling of being lighter than on Earth, making it possible to do one-handed push-ups and flips in the air. Martian and Lunar parabolas are done first to help passengers get used to the feeling of reduced gravity.

When the plane begins a parabola, it accelerates upward at a 45-degree angle. While lying on the floor, passengers feel like they are almost twice as heavy as on Earth. The engine thrust is then decreased so the plane begins to decelerate. Passengers feel the force of the floor on their bodies decrease. They can float, do flips, and try to catch floating water droplets squirted from the instructors' water bottles. Once the plane flies over the top of the parabola, it begins a descent toward Earth.

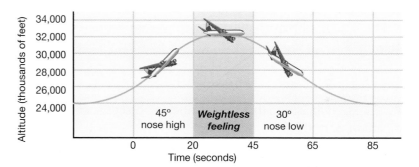

After approximately 25 seconds, passengers are alerted with the warning, "Feet down! Coming out!" Without the feeling of gravity pulling toward the bottom of the plane, it's impossible to tell up from down. As the plane begins to level out, the passengers slowly fall to the floor. Then the ride starts all over again, for a total of 15 fun-filled parabolas.

Questions:

1. Why does a person in a freely falling elevator feel weightless?

2. Are astronauts in orbit around Earth truly weightless? Explain.

3. List some ways parabolic flights are used.

4. Describe an experiment you could do on a parabolic flight to demonstrate one of Newton's laws. Explain which law the experiment demonstrates.

Astronaut training photo courtesy of NASA. Zero-G photo courtesy of www.gozerog.com.

Chapter 5 Assessment

Vocabulary

Select the correct term to complete each sentence.

balanced	compression	equilibrium
force	free-body diagram	friction
net force	newton	normal force
pound	sliding friction	static friction
tension	weight	

Section 5.1

1. A _____ is an action that can change an object's speed, direction, or both.

2. The English unit of force equal to 4.448 newtons is the _____ .

3. The metric unit of force needed to accelerate a 1-kg mass at 1 m/s^2 is the _____ .

4. A force that comes from the action of Earth's gravity is called _____ .

5. Squeezing creates _____ in a spring.

6. A pulling force carried by a rope is called _____ .

Section 5.2

7. _____ is a force that always resists the relative motion of objects or surfaces.

8. A frictional force that occurs when one surface slides over another is called _____ .

9. _____ is a frictional force between two non-moving surfaces.

Section 5.3

10. We sometimes say forces are _____ when they add up to make zero net force.

11. When all forces on an object are balanced, the object is in _____ .

12. The perpendicular force exerted by a surface on an object pressing against it is called the _____ .

13. A diagram representing all forces acting on an object is called a _____ .

14. The sum of all forces acting on an object is called the _____ .

Concepts

Section 5.1

1. Describe one situation in which forces are created.

2. Name the four fundamental forces of nature, the forces from which all others are derived.

3. Why is weight considered a force?

4. Forces cause changes to the motion of objects. Name a force and describe two changes it makes.

5. What two pieces of information do you need to describe a force?

6. Draw the following force vectors on a piece of paper and show the scale you use.
   a. 20 N west
   b. 4 N southeast

7. Name one contact force and one force that acts through a force field.

8. What happens to a spring's force if you stretch it more?

9. Compare and contrast tension, compression, and extension.

10. Which of the following is most often used to change the direction of a force, but not the strength of the force?

   a. a ball bearing   **c.** a spring

   b. a rope   **d.** a parachute

11. You know the relationship between weight and mass at the surface of the Earth. Describe this relationship on the Moon.

12. Identify which of the following are units of force (F) and which are units of mass (M).

   a. _____ kilogram   **c.** _____ pound

   b. _____ newton   **d.** _____ gram

Section 5.2

13. Give a reasonable explanation for why the friction is so low between an ice skate blade and the ice.

14. Does it require more force to start an object sliding or to keep it sliding? Explain your answer.

15. Why is it much easier to slide a cardboard box when it is empty compared to when it is full of heavy books?

16. Explain two ways friction can be reduced.

17. Explain how friction keeps a nail in place in a block of wood. If you try to pull out the nail, which way does the friction act?

18. Name two types of energy generated by friction and give an example of each.

19. Is friction something we always want to reduce? Explain.

Section 5.3

20. If the net force on an object is zero, can the object be moving? Explain.

21. Standing on Earth, gravity exerts a downward force on you, yet you don't fall toward the center of the planet.

   a. Name the other force that acts on you and keeps you in equilibrium.

   b. What is the direction of the other force?

   c. What do you know about the strength of this other force?

22. Describe the motion of the race car shown in the graphic to the right. Is it speeding up or slowing down?

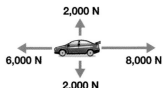

23. What are the four main forces acting on an airplane in flight? If the plane accelerates forward, which two forces must be out of balance? To fly on a level path, which two forces must be in balance?

24. Which of the following diagrams correctly shows the normal force on the block of wood sliding down the incline?

   a.   **b.**   **c.**   **d.**

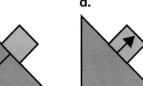

25. Draw a free-body diagram for the forces acting on the parachutist shown. Don't forget about air friction!

Problems

Section 5.1

1. Calculate the weight of a 66-newton bowling ball in pounds.

2. A frozen turkey bought in Canada is labeled "5.0 kilograms." This is a measurement of its mass. What is its weight in newtons?

3. What is the mass, in kilograms, of a large dog that weighs 441 newtons?

4. How much does a 40-kg student weigh on Earth in newtons?

5. How much mass, in kilograms, does a 50,000-N truck have?

6. An astronaut has a mass of 70 kilograms on Earth. What would her mass be on Mars? What would her weight be on Mars? The value of gravity (g) on Mars is 3.7 kg/N.

7. Using a scale of 1 cm = 5 N, draw force vectors representing a +20 N force and a -10 N force.

8. A spring is stretched 15 cm by a 45-N force. How far would the spring be stretched if a 60-N force were applied?

9. You and your friend pull on opposite ends of a rope. You each pull with a force of 10 newtons. What is the tension in the rope?

10. Two friends decide to build their strength by having a tug of war each day. They each pull with a force of 200 N.
    a. How much tension is in the rope?
    b. One day, one of the friends is sick and cannot work out. The other friend decides to build strength by tying the rope around a tree and pulling on the rope. How much must the single friend pull in order to get the same workout as he normally does? What is the tension on the rope? Explain.
    c. In both cases above, what is the net force on the rope if neither person is moving, and the tree stays put?

Section 5.2

11. Thomas pushes a 250-N box across a wooden floor using 75 N of force. If a second box of the same weight is stacked on top of the first, how much force would Thomas need to push the two boxes across the same floor?

12. Your backpack weighs 50 N. You pull it across a table at a constant speed by exerting a force of 20 N to the right. Draw a free-body diagram showing all four forces on the backpack. State the strength of each.

13. You exert a 50-N force to the right on a 300-N box that is on a table. However, the box does not move. Draw a free-body diagram for the box. Label all the forces and state their strengths. Explain why the box doesn't move.

Section 5.3

14. Find the net force on each box.

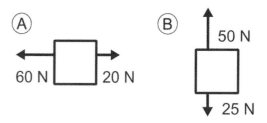

15. A 20-kilogram monkey hangs from a tree limb by both arms. Draw a free-body diagram showing the forces on the monkey. (*Hint*: 20 kilograms is not a force!)

16. The weight of a book resting on a stationary table is 9 N. How much is the normal force on the book? What would you need to do to increase the normal force on the book?

17. Is it possible to arrange three forces of 100 N, 200 N, and 300 N so they are in equilibrium? If so, draw a diagram.

18. You weigh a bear by making him stand on four scales as shown. Draw a free-body diagram showing all the forces acting on the bear. If his weight is 1,500 newtons, what is the reading on the fourth scale?

Applying Your Knowledge

Section 5.1

1. What is the weight of your favorite animal at different places in the universe?

  a. First find your favorite animal's mass in kilograms. (1 pound = 0.454 kilogram; 2.2 pounds = 1 kilogram)

  b. Then find the values of gravitational force (g) on five different planets or moons. The next page has values for g for the planets in our solar system in units of N/kg.

  c. Make a table that lists g for each planet or moon and your animal's weight on each of these.

2. Use the data in the table on the next page to answer the following questions.

  a. You know that mass is related to the strength of an object's gravitational force. Does the data in the table support this statement? Support your answer with an explanation.

  b. Is gravitational force related to the number of moons that a planet has?

  c. Is gravitational force related to how far a planet is from the Sun?

  d. Now come up with your own question and answer it using the data in the table.

Section 5.2

3. When an ice skater is on ice, a small amount of melting occurs under the blades of the skates. How does this help the skater glide? Your answer should discuss at least one kind of friction.

4. Joints like knees and elbows are designed to move freely. Find out how friction is reduced in a joint.

5. When on a mission, astronauts experience weightlessness.

  a. Research weightlessness. What is it in terms of the forces experienced by an astronaut?

  b. Research the effects of weightlessness on people and what astronauts do to counter those effects.

Section 5.3

6. Use this diagram to answer the following questions.

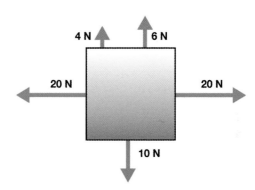

  a. Is the object shown above in equilibrium? Why or why not?

  b. Redraw this free-body diagram in a way that shows that the box will move to the right.

  c. Redraw this free-body diagram so that the box moves downward.

Comparing the Properties of the Planets (refer to *Applying Your Knowledge*)

	Mercury	Venus	Earth	Mars	Jupiter	Saturn	Uranus	Neptune
Diameter *(km)*	4,879	12,104	12,756	6,792	142,984	120,536	51,118	49,528
Mass *(kg)*	3.3×10^{23}	4.9×10^{24}	6.0×10^{24}	6.4×10^{23}	1.9×10^{27}	5.7×10^{26}	8.7×10^{25}	1.2×10^{26}
Density *(g/cm³)*	5.43	5.24	5.51	3.93	1.33	0.69	1.27	1.64
Average distance from the Sun *(km)*	58 million	108 million	150 million	228 million	779 million	1.43 billion	2.87 billion	4.50 billion
Moons *(confirmed number)*	0	0	1	2	53	53	27	14
Gravitational force *(N/kg)*	3.7	8.9	9.8	3.7	23.1	9.0	8.7	11.0
Surface temperature *(°C)*	−180 to +450	+465	−88 to +48	−89 to −31	−110	−139	−195	−200
Rotation period *(Earth days)*	59	243	1	1.03	0.41	0.44	0.72	0.67
Revolution period *(Earth years)*	0.24	0.62	1	1.90	12	29.50	84	165
Major gases in atmosphere	Trace He, H_2, O_2	CO_2	N_2, O_2	CO_2	H_2, He, CH_4, NH_3	H_2, He, CH_4, NH_3	H_2, He, CH_4, NH_3	H_2, He, CH_4, NH_3
Orbital velocity *(km/s)*	47.4	35.0	29.8	24.1	13.1	9.7	6.8	5.4

Planet photos courtesy of NASA/JPL. Planets not shown to scale.

Over the last 30 years, astronauts on different space missions have brought toys on board to compare how they work on Earth to how they work in "microgravity." During the missions, crew members take the toys out and play with them. Can you imagine trying to jump rope while floating around in the International Space Station? How do you think a spring toy will jump in space? You can learn how the toys behaved in space by doing an Internet search on "toys in space." But by reading this chapter first, you may be able to predict how toys might work in space. This chapter presents the laws of motion as stated by Sir Isaac Newton (1642–1727). Newton discovered answers to many questions about motion. Many historians believe Newton's ideas about motion were the beginning of modern science. Read this chapter and you will know all about motion too!

CHAPTER 6 INVESTIGATIONS

6A: Newton's First and Second Laws
What is the relationship between force and motion?

6B: Newton's Third Law
What happens when equal and opposite forces are exerted on a pair of Energy Cars?

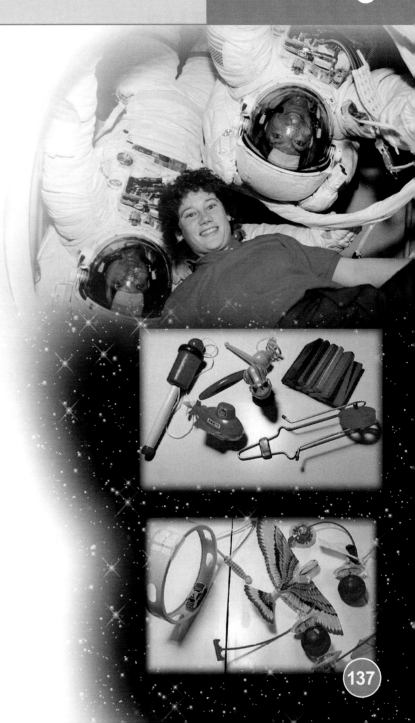

6.1 Newton's First Law

People who study science consider Sir Isaac Newton one of the most brilliant scientists who has ever lived. The three laws of motion are probably the most widely used natural laws in all of science. Newtons laws are not complicated math equations. They are brilliantly simple rules that show us an elegant way to make sense of how our world works.

Force changes motion

Force changes an object's motion   When playing miniature golf, what do you do to move the golf ball toward the hole? Do you tell the ball to move? Of course not! You hit the ball with the golf club to get it rolling. In physics, "hit the ball" means the golf club applies a force to the ball. This force is what changes the ball from being at rest to being in motion (Figure 6.1). *Motion can change only through the action of a force.* This statement is the beginning of Newton's first law.

Figure 6.1: *Force has the ability to change the motion of an object.*

Why do things stop moving after awhile?   Once moving, the ball rolls some, slows down, and eventually stops. For a long time, scientists thought the natural state of all things was to be at rest (stopped). They believed force had to be applied to keep an object moving and that constant motion required a constant force. *They were wrong!*

The real explanation   The golf ball stops because the force of friction keeps acting on it until there is no longer any motion. Suppose the golf course were perfectly level and had no friction. After being hit with the golf club, the ball would keep moving in a straight line at a constant speed *forever*. The ball would neither slow down nor change direction *unless another force acted on it*. Being stopped or moving with constant speed and direction are *both* natural states of motion and *neither one requires any force to sustain it.*

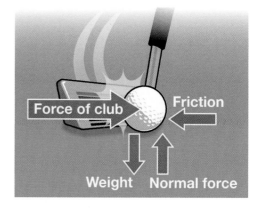

Net force   When you hit a golf ball, the force from the club is not the only force that acts on the ball (Figure 6.2). The ball's weight, the normal force from the ground, and friction are also acting. The ball moves according to the net force acting on the ball. The golf club causes the ball to move to the right because its force overcomes the friction force keeping the ball in place. Newton's first law is written in terms of the *net force* because that is what affects motion.

Figure 6.2: *Four forces act on a golf ball. The net force determines how it moves.*

The first law: The law of inertia

Newton's first law **Newton's first law** says that objects continue the motion they already have *unless* they are acted on by a net force (the sum of all forces acting on an object at any given time). When the net force is zero, objects at rest stay at rest, and objects that are moving keep moving in the same direction with the same speed.

> *When the net force is zero, objects at rest stay at rest and objects in motion keep moving with the same speed and direction.*

Force is required to change motion The first law says there can be no change in motion without a net force. *That includes slowing down!* It takes a net force (often friction) to make things slow down. If forces are truly balanced, a moving object will keep moving forever with the same speed, in the same direction.

Balanced and unbalanced forces Changes in motion come from **unbalanced forces**. Forces are "unbalanced" when the net force is NOT exactly zero. A rolling golf ball on a grassy golf course is not in equilibrium because friction is an unbalanced force. In the opposite situation, forces are "balanced" when they add up to zero net force. Forces are always balanced in equilibrium.

Inertia The first law is often called the "law of inertia" because **inertia** is the property of an object that resists changes in motion. Inertia comes from mass. Objects with more mass have more inertia. To understand inertia, imagine moving a bowling ball and a golf ball that are at rest (Figure 6.3). A golf ball has a mass of 0.05 kilogram, and suppose the bowling ball has a mass of 5 kilograms. The bowling ball has 100 times more mass than the golf ball, so it has 100 times more inertia too. Now ask yourself which needs more force to start moving. If you push for the same distance, the bowling ball takes MUCH more force to get it moving the same speed as the golf ball. The bowling ball needs more force because a bowling ball has more inertia than a golf ball. The greater an object's inertia, the greater the force needed to change its motion.

■■■■ **VOCABULARY** ■■■■

Newton's first law – a law of motion that states that an object at rest will stay at rest and an object in motion will stay in motion with the same velocity unless acted on by an unbalanced force

unbalanced forces – forces that result in a net force on an object and can cause changes in motion

inertia – the property of an object that resists changes in its motion

A bowling ball	A golf ball
5,000 grams	50 grams
5 kilograms	0.050 kilogram

Figure 6.3: *A bowling ball has more mass than a golf ball. The bowling ball is harder to move because it has more inertia.*

Solving Problems: Net Force and the First Law

A car drives along the highway at constant velocity. Find the car's weight and the friction force if the engine produces a force of 2,000 newtons between the tires and the road and the normal force on the car is 12,000 N.

1.	**Looking for:**	You are asked for the car's weight and the friction force.
2.	**Given:**	You are given the normal force and engine force. The normal force is 12,000 N and the engine force is 2,000 N. The car is moving at a constant velocity.
3.	**Relationships:**	Newton's first law states that if the car is moving at a constant velocity, the net force must be zero.
4.	**Solution:**	The weight of the car balances the normal force. Therefore, the weight of the car is a downward force: 12,000 N. The forward engine force balances the friction force, so the friction force is 2,000 N opposite the direction of the car's motion.

Your turn...

a. Identify the forces on the same car if it is stopped at a red light on level ground.

b. While the car is moving forward, a gust of wind gives it a big push from the back. Because most of the friction on a car (at highway speeds) is from the air, the friction force is reduced from 2,000 N to 1,500 N. What is the net force on the car if the engine force remains at 2,000 N? Does it still move at constant velocity?

c. What is normal force on the car if 1,000 N of luggage is added?

d. As you sit on the passenger seat of the car, the seat exerts a normal force of 550 N on you. If you weigh 600 N, what is the normal force of the car's floorboard on your feet?

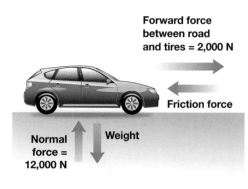

Figure 6.4: *The forces on the car.*

SOLVE FIRST LOOK LATER

a. When stopped, the car experiences a normal force of 12,000 N and its weight of 12,000 N.

b. The net force is 500 N. No, while the wind is blowing, it is not moving at constant velocity because it is experiencing a net force.

c. The normal force would be 13,000 N.

d. The normal force of the floorboard on your feet is 50 N.

Section 6.1 *Review*

1. For each of the following situations, identify what creates one of the forces that creates the motion described (there may be many).

    a. A flag flaps back and forth at the top of a flagpole.

    b. A soccer ball is passed from one player to another.

    c. A large piece of hail falls toward the ground.

    d. The ocean tide goes from high to low at the seashore (you might have to do a little research to get this one if you don't know already).

2. Which has more inertia—a shopping cart full of groceries or an empty shopping cart?

3. In the following situation, which diagram (A, B, C, or D) best illustrates the net force experienced by the cart when the weight pulls downward?

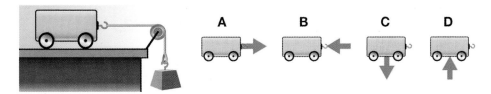

4. Forces contribute to the net force on a car rolling down a ramp.

    a. Which force supports the car's weight?

    b. Which force accelerates the car down the ramp?

    c. Which force acts against the motion of the car?

5. Imagine whirling a ball on a string over your head. Suppose the knot holding the ball comes loose and the ball is instantly released from the string. What path does the ball take after leaving the string? Use Newton's first law to explain your answer.

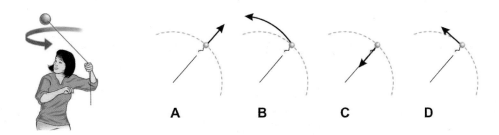

6.2 Newton's Second Law

What kind of change happens when forces are not balanced? The answer is *acceleration*. Acceleration is a change in velocity (speed or direction). Newton's *second* law describes how acceleration depends on both force and mass.

The three main ideas of the second law

What is the second law about?
Newton's first law tells us that motion cannot change without a net force. The second law tells us exactly what kind of change is caused by unbalanced forces. The second law answers questions like: "How much force does it take to change the speed of a 1,000-kg car from 0 to 80 km/h?" Anyone who does anything involving motion needs to understand the second law.

The three main ideas
Here are three big ideas in the second law.

1. Acceleration is the result of unbalanced forces.

2. A larger force makes a proportionally larger acceleration.

3. Acceleration is inversely proportional to mass.

Unbalanced forces cause acceleration
The first law tells us that things in motion can continue to move even without any net force. This is true as long as the motion is at a constant speed and in a straight line. The second law says that any unbalanced force results in acceleration. We know that acceleration causes changes in velocity (speed or direction). Putting these two ideas together tells us two things about force and motion: (1) Unbalanced forces cause changes in speed, direction, or both; and (2) any time there is a change in speed or direction, there *must be an unbalanced force acting*.

Force and motion connect through acceleration
The second law is the connection between force, mass, and motion. The connection occurs through *acceleration*, which results in *changes* in speed and/or direction. In fact, the unit of force (newton) is defined by the second law (Figure 6.5).

Newton
One newton (N) is the force it takes to change the speed of a 1-kg mass by 1 m/s in 1 second.

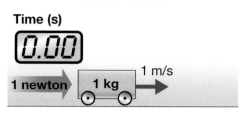

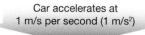

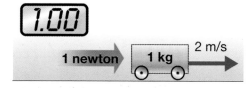

Figure 6.5: *The newton, a unit of force, is defined in terms of the acceleration it can create.*

Acceleration and force

Acceleration is proportional to force

The second law says that acceleration is *proportional* to force. What does that mean? It means that all other things being equal, if the force doubles, the acceleration also doubles. If the force is reduced by half, the acceleration is also reduced by half (Figure 6.6).

Example: A robot mail cart

Here is an example. Two engineers are each asked to design a battery-operated motor for a robot mail cart. The cart is supposed to drive around to people's offices and stop so they can collect their mail. One engineer chooses a motor that produces a force of 50 newtons. The other chooses a motor that produces a force of 100 newtons.

The acceleration of the mail cart

The robot with the smaller motor goes from rest to a top speed of 4 m/s in 4 seconds. The acceleration is 1 m/s². The robot with the larger motor accelerates to the same top speed in 2 seconds. Its acceleration is 2 m/s². Both robots reach the same top speed. The one with the bigger motor accelerates to its top speed twice as fast because it uses twice as much force. Of course, the one with the bigger motor drains its batteries faster too. There is also a trade-off between acceleration and energy!

Acceleration is in the direction of the net force

Another important factor of the second law is that the acceleration is always in the same direction as the net force. A force in the positive direction causes acceleration in the positive direction. A force in the negative direction causes acceleration in the negative direction. A sideways net force causes a sideways acceleration.

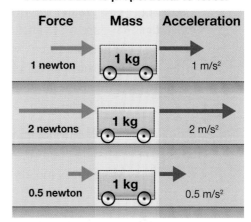

What it means to say "Acceleration is proportional to force."

Figure 6.6: *"Acceleration is proportional to force" means that if force is increased or decreased, acceleration will be increased or decreased by the same factor.*

STUDY SKILLS

Reviewing the newton

One newton is the force needed to change the speed of 1 kilogram by 1 m/s in one second. This means that:

$$1 \text{ N} = 1 \text{ kg} \cdot \text{m/s}^2$$

Or you can say 1 newton equals 1 kilogram-meter per second squared.

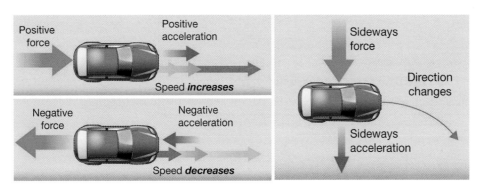

Acceleration and mass

Mass and acceleration
The greater the mass, the smaller the acceleration for a given force (Figure 6.7). That means acceleration is *inversely proportional* to mass. When the forces stay the same, increasing mass decreases the acceleration. For example, an object with twice the mass will have half the acceleration if the same force is applied. An object with half the mass will have twice the acceleration.

Why mass reduces acceleration
Acceleration decreases with mass because mass creates inertia. Remember, inertia is the property of matter that resists changes in motion (acceleration). More mass means more inertia, and therefore more resistance to acceleration.

Newton's second law
Force causes acceleration and mass resists acceleration. **Newton's second law** relates to the force on an object, the mass of the object, and the object's acceleration.

The acceleration caused by a net force is proportional to force and inversely proportional to mass.

The formula for the second law
The relationships between force, mass, and acceleration are combined in the formula for Newton's second law.

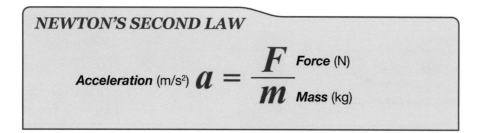

NEWTON'S SECOND LAW

$$\text{Acceleration (m/s}^2)\ a = \frac{F\ \text{Force (N)}}{m\ \text{Mass (kg)}}$$

Newton's second law – a law of motion that states that acceleration is force divided by mass

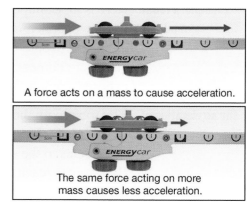

A force acts on a mass to cause acceleration.

The same force acting on more mass causes less acceleration.

Figure 6.7: *How acceleration is affected by mass.*

■■■■■ **SOLVE IT!** ■■■■■

Answer these questions to test your understanding of Newton's second law.

1. Force is tripled but mass stays the same. What happens to acceleration?

2. Acceleration decreases but the force is the same. What must have happened to the mass?

Summarizing the second law

Writing the second law You can use Newton's second law to calculate force, mass, or acceleration if two of the three values are known. As you solve problems, keep in mind the concepts shown below. Larger force leads to larger acceleration. Larger mass leads to smaller acceleration.

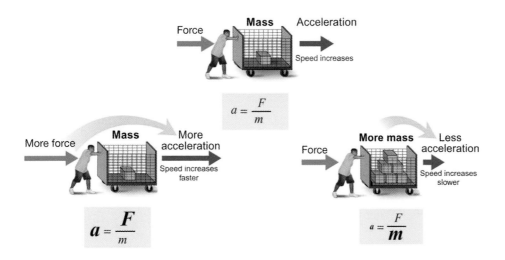

Net force and the second law Newton's second law explains the effect of the *net force* on motion. You must consider all the forces that are acting and add them up to find the net force. Then you use the net force to calculate any acceleration. You can also use the second law to work in the other direction, calculating net force from a given mass and acceleration.

To use Newton's second law properly, keep the following important ideas in mind.

1. The *net* force is what causes acceleration.

2. If there is *no* acceleration, the net force *must* be zero.

3. If there *is* acceleration, there *must* also be a net force.

4. The force unit of newtons is based on kilograms, meters, and seconds.

Solving Problems: Newton's Second Law

A car has a mass of 1,000 kilograms. If a net force of 2,000 N is exerted on the car, what is its acceleration?

1.	**Looking for:**	You are asked for the car's acceleration.
2.	**Given:**	You are given mass (kg) and net force (N).
3.	**Relationships:**	acceleration = force ÷ mass
4.	**Solution:**	acceleration = (2,000 N) ÷ (1,000 kg) = 2 m/s^2

Your turn...

a. As you coast down a hill on your bicycle, you accelerate at 0.5 m/s^2. If the total mass of your body and the bicycle is 80 kilograms, what is the net force pulling you down the hill (gravity − friction)?

b. What is the mass of an object that is experiencing a net applied force of 200 N and an acceleration of 500 m/s^2?

c. Recall that speed = distance ÷ time. The ratio of distance ÷ time is the same as the *slope* of a distance vs. time graph. That means speed is the slope of the distance vs. time graph. Acceleration is speed ÷ time. Use this graph of speed vs. time to find acceleration (the slope of this graph).

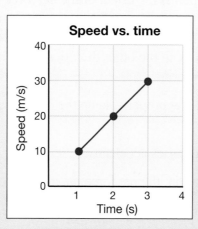

Speed vs. time

TECHNOLOGY

Race car design

Race cars are designed to have strong engines that produce large forces between the car and the road. They are also designed to be lightweight. Why is this combination of high forces and low mass useful for the design of a race car? Use Newton's second law to explain.

SOLVE FIRST LOOK LATER

a. 40 N

b. 0.40 kilogram

c. 10 m/s^2

Section 6.2 *Review*

1. What are the three main ideas associated with Newton's second law of motion? List these in your own words.

2. What conditions are necessary for acceleration to occur?

3. One kilogram-meter per second squared is also equal to what unit?

4. How much force would you need to cause a 20-kilogram object to accelerate in a straight line to 20 m/s²?

5. Different forces are applied to cars of different masses. The acceleration is measured for each combination of force and mass. Graph the data and determine the acceleration. Force goes on the *y*-axis and mass goes on the *x*-axis. Be sure to label each axis and give your graph a title.

Force (N)	Mass (kg)
5	1
10	2
15	3
20	4

6. A 2-kilogram rabbit starts from rest and is moving at 6 m/s after 3 seconds. What net force must be exerted on the rabbit (by the ground) to cause this change in speed (Figure 6.8)?

7. Explain how changing force or mass affects the acceleration of an object. Provide one example to support your answer.

8. A tow truck pulls a 1,500-kilogram car with a net force of 4,000 newtons. What is the acceleration of the car?

9. A potato launcher uses a spring that can apply a force of 20 newtons to potatoes. A physics student launched a 100-gram potato, a 150-gram potato, and a 200-gram potato with the launcher. Which potato had the greatest acceleration?

10. An experiment measures the speed of a 250-kilogram motorcycle every 2 seconds (Figure 6.9). The motorcycle moves in a straight line. What is the net force acting on the motorcycle?

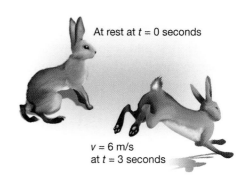

At rest at *t* = 0 seconds

v = 6 m/s
at *t* = 3 seconds

Figure 6.8: *Question 6.*

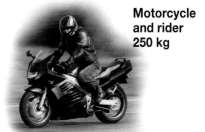

Motorcycle and rider 250 kg

Speed	Time
0 m/s	0 s
5 m/s	2 s
10 m/s	4 s
15 m/s	6 s
20 m/s	8 s

Figure 6.9: *Question 10.*

6.3 Newton's Third Law and Momentum

Newton's first and second laws apply to the motion of an *individual* object. Newton's third law applies to forces between interacting objects. Think about throwing a basketball (Figure 6.10). You feel the ball push back against your hand as you throw it. You apply a force to the ball to make it move. Where does the force against your hand come from? Can you predict your hand's motion and the basketball's motion after the throw?

Forces always come in matched pairs

An imaginary skateboard contest   Imagine a skateboard contest between Isaac Newton and an elephant. They can push against each other, but not against the ground. The one whose skateboard moves the fastest wins. The elephant is much stronger and pushes off Newton with a huge force, thinking he will surely win. But will he?

Your hand exerts a force on the ball (action)

The ball exerts an equal and opposite force on your hand (reaction)

Figure 6.10: *You experience Newton's third law (action-reaction) whenever you apply force to any object, such as a basketball.*

Reaction force   Action force

More acceleration   Less acceleration

JOURNAL

Think of three examples of action-reaction pairs that you experienced before class today. Write each one down and identify the action and reaction forces. Also write down what object each force acted on. Hint: The action and reaction forces never act on the same object.

The winner   Newton flies away with a great speed and the puzzled elephant moves backward with a much smaller speed. Newton wins—and will always win this contest against the elephant. No matter how hard the elephant pushes, Newton will always move away faster. Why?

Forces always come in pairs   It takes force to make both Newton and the elephant move. Newton wins because *forces always come in pairs*. The elephant pushes against Newton, and that *action* force pushes Newton away. The elephant's force against Newton creates a *reaction* force against the elephant. The action and reaction forces are equal in strength. Newton has much less mass, so he has much more acceleration, and therefore his speed is always greater.

The third law: Action and reaction

The first and second laws The first two laws of motion apply to individual objects. The first law says an object will remain at rest or in motion at a constant velocity unless acted upon by a net force. The second law states that acceleration equals the force on an object divided by the mass of the object.

The third law The third law of motion deals with pairs of objects. This is because *all forces come in pairs*. **Newton's third law** states that every action force creates a reaction force that is equal in strength and opposite in direction.

> ### *Every action force creates a reaction force that is equal in strength and opposite in direction.*

Force pairs There can never be a single force acting alone, without its action-reaction partner. Forces *only* come in action-reaction pairs. In the skateboard contest, the net force is the difference between the force created by the elephant in one direction and the force created by Newton in the opposite direction. The *action* of this force acts on Newton and moves Newton. The *reaction* of the same force acts on the elephant and moves the elephant. The combined strength of Newton and the elephant create two equal and opposite forces, an action and a reaction.

The labels *action* and *reaction* The words *action* and *reaction* are just labels. It does not matter which force is called action and which is called reaction. You simply choose one to call the action and then call the other one the reaction (Figure 6.11).

Why action and reaction forces do not cancel each other out Why don't action and reaction forces cancel each other out? The reason is *action and reaction forces act on different objects*. For example, think again about throwing a ball. When you throw a ball, you apply the action force to the ball, creating the ball's acceleration. The reaction is the ball pushing back against your hand. The action acts on the ball, and the reaction acts on your hand. The forces do not cancel each other out because they act on different objects. You can only cancel out forces acting on the same object (Figure 6.12).

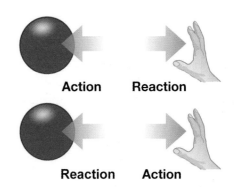

Figure 6.11: *It doesn't matter which force you call the action and which you call the reaction.*

Figure 6.12: *Action and reaction forces do not cancel each other out. One force acts on the ball, and the other force acts on the hand.*

Action and reaction forces

A skateboard example   Think carefully about propelling a skateboard with your foot. Your foot presses backward against the ground (Figure 6.13). The force acts *on* the ground. However, *you* move, so a force must act on you, too. Why do you move? What force acts on you? You move because the action force of your foot against the ground creates a reaction force of the ground against your foot. You "feel" the ground because you sense the reaction force pressing on your foot. The reaction force is what makes you move because it acts on *you*.

Draw diagrams   When sorting out action and reaction forces, it is helpful to draw diagrams. Draw each object apart from the other. Represent each force as an arrow in the appropriate direction. The illustration in Figure 6.13 is a good example of a diagram that shows a pair of action and reaction forces. The Solve It! box in the sidebar gives you an opportunity to think of your own example and draw a diagram.

Action and reaction guidelines   Here are some guidelines to help you sort out action and reaction forces:

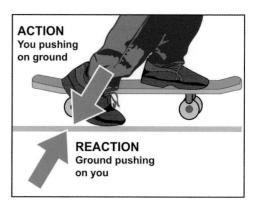

ACTION
You pushing on ground

REACTION
Ground pushing on you

Figure 6.13: *You move forward because of the reaction force of the ground on your foot.*

SOLVE IT!

Think of an action-reaction pair situation. Then draw a diagram illustrating the action-reaction pair. Use the tips in the text for drawing your diagram.

Guidelines for Action–Reaction Forces	Examples
Both forces are always there whenever any force occurs.	Your foot pushes (action) and the ground pushes back (reaction).
They always have the exact same strength.	The force arrows are the same length.
They always act in opposite directions.	The force arrows point in opposite directions.
They always act on different objects.	Your foot and the ground.
Both are real forces and can cause changes in motion.	You move forward on your skateboard.

Solving Problems: Action and Reaction

Action:
Sitting on a chair

A woman with a weight of 500 newtons is sitting on a chair (Figure 6.14). Describe one action-reaction pair of forces in this situation.

1. **Looking for:**   You are asked for a pair of action and reaction forces.

2. **Given:**   You are given an action force—the woman's force on the chair. Her force is 500 N.

3. **Relationships:**   Action-reaction forces are equal and opposite and act on different objects.

4. **Solution:**   The downward force of 500 N exerted by the woman on the chair is an action. Therefore, the chair acting on the woman provides an upward force of 500 N and is a reaction.

Figure 6.14: *An action is sitting on a chair.*

SOLVE FIRST LOOK LATER

a.   540 newtons

b.   The weight of the chair is 90 N. Action-reaction pairs include the cat-woman's lap, the woman-chair, the chair-strongman, and the strongman-ground.

c.   The force of the bat on the ball (action) accelerates the ball. The force of the ball on the bat (reaction) slows down the swinging bat.

d.   Earth attracts the Moon (action) and the Moon attracts Earth (reaction) in an action-reaction pair. Both action and reaction are due to gravity.

Your turn...

a. A cat jumps up and sits on the lap of the woman who is sitting in the chair in Figure 6.14. The cat's weight is 40 newtons. What is the reaction force provided by the chair now?

b. A strong man now picks up the chair with the woman and the cat and holds them all above his head. If the upward force from the strong man is 630 newtons, what is the weight of the chair in newtons? Describe the different action-reaction pairs in this scenario.

c. A baseball player hits a ball with a bat. Describe an action-reaction pair of forces in this situation.

d. Earth and the Moon are linked by what action-reaction pair?

Collisions and momentum

The effect of forces Newton's third law tells us that when two objects collide, they exert equal and opposite forces on each other. However, the *effect* of the force is not always the same. Imagine two hockey players moving at the same speed toward each other, one with twice the mass of the other. The force on each during the collision is the same strength, but they do not have the same change in motion after the collision.

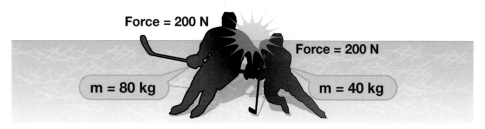

Force = 200 N

Force = 200 N

m = 80 kg

m = 40 kg

Momentum When studying motion related to collisions, we can predict how two colliding objects might move using Newton's third law of motion and *momentum*. **Momentum** is the mass of an object times its velocity. The units for momentum are kilogram-meter per second (kg · m/s).

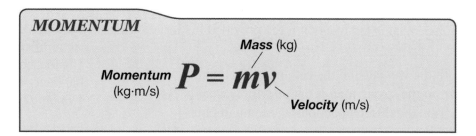

MOMENTUM

Mass (kg)

Momentum (kg·m/s) $P = mv$

Velocity (m/s)

The law of conservation of momentum Using this information, we can determine the momentum of each player in the example above. The **law of conservation of momentum** states that as long as the interacting objects are not influenced by outside forces (like friction) the total amount of momentum is constant (does not change). This means that the total amount of momentum for the colliding hockey players before the collision equals the total amount of momentum afterward. Also, any momentum lost by one player is gained by the other one.

VOCABULARY

momentum – the mass of an object times its velocity

law of conservation of momentum – a law that states that as long as interacting objects are not influenced by outside forces, the total amount of momentum is constant

SOLVE IT!

Calculate: Use the momentum formula to find the momentum of each hockey player before they collide.

Player 1: m = 80 kg; v = 2 m/s

Player 2: m = 40 kg; v = 3 m/s

Predict: Let's say the motion of Player 1 is in the positive direction and the motion of Player 2 is in the negative direction. Based on your momentum calculations, in which direction do you think the two combined players will move after the collision?

Understanding the law of conservation of momentum

Using positive and negative
The forces on each player or any two interacting objects are always equal and opposite. Similarly, the momentum of two interacting objects are equal and opposite. Therefore, it makes sense to use positive and negative values to tell the direction of motion (Figure 6.15). Momentum can be positive (moving to the right) or negative (moving to the left) (Figure 6.15).

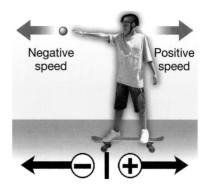

Figure 6.15: *The direction is important when using the law of conservation of momentum. We use positive and negative numbers to represent opposite directions.*

A ball example
Let's say a skateboarder is standing on a skateboard and has a ball. Before he throws the ball, his velocity (and the ball's) is zero. Since momentum is mass times velocity, the total momentum is also zero. The law of conservation of momentum says that after the ball is thrown, the total momentum still has to be zero. Here's where positive and negative values help us.

Conservation of momentum
If the ball has a mass of 1 kilogram and the skateboarder throws it at a velocity of –20 m/s to the left, the ball takes away –20 kg · m/s of momentum. To make the total momentum zero, the skateboarder must take away +20 kg · m/s of momentum. If his mass is 40 kilograms and you ignore friction, then his speed is +0.5 m/s to the right (Figure 6.16).

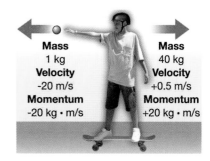

Figure 6.16: *The result of a skateboarder throwing a 1-kg ball at a speed of –20 m/s is that he and the skateboard with a total mass of 40 kg move backward at a speed of +0.5 m/s if you ignore friction. If you account for friction, would the calculation for velocity of the skateboarder on the skateboard end up being less or more than 0.5 m/s?*

More mass results in less acceleration
Because of his greater mass, the skateboarder will have a smaller velocity after he throws the ball. The ball, which has less mass, has the greater velocity. *They each have equal and opposite momentum after the throw.* The two objects, the skateboarder and the ball, have different velocities because they have different masses, *not because the forces are different!*

Jet planes and rockets
Rockets and jet planes use the law of conservation of momentum to move. In a process called jet propulsion, a jet moves forward when the engine pushes exhaust air at very high speed out of the back of the engine. The momentum lost by the air going backward is compensated by the momentum gained by the jet moving forward. Similarly, a rocket accelerates in space because it pushes mass at high speed out the end of the engine in the form of exhaust gases from burning fuel. The forward momentum of a rocket equals the momentum of the escaping mass ejected from the end of the engine.

➕ ➖ ✖ ➗ **Solving Problems: Conservation of Momentum**

An astronaut in space throws a 2-kilogram wrench away from her at a speed of –10 m/s. If the astronaut's mass is 100 kilograms, at what speed does the astronaut move backward after throwing the wrench?

1.	**Looking for:**	You are asked for the astronaut's speed. Since the astronaut is in space, we can ignore friction.
2.	**Given:**	You are given the mass and speed of the wrench and the mass of the astronaut.
3.	**Relationships:**	This is enough information to apply the law of conservation of momentum. The momentum of the wrench (m_1v_1) and the momentum of the astronaut (m_2v_2) add up to zero BEFORE the wrench is thrown.  $m_1 v_1 + m_2 v_2 = 0$
4.	**Solution:**	The momentum of the wrench and the astronaut also add up to zero AFTER the wrench is thrown. $$[2 \text{ kg} \times (-10 \text{ m/s})] + [(100 \text{ kg}) \times v_2] = 0; \ v_2 = +20 \div 100 = +0.2 \text{ m/s}$$ The astronaut moves backward to the right at a speed of +0.2 m/s.

Photo courtesy NASA

Your turn...

a. Two hockey players have a total momentum of +200 kg · m/s before a collision (+ is to the right). After their collision, they move together. In what direction do they move and what is their momentum?

b. When a large truck hits a small car, the forces are equal (Figure 6.17). However, the small car experiences a much greater change in velocity than the big truck. Explain why.

Figure 6.17: *Your turn, question b.*

SOLVE FIRST, LOOK LATER

a. The two hockey players move in the positive direction (or to the right). Their momentum after the collision is +200 kg · m/s.

b. The car has less mass and therefore less inertia, so it accelerates more and may become more damaged than the truck in this collision.

Section 6.3 *Review*

1. Emilio tries to jump to a nearby dock from a canoe that is floating in the water. Instead of landing on the dock, he falls into the water beside the canoe. Use Newton's third law to explain why this happened. Hint: First identify the action-reaction pair in this example.

2. You push backward against the ground to move a skateboard forward. The force you make acts against the ground. What force acts against you to move you forward?

3. Explain why action-reaction forces do not cancel each other out, resulting in zero net force.

4. The momentum of an object depends on what two factors?

5. The engine of a jet airplane pushes exhaust gases from burning fuel backward. What pushes the jet forward?

Forward motion of jet

Backward motion of exhaust

6. A small rubber ball is thrown at a heavier, larger basketball that is still. The small ball bounces off the basketball. Assume there are no outside forces acting on the balls.

   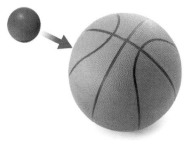

   a. How does the force on the small ball compare to the force on the basketball?

   b. Compare the total momentum of the two balls before and after the collision.

   c. The mass of the basketball is 600 grams, and its velocity before the small ball hits is 0 m/s. The mass of the small ball is 100 grams, and its velocity is +5 m/s before the collision and −4 m/s afterward. What is the velocity of the basketball after the collision?

Forensic Engineering: A Two-Part Science

We usually think of engineering as a science focused on designing and constructing things—like bridges, computers, automobiles, or sneakers. However, there is one branch of engineering that focuses on how things fail, collapse, or crash. It's called forensic engineering.

Forensic engineers are like time travelers, rewinding the clock to a point just before a bridge collapses or a car crashes. Their job is to gather and analyze information from the scene so they can reconstruct the event step by step. A forensic engineer's work is often used in court as evidence in personal injury or product liability cases. Forensic engineers must play two roles in their work: that of a detective, gathering clues and evidence, and that of an engineer, using this evidence to analyze the event.

Gathering information: The detective role

One task of a forensic engineer is to reconstruct automobile crashes. Working with law enforcement officials, forensic engineers act as detectives looking for clues about how the collision occurred. The vehicle or vehicles involved in the crash are the most important pieces of evidence. They can give the forensic engineer an idea of angle of impact, speeds involved, and seat belt usage.

John Kwasnoski has been a forensic engineer and physics professor for more than thirty years. He is often asked to testify in court when collisions result in criminal charges. Many of his cases involve a driver (often alcohol-impaired) losing control of a vehicle and colliding with another vehicle or pedestrians. In other cases, a driver may have hit a telephone pole, concrete barrier, or some other stationary object, resulting in injury to passengers in the vehicle.

Professor Kwasnoski explains, "As an investigator at the scene of a crash, I'm most often looking for evidence of the transfer of energy. Before the crash, the vehicle has a certain amount of kinetic energy. The police find the car at rest. The law of conservation of energy tells us the vehicle's kinetic energy had to be transferred somewhere. Often it's found in damage to roadside obstacles or the roadway itself, and in change to the vehicle's shape."

"I investigated the crash of the car in the photo [to the right]. This car hit a utility pole when the teenage driver lost control on a rural road. The front seat passenger was injured in the crash. The passenger was not wearing a seat belt that probably would have prevented her from striking the windshield."

This car hit a utility pole when the teenage driver lost control on a rural road.

After photographing the scene and making careful measurements of the damage to vehicles, the length of skid marks, and other evidence of transfer of energy, forensic engineers like Professor Kwasnoski head back to the lab.

Analyzing the information: The engineering role

The next step, explains Kwasnoski, is to figure out how much energy it took to cause the damage he observed. He looks at the results of crash tests where vehicles are crashed into concrete barriers at various speeds. The amount of damage depends on the specific properties (like stiffness) of the materials used to build the car, so it is important to analyze crash test records of the specific make and model of the vehicle involved in the crash. By comparing measurements of the vehicle's damage to the crash test records, the speed of the vehicle at the time of the collision can be inferred. This information is often a crucial piece of evidence in a criminal trial.

The study of how vehicles moved before, during, and after a collision is called *vehicular kinematics*. Another important part of a forensic engineer's job is to analyze how the passengers moved before, during, and after the collision. This is called *occupant kinematics*. From crash test data, the forensic engineer can calculate peak accelerations of an occupant. These accelerations, especially those of the head and neck, can be greater than the vehicle's peak acceleration. Calculating an occupant's peak acceleration can help determine the cause of his or her injuries.

Crash prevention through physics lessons

Investigating crashes has convinced Professor Kwasnoski that if people understood the physics of force and motion, they would be better equipped to make good decisions about driving and seat belt use. So he often speaks to high school and community groups about crash prevention.

Sometimes audience members will comment that they don't think it's important to wear seat belts when they are driving in town, where the speed limit is 35 miles per hour.

"I point out to them that if you calculate acceleration due to gravity, 35 miles per hour is the speed you would be going when you hit the ground after falling off a four-story building. I ask, 'Would you rather be strapped into a padded steel cage or just hurtling through the air on your own in a fall like that?'"

"We also talk about Newton's first law of motion—objects in motion stay in motion, unless acted on by a force. So if a car is traveling at 35 miles per hour and crashes, an unbelted occupant will collide with the interior of the car at 35 miles per hour. There are also secondary crashes—your organs collide with your rib cage, and your brain collides with your skull at 35 miles per hour."

While seat belts can't prevent every internal injury, Kwasnoski points out that all of the significant automobile safety advances in the past 50 years—air bags, padded dashboards, stronger frames—are designed to protect people who stay in the car. After investigating more than 650 collisions, Kwasnoski concludes, "You just do not want to be ejected from a vehicle in a crash." Human bodies are not designed to handle the impact of crashing into a stationary object after traveling through space at the speed of a car.

Questions:

1. What two roles do forensic engineers play?

2. How does a forensic engineer use the law of conservation of energy in a crash investigation?

3. Explain the difference between vehicular kinematics and occupant kinematics.

4. Project idea: Design a poster that uses a physics principle to encourage seat belt usage.

To learn more about Professor Kwasnoski's work, try this Internet keyword search: "Kwasnoski + legal sciences."

Car photo courtesy of John Kwasnoski.

Chapter 6 *Assessment*

Vocabulary

Select the correct term to complete each sentence.

inertia	momentum	Newton's first law
Newton's second law	Newton's third law	unbalanced forces

Section 6.1

1. _____ says that objects continue the motion they already have unless they are acted on by an unbalanced force.

2. If the net force acting on an object is not zero, then the forces acting on the object are _____ .

3. Objects with more mass have more _____ .

Section 6.2

4. The relationship between the force on an object, the mass of the object, and its acceleration is described by _____ .

Section 6.3

5. _____ states that every action force creates a reaction force that is equal in strength and opposite in direction.

6. The law of conservation of _____ can be used to predict motion of interacting objects after they collide.

Concepts

Section 6.1

1. Newton's first law states that no force is required to maintain motion in a straight line at constant speed. If Newton's first law is true, why must you continue to pedal a bicycle on a level surface to keep moving?

2. Two identical-looking, large, round balls are placed in front of you. One is filled with feathers and the other is filled with sand. Without lifting the balls, how could you use inertia to distinguish between them?

3. What are the natural states of motion? List all correct answers.
   a. being stopped
   b. moving with constant direction
   c. moving with changing speed
   d. moving with constant velocity

4. What happens to the inertia of an object if its mass is decreased?

5. Identify whether the following scenarios involve balanced (B) or unbalanced (U) forces.
   a. _____ a car stopped at a red light
   b. _____ a ball rolling down a hill
   c. _____ an airplane flying at constant speed at the same altitude in one direction
   d. _____ an airplane taking off
   e. _____ a person sitting motionless on a chair
   f. _____ a person running at constant speed around a circular track

Section 6.2

6. What is a newton?
   a. The time it takes to move 1 kilogram.
   b. The force it takes to change the speed of 1 kilogram by 1 m/s in 1 second.
   c. The speed it takes to move a 1-kilogram mass in one hour.

7. Explain the difference between "directly proportional" and "inversely proportional."

8. What does it mean to say that the "net force" determines an object's acceleration?

9. Describe three ways you could cause an acceleration of a moving car.

10. If you are applying the brakes on your bicycle, and you are slowing down, are you accelerating? Why or why not?

11. What is the formula that summarizes Newton's second law?

12. Which of the following is the equivalent unit to a newton?
   a. m/s^2
   b. m/s
   c. $kg \cdot m/s^2$

Section 6.3

13. Are these statements correct or incorrect? If incorrect, rewrite the sentence so that it is correct.
   a. In an action-reaction pair, the forces work on the same object.
   b. Every action force creates a reaction force, and the two forces are different in strength but act in the same direction.

14. A brick is sitting on a table. The force of gravity pushes down on the brick. What prevents the brick from accelerating downward and falling through the table?

15. When a bug traveling west collides with the windshield of a car traveling east, what can be said about the collision?
   a. The bug feels a stronger force than the car.
   b. The bug and the car feel the same size force.
   c. The car accelerates more than the bug.
   d. The bug does not accelerate due to the force.

16. Give an example of the law of conservation of momentum from everyday life.

Problems

Section 6.1

1. While an object is moving at a constant 20 m/s, a 5-N force pushes the object to the left. At the same time, a 5-N force is pushing the object to the right. What will the object's velocity be after 10 seconds?

2. A bowling ball has a mass of 6 kilograms. A tennis ball has a mass of 0.06 kilogram. How much inertia does the bowling ball have compared to the tennis ball?

3. A rider and motorcycle with a combined mass of 250 kilograms are driving down the road at a constant speed of 55 mph. The motorcycle's engine is producing a force of 1,700 newtons between the tires and the road.
   a. Find the weight of the motorcycle and rider in newtons.
   b. Find the normal force of the road on the motorcycle and rider.
   c. Find the frictional force of the road and air on the motorcycle and rider.

4. What is the net force on the refrigerator shown?

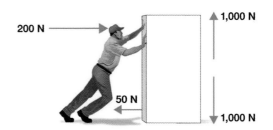

5. Make a free-body diagram of someone pushing a refrigerator that shows:
   a. A net force of 100 N with the refrigerator being pushed to the right.
   b. The refrigerator in equilibrium.

Section 6.2

6. Copy the following table and fill it in based on Newton's second law:

Force (N)	Mass (kg)	Acceleration (m/s²)
20	10	
50	10	
10	2	
10	5	
100		2
	100	5

7. What force is needed to accelerate a 1,000-kg car from a stop to 5 m/s²?

8. What is the acceleration of a truck with a mass of 2,000 kg when its brakes apply a force of 10,000 N?

9. A 20-N force accelerates a baseball at 140 m/s² (briefly). What is the mass of the baseball?

10. Gina is pushing a 10-kilogram box with 50 N of force toward the east. Dani is pushing the same box at the same time with 100 N of force toward the west. Assuming there is no friction, what is the acceleration of the box?

11. A cheetah can accelerate at 7 m/s², and the average cheetah has a mass of 40 kg. With what average force does the cheetah push against the ground?

12. A car speeds up from 5 m/s to 29 m/s over 4 seconds.

    a. What is the car's acceleration?

    b. If the car had started at 29 m/s and ended at 5 m/s after 4 seconds, what would its acceleration be? How is this different from the answer above?

Section 6.3

13. Jane has a mass of 40 kg. She pushes on a 50-kg rock with a force of 100 N. What force does the rock exert on Jane?

14. Look at the picture below.

    a. Identify at least three action-reaction pairs.

    b. Why might it be hard for the firefighter to hold the hose steady when the water gushes out of the hose? Think about the law of conservation of momentum.

15. A 3,000-kg car bumps into a stationary, 5,000-kg truck. The velocity of the car before the collision was +4 m/s and −1 m/s after the collision. What is the velocity of the truck after the collision?

Applying Your Knowledge

Section 6.1

1. The work of Sir Isaac Newton made this chapter on motion possible. You may already know about Newton because he is such a well-known scientist. Do some research to learn something you did not know about Newton. Write a paragraph describing your findings.

2. You are watching a magic show. For one trick, the magician rolls a ball down a hill. Suddenly the ball stops moving down the hill. It is as if the ball is defying gravity! Come up with an explanation for how the magician might have accomplished his trick. Hint: Think of all the forces that might be acting on the ball.

3. Answer the following motion questions for a hot-air balloon.

    a. List all the forces that are act on a hot-air balloon to keep it on the ground.

    b. List all the forces that act on a hot-air balloon when it is in the sky.

    c. Sketch a free-body diagram for a hot-air balloon that is rising straight off the ground. Indicate the magnitude of forces with the length of the force vectors.

    d. Sketch a free-body diagram for a hot-air balloon that is in a neutral position in the sky (neither rising nor sinking) but being blown eastward by the wind. Indicate the magnitude of forces with the length of the force vectors. What force might be opposing the wind?

Section 6.2

4. The text stated that anyone who does anything involving motion needs to understand Newton's second law. Think about a job or career that might involve using and understanding motion and answer the following.

    a. Name the job or career. Describe the types of motion-related tasks that are involved in this job or career.

    b. Pick one task listed in your answer for 4a and explain how understanding Newton's laws of motion might help accomplish the task better.

    c. Extension: Research and/or interview someone who has this career. Find out how they use their understanding of motion in their job. Write a paragraph about your findings.

5. Describe the design features you would incorporate into a battery-operated motor for a robot mail cart for the following situations. The design features to consider are the mass of the motor, rate of acceleration, and speed.

    a. A robot mail cart is needed to collect mail from offices located in a large warehouse. The warehouse has a lot of open space.

    b. A robot mail cart is needed in a small office space that has many offices that are close together.

    c. A robot mail cart is needed in an elementary school that has long hallways and many offices. However, many children are often in the hallways.

Section 6.3

6. You are playing a game of soccer. Describe as many action-reaction pairs in this situation as you can think of.

7. At the beginning of the chapter, you read about astronauts investigating how toys work in space. Describe how you think the following toys would work in space based on what you have learned in this chapter.

    a. a ball that can be thrown through a hoop

    b. building blocks

    c. a board game with game pieces for each player

    d. a deck of cards

8. Auto manufacturers design cars to withstand collisions. Research design features that allow a car and the people inside the car to survive a crash. Write a paragraph about one design feature that interests you.

9. If you push a very large object, like a building, it doesn't move before or after the interaction. Explain why.

10. For fun: Which one of the laws of motion is your favorite? Pick one and make a brochure explaining why it is your favorite law of motion.

Unit 3

Work and Energy

Try this at home

Tie a string between two chairs. Now tie another string to the first so it hangs straight down. Tie something heavy to the second string, like a small toy action figure or a metal washer. You created a pendulum! Pull the object back so it is parallel with the floor and release it, but don't move your hand. Does the object swing back and hit your hand? Why or why not? How many times does the object swing before it stops moving? Why doesn't it go on forever? What would you have to do to make it go again? Write down your observations and ideas in a paragraph.

Energy

Look around you. Do you see any changes taking place? When the lights came on in your classroom, for example, the light bulbs gave off light and heat. Outside, the Sun may be shining and causing changes in plants. And right now your eyes are moving across the page while you read this introduction. Energy is at the heart of all these events. When an object falls toward Earth, when you play a sport or a musical instrument, when your alarm clock wakes you up in the morning, and when a bird flies through the air, changes are taking place thanks to the presence of energy.

Energy is everywhere! As you read this chapter, think about how energy is responsible for the changes that take place around you and even inside your body! For starters, can you identify the different forms of energy in the picture on this page?

CHAPTER 7 INVESTIGATIONS

7A: Energy in a System
How is energy related to motion?

7B: Conservation of Energy
What limits how much a system may change?

7.1 What Is Energy?

Unlike matter, pure energy cannot be smelled, tasted, touched, seen, or heard. However, energy does appear in many forms, such as motion and heat. Energy can also travel in different ways, such as in light and as electricity. Without energy, nothing could ever change. In fact, the workings of the entire universe (including all of our technology) depend on energy flowing and changing back and forth from one form to another.

Defining energy

What is energy? **Energy** describes the ability of things to change themselves or to cause change in other things. What types of changes are we talking about? Some examples are changes in temperature, speed, position, pressure, or any other physical variable. Energy can also cause changes in materials, such as when burning wood changes into ashes and smoke.

What has energy? The list below describes objects that have energy. Read through this list and notice how many different forms of energy exist. We will talk more about these different forms in this chapter.

- A gust of wind has energy because it can move objects in its path.
- A piece of wood burning in a fireplace has energy because it can produce heat and light.
- You have energy because you can change the motion of your body.
- Batteries have energy; they can be used in a radio to make sound.
- Gasoline has energy; it can be burned in an engine to move a car.
- A ball at the top of a hill has energy because it can roll down the hill and move objects in its path.

Measuring energy A **joule (J)** is the unit of measurement for energy. One joule is the energy needed to push with a force of 1 newton for a distance of 1 meter (Figure 7.1). So, 1 joule is equivalent to 1 newton multiplied by 1 meter (or 1 newton-meter). If you push a toy car forward with a force of 1 newton over a distance of 1 meter, you have applied 1 joule of energy to the car. One joule is a pretty small amount of energy. An ordinary 100-watt electric light bulb uses 100 joules of energy *every second*!

energy – a quantity that describes the ability of an object to change or cause changes

joule (J) – a unit of energy; 1 joule is enough energy to push with a force of 1 newton for a distance of 1 meter

joule (J)

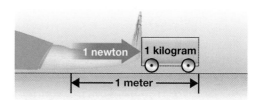

Figure 7.1: *Pushing a 1-kilogram object with a force of 1 newton for a distance of 1 meter uses 1 joule of energy.*

Units related to the joule

1 joule = 1 newton-meter

1 newton = 1 kg·m/s^2

therefore . . .

1 joule = 1 kg·m^2/s^2

Some forms of energy

Understanding energy One way to understand energy is to think of it as nature's money. Energy can be spent and saved in a number of different ways. It takes energy to "buy" changes like going faster, moving higher, or getting hotter. These three changes *use* energy. The opposite changes, such as slowing down, falling, or cooling off, *release* energy. Just like a bank account, nature keeps perfect track of energy. What you "spend" diminishes what you have left. You can only "buy" as much change as you have energy to "pay for."

Mechanical energy **Mechanical energy** is the energy possessed by an object due to its motion or its position. Energy of motion is called kinetic energy, and energy of position is called potential motion.

Chemical energy **Chemical energy** is a form of energy stored in molecules. Batteries are really storage devices for chemical energy. For example, the chemical energy in a battery changes to electrical energy when you connect wires and a light bulb to the battery. Your body also uses chemical energy when it converts food into energy so that you can walk or think. A car and many other types of machines use chemical energy when they burn fuel to operate.

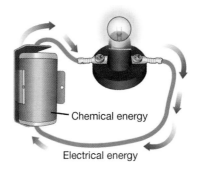

Chemical energy

Electrical energy

Electrical energy Electrical energy comes from electric charge, which is one of the fundamental properties of all matter. You will learn more about electricity and electric charge in Unit 7. The electrical energy we use in our homes is transformed from other forms of energy, such as the chemical energy released by burning oil and gas, or the mechanical energy released by falling water in a *hydroelectric* dam or power plant.

Pressure energy Pressure in gases and liquids is also a form of energy. An inflated bicycle tire has more energy than a flat tire. An inflated tire can hold up a bicycle (with you on it) against the force of gravity, whereas a flat tire cannot.

More forms of energy

Elastic energy   Elastic energy is energy that is stored or released when an object changes shape (or *deforms*). For example, you use energy to stretch a rubber band. Some of the energy from your muscles is stored as elastic energy in the stretched (changed) shape of the rubber band. The energy is released again when the rubber band changes back to its original (unstretched) shape. Objects that are commonly used to store and release elastic energy include rubber bands, springs, and archery bows (Figure 7.2).

nuclear energy – a form of energy that is stored in the nuclei of atoms

radiant energy – a form of energy that is represented by the electromagnetic spectrum

Nuclear energy and radiant energy   Every second, about 5 million tons of mass are converted to energy through nuclear reactions in the core of the Sun. In the Sun, *nuclear energy* is transformed to heat that eventually escapes the Sun as *radiant energy*. **Nuclear energy** is a form of energy stored in the nuclei of atoms (particles of matter). You will read more about nuclear energy and nuclear reactions in Chapter 18. **Radiant energy** is energy that is carried by electromagnetic waves. Light is one form of radiant energy, and so are radio waves that carry music through the air.

The electromagnetic spectrum   Light and radio waves are a traveling form of pure energy. In fact, they are only two of a whole family of energy waves called the *electromagnetic spectrum*. The electromagnetic spectrum includes infrared radiation (heat), visible light (what we see), and ultraviolet light. In other words, *light energy* and *heat energy* are included in the electromagnetic spectrum. You will recognize other components of the spectrum as well. You have listened to radio waves, may have cooked with microwaves, and maybe you have had an image made of a part of your body with X-rays.

Figure 7.2: *A stretched bowstring on a bent bow has elastic energy, so it is able to create change in itself and in the arrow.*

The electromagnetic spectrum

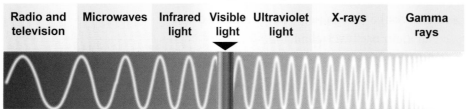

| Radio and television | Microwaves | Infrared light | Visible light | Ultraviolet light | X-rays | Gamma rays |

The Sun and gravity

The Sun and energy Both living creatures and human technology derive virtually all of their energy from the Sun. Without the Sun's energy, Earth would be a cold, icy place with a temperature of –273 degrees Celsius. The Sun's energy not only warms the planet, it also drives the entire food chain (Figure 7.3). Plants store the energy as carbohydrates, like sugar. Animals eat the plants to get energy. Other animals eat *those* animals for their energy. It all starts with the Sun.

Life on Mars and other planets A very important question in science today is whether there is life on other planets such as Mars. Mars is farther from the Sun than Earth. For this reason, Mars receives less energy from the Sun than does Earth. In fact, the average temperature on Mars is well below the freezing point of water. Can life exist on Mars? Recent research suggests that it may be possible. Scientists have found bacteria in the Antarctic ice living at a temperature colder than the average temperature of Mars.

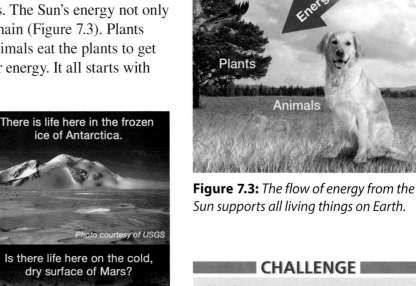

There is life here in the frozen ice of Antarctica.

Photo courtesy of USGS

Is there life here on the cold, dry surface of Mars?

Photo courtesy of NASA/NSSDC

Figure 7.3: *The flow of energy from the Sun supports all living things on Earth.*

Gravity and energy A falling rock gains speed as it falls. Energy must be supplied to increase speed. The falling water that turns a hydroelectric turbine must also have energy, otherwise no electrical energy could be produced. Where does this energy come from?

The answer has to do with Earth's gravity. If an object, or any matter, is lifted against gravity, energy is stored. This stored energy is transformed into energy of motion, such as the object falling back down. Many forms of human technology, including roller coasters, swings, water wheels, hydroelectric power plants, and even a kind of medieval catapult called a *trebuchet*, rely on gravity.

CHALLENGE

The planet Venus is closer to the Sun than Earth. Should this make Venus warmer or colder than Earth? Research your answer to see what scientists think Venus is like on its surface.

Energy and work

What *work* means in physics   In physics, the word *work* has a very specific meaning. *Work* is the transfer of energy that results from applying a force over a distance. Work is a product of the force applied times the distance traveled (work = force × distance). For example, if you push a block with a force of 1 newton for a distance of 1 meter, you do 1 joule of work. Both work and energy are measured in the same units (joules) because work is a form of energy.

Work and potential energy   Doing work always means transferring energy. The energy may be transferred to the object to which force is applied, or it may go somewhere else. For example, you can increase the energy of a rubber band by exerting a force that stretches it. The work you do stretching the rubber band is stored as elastic potential energy by the rubber band. The rubber band can then use that stored energy to do work on a paper airplane, giving it energy (Figure 7.4).

Work is done on objects   When thinking about work, you should always be clear about which force is doing the work on which object. Work is done *on* objects. If you lift a block 1 meter with a force of 1 newton, you have done 1 joule of work *on the block*.

Energy is needed to do work   An object that has energy is able to do work; without energy, it is impossible to do work. In fact, energy can sometimes be thought of as *stored work*. As the block you lifted earlier falls, it has energy that can be used to do work. If the block hits a ball, it will do work on the ball and change the ball's motion. Some of the block's energy is transferred to the ball during the collision (below right). You will learn more about the concept of work in the next chapter.

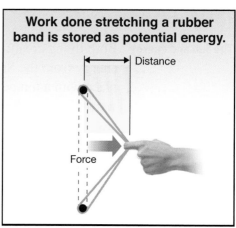

Work done stretching a rubber band is stored as potential energy.

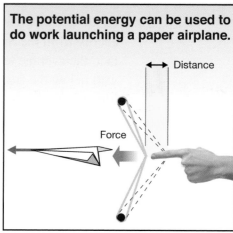

The potential energy can be used to do work launching a paper airplane.

Figure 7.4: *You can do work to increase an object's energy. Then that energy can do work on another object, giving it energy.*

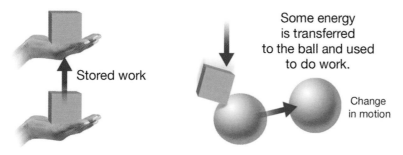

Stored work

Some energy is transferred to the ball and used to do work.

Change in motion

Potential energy

What is potential energy?
Potential energy is energy due to *position*. The word *potential* means that something is capable of becoming active. Systems or objects with potential energy are able to exert forces (exchange energy) as they change to other arrangements. For example, a stretched spring has potential energy. If released, the spring will use this energy to move itself (and anything attached to it) back to its original length.

Gravitational potential energy
A block suspended above a table has potential energy. If released, the force of gravity moves the block down to a position of lower energy. The term *gravitational potential energy* describes the energy of an elevated object. The term is often shortened to just *potential energy* because the most common type of potential energy in physics problems is gravitational. Unless otherwise stated, you can assume that *potential energy* means gravitational potential energy.

How to calculate potential energy
How much potential energy does a raised block have? The block's potential energy is exactly the amount of work it can do as it goes down. Work is force multiplied by distance. The force is the weight (*mg*) of the block in newtons. The distance the block can move down is its height (*h*) in meters. Multiplying the weight by the distance gives you the block's potential energy at any given height (Figure 7.5).

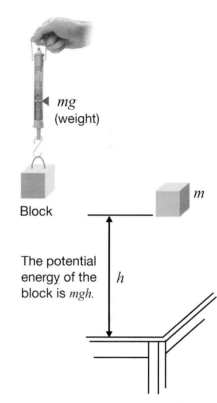

Figure 7.5: *The potential energy of the block is equal to the product of its mass, the strength of gravity, and the height from which the block can fall.*

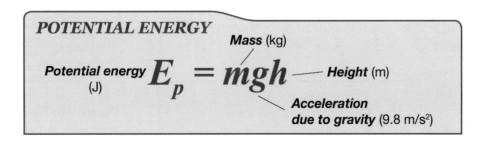

POTENTIAL ENERGY

Potential energy (J) $E_p = mgh$ — Mass (kg), Height (m), Acceleration due to gravity (9.8 m/s²)

Kinetic energy

Kinetic energy is energy of motion Objects that are moving also have the ability to cause change. Energy of *motion* is called **kinetic energy**. A moving billiard ball has kinetic energy because it can hit another ball and change its motion. Kinetic energy can easily be converted into potential energy. The kinetic energy of a basketball tossed upward converts into potential energy as the height increases.

Kinetic energy can do work The amount of kinetic energy an object has equals the amount of work the object can do by exerting force as it stops. Consider a moving skateboard and rider (Figure 7.6). Suppose it takes a force of 500 newtons applied over a distance of 10 meters to slow the skateboard down to a stop (500 N × 10 m = 5,000 joules). The kinetic energy of the skateboard and rider is 5,000 joules because that is the amount of work it takes to stop the skateboard.

Kinetic energy depends on mass and speed If you had started with twice the mass—say, two skateboarders—you would have to do twice as much work to stop them both. Kinetic energy increases with mass. If the skateboard and rider are moving faster, it also takes more work to bring them to a stop. This means kinetic energy also increases with speed. Kinetic energy is related to *both* an object's speed and its mass.

The formula for kinetic energy The kinetic energy of a moving object is equal to one half its mass multiplied by the square of its speed. This formula comes from a combination of relationships, including Newton's second law, the distance equation for acceleration ($d = \frac{1}{2}at^2$), and the calculation of energy as the product of force and distance.

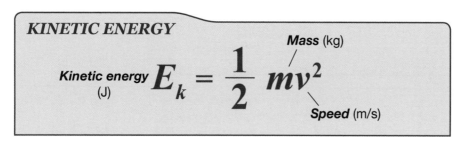

KINETIC ENERGY

$$E_k = \frac{1}{2}mv^2$$

Kinetic energy (J)   Mass (kg)   Speed (m/s)

kinetic energy – energy of motion

Moving skateboard and rider

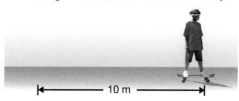

A force of 500 N applied for 10 m . . .

500 N

. . . brings the skateboard and rider to a stop.

|← 10 m →|

Work done = 500 N × 10 m
= 5,000 J

Therefore . . .
The kinetic energy is 5,000 joules because that is the amount of work the skateboard can do as it stops.

Figure 7.6: *The amount of kinetic energy the skateboard has is equal to the amount of work that must be done to stop the skateboard.*

Solving Problems: Potential and Kinetic Energy

A 2-kilogram rock is at the edge of a cliff 20 meters above a lake. The rock becomes loose and falls toward the water below. Calculate its potential and kinetic energy when it is at the top and when it is halfway down. Its speed is 14 m/s at the halfway point.

1. **Looking for:**	You are asked for the potential and kinetic energy at two locations.	
2. **Given:**	You are given the mass in kilograms, the height at each location in meters, and the speed halfway down in m/s. You can assume the initial speed is 0 m/s because the rock starts from rest.	
3. **Relationships:**	$E_p = mgh$ and $E_k = \dfrac{1}{2}mv^2$	
4. **Solution:**	Potential energy at the top:	$m = 2$ kg, $g = 9.8$ N/kg, and $h = 20$ m $E_p = (2 \text{ kg})(9.8 \text{ N/kg})(20 \text{ m}) = 392$ J
	Potential energy halfway down:	$m = 2$ kg, $g = 9.8$ N/kg, and $h = 10$ m $E_p = (2 \text{ kg})(9.8 \text{ N/kg})(10 \text{ m}) = 196$ J
	Kinetic energy at the top:	$m = 2$ kg and $v = 0$ m/s $E_k = (1/2)(2 \text{ kg})(0^2) = 0$ J
	Kinetic energy halfway down:	$m = 2$ kg and $v = 14$ m/s $E_p = (1/2)(2 \text{ kg})(14 \text{ m/s})^2 = 196$ J

Your turn...

a. Calculate the potential energy of a 4-kilogram cat crouched 3 meters off the ground.

b. Calculate the kinetic energy of a 4-kilogram cat running at 5 m/s.

SCIENCE FACT

Kinetic energy and speed

Kinetic energy increases as the square of the speed. This means that if you go twice as fast, your energy increases by four times ($2^2 = 4$). If your speed is three times as fast, your energy is nine times bigger ($3^2 = 9$). A car moving at a speed of 100 km/h (62 mph) has four times the kinetic energy it had when going 50 km/h (31 mph). At a speed of 150 km/h (93 mph), it has nine times as much energy as it did at 50 km/h. The stopping distance of a car is proportional to its kinetic energy. A car going twice as fast has four times the kinetic energy and needs four times the stopping distance. This is why driving at high speeds is so dangerous.

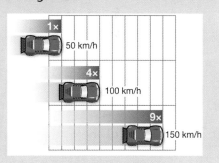

SOLVE FIRST LOOK LATER

a. 117.6 J

b. 50 J

Section 7.1 *Review*

1. Imagine you are holding an apple.

   a. Does this apple have energy? How do you know?

   b. How could you increase the potential energy of this apple?

   c. How could you increase the kinetic energy of this apple?

2. Do a stretched spring and a box on a high shelf both have potential energy? Why or why not? Explain your answer.

3. A book on a 2-meter-high shelf has 20 joules of potential energy. What is the mass of this book?

4. A 1-kilogram ball has 8 joules of kinetic energy. What is its speed?

5. If the speed of a ball increased from 1 m/s to 4 m/s, by how much would kinetic energy increase?

6. Which of these graphs illustrates the relationship between speed and the amount of kinetic energy for a 1-kilogram object?

   A 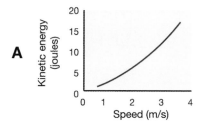   B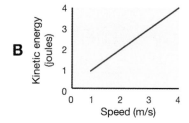

7. List two forms of mechanical energy.

8. Does a rubber band have more or less elastic energy when it is stretched?

9. Name a form of energy that is part of the electromagnetic spectrum.

SOLVE IT!

Energy from food

We get energy from eating food. The Calorie is a unit of energy often used for food. One food Calorie is equal to 4,187 joules. One calorie (lowercase "c") equals 4.187 joules.

1. If you push a box a distance of 2,000 meters with a force of 1 newton, how many Calories have you used?

2. If you push a box for a distance of 1 meter with a force of 4.187 newtons, how many calories have been used?

7.2 Energy Transformations

Systems change as energy flows and changes from one part of the system to another. Parts of the system may speed up or slow down, get warmer or colder, or change in other measurable ways. Each change transfers energy or transforms energy from one form to another.

Transforming energy

An example of energy flow

An example of a flow of energy is illustrated below. This example involves transforming chemical energy into electrical energy. The chemical energy (a fuel) is a gas called methane. It is burned in a chemical reaction and heat energy is released. The heat energy makes hot steam. The steam turns a device called a turbine, making mechanical energy. Finally, the turbine turns an electric generator, producing electrical energy. You can obtain this electrical energy by "plugging in" to an electrical outlet!

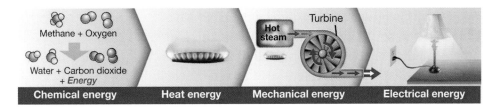

High potential energy, *Low* kinetic energy

Lower potential energy, *Higher* kinetic energy

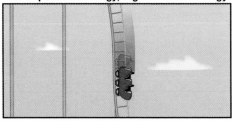

Lower kinetic energy

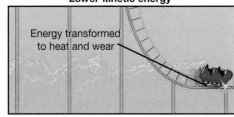

From high to low energy

How can we predict how energy will flow? One thing we can always be sure of is that systems tend to move from higher to lower energy. For example, at the top of a roller coaster hill, the car has more potential energy (Figure 7.7). The potential energy is transformed to kinetic energy as the car rolls down the hill. Once it reaches the bottom, the car has less potential energy and is more stable.

Friction and the law of conservation of energy

At the bottom of a hill, a roller coaster car has more kinetic energy. Without friction, due to Newton's first law of motion, the car would roll on a straight path forever. However, on a straight path, the kinetic energy of the car eventually decreases due to friction slowing it down. Friction transforms energy of motion to energy of heat or to the wearing away of the material of the wheels. The energy converted to heat or wear is no longer available as potential energy or kinetic energy, but it was not destroyed!

Figure 7.7: *This roller coaster car illustrates how systems go from high to low energy to become more stable. Potential energy decreases as the car rolls down the hill. Kinetic energy eventually decreases due to friction along the track and is transformed to heat and the wear of the wheels.*

Following an energy transformation

An example of energy transformation   Suppose you are skating and you come to a steep hill. You know skating up the hill requires energy. From your mass and the height of the hill you can calculate how much more potential energy you will have on the top (Figure 7.8, top). You need at least this much energy, plus some additional energy, to overcome friction.

Chemical energy to potential energy   The energy you use to climb the hill comes from food. The chemical potential energy stored in the food you ate is converted into simple sugars that are burned as your muscles work against gravity as you climb the hill. Upon reaching the top of the hill, some of the energy you spent is now stored as potential energy because your position is higher than when you began. Some of the energy was also converted by your body into heat. Can you think of any other places the energy might have gone?

How does potential energy get used?   Once you get over the top of the hill and start to coast down the other side, your speed increases. An increase in speed implies an increase in kinetic energy that comes from the potential energy gained from climbing up the hill. Energy was saved and used to "purchase" greater speed as you descend down the other side of the hill (Figure 7.8, bottom).

Kinetic energy is used up in the brakes   If you are not careful, stored up potential energy can generate too much speed! Assuming you want to make it down the hill with no injuries, some of the kinetic energy must change into some other form. Brakes on your skates slow you down and use up the extra kinetic energy. Brakes convert kinetic energy into heat and the wearing away of the brake pads. As you slow to a stop at the bottom of the hill, you should notice that your brakes are very hot, and some of the rubber is worn away.

The flow of energy   During the trip up and down the hill, energy flowed through many forms. Starting with chemical energy, some energy appeared in the form of potential energy, kinetic energy, heat, air friction, sound, evaporation, and more. During all these transformations, no energy was lost because energy can never be created or destroyed. All the energy you started with went somewhere (Figure 7.9).

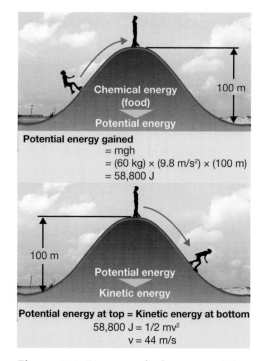

Potential energy gained
= mgh
= (60 kg) × (9.8 m/s²) × (100 m)
= 58,800 J

Potential energy at top = Kinetic energy at bottom
$$58,800 \text{ J} = 1/2 \, mv^2$$
$$v = 44 \text{ m/s}$$

Figure 7.8: *How to calculate potential energy needed, and speed on the way down.*

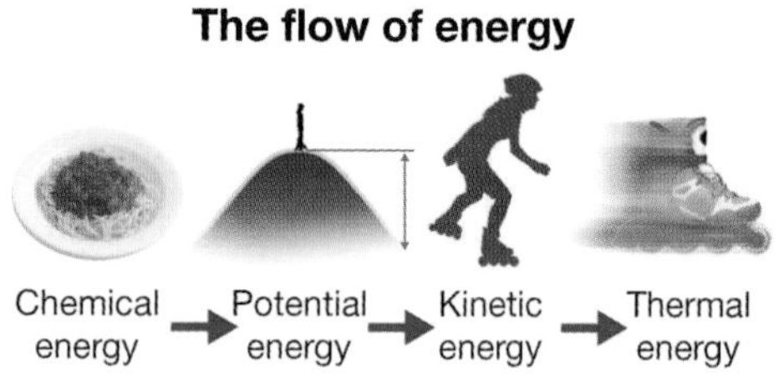

The flow of energy

Chemical energy → Potential energy → Kinetic energy → Thermal energy

Figure 7.9: *A few of the forms the energy goes through during the skating trip.*

Energy in your life

Common units of energy   A joule is a tiny amount of energy compared to what you use every day. One joule is just enough energy to lift a pint of ice cream 21 centimeters off the table. That same pint of ice cream releases 3 million times as much energy when it is digested by your body! Some units of energy that are more appropriate for everyday use are the kilowatt-hour (kWh), food Calorie, and British thermal unit (Btu) (Figure 7.10).

Daily energy use   The table below gives some average values for the energy used by humans in daily activities.

Table 7.1: Daily energy use in different energy units

Activity	kWh	Joules	Gallons of gas
Climb a flight of stairs	0.017	60,000	0.0005
Use an electric light for 1 hour	0.1	360,000	0.003
Cook an average meal	1	3,600,000	0.03
Cut the grass	18	65,000,000	0.5
Drive 30 miles to the mall and back in a small, efficient car	36	130,000,000	1
Drive 30 miles to the mall and back in a large SUV	72	260,000,000	2

Electrical energy

1 kilowatt-hour **(kWh)** = 3,600,000 J

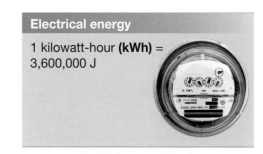

Food energy

1 food Calorie **(kcal)** is the energy needed to raise the temperature of 1 kg of water by 1°C.

Heat energy

1 **Btu** is the energy needed to raise the temperature of 1 lb. of water by 1°C.

1 British thermal unit = 1,055 J

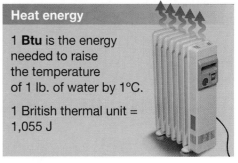

Figure 7.10: *Energy units you might use in daily life.*

7.2 Energy Transformations   **175**

Section 7.2 *Review*

1. When you rub your hands together, you produce a little heat. Describe the flow of energy that causes the heat to be produced. Use the terms *chemical energy, kinetic energy, friction*, and *heat* in your answer.

2. Arrange the four energy units from largest to smallest.
   a. joule (J)
   b. kilowatt-hour (kWh)
   c. British thermal unit (Btu)
   d. Calorie (kcal or C)

3. Martha wakes up at 5:30 a.m. and eats a bowl of cereal. It's a nice day, so she decides to ride her bicycle to work, which is uphill from her house. It is still dark outside. Martha's bike has a small electric generator that runs from the front wheel. She flips on the generator so that her headlight comes on when she starts to pedal. She then rides her bike to work. Write a paragraph to describe all the energy transformations that occur in this situation.

CHALLENGE

At the end of a ride up a steep hill, Ken was at an elevation of 1,600 meters above where he started. He figured out that he and his bicycle had stored 1,000,000 joules of energy. If Ken has a mass of 54 kg, what is the mass of Ken's bicycle?

(Hint: $g = 9.8$ m/s^2)

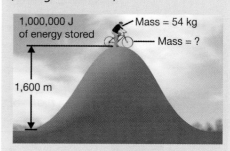

7.3 Conservation and Forms of Energy

What happens when you throw a ball straight up in the air (Figure 7.11)? The ball leaves your hand with kinetic energy it gained while your hand accelerated it from rest. As the ball goes higher, it gains potential energy. However, the ball slows down as it rises so its kinetic energy decreases. The increase in potential energy is exactly equal to the decrease in kinetic energy. The kinetic energy converts into potential energy, and the ball's total energy stays the same!

VOCABULARY

law of conservation of energy – energy can never be created or destroyed, only transformed into another form; the total amount of energy in the universe is constant

The law of conservation of energy

Law of conservation of energy
The idea that energy transforms from one form into another without a change in the total amount is called the **law of conservation of energy**. The law states that energy can never be created or destroyed, just transformed from one form into another. The law of conservation of energy is one of the most important laws in physics. It applies to not only kinetic and potential energy, but to all forms of energy.

> *Energy can never be created or destroyed, just transformed from one form into another.*

Using energy conservation
The law of conservation of energy explains how a ball's launch speed affects its motion. As the ball in Figure 7.11 moves upward, it slows down and loses kinetic energy. Eventually, it reaches a point where all the kinetic energy has been converted to potential energy. The ball has moved as high as it will go, and its upward speed has been reduced to zero. If the ball had been launched with a greater speed, it would have started with more kinetic energy. It would have had to climb higher for all of the kinetic energy to be converted into potential energy. If the exact launch speed is given, the law of conservation of energy can be used to predict the height the ball reaches.

Energy transforms between potential and kinetic
The ball's conversion energy on the way down is opposite what it was on the way up. As the ball falls, its speed increases and its height decreases. The potential energy decreases as it converts to kinetic energy. If gravity is the only force acting on the ball, it returns to your hand with exactly the same speed and kinetic energy it started with—except that now it moves in the opposite direction.

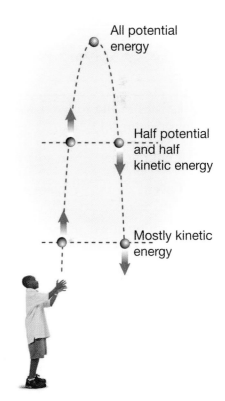

Figure 7.11: *When you throw a ball in the air, the energy transforms from kinetic to potential and then back to kinetic.*

Using energy conservation to solve problems

How to use energy conservation   Energy conservation is a direct way to find out what happens before and after a change from one form of energy to another (Figure 7.12). The law of energy conservation says that the total energy before the change equals the total energy after it. In many cases (with falling objects, for instance), you need not worry about force or acceleration. Applying energy conservation allows you to find speeds and heights easily.

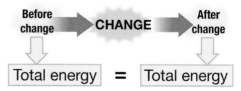

Figure 7.12: *Applying energy conservation.*

   Solving Problems: Energy Conservation

A 2-kg car moving with a speed of 2 m/s starts up a hill. How high does the car roll before it stops (Figure 7.13)?

1. **Looking for:**	You are asked for the height.	
2. **Given:**	You are given the mass in kg and starting speed in m/s.	
3. **Relationships:**	$E_K = \dfrac{1}{2}mv^2$, $E_P = mgh$	
4. **Solution:**	Find the kinetic energy at the start: $E_K = (1/2)(2 \text{ kg})(2 \text{ m/s})^2 = 4$ J Use the potential energy to find the height: $mgh = 4$ J; therefore: $h = (4 \text{ J}) \div (2 \text{ kg})(9.8 \text{ N/kg})$ $\quad = 0.2$ m The car rolls upward to a height of 0.2 m above where it started.	

Figure 7.13: *How high does the car roll before it stops?*

SOLVE FIRST LOOK LATER

a. 147,000 J

b. 20 m

Your turn...

a. A 500.-kg roller coaster car starts from rest at the top of a 60.0-meter hill. Find its potential energy when it is halfway to the bottom.

b. A 1-kg ball is tossed straight up with a kinetic energy of 196 J. How high does it go?

"Using" and "conserving" energy in the everyday sense

"Conserving" energy Almost everyone has heard that it is good to "conserve energy" and not waste it. This is useful advice because energy from gasoline or electricity costs money and uses resources. But what does it mean to "use energy" in the everyday sense? If energy can never be created or destroyed, how can it be "used up"? Why do people worry about "running out" of energy?

"Using" energy When you "use" energy by turning on a light, you are really converting energy from one form (electricity) to other forms (light and heat). What gets "used up" is the amount of energy *in the form of electricity*. Electricity is a valuable form of energy because it is easy to move over long distances (through wires). In the "physics" sense, the energy is not "used up" but converted to other forms. The total amount of energy stays constant.

Power plants Electric power plants don't *make* electrical energy. Energy cannot be created. What power plants do is convert other forms of energy (chemical, solar, nuclear) to electrical energy. When someone asks you to turn out the lights to conserve energy, they are asking you to use less electrical energy. If people used less electrical energy, power plants would burn less oil, gas, or other fuels in "producing" the electrical energy they sell.

"Running out" of energy Many people are concerned about "running out" of energy. What they worry about is running out of certain *forms* of energy that are easy to use, such as fossil fuels like oil and gas. It took millions of years to accumulate these fuels because they are derived from decaying, ancient plants that obtained their energy from the Sun when they were alive. Because it took a long time for these plants to grow, decay, and become oil and gas, fossil fuels are a limited resource.

Transitioning to new resources When you use gas in a car, the chemical energy in the gasoline mostly becomes heat energy. It is impractical to put the energy back into the form of gasoline, so we say the energy has been "used up," even though the energy itself is still there, only in a different form. Other forms of energy, such as flowing water, wind, and solar energy, are not as limited. They don't get used up. Many scientists hope our society will make a transition to these forms of energy over the next 100 years.

Switch to fluorescent bulbs

75 W Incandescent bulb

Same amount of light!

20 W Compact fluorescent bulb

There are about 300,000,000 people in the United States. If an average house has 4 light bulbs per person, it adds up to 1,200,000,000 light bulbs. One kWh of electrical energy will light a bulb for 10 hours. Multiplying by 4 bulbs per person totals 120,000,000 kWh every hour just for light bulbs!

An average electric power plant puts out 1,000,000 kWh of electrical energy per hour. That means 120 power plants are burning up resources each hour just to run light bulbs! Incandescent light bulbs convert only 10 percent of electrical energy to light. Fluorescent bulbs make the same amount of light with one quarter the electrical energy. LED bulbs are even more energy efficient, saving 75-80% of the electricity used by incandescent bulbs.

Energy and running

Humans have high endurance You know that you cannot run as fast as a dog or many other animals, like the cheetah. Human beings get tired and have to rest after running fast. Although humans are not the best sprinters on the planet, they are the best runners in terms of endurance. Scientists are learning that the human body is ideal for running long distances.

Heat production Machines, including the human body, always lose some of the converted energy to heat. Car engines and computers all produce heat that can cause damage unless it is removed. This is why cars have radiators and computers have fans.

Humans keep cool The human body works a little like a radiator by directing blood toward the skin's surface. Blood flowing near the surface can lose some heat to the relatively cooler air. A more effective way of removing heat is sweating. As sweat leaves the body, it evaporates from the skin and carries away heat. This one mechanism—sweating—makes it possible for human beings to run for long periods of time. Humans can continuously cool down while performing strenuous exercise like running. Animals with fur, like cheetahs, quickly get overheated and need to rest (see sidebar at the right). Scientists believe that sweating has allowed mankind to be successful at hunting large game throughout human history.

Energy conservation and the Achilles tendon The Achilles tendon is a good example of energy conversion between kinetic and potential energy (Figure 7.14). When the heel is down, the Achilles tendon stretches like a rubber band and potential energy is stored. Let's say 100 units of energy are stored. As the foot moves through the running stride, the tendon shortens and pulls up the heel, using about 90 units of this stored energy. In effect, the energy transformation by the Achilles tendon and the associated muscles in the foot is 90 percent. Only 10 percent of the stored energy is lost as heat!

SCIENCE FACT

Speed vs. endurance

The top speed of a cheetah is 30 m/s, and the top speed of a human is 10 m/s. A human cannot outrun a cheetah over a short distance. However, a human could win a long-distance race. Because the furry body of a cheetah does not effectively release heat, it gets overheated quickly and is exhausted after a high-speed sprint. Humans, on the other hand, constantly release heat from the skin's surface by sweating and have greater endurance as a result.

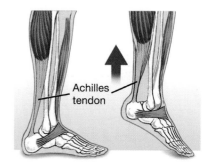

Fully stretched with stored potential energy

Shortened as potential energy converts to kinetic energy

Figure 7.14: *The Achilles tendon illustrates energy conversion.*

Section 7.3 *Review*

1. Explain what it means to say that energy is conserved.

2. Imagine you are the teacher of a science class. A student brings in a newspaper article that claims the world will run out of energy by the year 2050 because all the oil will be pumped out of the planet. The student is confused because she has learned in your class that energy can never be created or destroyed. How would you explain to her what "running out of energy" means in the article?

3. Explain what it means to say that energy is conserved as a ball falls toward the ground.

4. Some but not all of the gasoline used by a car's engine is transformed into kinetic energy. Where might some of the energy go in this system?

5. A 0.5-kg ball moving at a speed of 3 m/s rolls up a hill. How high does the ball roll before it stops?

6. Explain in your own words why energy is considered to be "nature's money." Give an example to support your explanation.

7. The table below lists normal and abnormal events. Explain why some events would never happen normally.

Normal events	Abnormal events
A ball rolls downhill	A ball at rest begins to roll uphill
An apple falls off a tree and lands on the ground	An apple on the ground flies up into the tree
A stretched rubber band snaps back to its original shape	A rubber band fully stretches on its own

CHALLENGE

Energy projects

Conduct Internet research on energy conservation. Use your favorite search engine and the following keywords to help you find information: "green communities," "energy conservation local," and "local electricity costs." The United States Environmental Protection Agency is another good resource (www.epa.gov).

1. Research what is going on in your community regarding energy conservation. Write about a project designed to save energy that is being planned or is already implemented. How much energy has been or might be saved?

2. Every month your family pays an electric bill for energy you have used. Research the cost of electricity in your area. How much does it cost for 1 million joules? This is the amount of energy used by a single electric light bulb in 3 hours.

STEM Electric Wind

AFRICA

Malawi

Fourteen-year-old William Kamkwamba pushed aside the tall grass that grew among the decaying tractor parts and broken machinery of the junkyard near his hometown of Wimbe, Malawi. He smiled as he spotted a few nuts and bolts that weren't too rusty. To most of his neighbors, this stuff was just trash. But to William, the junkyard contained hidden treasure: a tractor fan, a shock absorber, some ball bearings. He had big plans for this stuff: He would build a windmill.

Students from the Kachokolo Secondary School across the street teased him from the schoolyard, calling out "Hey, look, it's William, digging in the garbage again." Despite the taunts, William wished he, too, was in school. But a severe drought had struck Malawi the previous year (2002), and corn crops failed. His family could no longer afford the $80 per year tuition. He was forced to drop out.

William didn't want to fall behind in his studies. When he learned that a small public library had opened in his old primary school, he decided to see if he could find any books to help him keep up.

Library book sparks idea

In a textbook called *Explaining Physics*, William read that if you spin a coil of wire inside a magnetic field, an electric current is created. This is called *electromagnetic induction*. William learned that electromagnetic induction enables pedal-powered bicycle lamps to light up. He had always wondered how those worked.

Back at the library, William stumbled across another textbook called *Using Energy*. It had a row of windmills on its cover. He'd never seen a windmill before, but he had made pinwheels from plastic lids as a little boy. What were these giant pinwheels good for? The book said that windmills could be used to generate electricity.

An idea began to take shape in William's mind. If he could build a windmill, he could have light in the evenings. Electric lights were a luxury enjoyed by only two percent of Malawians. And even better, maybe he could build a windmill to operate a water pump. Then, his family could irrigate their crops. They could have a good harvest, even in a dry year. Perhaps there would be extra food to sell, and then he could go back to school!

Slowly William gathered the things he needed to build his windmill. He salvaged parts of his father's broken bicycle to serve as the body of the windmill. He cut and flattened strips of plastic pipe to make blades. His best friend purchased a beat-up second hand bicycle lamp with its pedal-powered generator.

William had to pay a welder to attach the shaft from the junkyard shock absorber to the bicycle sprocket so it could spin, and to melt holes in the tractor fan so he could attach the long blades. To earn the money, he worked long hours loading wood into a truck for a driver at the local market. Finally, he was ready to assemble his windmill.

William made a drill using a corncob handle and a sharp nail. He heated the point in a fire and then melted holes in the plastic blades. He gathered bottle tops and hammered them flat to use as washers.

He bolted the pieces together. Finally, he attached the bicycle lamp generator to the bike tire. Now it could be turned by the spinning motion of the windmill blades instead of pedals.

Next, William and two friends built a five-meter-tall tower out of tree trunks and branches they cut by hand.

The next day, they hoisted the windmill with a rope and pulley system created from William's mother's clothesline wire. They bolted the windmill to the tower.

Soon, a dozen or so people from the village wandered down to see what this tall contraption could be. Some of them laughed at the sight of all this useless junk lashed together. A man who worked in the market asked William what he called his machine. Since there is no word in his native language for windmill, William answered, "Electric wind."

William had jammed a bent bicycle spoke into the wheel to prevent it from spinning before he was ready. He ran to get the bicycle lamp bulb, then attached it to the generator wires hanging from the windmill. Meanwhile, the crowd grew to about 60. William took a deep breath and pulled the spoke away. The windmill began to turn, faster and faster. The bicycle lamp flickered, and then glowed with a steady bright light. All the people began to clap and cheer.

Windmill power transforms lives

Afterward, William installed some lights in his family home. He began making a little money using his generator to charge cell phones. Then, in 2006, some people from the organization that built the library came to inspect it. They noticed the windmill and asked the librarian who made it. A few days later they sent a colleague to see it. He asked William to tell him all about how he built the windmill.

William installs lights in his family's home.

A few days later, the man returned with some reporters. Several radio stations and newspapers ran stories about William's windmill. A man in Lilongwe, Malawi's capital, read about the windmill and shared William's story on his blog. As a result, William was invited to speak at a conference called TEDglobal 2007, in Tanzania. Many global investors there chose to support William's work, enabling him to build a second windmill to power a water pump, and to return to school.

William graduated from the African Leadership Academy in 2010 and graduated from engineering school in the United States. His goal is to start his own windmill company in Malawi, bringing a reliable, affordable source of electricity to his fellow citizens.

William at the TEDglobal conference in 2007

Questions:

1. Name three ways the windmill benefited William and his family.

2. What challenges might have prevented William from building the windmill? How did he overcome them?

3. What simple machine can you see in the windmill photo? Name the input and output forces.

To learn more about William Kamkwamba, visit www.movingwindmills.org and read his book: Kamkwamba, William, and Bryan Mealer. (2009). The Boy Who Harnessed the Wind. New York: HarperCollins Publishers.

Photos by Tom Rielly

Chapter 7 Assessment

Vocabulary

Select the correct term to complete each sentence.

Calorie	kinetic energy	mechanical energy
chemical energy	law of conservation of energy	potential energy
joule	nuclear energy	radiant energy

Section 7.1

1. This energy is related to Earth's gravity: _____ .

2. Energy that is due to motion is called _____ .

3. The _____ is the SI unit of energy.

4. A fossil fuel is a good example of this kind of energy: _____ .

5. Potential and kinetic are types of this kind of energy: _____ .

6. _____ from the Sun depends on this kind of energy: _____ .

Section 7.2

7. This unit is often used to measure the amount of energy in food: _____ .

Section 7.3

8. The _____ states that in a closed system, the total amount of energy does not change over time.

Concepts

Section 7.1

1. What does energy give objects the ability to do?

2. Identify at least one way that energy is involved in these situations:

   a. An ocean wave at the beach knocks over a sand castle.
   b. Your houseplant grows better when it is placed in sunlight.
   c. When you eat breakfast in the morning, you have more energy for your school day.
   d. When you drop a plate, it breaks into pieces.
   e. Your hair dryer works when you plug it into an electrical outlet.

3. In this chapter, you learned that you can increase the pressure energy of a tire by blowing it up. Give another example of an object that has pressure energy.

4. What provides and has always provided most of Earth's energy for living things and technology?

5. Explain how work and energy are related.

6. Describe the difference between potential and kinetic energy.

7. Copy the following table onto a piece of paper and fill it in based on your understanding of potential and kinetic energy.

	Potential energy	Kinetic energy
What's the formula?		
What happens to energy when the mass of an object increases?		
What happens when the object is lifted to a higher height (without a change in speed)?		
What happens when the speed of an object increases (without a change in height)?		

8. Which of the following is equivalent to 2 joules?

    a. 2 newtons

    b. $2 \text{ kg·m}^2/\text{s}^2$

    c. 2 kilograms

    d. 2 meters

Section 7.2

9. Give an example of how energy flows in a system. Come up with an example that was not explained in the text.

10. A sled going down a hill covered in snow eventually comes to a stop. Explain why this happens in terms of energy. Use the terms *potential energy*, *kinetic energy*, and *friction*.

Section 7.3

11. A roller coaster track is a good example of the law of conservation of energy. Use this law to explain these facts about a roller coaster track.

    a. The largest hill for a roller coaster track is the first hill on the track. The hills after the first are smaller and smaller.

    b. To get to the top of the first (highest) hill, a motor pulls the cars up to the top. After the top of the first hill, a motor is not needed to keep the cars going.

    c. The roller coaster car moves really fast at the bottom of a hill on the track but slows down as it moves up a hill (not including the first hill).

12. Draw a pie chart showing the relative amounts of potential for a book in each of the following situations.

    a. The book on a high shelf that is 2 meters off the ground.

    b. The book after it has fallen off the 2-meter shelf and is now 1 meter off the ground.

    c. The book just before it hits the ground.

13. What is the difference between a resource that is limited and one that is not limited? Give an example of each.

Problems

Section 7.1

1. What is the minimum energy required to lift an object weighing 200 newtons to a height of 20 meters?

2. How many joules of energy would it take for Jolene to lift a 49-newton platter of food 1 meter off the table?

3. Calculate the potential energy of a bird sitting on a tree limb. The mass of the bird is 0.1 kilogram, and it is 5 meters off the ground.

4. How high is a 0.1-kilogram bird from the ground when its potential energy is 3 joules?

5. What is the kinetic energy of a 2,000-kg car that is traveling 10 m/s?

Section 7.2

6. On a typical day, let's say you do the following: cook three average meals, climb two flights of stairs, use an electric light for six hours, and ride in a small, efficient car for 15 miles. What is the total amount of energy that has been used in these activities? Record your answer in kilowatt-hours, joules, and gallons of gas.

Section 7.3

7. A 2-kg ball is released from rest at the top of a track and reaches a speed of 10 m/s at the bottom.

    a. How much kinetic energy does the ball have?

    b. How much potential energy did it have at the top of the hill (assuming no energy was lost)?

    c. What was the height of the hill?

8. An 80-kilogram cliff diver is standing on a cliff that is 30 meters high.

   a. What is the potential energy of the cliff diver?

   b. The diver makes his dive. What is his potential energy when he is 10 meters above the water surface?

   c. What is the kinetic energy of the diver when he is traveling at 19.6 m/s during his dive?

   d. What is the potential energy of the diver when he is traveling at 19.6 m/s?

Applying Your Knowledge

Section 7.1

1. Solar energy and hydroelectric energy are important sources of energy. Find out more about either one of these forms of energy. What is being done to make it a more efficient source of energy, and where it is being used in the United States?

2. Gas and oil are nonrenewable resources. Make a list of renewable resources (e.g., solar and hydroelectric energy). Find data that indicates how much nonrenewable resources are used in the United States.

3. Nuclear energy is a controversial energy resource. Find out why. List two pros and two cons for this form of energy.

Section 7.2

4. The images below show what happens when a child rides a swing. Match the following descriptions to the images.

  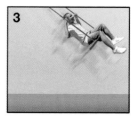

   a. High potential energy

   b. High kinetic energy

   c. Equal amounts of potential and kinetic energy

Section 7.3

5. Here is some data for kinetic energy versus speed for a moving object. Make a graph of this data and answer the following questions. Place kinetic energy on the y-axis and speed on the x-axis.

Speed (m/s)	Kinetic energy (joules)
12	720
24	2,880
48	11,520
60	18,000

   a. What is the mass of the object represented by this data set?

   b. Use your graph to find the kinetic energy at 30 m/s. Then use the kinetic energy formula to check yourself.

   c. In the Section 7.1 review, you saw two graphs. One could be described as *linear* and the other could be described as *exponential*. Find out which is which. Identify which of these terms describes the relationship of kinetic energy versus speed.

Work and Power

CHAPTER 8

We all have our ideas about what the word *work* means. Sometimes we really want to work on something we enjoy—like learning how to play guitar. Other times, we would like to be lazy and do no work. You may also have ideas about the words *efficiency* and *power*. In science, work, efficiency, and power have special meanings. Work means that something is moved by a force. Look at the excavator on this page. What is being moved? What force is involved? Machines like excavators have a lot of power because they help people move more dirt faster. However, is this machine as efficient at moving dirt as the human body machine? Think about these questions as you get to work reading this chapter!

CHAPTER 8 INVESTIGATIONS

8A: Manipulating Forces
How do simple machines work?

8B: Work
How can a machine multiply forces?

8.1 Work

Energy is a measure of an object's ability to do work. If you have energy, then you can do work. That means you can make forces that act to move things. Suppose you lift your book over your head. Your arm muscles make forces, and those forces cause the book to move, therefore you do work. Now suppose you lift your book fast, then lift it again slowly. The work is the same because the force it takes to lift the book (its weight) is the same, and the distance (height) is the same. But it feels different to do the work fast or slow. The difference between doing work fast or slow is described by *power*. Power is the rate at which energy flows or at which work is done. This section is about work and power.

Reviewing the definition of work

What *work* means in physics   In the last chapter, you learned that work has a very specific meaning in physical science. **Work** is the transfer of energy that results from applying a force over a distance (Figure 8.1). If you push a box with a force of 1 newton for a distance of 1 meter, you do 1 joule of work. Both work and energy are measured in the same units (joules) because work is a form of energy.

Work is done by forces that cause movement   When thinking about work, remember that work is *done by forces that cause movement*. If nothing moves (distance is zero), then no work is done, even if a huge force is applied. For example, in the scientific sense, you don't do any work if you push a box that stays glued to the table. However, if you push the box 1 meter with a force of 1 newton, you have done 1 joule of work.

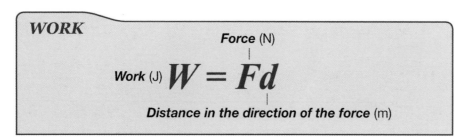

WORK

$$W = Fd$$

Work (J)   Force (N)   Distance in the direction of the force (m)

Figure 8.1: *Work is a form of energy you either use or get when a force is applied over a distance.*

VOCABULARY

work – a form of energy that comes from force applied over distance; a force of 1 newton does 1 joule of work when the force causes 1 meter of motion in the direction of the force

STUDY SKILLS

When studying, remember that the definition for a joule, the unit of energy, is the same as the definition for work!

Definition of **joule** = Definition of **work**

Reviewing work and energy

Energy is needed to do work   Recall that energy is always needed to do work. An object that has energy is able to do work; without energy, it is impossible to do work. The more energy you have, the more work you can do. For example, a ball rolling across a table has kinetic energy that can be used to do work. If the ball collides with a toy car, it will do work on the car and change its motion. Some of the ball's kinetic energy is transferred to the car. Collision is a common method of doing work.

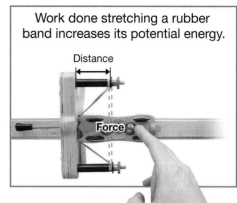

Work done stretching a rubber band increases its potential energy.

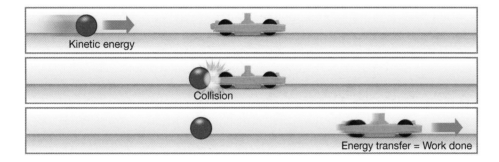

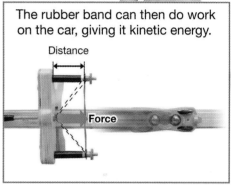

The rubber band can then do work on the car, giving it kinetic energy.

Work and energy transfer   Doing work always means *transferring* energy. The energy may be transferred to the object to which the force is applied, or it may go somewhere else. For example, you can increase the potential energy of a rubber band by exerting a force that stretches it. The work you do stretching the rubber band is stored as elastic energy in the rubber band. In this case, the work you do stretching the rubber band is partially transferred to the rubber band itself. The rubber band can then use the elastic energy to do work on a toy car, giving it kinetic energy (Figure 8.2).

Figure 8.2: *You can do work to increase an object's potential energy. Then the potential energy can be converted to kinetic energy.*

Work may not increase the energy of an object   The exact amount of energy used to do work is always transferred somewhere. But not all work is transformed to the kind of energy you might initially think about. For example, you can do work on a block by sliding it across a level table. In this example, the work you do does *not* increase the energy of the block. Because the block will not slide back all by itself, it does not gain the ability to do work *itself*; therefore, it gains no energy. Your work is done to overcome *friction* and eventually becomes heat and wear.

Work done against gravity

What is the most effective force to do work?

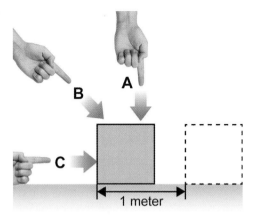

1 meter

Work is *only* done by the part of a force that acts in the same direction as the resulting motion. Force A in the diagram does no work at all because it does not *cause* the block to move sideways. Force B is applied at an angle to the direction of motion of the block. Only part of force B (in the direction the block moves) does work. The most effective force is force C. All of force C does work because force C acts in the same direction the block moves.

Lifting force equals the weight Many situations involve work done by or against the force of gravity. To lift something off the floor, you must apply an upward force with a strength equal to the object's weight. The work done while lifting an object is equal to its change in potential energy. It does not matter whether you lift the object straight up or you carry it up the stairs in a zig-zag pattern. The work is the same in either case.

Work done against gravity is equal to weight multiplied by the change in height.

Why the path does not matter The reason the path does not matter is because work is only done by the part of a force that acts *in the direction of the motion*. Gravity acts vertically so only vertical motion counts toward work. If you move an object on a diagonal, only the vertical distance matters, because the force of gravity is vertical. It is much easier to climb stairs or go up a ramp, but the work done *against gravity* is the same as if you jumped straight up. Stairs and ramps are easier because you need less force, but you have to apply the force over a longer distance. In the end, the total work done against gravity is the same, no matter what path you take.

÷ Solving Problems: **Work**

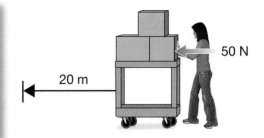

How much work is done by a person who pushes a cart with a force of 50 newtons if the cart moves 20 meters in the direction of the force (Figure 8.3)?

1.	**Looking for:**	You are asked for work.
2.	**Given:**	You are given values for force and distance.
3.	**Relationships:**	Work = force × distance.
4.	**Solution:**	The work done is: 50 N × 20 m = 1,000 J.

Figure 8.3: *How much work is this person doing by pushing the cart?*

Your turn...

a. How far does a 100-newton force have to move to do 1,000 joules of work?

b. An electric hoist does 500 joules of work lifting a crate 2 meters. How much force does the hoist use?

c. An athlete does one push-up. In the process, she moves half of her body weight, 250 newtons, a distance of 20 centimeters. This distance is the distance her center of gravity moves when she fully extends her arms. How much work did she do after one push-up?

d. You decide to push on a brick wall with all your might for 5 minutes. You push so hard that you begin to sweat. However, the wall does not move. If you end up pushing with a force of 500 newtons, how much work did you do?

SOLVE FIRST / LOOK LATER

a. 10 meters

b. 250 newtons

c. 50 joules

d. You didn't do any work because the wall did not move.

Section 8.1 *Review*

1. What is the best way to define work?

    a. applying force for a period of time

    b. moving a certain distance

    c. applying a force over a distance

    d. applying force at a given speed

2. Push a box across a table with a force of 5 newtons and the box moves 0.5 meters. How much work has been accomplished?

3. If you do 200 joules of work using a force of 50 newtons, over what distance was the force applied?

4. A cart was pulled for a distance of 1 kilometer, and the amount of work accomplished equaled 40,000 joules. With what force was the work accomplished?

5. In which of these cases is a waiter doing work on the object (Figure 8.4)? Explain your answer.
    Situation 1: The waiter is carrying a tray of glasses across a room.
    Situation 2: The waiter is pushing a cart across a room.

6. When you climb a flight of stairs, you are moving your body weight (a force) up a certain distance (the vertical height from the bottom to the top stair). In which case does the amount of work you do increase? Explain your answer.

    a. You run up 10 stairs then you run up 50 stairs.

    b. You walk slowly up 10 stairs and then you run up 10 stairs really fast.

7. How is work related to potential and kinetic energy?

8. A 2-kilogram object falls 3 meters.

    a. How much potential energy did the object have before it fell?

    b. How much work was accomplished by the fall?

9. It takes 300 newtons of force and a distance of 20 meters for a moving cart to come to a stop.

    a. How much work is done on the cart?

    b. How much kinetic energy did this cart have?

Situation 1

Situation 2

Figure 8.4: *Question 5.*

▬▬▬ SOLVE IT! ▬▬▬

Design your own work problem

To design your work problem, describe a situation involving work and provide the force and distance values. Solve the problem to find the correct answer.

Then give your problem to a friend and have them solve it. Check and see if they got the right answer.

8.2 **Efficiency and Power**

One day your science teacher declares, "Today we are going to do our work with greater *efficiency* and greater *power*." That sounds like a good idea, but what does your teacher mean? Read on and you will find out!

Work input and output

Input work and output work Every process that transforms energy can be thought of as a machine. Work or energy goes in one end and work or energy comes out the other end. The "machine" may be a toaster heating bread, which transforms electrical energy into heat, or even a human consuming food in order to have the energy to exercise. Using this concept, the **work input** is the work or energy supplied to the process (or machine). The **work output** is the work or energy that comes out of the process (or machine).

A rope and pulley example As an example, consider using a rope and pulley machine to lift a load weighing 10 newtons (Figure 8.5). If you lift the load a distance of 1 meter, the machine has done 10 joules of work and the work output is 10 joules. For this particular machine, you only need to pull with a force of 5 newtons, but you need to pull the rope a distance of 2 meters. Your work input is 5 newtons × 2 meters, or 10 joules.

How work input and output are related The example of a rope and pulley machine illustrates a rule that is true for all machines and all processes that transform energy. The total energy of work output can never be greater than the total energy of work input.

The energy output of a process or machine can never exceed the energy input.

You may recognize this statement as just another way of saying the law of conservation of energy. You are right! If you carefully account for all the work and energy in any process, you find that the total work and energy output of the process is exactly equal to the total work and energy input.

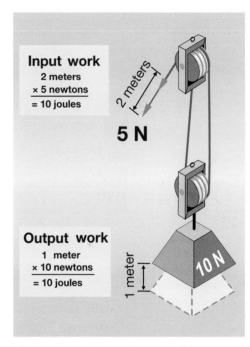

Figure 8.5: *The work input of the rope and pulley machine is the same as the work output.*

Efficiency

Real machines Suppose you measure the forces on an actual rope and pulley machine. Figure 8.6 shows what you find. Notice that the work input is a little more than the work output! It took 11 joules of input work applied to the rope to produce 10 joules of output work lifting the weight. This kind of behavior is true of all real machines. The work output is less because some work is always converted to heat and other kinds of energy by *friction*.

VOCABULARY

efficiency – the ratio of usable output work divided by total input work, efficiency is often expressed as a percent, with a perfect machine having 100 percent efficiency

Everyday machines

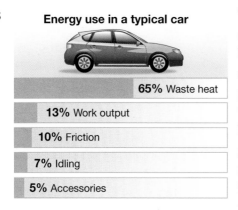

Energy use in a typical car

65% Waste heat	
13% Work output	
10% Friction	
7% Idling	
5% Accessories	

The diagram at the left shows how the chemical energy (input) released by burning gasoline is used in a typical car. Only 13 percent of the energy in a gallon of gas is transformed into output work! Car engines in use get hot. That's because 65 percent of the energy in gasoline is converted to heat. As far as moving the car goes, this heat energy is "lost." The energy doesn't vanish; it just does not appear as useful output work.

Efficiency Now we can talk about efficiency. The **efficiency** of a machine is the ratio of usable output work divided by total input work. Efficiency is usually expressed in percent. The car in the diagram has an efficiency of 13 percent. That means 13 joules go to making the car move out of every 100 joules released from gasoline. A "perfect" car would have an efficiency of 100 percent. Since all real machines have some friction, perfect machines are technically impossible.

Calculating efficiency You calculate efficiency by dividing the usable output work by the total input work. The rope and pulley machine in Figure 8.6 has an efficiency of 91 percent. That means that 1 joule out of every 11 (9 percent) is "lost" to friction. The work isn't really "lost," but is converted to heat and other forms of energy that are not useful in doing the job the rope and pulley machine is designed to do.

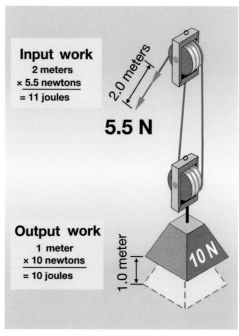

Input work
2 meters
× 5.5 newtons
= 11 joules

2.0 meters

5.5 N

Output work
1 meter
× 10 newtons
= 10 joules

1.0 meter

10 N

Figure 8.6: *If the input work is 11 joules, and the output work is 10 joules, then the efficiency is 91 percent.*

Efficiency in natural systems

The meaning of efficiency Energy drives all the processes in nature, from winds in the atmosphere to nuclear reactions occurring in the cores of stars. In the environment, efficiency is interpreted as the fraction of incoming energy that goes into a process. For example, Earth receives energy from the Sun. Earth absorbs this solar energy with an average efficiency of 78 percent. The energy that is not absorbed is reflected back into space.

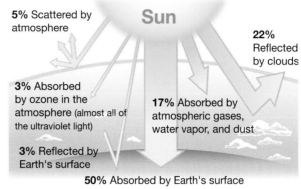

What happens to the incoming solar radiation?

5% Scattered by atmosphere

22% Reflected by clouds

3% Absorbed by ozone in the atmosphere (almost all of the ultraviolet light)

17% Absorbed by atmospheric gases, water vapor, and dust

3% Reflected by Earth's surface

50% Absorbed by Earth's surface

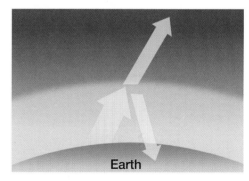

Figure 8.7: *Dust and clouds reflect light back into space, decreasing the efficiency with which Earth absorbs energy from the Sun.*

Earth's temperature Earth's efficiency at absorbing solar energy is critical to living things. If the efficiency decreased by a few percent, Earth's surface would become too cold for life. Some scientists believe that many volcanic eruptions or nuclear war could decrease the absorption efficiency by spreading dust in the atmosphere. Dust reflects solar energy (Figure 8.7). On the other hand, if the efficiency increased by a few percent, it would get too hot to sustain life. Too much carbon dioxide in the atmosphere increases absorption efficiency (Figure 8.8). Scientists are concerned that the average annual temperature of Earth has already warmed 1°C degree since the 1880s as a result of carbon dioxide released by human technology.

Figure 8.8: *Carbon dioxide and other greenhouse gases in the atmosphere absorb some energy that otherwise would have been radiated back into space. This increases the efficiency with which Earth absorbs energy from the Sun.*

Conservation of energy In any system, all of the energy goes somewhere. Another way to say this is that energy is conserved. For example, rivers flow downhill. Most of the potential energy lost by water moving downhill becomes kinetic energy in the motion of the water. Erosion takes some of the energy and slowly changes the land by wearing away rocks and dirt. Friction takes some of the energy and heats up the water. If you could add up the efficiencies for every single process in which water is involved, the total would be 100 percent.

Solving Problems: Efficiency

You see a newspaper advertisement for a new, highly efficient machine. The machine claims to produce 2,000 joules of output work for every 2,100 joules of input work. What is the efficiency of this machine? Is it as efficient as a bicycle (see sidebar)? Do you believe the advertisement's claim? Why or why not?

1.	**Looking for:**	You are asked to calculate efficiency.
2.	**Given:**	You are given the input work and output work.
3.	**Relationships:**	Efficiency is calculated by dividing output work by input work and then multiplying by 100 to get a percentage. $$\text{Efficiency} = \frac{\text{Output work}}{\text{Input work}} \times 100$$
4.	**Solution:**	$$\text{Efficiency} = \frac{2,000 \text{ J}}{2,000 \text{ J}} \times 100 = 95\%$$ The efficiency of the machine is 95 percent, which is as efficient as a bicycle. Because a bicycle is the most efficient machine ever invented, I won't believe the advertisement until I see actual scientific data that proves its amazing efficiency.

Your turn...

a. Suppose 1,000 joules of input work were applied to a machine with only 8 percent efficiency. What would be its output work?

b. You do 32 joules of work using a pair of scissors. The scissors do 25 joules of work cutting a piece of fabric. What is the efficiency of the scissors?

SOLVE FIRST LOOK LATER

a. 80 joules

b. About 78 percent

Power

Energy vs. power
If you lift a book over your head, the book gets potential energy from your action. Even if you lift the book faster, it has the same amount of potential energy. This is because the height is the same. But it feels different to transfer the energy to the book at different speeds. *Power* describes how fast energy is transferred to an object.

What is power?
Power is the rate at which work is done. Here's an example. Suppose Michael and Jim each lift a barbell weighing 100 newtons from the ground to a height of 2 meters (Figure 8.9). Michael lifts quickly and Jim lifts slowly. Michael and Jim do the same amount of work (100 N × 2 m = 200 joules of work). However, Michael's *power* is greater because he gets the work done in less time!

Watts and horsepower
Power is calculated in watts. One **watt (W)** is equal to 1 joule of work per second. A *kilowatt*, which you may have heard of, equals 1,000 watts. The watt was named after James Watt, the Scottish engineer who invented the steam engine. Another unit of power is the **horsepower**. Watt expressed the power of his engines as the number of horses an engine could replace. One horsepower equals 746 watts, or 746 joules of work per second!

> **POWER**
>
> Power (W) $P = \dfrac{W}{t}$ Work (J)
>
> Time (s)

Calculating power
Now, let's calculate and compare the power output of Michael and Jim. Michael's power is 200 joules divided by 1 seconds, or 200 watts. Jim's power is 200 joules divided by 10 seconds, or 20 watts. Jim takes 10 times as long to lift the barbell, so his power is one-tenth as much. The maximum power output of an average person is a few hundred watts.

> ### ■ VOCABULARY ■
>
> **power** – the rate of doing work or moving energy; power is equal to energy (or work) divided by time
>
> **watt (W)** – a unit of power equal to 1 joule per second
>
> **horsepower (hp)** – a unit of power equal to 746 watts

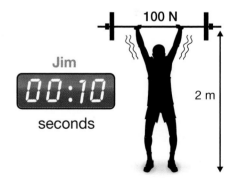

Figure 8.9: *Michael and Jim do the same amount of work but do not have the same power.*

Solving Problems: Power

Allen lifts his weight (500 newtons) up a staircase that is 5 meters high in 30 seconds (Figure 8.10). How much power does he use? How does his power compare with a 100-watt light bulb?

1. **Looking for:**	You are asked to calculate Allen's power.
2. **Given:**	You know force, distance, and time.
3. **Relationships:**	Relationships that apply include the formulas for work and power.
	Work = force × distance Power = work ÷ time
4. **Solution:**	Solve for work by multiplying "force × distance."
	Then solve for power by replacing the value for "work" in the formula.
	Power = work ÷ time
	Remember: 1 joule = 1 N · m, and 1 watt = 1 N-m/s
	Power = (500 N × 5 m) ÷ 30 s = 2,500 N · m ÷ 30 s = 83 watts
	Allen's power is less than a 100-watt light bulb. Most human activities use less power than a light bulb.

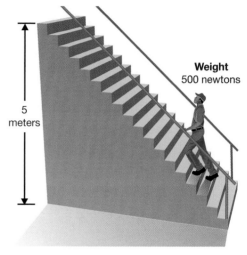

Figure 8.10: *How much power does this 500-newton person use?*

SOLVE FIRST LOOK LATER

a. He would need to climb the stairs in 25 seconds. To answer this question, you need to divide work by power (work ÷ power = time).

b. 2.5 seconds

Your turn...

a. Let's say that Allen did the same amount of work as in the problem above, but he wanted to have the same amount of power as a 100-watt light bulb. How fast would he have to climb the stairs?

b. What is the minimum time needed to lift a 2,000-newton weight 10 meters using a motor with a maximum power rating of 8,000 watts?

Section 8.2 *Review*

1. You read about a rope and pulley machine that was able to produce equal amounts of output work and input work. Was this a realistic example? Why or why not?

2. What do you need to do to calculate the efficiency of any machine?

3. A car's efficiency is only 13 percent.
   a. If the input work for a car is 200 joules, what is the output work?
   b. List two things that car manufacturers do to improve a car's efficiency.

4. A simple machine produces 25 joules of output work for every 50 joules of input work. What is the efficiency of this machine?

5. How is work related to power?

6. If you know the power for a machine and the time it takes to produce that power, what value can you calculate?

7. How does 1 horsepower compare to 1 watt of power?

8. A gallon of gasoline contains about 36 kilowatt-hours of energy. Suppose a gallon of gas costs $2.50 and a kilowatt-hour of electricity costs 12 cents. Which form of energy is less expensive?

9. A 100-newton object is lifted 100 meters in 100 seconds. What is the power generated in this situation?

10. Which situation would produce 200 watts of power?
    a. 100 J of work done in 2 s
    b. 400 J of work done in 2 s
    c. 2,000 J of work done in 5 s
    d. 2 J of work done in 100 s

11. An average car engine can produce about 100 horsepower. How many 100-watt light bulbs does it take to use the same amount of power?

12. A half-cup of ice cream contains about 200 food Calories. How much power can be produced if the energy in a cup of ice cream is expended over a period of 10 minutes (600 seconds)? Each food Calorie is equal to 4,187 joules. Write your answer in watts and then in horsepower.

JOURNAL

Are you really doing work when you do your homework?

This question was posed at the beginning of the chapter. Answer it in your own words based on what you know about work.

Think about this question from different angles before you answer it!

KEYWORDS

Energy-efficient technologies

Engineers are always trying to improve efficiency of the machines we use every day. Do an Internet search using the key phrase "energy-efficient technologies" and see what you find. Pick a topic and present your findings to your class. If requested, cite your sources in a required format in your report.

Extension: Be a roving reporter within your home and see how many energy-efficient appliances you can find. Use technology to present your findings.

Human-Powered Transportation

How many ways can you think of to get around using only your own energy? Some possibilities include walking, running, skateboarding, ice-skating, canoeing, swimming, cross-country skiing, kayaking, roller-skating, riding in a wheelchair, and cycling. Until 200 years ago, people had few choices for transportation and relied on human or animal power to get from place to place. Today, people travel most of their miles in cars, planes, buses, and trains. Yet there is a growing movement for people to use more efficient and cleaner forms of transportation. Human power is environmentally friendly and also has health benefits.

Work and power

When we move our bodies along, whether by walking, swimming, or skiing, we exert forces over a distance and do work. Our leg muscles are stronger than those in our arms, so methods of transportation that rely on the legs are usually faster than ones in which we use only the arms. Factors such as friction and air resistance also greatly impact how quickly we can get from place to place. If given a push, you can easily slide across an ice-skating rink. Try the same thing on a basketball court and you won't go anywhere because of friction.

The rate at which you do work or use energy is your power. On average, the human body uses 100 joules per second (100 watts). The body converts chemical energy from food into other forms of energy to generate heat, pump blood, digest food, move around, and conduct other bodily functions.

Your body requires more energy during physical activity. You may have heard that marathon runners often eat a large pasta dinner the night before a race to ensure that they have enough energy for the next day. Professional athletes calculate the amount of food they need to eat each day based on the work their bodies do while running, swimming, or participating in other sports.

Ways to get around

Walking is the most common way for people to get from place to place using their own power. For short trips, walking is often faster and easier than other modes of transportation. An average person can walk a mile (1.6 km) in 20 minutes, so it is convenient, especially for people living in cities, to do many errands on foot.

Scientists and engineers have developed equipment and machines to help us get around when walking isn't ideal. Cross-country skis and snowshoes make it possible to travel in conditions that would be impossible on foot. Ice skates and sea kayaks are also designed for travel in specific environments. Bicycles increase the speed and distance we can move under our own power. People have even built human-powered machines such as submarines, airplanes, and helicopters!

Human-powered aircraft flight over Rogers Dry Lake at NASA Dryden Flight Research Center, California. For more information, try this Internet keyword search: "NASA + human powered flight"

Cycling science

The most popular human-powered vehicle is the bicycle, with over 1 billion in use worldwide. Bicycles are affordable, easy to ride and park, and make commuting fun rather than a chore. Riding at a comfortable pace of 12 mph, a cyclist can travel a mile in only five minutes. In cities, cycling is often faster than driving a car. This is why courier companies often use bicycle messengers to make deliveries.

What makes the bicycle unique among transportation methods is its efficiency. If you want to get from one place to another using the least amount of energy, cycling is the way to go. It takes only 35 calories, or 150,000 joules, to cycle one mile. If you were walking, you could travel only one-third of a mile using the same energy. And a typical car could travel only 100 feet using an amount of gasoline with 150,000 joules of energy.

In a well-designed bicycle, up to 98 percent of the work done by applying force to the pedals is transmitted to the wheels. But not all of the energy in the spinning wheels is converted into the kinetic energy of the bicycle and rider. Energy is lost as tires change shape and create heat. Keeping tires inflated to the correct pressure minimizes this energy loss in bicycles as well as in automobiles.

Air resistance is often the largest source of energy loss while cycling. When moving at low speeds, the force of air resistance is small because the cyclist does not have to force a large amount of air out of the way during each second of motion. However, as speed increases, the force of air resistance increases rapidly. Doubling a bicycle's speed causes the air resistance to approximately quadruple.

The shape of the bicycle and position of the rider are major factors that determine air resistance. Recumbent bicycles, in which a rider's body is low and the legs are extended forward, are much more aerodynamic than upright bicycles. Many people also find recumbent bicycles to be more comfortable, especially for long trips.

A streamliner is a specially designed recumbent bicycle with an aerodynamic shell. Streamliners have been used to set world records for the fastest human-powered vehicles. Canadian Todd Reichert set a record of 89.6 mph for the fastest speed achieved in a 200-meter flat stretch of track. Another Canadian, Greg Kolodziejzyk, broke the world record for the longest distance traveled in 24 hours using human power when he pedaled a whopping 650 miles in a single day!

Energy needed to travel one mile

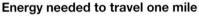
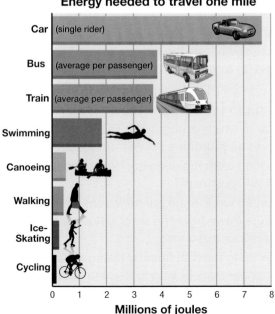

| Car (single rider) |
| Bus (average per passenger) |
| Train (average per passenger) |
| Swimming |
| Canoeing |
| Walking |
| Ice-Skating |
| Cycling |

Millions of joules (0 1 2 3 4 5 6 7 8)

Questions:

1. Which forms of human-powered transportation do you use most often?

2. Compare the energy needed to ice skate and swim one mile. Which requires more energy? Why?

3. Is 100 percent of the work you do to pedal a bicycle converted into kinetic energy? Explain.

4. Why are recumbent bicycles more efficient than upright bicycles?

Chapter 8 *Assessment*

Vocabulary

Select the correct term to complete each sentence.

efficiency	watt	work output
horsepower	work	work input
power		

Section 8.1 and Section 8.2

1. The rate at which work is done is called _____ .

2. In physics, _____ is the product of the force applied and the distance moved in the direction of the force.

3. A unit of power equal to 746 watts is a(n) _____ .

4. The unit for one joule per second is one _____ .

5. You calculate the _____ of a machine by dividing its _____ by its work input and multiplying by 100.

6. The work output of a machine can never be greater than the _____ .

Concepts

Section 8.1

1. For each situation, explain whether work (**W**) is done or not (**N**) done.

    a. _____ standing still while holding a box of heavy books

    b. _____ hitting a baseball with a bat

    c. _____ picking up a suitcase

    d. _____ pushing hard against a stationary stone wall for an hour

    e. _____ falling toward Earth while sky diving

2. Why are energy and work measured using the same units?

3. Copy the table below onto a piece of paper. Then use the graphic to fill it in. In the Work Done? column, write *yes*, *no*, or *some*. In the Motion of the Block column, describe how the block would move under each force.

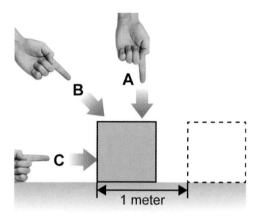

Force	Work done?	Motion of the block
A		
B		
C		

4. It's moving day and you need to move boxes and furniture from your old second-floor apartment on Main Street to your new fifth-floor apartment on Harmony Street. You have to take the stairs to move your furniture down from the old apartment, but you can use the elevator to get up to your new apartment on the fifth floor.

    a. Describe one way in which your muscles DO do work while moving your boxes and furniture down from the second floor.

    b. Describe one way in which your muscles DON'T do work while moving your boxes and furniture down from the second floor.

    c. Does the elevator do work moving your boxes and furniture up to the fifth-floor apartment?

    d. Which involves more work in the scientific sense: moving the boxes and furniture down from the second floor or up to the fifth floor? Explain your reasoning.

    e. You take one box up to the fifth floor by taking the stairs. If the elevator had taken the same box up to the fifth floor, would it have done more, less, or the same amount of work as you? Explain your reasoning.

Section 8.2

5. Your lab partner shows you results from an experiment with a simple machine. The output work is 10 joules, and the input work is 8 joules. She asks, "Does this data look correct?" What would be your response and why?

6. A bicycle is considered to be one of the most efficient human-powered machines. Explain why.

7. At the beginning of the chapter, there was the question: How can you produce more power than an excavator? Answer this question using your understanding of work and power. Give an example that illustrates your answer.

8. Mikhail lifts a 500-newton weight 2 meters in 2 seconds. Tobias lifts the same 500-newton weight 2 meters in 4 seconds.

   a. Which boy does more work?

   b. Which boy uses greater power?

   c. The human body is only 8 percent efficient. To obtain the amount of work accomplished by Mikhail or Tobias, how much input work was required?

Problems

Section 8.1

1. How much work can be done with 10 joules of energy?

2. A 2-kilogram object falls a distance of 5 meters. How much potential energy does this object have before it falls? How much work is done on it by gravity as it falls?

3. Sara's mother gets a flat tire on her car while driving Sara to school. They use a jack to change the tire. It exerts a force of 5,000 newtons to lift the car 0.25 meters. How much work is done by the jack?

4. How far does Isabella lift a 50-N box if she does 40 joules of work in lifting the box from the floor to a shelf?

5. A man pushes a television crate across the floor with a force of 200 newtons. How much work does he do if the crate moves 20 meters in the same direction as the force?

6. A bottle rocket is a toy that is made from an empty soda bottle. A bicycle pump is used to pump air into the bottle. The rocket shoots upward when it is released from the launcher, allowing the high-pressure air to come out.

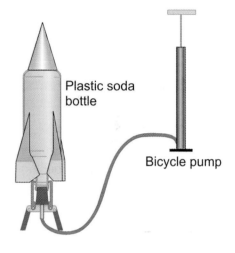

Plastic soda bottle

Bicycle pump

   a. Work is done as the pump is pushed, forcing air into the bottle. What happens to this work? Does it just disappear?

   b. Suppose a person does 2,000 joules of work using the pump. What is the maximum kinetic energy the rocket can have after it is launched?

   c. Do you think the rocket could actually have this much kinetic energy? Explain why or why not.

Section 8.2

7. A certain battery contains 20 joules of energy. The battery is connected to a perfect motor that uses 100 percent of the energy to make force.

   a. Suppose the motor made a 2-newton force. Over how much distance could this force be applied?

   b. How large a force can be sustained for 5 meters?

   c. If the battery was 80 percent efficient, what would the answers be for questions 7a and 7b?

8. A machine is used to lift an object a distance of 2 meters. If the power of the machine is increased, what happens to the time it takes for the object to be lifted 2 meters?

9. During construction, a crane lifts a 2,000-newton weight to the top of a 50-meter-tall building. How much power must the crane have to perform this task in 5 seconds? Give your answer in watts, kilowatts, and horsepower.

10. What is the minimum time needed to lift a 1,000-newton weight 20 meters using a motor with a maximum power rating of 8,000 watts?

Applying Your Knowledge

Section 8.1

1. Spend one day recording a variety of tasks that you do that involve doing work in the scientific sense. Also record the machines that allow you do certain tasks. Then spend the next day doing one or two of these tasks without using the machine. Answer the following questions.

   a. Was more or less work done using the machine? How do you know?

   b. Was the power output more or less with the machine? How do you know?

   c. What are your thoughts about using machines to accomplish work?

Section 8.2

2. A water-powered turbine makes electricity using the energy of falling water. At the location of one turbine, 100 kilograms of water falls every second from a height of 20 meters.

   a. How much potential energy does 100 kilograms of water have at a height of 20 meters?

   b. How much power in watts could you get out of the turbine if it was perfectly efficient?

   c. Research the efficiency of modern water-powered turbines. How efficient are these devices?

3. In this chapter, you learned the scientific meanings of *work*, *efficiency*, and *power*. Create a superhero cartoon character and draw a comic strip story that illustrates the scientific meanings of these words.

4. Choose an appliance in your house that you use on a regular basis. It may be a television, stereo, fan, or other device.

   a. The power of an appliance is a measure of the number of joules of electrical energy it converts to other forms of energy each second. What type or types of energy are converted from electrical energy by your appliance?

   b. There should be a label on the appliance that indicates its power in watts. Record the power. Then estimate the amount of time you use the appliance in an average day. Convert this time to seconds. Now calculate the number of joules of energy used by the appliance in a day.

Simple Machines

The Great Pyramid of Egypt, built around 2570 BCE and originally 147 meters tall, is one of the Seven Wonders of the Ancient World. Wind and weather over thousands of years have eroded the height of the pyramid. Yet the Great Pyramid is still considered one of the largest structures ever built. You might be amazed to find out that simple machines and not engines powered by fossil fuels were used to construct the pyramid. What are simple machines? Common items like a seesaw, a bottle opener, and a broom are all simple machines called levers. A screw is a type of simple machine. More complex machines such as bicycles are arrangements of simple machines. Surprisingly, your arms, legs, and other parts of your body are simple machines.

Simply put, simple machines are key parts of daily life and have been for thousands of years. In this chapter, you will learn why they are so useful.

CHAPTER 9 INVESTIGATIONS

9A: Levers
How does a lever work?

9B: Levers and the Human Body
What types of levers does your body have?

9.1 Types of Simple Machines

How do you move something that is too heavy to carry? How do you pry open a tight lid? How do you transfer pedaling motion to the wheels on a bicycle? The answer to each of these questions is "use a simple machine." In this section, you will learn how simple machines multiply forces to accomplish tasks.

Using machines

What technology allows us to do   Machines allow us to do incredible things. Moving huge steel beams, digging tunnels that connect two islands, and building 100-story skyscrapers are examples. What makes these things possible? Do we have super powers?

What is a machine?   In a way, we do have super powers. Our powers come from the clever human invention of machines. A **machine** is a device, like a bicycle, with moving parts that work together to accomplish a task (Figure 9.1). All the parts of a bicycle work together to transform forces from your muscles into motion. A bicycle allows you to travel at faster speeds and for greater distances than you could on foot.

Work output is forward motion

Work input is force applied to pedals

The concepts of input and output   In Chapter 8, you learned that a machine accomplishes *output work* when *input work* is applied to it. For the machines in this chapter, the **input** includes everything you do to make the machine accomplish a task, like pushing on the bicycle pedals. The **output** is what the machine does for you, like going fast or climbing a steep hill. In other words, the input and output may be force, energy, or power.

Parts of a Bicycle

Wheels   Gears   Pedals

Figure 9.1: *A bicycle is a machine that allows you to travel faster than you can on foot.*

Simple machines

The beginning of technology   The development of cars, airplanes, and other modern machines began with the invention of simple machines like levers. A **simple machine** is an unpowered mechanical device that accomplishes a task with only one movement. For example, a lever allows you to open a paint can, sweep the floor, or move a heavy rock (Figure 9.2). The variety of simple machines is shown below.

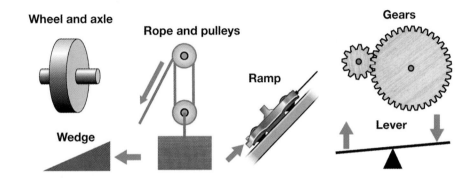

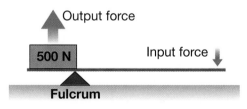

Figure 9.2: *Levers accomplish a task with one motion.*

Input force and output force   Simple machines work with forces. The *input force* is the force you apply to the machine. The *output force* is the force the machine applies to what you are trying to move. Figure 9.3 shows how a lever is arranged to create a large output force from a small input force. A **lever** is a stiff structure that rotates around a fixed point called a *fulcrum.*

Machines within machines   Most of the machines we use today are made up of combinations of different types of simple machines. For example, a bicycle is a complex machine made up of simple machines. A bicycle uses wheels and axles, levers (the pedals and kickstand), and gears. A **gear** is a rotating wheel with teeth that receives or transfers motion and forces to other gears or objects. If you take apart a complex machine such as a clock, a food processor or blender, or a car engine, you will find it is made of simple machines like gears.

Figure 9.3: *If arranged like this, a lever can create a large output force.*

What do simple machines do for us?

A review   In Chapter 8, you learned about work, efficiency, and power. Let's review these concepts because they illustrate why simple machines are so useful.

Machines do work   A simple machine does work because it applies a force over a distance. If you are using the machine, you also do work, because you apply force to the machine to make it move. By definition, a simple machine has no source of energy except the immediate forces you apply. That means the only way to get output work *from* a simple machine is to do input work *on* the machine.

Output and input work   Remember, the output work done by a simple machine can never exceed the input work done on the machine. In a *perfect* machine, the output work exactly equals the input work. Of course, there are no perfect machines. Friction always converts some of the input work to heat and wear, so the output work is always *less than* the input work. However, for some machines, the effect of friction is so small that we can often assume input and output work are approximately equal.

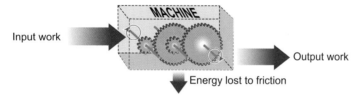

Machines have efficiency and power   The efficiency of a machine equals output work divided by input work multiplied by 100. For a perfect machine, efficiency is 100 percent. A bicycle is a highly efficient machine at 95 percent, but the human body is relatively inefficient at 8 percent. Power is how quickly work is accomplished (work ÷ time).

Machines multiply forces   A rope and pulley system, for example, allows you to lift more weight (force) than you could lift on your own without a machine. In other words, the rope and pulley system multiplies force. To compensate for multiplying force, you need to apply it over a greater distance than when the output work is done. For example, Figure 9.4 shows that to multiply force from 5 newtons to 10 newtons and to lift the weight by 1/2 meter, you need to pull the rope twice as far as the weight is lifted—you need to pull 1 meter of rope.

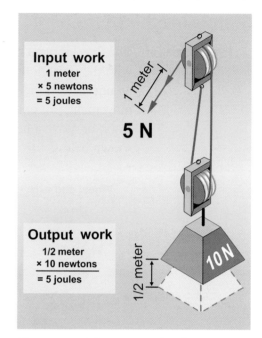

Input work
1 meter
× 5 newtons
= 5 joules

5 N

Output work
1/2 meter
× 10 newtons
= 5 joules

Figure 9.4: *This rope and pulley system illustrates that in order to multiply force in the output work, you need to compensate by applying the input force over a greater distance. In this example, friction is ignored.*

SOLVE IT!

Solve the following for the rope and pulley system in Figure 9.4.

a. What is the efficiency of this machine? Is this possible? Why or why not?

b. If the output work was accomplished in 2 seconds, what is the power of this machine?

Section 9.1 *Review*

1. In your own words, explain the difference between a simple machine and a complex machine.

2. List two reasons from this section that explain why a simple machine is a useful device.

3. In order to get work output from a machine, what do you need to do?

4. You have learned that a bicycle is a very efficient machine. Fill in the table with the words that explain how each factor compares as you use this machine. Input is applied to the pedals of the bicycle, and output is what the wheels do for you. Words to use: *similar*, *smaller*, or *larger*. Some words are filled in for you.

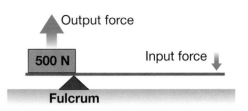

Figure 9.5: *Question 5.*

Comparing input and output for a bicycle		
Work	Input	
	Output	
Force	Input	
	Output	Smaller
Distance	Input	Smaller
	Output	

5. The arrangement of the lever in Figure 9.5 is similar to the arrangement you need to pry open the lid of a paint can. For the diagram to the right, label each part as *fulcrum*, *output force*, or *input force*.

6. You have a rope and pulley system for lifting objects. You need to lift a 20-newton object a distance of 3 meters. If you provide an input force of 10 newtons to pull on the rope, what length of rope would you need to pull before you accomplished this task?

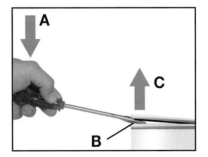

JOURNAL

Why use simple machines?

In the introduction of this chapter, you were asked: Is is possible to get through a day without using simple machines?

Write your answer to this question as a paragraph or short essay.

Extension: You need to educate others that simple machines are part of our lives. Write a television ad or design a magazine advertisement that illustrates how and why we use simple machines every day.

9.2 **Mechanical Advantage**

The human body is a machine, but is it capable of building a huge skyscraper on its own? Probably not. It would be impossible for a single person to lift and carry all the materials needed to build a tall building. We need machines to do this job. Machines help us lift more than we can lift on our own. This section continues our discussion about why simple machines are so useful and introduces the concept of mechanical advantage.

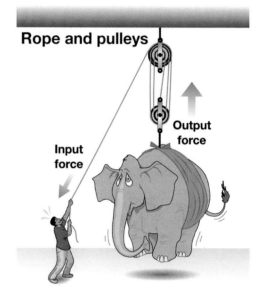

What is mechanical advantage?

Multiplying forces Another way to say machines help us lift more than we can lift is to say "machines multiply forces." Remember that a force is an action that has the ability to change motion, like a push or a pull. Figure 9.6 illustrates how a rope and pulley system can multiply forces. You apply the input force to the rope, and the output force is applied to the *load* you are lifting. A load is the amount of force or weight that a machine lifts or moves. One person could lift an elephant—quite a heavy load—with a properly designed system of ropes and pulleys!

Mechanical advantage **Mechanical advantage** is the ratio of output force to input force. If the mechanical advantage is greater than 1, the output force is bigger than the input force (Figure 9.6). A mechanical advantage less than 1 means the output force is smaller than the input force. What does it mean when mechanical advantage equals 1? In this case, the output force equals the input force. On the next pages, you will use this definition of mechanical advantage and discover that there are other ways to calculate this important value.

> **VOCABULARY**
>
> **mechanical advantage** – the ratio of output force divided by input force

Rope and pulleys

Figure 9.6: *A rope and pulley system can multiply forces.*

MECHANICAL ADVANTAGE

$$\underset{\text{advantage}}{\overset{\text{Mechanical}}{}} MA = \frac{F_o}{F_i} \quad \begin{array}{l} \textit{Output force (N)} \\ \\ \textit{Input force (N)} \end{array}$$

> **STUDY SKILLS**
>
> **If mechanical advantage Is...**
>
> \> 1, then output force is > input force
>
> < 1, then output force is < input force
>
> = 1, then output force is = input force

Levers

Parts of the lever All levers include a stiff structure that rotates around a fulcrum. For example, you can make a lever by balancing a board on a log. The log is the fulcrum. The side of the lever where the input force is applied is called the *input arm*. The *output arm* is the end of the lever that applies the output force.

Mechanical advantage Levers are useful because you can arrange the fulcrum and the input and output arms to adjust the mechanical advantage of the lever. By changing the position of the fulcrum, you can alter the amount of input force needed compared to output force desired. The length of the lever arm is indirectly related to the corresponding force. For example, if the input arm is 3 times longer than the output arm, the output force is 3 times greater than the input force. This lever has a mechanical advantage of 3. *Using the length of the lever arms, mechanical advantage can also be calculated by dividing the length of the input arm by the length of the output arm.*

The three classes of levers Pliers, a wheelbarrow, and your arm represent each of three classes of levers. These objects look different from each other, so how are they similar? For starters, they accomplish a task with one movement. They also each operate using a fulcrum and lever arms. Each class of levers is defined by the location of the input and output forces relative to the fulcrum (Figure 9.7).

First-class lever

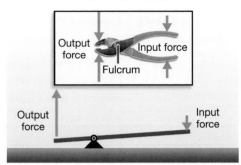

Second-class lever

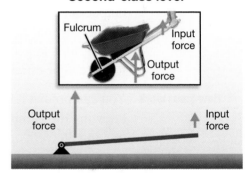

Third-class lever

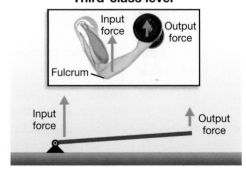

Figure 9.7: *These diagrams show the three classes of levers. What is the mechanical advantage of each of these levers: 1, > 1, or < 1?*

Class of lever	Fulcrum	Force	Length of arms
1st	Between input and output forces	Vary in magnitude	Vary in length
2nd	One end of lever	Output > input	Input > output
3rd	One end of lever	Input > output	Output > input

Gears

Gears transfer motion and force
Gears can transfer motion and force when the teeth of one gear press on the teeth of another gear as each gear rotates around a shaft (Figure 9.8). Connected in this way, the two gears turn in different directions. You can think of a gear as a rotating lever. The tip of a tooth of a gear is like the end of a lever, and the shaft of the gear is like the fulcrum. This means that forces are applied where the teeth press against each other.

Gears change force and speed
Gears can also multiply forces and change rotating speeds. Like other simple machines, gears have an input and an output. The *input gear* is the one you turn, or apply forces to. The *output gear* is the one that is connected to the output of the machine. Force is multiplied when the input gear in a pair is smaller and has fewer teeth than the output gear. Speed is increased when the input gear is larger than the output gear. Because gear teeth fit together, large forces can be efficiently transferred at high rotating speeds without slipping.

How gears work
You can predict how force and speed are affected when gears turn by knowing the number of teeth for each gear. Because the teeth don't slip, moving 36 teeth on one gear means that 36 teeth have to move on any connected gear. If the output gear has 36 teeth, it turns once to move 36 teeth. If the input gear has only 12 teeth, it has to turn 3 times to move 36 teeth ($3 \times 12 = 36$). In this example, the output gear is larger so force is multiplied.

Gear ratio and mechanical advantage
The **gear ratio** is the ratio of output turns to input turns. The gear ratio can also be calculated as the ratio of the number of teeth on the input gear versus the number on the output gear. The mechanical advantage of a pair of gears is the inverse of the gear ratio.

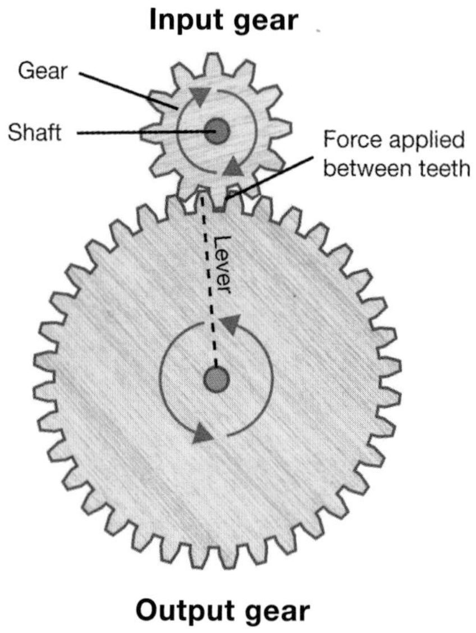

Figure 9.8: *A gear rotates around a shaft. Force is applied between the teeth of two gears. What is the mechanical advantage of this gear combination? Answer: Greater than 1 and force is multiplied. Do you see why?*

> ### GEAR RATIO
>
> Turns of output gear
> Turns of input gear
> $$\frac{T_o}{T_i} = \frac{N_i}{N_o}$$
> Number of teeth on input gear
> Number of teeth on output gear

Rope and pulley systems

Tension in ropes and strings
Ropes and strings carry *tension* forces along their length. The tension is the same at every point in a rope. If the rope is not moving, its tension is equal to the force pulling on each end (below). Ropes or strings do *not* carry pushing forces.

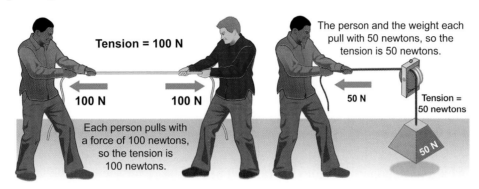

Tension = 100 N

100 N   100 N

Each person pulls with a force of 100 newtons, so the tension is 100 newtons.

The person and the weight each pull with 50 newtons, so the tension is 50 newtons.

50 N

Tension = 50 newtons

50 N

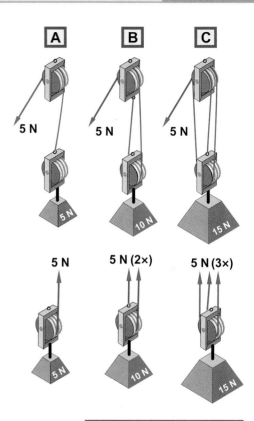

The forces in a rope and pulley system
Figure 9.9 shows three different configurations of rope and pulley systems. Imagine pulling with an input force of 5 newtons. In case A, the load feels a force equal to your input force. In case B, there are two strands of rope supporting the load, so the load feels two times your input force. In case C, there are three strands, so the output force is three times your input force.

Mechanical advantage
The mechanical advantage of a pulley system depends on the number of strands of rope directly supporting the load. In case C, three strands directly support the load, so the output force is three times the input force. The mechanical advantage is 3. To make a rope and pulley system with a greater mechanical advantage, you can increase the number of strands directly supporting the load.

Work
To raise the load 1 meter in case C, the input end of the rope must be pulled for 3 meters. This is because *each* of the three supporting strands must shorten by 1 meter. The mechanical advantage is 3, but the input force must be applied for three times the distance of the output force. In other words, to compensate for multiplying force, you need to apply the input force over a greater distance than when the output work is done.

	A	B	C
Input force	5 N	5 N	5 N
Output force	5 N	10 N	15 N
Mechanical advantage	1	2	3

Figure 9.9: *A rope and pulley system can be arranged to have different mechanical advantages.*

Ramps, wedges, and wheels

Ramps The mechanical advantage of a ramp is the ramp length divided by the height of the ramp. A ramp is a simple machine that allows you to raise a heavy object, such as a wheeled cart, with less force than you would need to lift it straight up (Figure 9.10). Ramps reduce the input force by increasing the distance over which the input force acts. For example, suppose a 10-meter ramp is used to lift a cart 1 meter. The input distance is 10 times the output distance. If the ramp were frictionless, the input force would therefore be 1/10 the output force. The mechanical advantage would be 10.

A screw is a type of ramp A screw is a simple machine that turns rotating motion into linear motion (Figure 9.11). A screw works just like a ramp that curves as it gets higher. The "ramp" on a screw is called a thread. Imagine unwrapping one turn of a thread to make a straight ramp. Each turn of the screw advances the nut the same distance it would have gone sliding up the ramp. The *lead* of a screw is the distance it advances in one turn. A screw with a lead of 1.2 millimeters advances 1.2 mm for each turn. You find the mechanical advantage of a screw by dividing its circumference by the lead.

Wedge A wedge is like a ramp that can work while in motion (a ramp is always stationary). A wedge has a side that slopes down to a thin edge. The mechanical advantage for a wedge is inversely related to the size of the wedge angle. For example, sharp wedges (small angles) produce large forces. The head of an axe is a wedge that is used to split wood. When the surface of the wedge is rough (producing high friction), a wedge provides a large holding or stopping force, thus preventing other objects from moving.

Wedge

Wheel and axle A wheel rotates around a rod called an axle. The wheel and axle move together to move or lift loads. The mechanical advantage of a wheel and axle is the ratio of the radius of the wheel to the radius of the axle. By applying a force to the wheel to turn the axle, output force is increased. In this case, distance and speed are decreased. If you want to increase distance and speed, you apply a force to the axle in order to turn the wheel. In this case, force output is decreased. Cars, bicycles, and household devices like a rolling pin and a door knob are wheel-and-axle machines.

Wheel
Axle

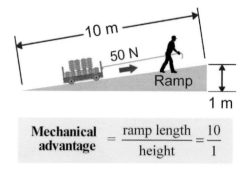

$$\text{Mechanical advantage} = \frac{\text{ramp length}}{\text{height}} = \frac{10}{1}$$

Figure 9.10: *A ramp allows you to raise a heavy object with less force than you would need if you lifted it upward.*

Mechanical advantage of a screw

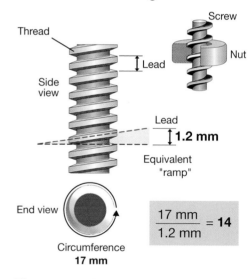

$$\frac{17 \text{ mm}}{1.2 \text{ mm}} = 14$$

Figure 9.11: *A screw is a rotating ramp.*

Solving Problems: Mechanical Advantage

A crowbar is a type of lever that you use to pull a nail out of a piece of wood. If the handle of a crowbar is 40 centimeters and the foot is 2 centimeters, what is its mechanical advantage (Figure 9.12)?

1. **Looking for:**	You are asked to find the mechanical advantage of a lever.	
2. **Given:**	The point between the foot and handle of the crowbar is the fulcrum. The input arm is 40 centimeters long and the output arm is 2 centimeters.	
3. **Relationships:**	Mechanical advantage $= \dfrac{\text{Length of input arm}}{\text{Length of output arm}}$ The formula above illustrates one way to calculate the mechanical advantage of a lever using the length of the arms.	
4. **Solution:**	Mechanical advantage $= \dfrac{40\ \text{cm}}{2\ \text{cm}} = 20$ The mechanical advantage of the crowbar is 20.	

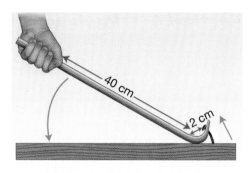

Figure 9.12: *What is the mechanical advantage of this crowbar?*

SOLVE FIRST/LOOK LATER

a. The output force would be 100 newtons.

b. The mechanical advantage is 1/4 or 0.25. Speed is multiplied.

c. The output gear turns 6 times. The mechanical advantage is 1/3.

d. You need to pull 4 m of rope.

Your turn...

a. What is the output force of the crowbar if the input force is 5 newtons (Figure 9.12)?

b. An output gear has 10 teeth and an input gear has 40 teeth. What is the mechanical advantage of this gear combination? Is force or speed multiplied in this example?

c. An input gear turns twice. How many times will the output gear turn if the gear ratio is 3? What is the mechanical advantage?

d. The mechanical advantage of a rope and pulley system is 2. The input force is 6 N and the output force is 12 N. If a 12-N object is lifted 2 m by the system, what length of rope do you need to pull?

Section 9.2 *Review*

1. If you know the lengths of the lever arms for a lever, can you calculate its mechanical advantage? If so, what is the formula?

2. Describe the term *mechanical advantage*. Why is it an important value to know when working with machines?

3. For a wheelbarrow, the output force is usually greater than the input force. Therefore, what can you say about the mechanical advantage of a wheelbarrow?

4. You might be surprised to learn that a broom is a lever. What kind of lever is it: first, second, or third class? Explain your answer.

5. A input gear with 24 teeth is connected to an output gear with 12 teeth (Figure 9.13).
   a. If the output gear turns twice, how many times does the input gear turn? What is the gear ratio? What is the mechanical advantage?
   b. What is being multiplied in this gear combination, force or speed?

6. Jane designed a rope and pulley system to lift a log. The log weighs 2,000 newtons. If she pulls 10 meters of rope to lift the log 2 meters, what force does she apply to the rope? You may assume a perfect machine (no friction). What is the mechanical advantage of this system?

7. You can predict the mechanical advantage of a rope and pulley system:
   a. by looking at the size of the load being lifted by the system.
   b. by seeing how fast you can lift a load with the system.
   c. by weighing the whole system.
   d. by counting the number of supporting strands in the system.

8. What is the mechanical advantage of a 15-meter ramp that rises 3 meters?

9. A screw is similar to another simple machine. Name that machine and explain why a screw resembles it.

10. How is force multiplied in a machine that uses a wheel and axle?

11. What is the difference between a wedge and a ramp?

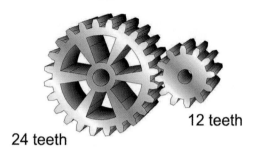

12 teeth

24 teeth

Figure 9.13: *Question 5.*

9.3 Levers in the Human Body

In this section, we will look at how the human body system uses levers. Here's one reason you need simple machines each day—you use the simple machines of your body on a daily basis!

Levers in your arms and legs

The human body machine   The human body is a complex machine that includes a number of simple machines—levers. Levers provide the body with the ability to accomplish tasks. Your arms and legs, for example, work as levers to move and lift objects. In the last section, you learned that there are three classes of levers. Your limbs are examples of third-class levers. What does this mean for the ability of the arms and legs to accomplish tasks?

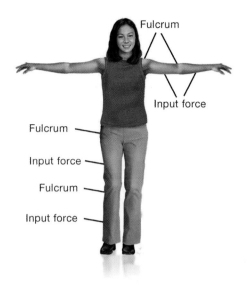

The usefulness of third-class levers   A classic example of a third-class lever is your forearm. Consider the act of lifting a book with your hand. Your forearm is a lever that rotates around a fulcrum at the elbow. The biceps muscle provides the input force. Notice that the input force is closer to the fulcrum than the output force. In fact, the ratio of lengths is about 12 to 1. That means your biceps muscle has to make 12 N of force for every 1 N of force exerted by your hand. But it also means your hand can move 12 times as far as the end of your biceps muscle! That's a good thing because the biceps muscle can only contract a few centimeters. Imagine how limited you would be if your hand could only move up or down a few centimeters!

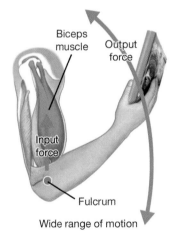

Figure 9.14: *Human arms and legs are all examples of third-class levers.*

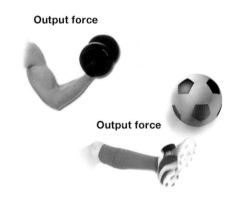

Your arms and legs   Figure 9.14 illustrates the locations of some of the levers of the human body. Human arms and legs are examples of third-class levers because the input forces are between a fulcrum and the output force. The output force is what you accomplish with your hands and feet (Figure 9.15). For the human body, your skeleton is the stiff structure that pivots or rotates around a fulcrum. Joints act as fulcrums. The input force is provided by your muscles.

Figure 9.15: *The output force for arms and legs is what you accomplish with your hands and feet.*

Levers from head to foot

Your neck is a lever   Stop reading for a moment. Relax your neck so that your head drops slowly forward. At about 4.5 kilograms, the head is a heavy object. Your head drops forward when you relax your neck because your head and neck work like a first-class lever (Figure 9.16). The fulcrum is at the top of the neck. The muscles in the neck provide an input force that allows you to raise your head. When you relax these muscles, gravity causes your head to fall forward.

Your lower jaw is a lever   Like your arms and legs, your lower jaw is a lever. However, the lower jaw can work both as a second-class and as a third-class lever.

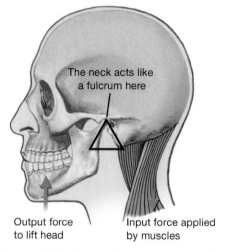

Figure 9.16: *The neck is an example of a first-class lever.*

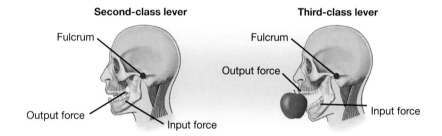

Chewing with a second-class lever   Whether you are chewing gum or your lunch, you are using a second-class lever. Think about this for a moment. Where is the fulcrum? Where is the input force? Where is the output force? The fulcrum is the hinge of your jaw. The output force is the piece of gum or food that you are chewing with your molars. The input force is provided by your jaw muscles.

Biting with a third-class lever   When you take a bite out of an apple, your lower jaw becomes a third-class lever. The hinge of your jaw, the fulcrum, doesn't change. However, the location of the output force does. You use your jaw muscles as input force, to provide an output force for biting with your front teeth.

Your foot is a lever   When you stand on your toes, the feet act as second-class levers (Figure 9.17). Your toes are the fulcrum. The input force is provided by your calf muscles. The output force is the weight of your foot being lifted.

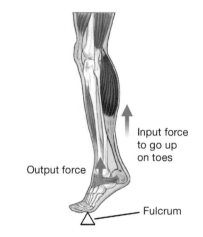

Figure 9.17: *The foot is an example of a second-class lever.*

Section 9.3 *Review*

1. The human body is often called a machine because of the way the bones and muscles work together. Is the body a "simple" machine? Why or why not?

2. Which of the following are levers in your body? To justify your answer, identify the output force, input force, and fulcrum for your choices. Also, describe a task that each lever accomplishes.
    a. your kneecap
    b. your lower leg
    c. your brain
    d. the top of your foot

3. Identify which kind of lever is represented in the following scenarios.
    a. A girl bites into a sandwich.
    b. A baseball pitcher throws a fastball with his arm.
    c. A swimmer does a flutter kick while swimming the length of a pool.
    d. Once you take a bite of food, you need to chew it well with your molars.
    e. I knew James was asleep because his head nodded forward.

4. The mechanical advantage of a lower jaw when it is used as a third-class lever is 0.7.
    a. What is being multiplied by the jaw when it is used as a third-class lever?
    b. If the input force is 500 newtons, what would the output force be for the jaw used in this way?
    c. Why do you think it is more useful to have the jaw act as a third-class lever for biting versus a second-class lever?

5. The mechanical advantage of a jaw when it is used as a second-class lever is 1.4.
    a. If the input force is 100 newtons, what is the output force?
    b. How does the input lever arm compare to the output lever arm when the jaw is used as a second-class lever? Draw a diagram to illustrate your answer.

▰▰▰▰ BIOGRAPHY ▰▰▰▰
Giovanni Alfonso Borelli

A long time ago, Giovanni Alfonso Borelli, an Italian physicist and mathematician, understood that our bodies work using mechanical forces. He a wrote a book about this—*On the Movement of Animals*—in the 17th century. This book is considered to be the first book on the subject of biomechanics.

As the name suggests, biomechanics is the study of mechanics as applied to biological systems. Scientists who study biomechanics ask questions such as: How do insect wings work? and How do animals living on coastlines withstand the battering of ocean waves? Biomechanics is also beneficial for understanding and preventing sports injuries.

In writing his book, Borelli led the way for recognizing that arms and legs work as levers. He used the scientific method in his work and supported his theories with mathematics.

⚙ STEM Prosthetics in *Action!*

The human leg is a complex and versatile machine. Designing a prosthetic (artificial) device to match the leg's capabilities is a serious challenge. Teams of scientists, engineers, and designers around the world use different approaches and technologies to develop prosthetic legs that help the user regain a normal, active lifestyle.

Studying the human gait cycle

Each person has a unique way of walking. But studying the way humans walk has revealed that some basic mechanics hold true for just about everyone. Scientists analyze how we walk by looking at our "gait cycle." The gait cycle consists of two consecutive strides while walking, one foot and then the other. By breaking the cycle into phases and figuring out where in the sequence prosthetic devices could be improved, designers have added features and materials that let users walk safely and comfortably with their own natural gait.

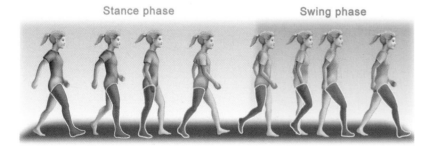

Stance phase   Swing phase

Designing a better prosthetic leg

In many prosthetic leg designs, the knee is the component that controls how the device operates. In the past, most designs were basic and relied on the user learning how to walk properly. This effort required up to 80 percent more energy than a normal gait and often made walking with an older prosthetic leg quite a workout!

The knee joint in those older designs was often a hinge that let the lower leg swing back and forth. The hinge could also lock in place to keep the leg straight and support the user's weight to make standing easier. This type of system worked relatively well on level surfaces, but could be difficult to use on inclines, stairs, irregular terrain (like a hiking trail), or slippery surfaces.

Current prosthetic legs have improved upon old designs by employing hydraulics, carbon fiber, mechanical linkages, motors, computer microprocessors, and innovative combinations of these technologies to give more control to the user. For example, in some designs, a device called a damper helps to control how fast the lower leg can swing back and forth while walking. The damper accomplishes this by changing the knee's resistance to movement as needed.

New knee designs allow users to walk, jog, and, with some models, even run with a more natural gait. In fact, in 2003, Marlon Shirley became the first above-the-knee amputee in the world to break the 11-second barrier in the 100-meter dash with a time of 10.97 seconds! He accomplished this feat with the aid of a special prosthetic leg designed specifically for sprinting.

Designs that learn

By continuously monitoring the velocities of the upper and lower leg, the angle of the bend of the knee, changes in the terrain, and other data, computer microprocessors in the knee calculate and make adjustments to changing conditions in milliseconds. This makes the prosthetic leg more stable and efficient, allowing the knee, ankle, and foot to work together as a unit. Some designs have built-in memory systems that store information from sensors about how the user walks. These designs "learn" how to make fine-tuned adjustments based on the user's particular gait pattern.

New foot designs

New foot designs also reduce the energy required to walk with prosthetic leg systems. They also smooth out the user's stride. Using composite materials, these designs allow the foot to flex in different ways during the gait cycle. Both the heel and the front part of the foot act like springs to store and then release energy. When the foot first strikes the ground, the heel flexes and absorbs some of the energy, reducing the impact. Weight gets shifted toward the front of the foot as the walker moves through the stride.

As this happens, the heel springs back into shape and the energy released helps to flex the front part of the foot, once again storing energy. When the foot leaves the ground in the next part of the gait cycle, the flexed front part of the foot releases its stored energy and helps to push the foot forward into the next stride.

The engineering cycle

Engineers have realized the advantage of making highly specialized feet that match and sometimes exceed the capabilities of human feet. Distance running and sprinting feet are built to different specifications to efficiently deal with the forces and demands related to these activities.

Questions:

1. What are some technologies used by designers of prosthetic legs to improve their designs?

2. How are computers used to improve the function of prosthetic devices?

3. Explain how new foot designs reduce the amount of energy required to walk with a prosthetic leg.

4. Research the field of *biomechanics*. In a paragraph:

   (1) describe what the term biomechanics means; and

   (2) write about a biomechanics topic that interests you.

Chapter 9 **Assessment**

Vocabulary

Select the correct term to complete each sentence.

gear	lever	output
gear ratio	machine	simple machine
input	mechanical advantage	

Section 9.1 and Section 9.2

1. A(n) _____ is a stiff structure that pivots on a fulcrum.

2. If the output force is 2 newtons and the input force is 1 newton, the _____ for a machine is 2.

3. A(n) _____ is a wheel with teeth that transfers motion or force.

4. A rope and pulley system or a lever is an example of a(n) _____ .

5. To travel 150 kilometers in less than 2 hours, I need a(n) _____ , a device that has moving parts that work together to help me travel that far in that amount of time.

6. The force you use when you pedal a bicycle is the _____ and the motion of the wheels and distance traveled is the _____ .

7. The number of turns of an output gear divided by the number of turns of the input gear is called the _____ .

Concepts

Section 9.1

1. Name two simple machines that are found on a bicycle.

2. Explain the difference between input work and output work for a machine.

3. Is a gas-powered lawn mower a simple machine? Explain why or why not.

4. An inventor claims to have created a new unpowered machine. He says the machine can push with an output force of 100 newtons for 1 meter if you apply an input force of 50 newtons for 0.5 meter. Could this machine work? Explain why or why not.

5. List the simple machines that you have learned about and provide an everyday example of each. Think of an example that was not mentioned in the text.

Section 9.2

6. Correct or incorrect? Explain your answer in each case.
   a. You know the input arm length and the output arm length of a lever. Therefore, you can calculate the mechanical advantage.
   b. You know the input arm length, output arm length, and the output force for a lever. Therefore, you can determine the input force.

7. Explain how mechanical advantage is calculated for each simple machine:
   a. a pair of gears
   b. a lever

8. How does the mechanical advantage of a second-class lever compare to the mechanical advantage of a third-class lever?

9. Identify whether the items in this list are first-, second-, or third-class levers when they are in use: a nutcracker, a baseball bat, a golf club, a hammer used to hit a nail, a pair of scissors.

10. Two people are pulling on opposite ends of a rope so that it has a tension of 150 newtons. If the rope is not moving, with what pulling force is each of the two people pulling?

11. A wheeled cart is used to bring a large load of luggage up a ramp onto a cruise ship. What simple machines are being used in this situation? Why are they being used? Include the term *mechanical advantage* in your answer.

Section 9.3

12. Which classes of levers are represented in the human body?
 a. first and second class only
 b. third class only
 c. all three classes
 d. second and third class only

13. Look at this diagram of a sweeping broom.

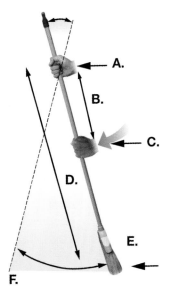

 a. Label the parts of this diagram using these terms: *input force, output force, input arm, output arm, fulcrum,* and *wide range of motion.*
 b. What kind of lever is a broom?
 c. What are the advantages to using this kind of lever to sweep a floor?
 d. If you had to move large rocks in your yard, would you still want to use a broom? Why or why not? If not, what kind of lever might you use?

14. Pick an animal and name one motion that this animal does. State whether or not the motion represents the motion of a lever. If a lever is represented, explain whether it is a first-, second-, or third-class lever. Justify your answer.

Problems

Section 9.1

1. If you lift a 200-newton box 1 meter with a rope and pulley system and you apply 20 newtons to lift this box, what is the mechanical advantage of the system? How far do you have to pull the rope when you are applying the input force? If you do this work in 5 seconds, what is the power?

Section 9.2

2. Look at the diagrams at the right. They show three first-class levers, each arranged in a different way. For each, answer the following questions.

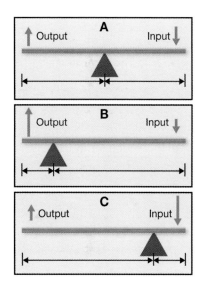

 a. Is the input arm larger, smaller, or the same as the output arm?
 b. Is the input force larger, smaller, or the same as the output force?
 c. Is the mechanical advantage greater than, less than, or equal to 1?

3. A lever has an input arm that is 2 meters long and an output arm that is 3 meters long. What is the mechanical advantage? Does this lever multiply force? Why or why not?

4. Betsy wants to use her own weight to lift a 1,500-newton box. She weighs 500 newtons. Suggest input and output arm lengths that would allow Betsy to lift the box with a lever. Draw a lever and label the input and output arms with the lengths and forces.

5. You need a wheelbarrow to transport some soil for your garden. The wheelbarrow you have gives you a mechanical advantage of 3.5. If you use 65 newtons of force to lift the wheelbarrow so that you can roll it, how much soil can you carry with this wheelbarrow? Give the weight of the soil in newtons and be sure to show your work.

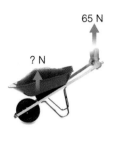

65 N

? N

6. An output gear has 12 teeth and an input gear has 48 teeth. What is the mechanical advantage of this gear combination? Is force or speed multiplied in this example?

7. An input gear turns four times. How many times will the output gear turn if the gear ratio is 0.5? Does the input gear have more or fewer teeth than the output gear?

8. You plan to use a rope and pulley system to lift a 100-newton box. How much input force is needed if the pulley has:

a. 1 supporting string?

b. 2 supporting strings?

c. 5 supporting strings?

d. 10 supporting strings?

9. The block and tackle machine (a rope and pulley system) on a sailboat can help a sailor raise her mainsail. She needs 500 newtons of force to raise the sail. If the block and tackle gives her a mechanical advantage of 5, how much input force must be applied to raise the sail?

10. Use the input and output forces listed in the table below to calculate the mechanical advantage.

Input force (N)	Output force (N)	Mechanical advantage
10	100	
30	30	
500	1,350	
625	200	

11. One of the examples in the table in problem 10 has a very low mechanical advantage. Identify this example and explain why you might or might not want to use this machine to lift something that weighs 200 newtons.

Section 9.3

12. Using the diagrams below, answer the following questions (on the next page).

Jaw

Fulcrum (pivot)

10 cm

7 cm

Equivalent lever

10 cm

7 cm

Input force (muscle)

Output force (bite)

Arm (Biceps muscle)

Biceps muscle

Fulcrum (pivot)

5 cm

35 cm

Equivalent lever

35 cm

5 cm

Input force (muscle)

Output force (lift)

a. Using the distances shown in the diagrams on the previous page, calculate the mechanical advantage of the jaw and arm. Which is larger?

b. Suppose the jaw and biceps muscle produce equal input forces of 800 N. Calculate and compare the output forces in biting (jaw) and lifting (arm). Which is larger?

c. Suppose you need an output force of 500 N. Calculate and compare the input forces of the jaw and biceps muscle required to produce 500 N of output force.

Applying Your Knowledge

Section 9.1

1. Think about a task that you need to do. Now invent and design a complex machine to do this task. Be sure to include two or more simple machines in your design. On a separate piece of paper, make a sketch of your design and describe how it works.
Extension: Present your complex machine to your class. Can you convince your classmates that your machine will work? Why or why not? (STEM)

Section 9.2

2. Does mechanical advantage have units? Why or why not?

3. Write a mechanical advantage question for each of the following simple machines. Be sure to work out the answers to the questions. Share your questions with a classmate and then check his or her work.

a. ramp
b. screw
c. wedge
d. wheel and axle

4. Here is a table with sample data for lifting (input) force versus the number of supporting strings in a rope and pulley system. Use the data to answer the following questions.

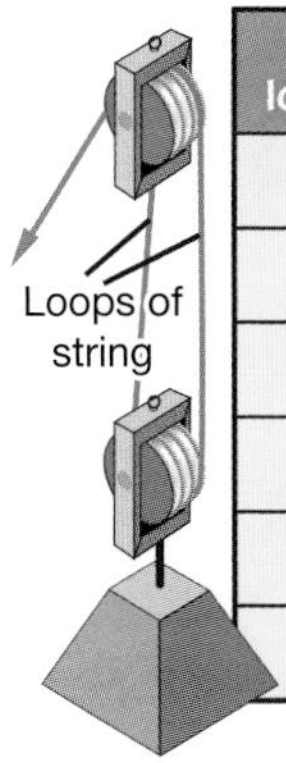

Supporting loops of string	Input force	Output force	Mechanical advantage
2	5 N	10 N	
4	2.5 N	10 N	
6	1.7 N	10 N	
1	10 N	10 N	
3	3.3 N	10 N	
5	2 N	10 N	

Loops of string

a. Describe the relationship between the lifting (input) force and the number of supporting strings in the pulley.

b. Make a graph that shows the relationship between lifting (input) force and the number of supporting strings. Which variable is dependent and which is independent?

c. Calculate the mechanical advantage for each number of supporting strings.

Section 9.3

5. In the text, you learned about the scientific field of biomechanics. Find out more about this fascinating field. What do biomechanics scientists do? What kind of research questions do they ask? Do they work indoors or outdoors or both? Write an essay that describes your findings.

Unit 4

Matter and Energy

Air: 78% nitrogen, 21% oxygen, 1% trace gases

H_2O

energy

Solid

Buoyancy

volume

$$\frac{V_1}{T_1} = \frac{V_2}{T_2}$$

temperature

Phase change

inside 50°C less density

outside 15°C more density

Liquid

Try this at home → Get a metal spoon and a plastic spoon, a cup, some ice, and water. Fill the cup with ice and then add water until the cup is about half full with water. Place both spoons in the cup and wait for two minutes. Touch the metal spoon and then the plastic spoon. Record your observations on a piece of paper. Which spoon feels colder? Warmer? Why do the spoons feel like they do? Write a paragraph about what you felt and why you think that happened.

226

Matter and Temperature

Have you ever imagined what it would be like to live in an atom-sized world? You may have seen movies where the characters are suddenly shrunk to the size of a flea or an even tinier animal. If you were that small, what would the matter around you look like? What if you were even smaller, say the size of an atom? In this world, even the air around you could be dangerous. Everywhere you looked, you would see atoms and molecules whizzing around at amazingly fast speeds and occasionally colliding with one another. Watch out! One of those particles might collide with you!

If you were the size of an atom, you would notice that the particles that make up everything are in constant motion. In liquids, the particles slide over and around each other. In solids, the particles vibrate in place. In gases, the particles move around freely. Ordinary air would look like a crazy, three-dimensional bumper-car ride where you are bombarded from all sides by giant beach balls. It will be helpful to imagine life as an atom as you study this chapter.

CHAPTER 10 INVESTIGATIONS

10A: Pure Substance or Mixture?
How can observing the melting point identify a pure substance or a mixture?

10B: Determining Freezing/Melting Point
How do you determine the freezing/melting point of cetyl alcohol?

10.1 The Nature of Matter

From a distance, a sugar cube looks like a single piece of matter. But up close, you can see it is made up of tiny, individual crystals of sugar fused together. Can those sugar crystals be broken into even smaller particles? What is the smallest particle of sugar that is still sugar (Figure 10.1)?

Matter is made of tiny particles in constant motion

The idea of atoms   *Matter* is a term used to describe anything that has mass and takes up space. The idea that matter is made of tiny particles goes back to 430 BCE. The Greek philosophers Democritus and Leucippus proposed that matter is made of tiny particles called *atoms*. For 2,300 years, few people believed this theory. In 1803, John Dalton revived the idea of atoms, but he lacked proof.

Brownian motion provides evidence for particles   In 1827, Robert Brown, a Scottish botanist, was looking through a microscope at tiny grains of pollen in water. He saw that the grains moved in an irregular, jerky manner. After observing the same motion in tiny dust particles, he theorized that all tiny particles move in the same way. The irregular, jerky motion was named *Brownian motion* in Brown's honor.

A human-sized comparison   Imagine throwing marbles at a tire tube floating in the water. The impact of any single marble is too small to make the tire tube move. If you throw enough marbles, the tube will start moving slowly. The motion of the tire tube will appear smooth because the mass of a single marble is tiny compared to the mass of the tire tube (Figure 10.2).

Why Brownian motion is jerky, not smooth   Now imagine throwing marbles at a foam cup floating in water. The motion is jerky, and the impact of individual marbles can be seen. The mass of the cup is not huge compared to the mass of a marble. A pollen grain in water moves around in a jerky manner much like the foam cup. That motion is caused by the impact of individual water molecules on the pollen grain. Like the cup, the mass of the pollen grain, while larger than a water molecule, is not so much larger that impacts are completely smoothed out.

Matter is made of atoms   In 1905, Albert Einstein proposed that Brownian motion is caused by collisions between visible particles like pollen grains, and smaller, invisible particles. This was strong evidence that matter was indeed made of *atoms*.

Figure 10.1: *What is the smallest particle of sugar that is still sugar?*

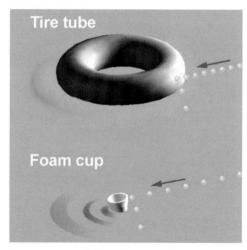

Figure 10.2: *Throwing marbles at a tire tube moves the tube smoothly. Throwing the same marbles at a foam cup moves the cup in a jerky manner, like Brownian motion.*

Atoms and elements

Elements   An **element** is defined as a pure substance that cannot be broken down into simpler substances by physical or chemical means. For example, water is made from the elements hydrogen and oxygen. If you add energy, you can break water down into hydrogen and oxygen, but you cannot break the hydrogen and oxygen down into simpler substances (Figure 10.3).

Defining atoms   A single **atom** is the smallest particle of an element that retains the chemical identity of the element. For example, you can keep cutting a piece of the element gold into smaller and smaller pieces until you cannot cut it any more. That smallest particle you can divide it into is one atom. A single atom of gold is the smallest piece of gold you can have. If you split the atom, it will no longer be gold.

How small are atoms?   A single atom has a diameter of about 10^{-10} meters. This means that you can fit 10,000,000,000 (10^{10}) atoms side-by-side in a one-meter length. You may think a sheet of aluminum foil is thin, but it is actually more than 200,000 atoms thick!

Atoms of an element are similar to each other   Each element has a unique type of atom. Sodium atoms are different from carbon atoms, carbon atoms are different from aluminum atoms, etc. But all atoms of a given element are similar to each other. If you could examine a million atoms of carbon, you would find them all to be similar. You will learn much more about atoms in Chapter 14.

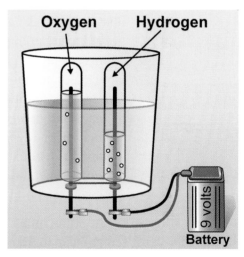

Figure 10.3: *You can break water down into oxygen and hydrogen by adding energy.*

Compounds contain two or more elements

Compounds Sometimes elements are found in their pure form, but more often they are combined with other elements. Most substances contain several elements combined together. A **compound** is a substance that contains two or more different elements chemically joined and has the same composition throughout. For example, water is a compound that is made from the elements hydrogen and oxygen. Figure 10.4 shows some familiar compounds.

Molecules If you could magnify a sample of pure water so you could see its atoms, you would notice that the hydrogen and oxygen atoms are joined together in groups of two hydrogen atoms to one oxygen atom. These groups are called molecules. A **molecule** is a group of two or more atoms joined together by *chemical bonds*. A compound is made up of only one type of molecule. Some compounds, like table salt (sodium chloride), are made of equal combinations of different atoms instead of individual molecules.

Mixtures Most of the things you see and use in everyday life are mixtures. A *mixture* contains more than one kind of atom, molecule, or compound.

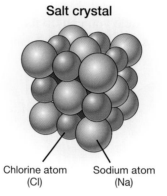

Salt crystal

Chlorine atom (Cl)   Sodium atom (Na)

COMPOUNDS contain more than one type of atom joined together.

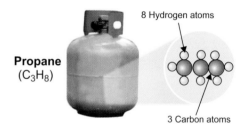

Propane (C₃H₈)

8 Hydrogen atoms

3 Carbon atoms

Water (H₂O)

2 Hydrogen atoms

1 Oxygen atom

Figure 10.4: *Examples of compounds.*

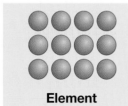

Element
One single kind of atom

Compound
One type of molecule or equal combinations of different atoms

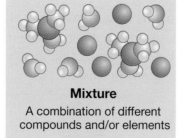

Mixture
A combination of different compounds and/or elements

Classifying matter

Pure substances Matter can be divided into two categories: pure substances and mixtures. A **pure substance** cannot be separated into different kinds of matter by physical means such as sorting, filtering, heating, or cooling. Elements and compounds are pure substances. Examples include water, table salt, gold, and oxygen.

Mixtures contain more than one kind of matter A **mixture** contains a combination of different elements and/or compounds. All mixtures share one common property: They can be separated into different types of matter by physical means such as sorting, filtering, heating, or cooling. For example, cola is a mixture that can be separated into carbonated water, corn syrup, caramel color, phosphoric acid, natural flavors, and caffeine.

Homogeneous mixture is the same throughout

A **homogeneous mixture** is the same throughout. In other words, all samples of a homogeneous mixture are the same. For example, an unopened can of cola is a homogeneous mixture. The cola in the top of the unopened can is the same as the cola at the bottom. Once you open the can, however, carbon dioxide will escape from the cola making the first sip a little different from your last sip. Brass is another example of a homogeneous mixture. It is made of 70 percent copper and 30 percent zinc. If you cut a brass candlestick into 10 pieces, each piece would contain the same percentage of copper and zinc.

Two samples of a heterogeneous mixture could be different A **heterogeneous mixture** is one in which different samples are not necessarily made up of exactly the same proportions of matter. One common heterogeneous mixture is chicken noodle soup (Figure 10.5). One spoonful might contain broth, noodles, and chicken, while another contains only broth. Can you think of a way to separate this mixture?

VOCABULARY

pure substance – matter that cannot be separated into other types of matter by physical means; includes all elements and compounds

mixture – matter that contains a combination of different elements and/or compounds and can be separated by physical means

homogeneous mixture – a mixture that is the same throughout; all samples of a homogeneous mixture are the same

heterogeneous mixture – a mixture in which different samples are not necessarily made up of the same proportions of matter

Figure 10.5: *Chicken soup is a heterogeneous mixture.*

Summary

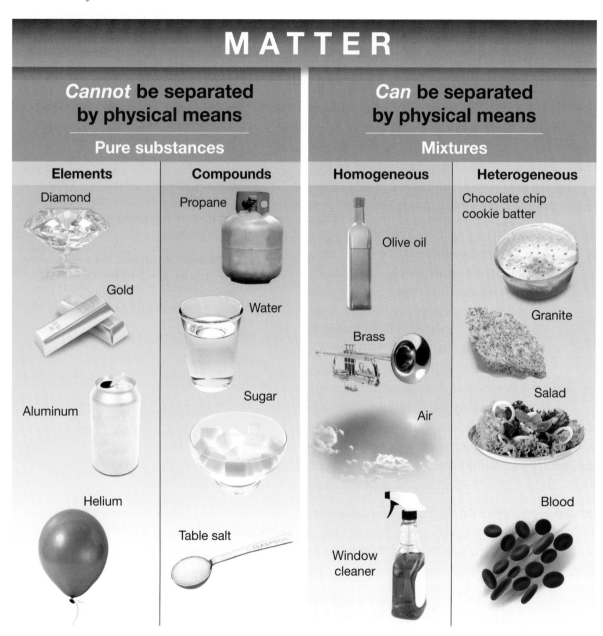

MATTER

Cannot be separated by physical means		*Can* be separated by physical means	
Pure substances		Mixtures	
Elements	Compounds	Homogeneous	Heterogeneous
Diamond	Propane	Olive oil	Chocolate chip cookie batter
Gold	Water	Brass	Granite
Aluminum	Sugar	Air	Salad
Helium	Table salt	Window cleaner	Blood

Section 10.1 *Review*

1. Explain why Brownian motion provides evidence for the existence of atoms and molecules.

2. Describe the difference between elements, compounds, and mixtures.

3. Which would be easier to separate, a mixture or a compound? Explain your answer.

4. Give an example of each: element, compound, and mixture.

5. Identify each of the following as element, compound, homogeneous mixture, or heterogeneous mixture. Explain your reasoning for each.

    a. milk
    b. iron nail
    c. glass
    d. sugar
    e. bottled spring water
    f. distilled water
    g. air
    h. alloy bicycle frame
    i. propane
    j. baking soda

6. Most things you use every day are:

    a. compounds
    b. elements
    c. mixtures

7. Your teacher has mixed salt, pepper, and water. Describe a procedure that you could use to separate this mixture. Be sure to list all of the materials you would need, your setup, and your expected results.

▰▰▰ BIOGRAPHY ▰▰▰

Edouard Benedictus

In 1903, a French chemist named Edouard Benedictus dropped a glass flask in the lab. The flask was full of cracks, but surprisingly, the pieces did not scatter across the floor. The shape of the flask remained intact. The flask had been used to store a compound called *cellulose nitrate*. Although the chemical had evaporated, it left a plastic film on the inside of the glass. Initially, Benedictus tried to sell his shatter-resistant glass to automobile manufacturers, but they weren't interested. During World War I, he sold it for use in gas mask lenses. Soon after the war, the auto industry began using his glass.

10.2 **Temperature**

You have probably used a thermometer. However, did you ever stop to think about how it works? In this section, you will learn what temperature is, how it is measured, and how the devices we use to measure temperature work.

Temperature scales

Fahrenheit There are two common temperature scales. On the **Fahrenheit** scale, water freezes at 32 degrees and boils at 212 degrees (Figure 10.6). There are 180 Fahrenheit degrees between the freezing point and the boiling point of water. Temperature in the United States is commonly measured in Fahrenheit; 68°F (68 degrees Fahrenheit) is a comfortable room temperature.

Celsius The **Celsius** scale divides the interval between the freezing and boiling points of water into 100 degrees (instead of 180). Water freezes at 0°C (0 degrees Celsius) and boils at 100°C. Most scientists and engineers use Celsius because 0 and 100 are easier to work with than 32 and 212.

Converting between the scales A weather report of 21°C in London, England, predicts a pleasant day, good for shorts and a T-shirt. A weather report of 21°F in Minneapolis, Minnesota, means a heavy winter coat, gloves, and a hat will be needed. Because the United States is one of only a few countries that use the Fahrenheit scale, it is useful to know how to convert between Fahrenheit and Celsius.

> **CONVERTING BETWEEN FAHRENHEIT AND CELSIUS**
>
> $$T_{Fahrenheit} = \frac{9}{5} T_{Celsius} + 32 \quad \Big| \quad T_{Celsius} = \frac{5}{9} (T_{Fahrenheit} - 32)$$

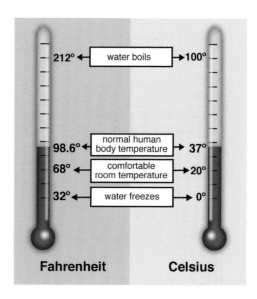

Figure 10.6: *The Fahrenheit and Celsius temperature scales.*

Solving Problems: Temperature Conversions

A friend in Paris sends you a recipe for a cake. The French recipe says to bake the cake at a temperature of 200°C for 45 minutes. At what temperature should you set your oven, which uses the Fahrenheit scale (Figure 10.7)?

1. **Looking for:** You are asked for the temperature in degrees Fahrenheit.

2. **Given:** You are given the temperature in degrees Celsius.

3. **Relationships:** Use the conversion formula: $T_F = {}^9\!/_5\, T_C + 32$.

4. **Solution:** $T_F = ({}^9\!/_5)\,(200) + 32 = 392°F$

Your turn...

a. You are planning a trip to Iceland, where the average July temperature is 11.2°C. What is this temperature in Fahrenheit?

b. You are doing a science experiment with a Fahrenheit thermometer. Your data must be in degrees Celsius. If you measure a temperature of 125°F, what is this temperature in degrees Celsius?

c. The temperature on the Moon varies from −230°C, at night, to 120°C during the day. What is the range in temperatures on the Moon in degrees Fahrenheit?

Photo courtesy of the Image Science & Analysis Laboratory, NASA Johnson Space Center

Figure 10.7: *A French recipe says to bake a cake at 200°C. At what temperature would you set the oven in degrees Fahrenheit?*

SOLVE FIRST LOOK LATER

a. 52.2°F

b. 51.7°C

c. −382°F to 248°F

Defining temperature

VOCABULARY

thermal energy – energy due to temperature

temperature – a quantity that measures the kinetic energy per molecule due to random motion

Atoms are always in motion   Imagine you had a microscope powerful enough to see individual molecules in a compound (or atoms, in the case of an element). You would see that the molecules are in constant motion, even in a solid object. In a solid, the molecules are not fixed in place, but act like they are connected by springs (Figure 10.8). Each molecule stays in the same average place, but constantly jiggles back and forth in all directions. As you might guess, the "jiggling" means motion, and motion means *energy*. The back-and-forth jiggling of molecules is caused by **thermal energy**, which is a kind of kinetic energy.

Temperature and energy   Thermal energy is proportional to temperature. When the temperature goes up, the energy of motion increases. That means the molecules jiggle around more vigorously. The higher the temperature, the more thermal energy molecules have and the faster they move around. Temperature measures a particular kind of kinetic energy per molecule.

Temperature measures the kinetic energy per molecule due to random motion.

Random versus average motion   If you throw a rock, the rock gets more kinetic energy, but the temperature of the rock does *not* go up. How can temperature measure kinetic energy then? The answer is the difference between *random motion* of the molecules and *average motion* of the object. For a collection of many molecules (like a rock), the kinetic energy has two parts. The kinetic energy of the thrown rock comes from the average motion of the whole collection—the whole rock. This kinetic energy is *not* what temperature measures.

Random motion   Each molecule in the rock is also jiggling back and forth independently of the other molecules in the rock. This jiggling motion is random motion. Random motion is motion that is scattered equally in all directions. On average, there are as many molecules moving one way as there are moving the opposite way. *Temperature measures the kinetic energy of the random motion.* Temperature is not affected by any kinetic energy associated with average motion. That is why throwing a rock does not make it hotter (Figure 10.9).

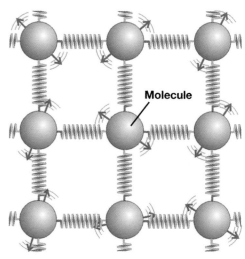

Figure 10.8: *Molecules in a solid are connected by bonds that act like springs.*

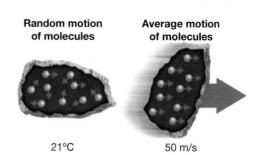

Figure 10.9: *A collection of molecules can have both average motion and random motion. That is why a rock has both a velocity and a temperature.*

Thermometers

Thermometers If you touch an object, you can sense whether it is hot or cold, but you cannot tell the exact temperature. A **thermometer** is an instrument that measures the exact temperature. The most common thermometers contain either a red fluid, which is alcohol containing a small amount of red dye, or a silvery fluid, which is mercury. You may have also used a thermometer with a digital readout.

Using a liquid to sense the temperature Thermometers can detect the physical changes in materials due to change in temperature. Different types of thermometers measure different physical changes. In a thermometer that uses a liquid to sense temperature, the expansion of the liquid is directly proportional to increase in temperature. As the temperature increases, the liquid expands and rises up a long, thin tube (Figure 10.10). You tell the temperature by the height the liquid rises. The tube is long and thin, so a small change in volume makes a large change in the height.

Digital thermometers Another physical property that changes with temperature is electrical resistance. The resistance of a metal wire will increase with temperature. Because the metal is hotter, and the metal atoms are shaking more, there is more resistance to electrons passing through the wire. A *thermistor* is a device that changes its electrical resistance as the temperature changes. Some digital thermometers sense temperature by measuring the resistance of a thermistor.

Liquid-crystal thermometers Some thermometers, often used on the outside of aquariums, contain liquid crystals that change color based on temperature. As temperature increases, the molecules of the liquid crystal bump into each other more and more. This causes a change in the structure of the crystals, which in turn affects their color. These thermometers are able to accurately determine the temperature between 65°F and 85°F.

Liquid-crystal thermometer

How a thermometer works

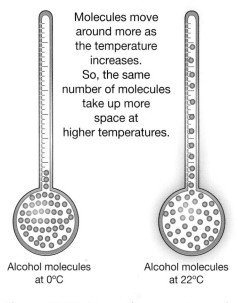

Molecules move around more as the temperature increases. So, the same number of molecules take up more space at higher temperatures.

Alcohol molecules at 0°C

Alcohol molecules at 22°C

Figure 10.10: *How a thermometer works.*

Absolute zero and the Kelvin temperature scale

Absolute zero   There is a limit to how cold matter can get. As the temperature is reduced, molecules move more and more slowly. When the temperature gets down to **absolute zero**, molecules have the lowest energy they can have and the temperature cannot get any lower. You can think of absolute zero as the temperature where molecules are completely frozen, with no motion. Technically, molecules never become absolutely motionless, but the kinetic energy is so small it might as well be zero. Absolute zero occurs at minus 273°C (−459°F).

You cannot have a temperature lower than absolute zero.

The Kelvin scale   A temperature in Celsius measures only *relative* thermal energy, relative to zero Celsius. The **Kelvin scale** is useful in science because it starts at absolute zero. A temperature in Kelvins measures the actual energy of molecules relative to zero energy.

Converting to Kelvin   The Kelvin (K) unit of temperature is the same size as the Celsius degree. However, water freezes at 273K and boils at 373K. Most of the outer planets and moons have temperatures closer to absolute zero than to the freezing point of water (Figure 10.11). To convert from Celsius to Kelvins, you add 273 to the temperature in Celsius. For example, a temperature of 21°C is equal to 294K (21 + 273).

High temperatures   While absolute zero is the lower limit for temperature, there is no practical upper limit. Temperature can go up almost indefinitely. As the temperature increases, exotic forms of matter appear. For example, at 10,000°C, atoms start to come apart and become a *plasma*. In a plasma, atoms are broken apart into separate positive ions and negative electrons. Plasma conducts electricity and is formed in lightning and inside stars. You'll read more about plasma in the next section.

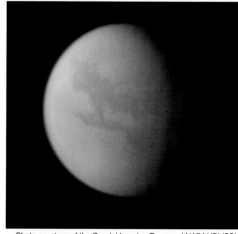

Photo courtesy of the Cassini Imaging Team and NASA/JPL/SSI

Figure 10.11: *The average surface temperature of Saturn's largest moon, Titan, is 93K.*

Section 10.2 *Review*

1. People in the United States know that water boils at 212°F. In Europe, people know that water boils at 100°C. Is the water in the United States different from the water in Europe? What explains the two different temperatures?

2. A comfortable room temperature is 20°C. What is this temperature in degrees Fahrenheit?

3. Which is colder, 0°C or 20°F?

4. Explain the scientific meaning of the word *random*.

5. Temperature measures:
   a. the kinetic energy of the random motion of molecules in an object.
   b. the kinetic energy of the average motion of molecules in an object.
   c. the potential energy of an object.
   d. the motion of an object.

6. A thermometer that uses a liquid to measure temperature works because:
   a. the electrical resistance in the liquid changes with temperature.
   b. the liquid changes color as temperature changes.
   c. the expansion of the liquid is directly proportional to increase in temperature.

7. Which statement best describes the relationship between temperature and thermal energy?
   a. Temperature is inversely related to thermal energy.
   b. Temperature is directly proportional to thermal energy.
   c. As temperature goes up, thermal energy goes down.
   d. Thermal energy is not related to temperature.

8. Would thermal energy be greater at 0°C or 48°F? Explain your answer.

9. Why can't there be a temperature lower than absolute zero?

SOLVE IT!

What is absolute zero in degrees Fahrenheit?

10.3 The Phases of Matter

You will notice that on a hot day, a glass of iced tea (or any cold beverage) has liquid water on the outside (Figure 10.12). The water does not come from inside the glass. The ice (the *solid* form of water) and cold liquid inside make the outside of the glass cold, too. This "outside" cold temperature causes water vapor in the air—a *gas*—to condense into *liquid* water on the exterior of the glass. What is happening at the level of atoms and molecules? Why can water take the form of solid, liquid, or gas?

Solid, liquid, and gas

Phases of matter On Earth, pure substances are usually found as solids, liquids, or gases. These are called *phases of matter*. Another phase of matter called *plasma* is discussed later in the section.

Solids A **solid** holds its shape and does not flow. The molecules in a solid vibrate in place but, on average, don't move far from their places.

Liquids A **liquid** holds its *volume* but does not hold its shape—it flows. The molecules in a liquid are about as close together as they are in a solid. But they have enough energy to change positions with their neighbors. Liquids flow because the molecules can move around.

Gases A **gas** flows like a liquid, but can also expand or contract to fill a container. A gas does not hold its volume. The molecules in a gas have enough energy to completely break away from each other and are much farther apart than molecules in a liquid or a solid.

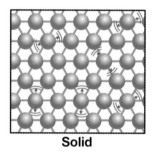

Solid

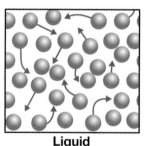

Liquid

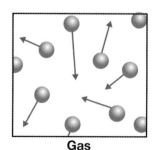

Water vapor in air (gas)

Liquid water

Solid water

Figure 10.12: *Why can water take the form of solid, liquid, and gas?*

Gas

Intermolecular forces

Intermolecular forces When they are close together, molecules are attracted through **intermolecular forces**. These forces are not as strong as the chemical bonds between atoms, but are strong enough to attach neighboring molecules to each other. Intermolecular forces have different strengths in different elements and compounds. Iron is a solid at room temperature. Water is a liquid at room temperature. This tells you that the intermolecular forces between iron atoms are stronger than those between water molecules.

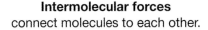

Chemical bonds connect atoms in a compound.

Intermolecular forces connect molecules to each other.

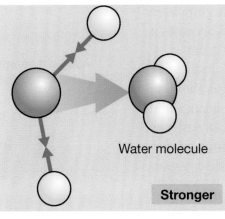

Water molecule

Stronger

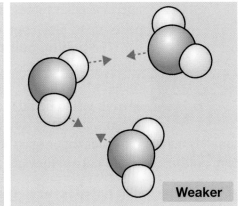

Weaker

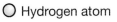 Hydrogen atom 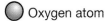 Oxygen atom

Temperature vs. intermolecular forces Within all matter, there is a constant competition between temperature and intermolecular forces. The kinetic energy from temperature tends to push atoms and molecules apart. When temperature wins the competition, molecules break away from each other and you have a gas. Intermolecular forces tend to bring molecules together. When intermolecular forces win the competition, molecules clump tightly together and you have a solid. Liquid is somewhere in the middle. Molecules in a liquid are not stuck firmly together, nor can they escape and break away from each other (Figure 10.13).

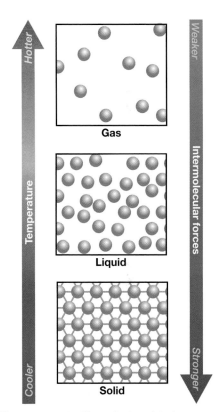

Figure 10.13: *The relationship between temperature, intermolecular forces, and phase of matter.*

10.3 The Phases of Matter **241**

Changing phase

Melting and freezing
The **melting point** is the temperature at which a substance changes from solid to liquid (melting) or from liquid to solid (freezing). Different substances have different melting points because intermolecular forces vary. When these forces are strong, it takes more energy to separate molecules from each other. Water melts at 0°C. Iron melts at a much higher temperature, about 1,500°C. The difference in melting points tells us the intermolecular forces between iron atoms are stronger than between water molecules.

Boiling and condensing

When enough energy is added, the intermolecular forces are completely pulled apart and a liquid becomes a gas. The **boiling point** is the temperature at which a substance changes from liquid to gas (boiling) or from gas to liquid (condensation). When water boils, you can easily see the change within the liquid as bubbles of water vapor (gas) form and rise to the surface. The bubbles in boiling water are not air, they are water vapor.

Changes in phase require energy
It takes energy to break the intermolecular forces between particles. This explains a peculiar thing that happens when you heat an ice cube. As you add heat energy, the temperature increases. Once it reaches 0°C, *the temperature stops increasing* as ice starts to melt and form liquid water (Figure 10.14). As you add more heat energy, more ice becomes liquid but the temperature stays the same. This is because the energy you are adding is being used to break the intermolecular forces and change solid into liquid. Once all the ice has become liquid, the temperature starts to rise again as more energy is added. Figure 10.14 shows the temperature change in an experiment. When heat energy is added or subtracted from matter, either the temperature changes or the phase changes, but usually not both at the same time.

VOCABULARY

melting point – the temperature at which a substance changes from solid to liquid (melting) or liquid to solid (freezing)

boiling point – the temperature at which a substance changes from liquid to gas (boiling) or from gas to liquid (condensation)

Start with ice at –20°C

Add heat energy at a constant rate

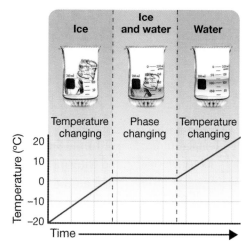

Figure 10.14: *Note how the temperature stays constant as the ice is melting into water.*

Melting and boiling points, sublimation, and plasmas

All substances can exist as a solid, liquid, or gas
On Earth, elements and compounds are usually found as solids, liquids, or gases. Each substance can exist in each of the three phases, and each substance has a characteristic temperature and pressure at which it will undergo a phase change. Figure 10.15 lists some examples.

Sublimation and deposition
Sometimes a solid can change directly to a gas with no liquid phase when heat energy is added. This process is called *sublimation*. Solid iodine is a substance that readily undergoes sublimation at room temperature. This is evident by the formation of a purple cloud above the crystals (Figure 10.16). A more common example is the shrinking of ice cubes (solid water) over time in the freezer. The opposite of sublimation is called *deposition*. One example of deposition is when water vapor changes directly into a solid—such as frost on a window on a cold winter night.

Plasma is a fourth phase of matter
At temperatures greater than 10,000°C, the atoms in a gas start to break apart. In the **plasma** phase, matter becomes ionized as electrons are broken loose from atoms. Because the electrons are free to move independently, plasma can conduct electricity. The Sun is made of plasma, as is most of the universe, including the Orion nebula (shown right).

Image courtesy of NASA, ESA, M. Robberto (STScI/ESA) and the Hubble Space Telescope Orion Treasury Project Team

Where else do you find plasma?
A type of plasma is used to make neon and fluorescent lights. Instead of heating the gases to an extremely high temperature, an electrical current is passed through them. The current strips the electrons off the atoms, producing plasma. You also see plasma every time you see lightning!

Substance	Melting point	Boiling point
Helium	–272°C	–269°C
Oxygen	–218°C	–183°C
Mercury	–39°C	357°C
Water	0°C	100°C
Lead	327°C	1,749°C
Aluminum	660°C	2,519°C

Figure 10.15: *Solid iodine readily undergoes sublimation at room temperature.*

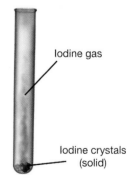

Iodine gas

Iodine crystals (solid)

Figure 10.16: *The melting and boiling points of some common substances.*

Summarizing the phases of matter

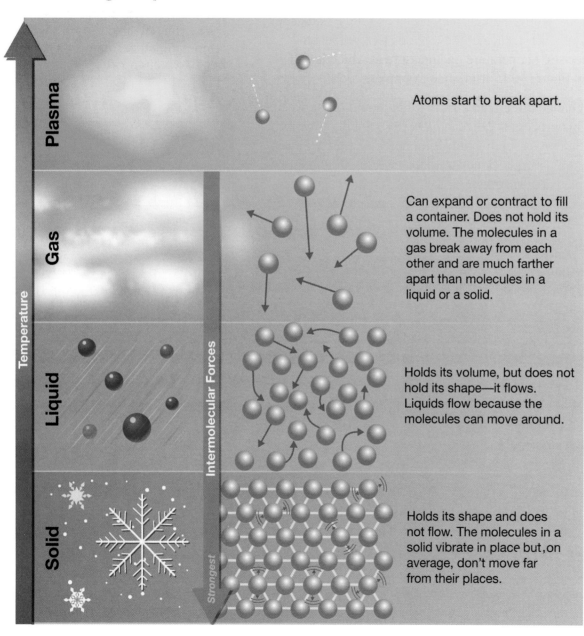

Atoms start to break apart.

Can expand or contract to fill a container. Does not hold its volume. The molecules in a gas break away from each other and are much farther apart than molecules in a liquid or a solid.

Holds its volume, but does not hold its shape—it flows. Liquids flow because the molecules can move around.

Holds its shape and does not flow. The molecules in a solid vibrate in place but, on average, don't move far from their places.

SCIENCE FACT

Evaporation

If you leave a pan of water in a room, eventually it dries out. Why does this happen? *Evaporation* occurs when molecules go from liquid to gas at temperatures below the boiling point. Remember, temperature measures the *average* random kinetic energy of molecules. Some molecules have energy above the average and some below the average. Some of the highest-energy molecules have enough energy to overcome the intermolecular forces with their neighbors and become a gas if they are near the surface of the liquid. Molecules with higher than average energy are the source of evaporation.

Evaporation takes energy away from a liquid. The molecules that escape are the ones with the most energy. The average energy of the molecules left behind is lowered. Evaporation cools the surface of a liquid because the fastest molecules escape and carry energy away. This is how your body cools off on a hot day. The evaporation of sweat from your skin cools your body.

Section 10.3 *Review*

1. Identify the phase or phases of matter (solids, liquids, and gases) that apply to each statement. More than one phase of matter may apply to each statement.

   a. Molecules do not move around, but vibrate in place.

   b. Has volume but no particular shape.

   c. Flows.

   d. Molecules break free of intermolecular forces.

   e. Does not have a volume or shape.

   f. Molecules can move around and switch places, but remain close together.

2. Explain why particles in a gas are free to move far away from each other.

3. Explain why liquids flow but solids do not.

4. Would you expect a substance to be a solid, liquid, or gas at absolute zero? Explain your answer.

5. Describe what happens, at the molecular level, during melting.

6. Describe what happens, at the molecular level, when a substance boils.

7. What is the most common phase of matter in the universe?

8. What is plasma? Where can you find plasma?

9. Matter has four phases that we experience. List the four phases in order of increasing temperature (lowest to highest).

10. Put the following terms in order from greatest intermolecular forces to weakest intermolecular forces: liquid, gas, solid.

11. Which would you expect to have stronger intermolecular forces:

    a. hydrogen, which exists as a gas at room temperature

    b. iron, which exists as a solid at room temperature

12. Identify the segment of the graph (A to B, B to C, C to D, D to E) in Figure 10.17 where a phase change is occurring. There could be more than one place. Explain your reasoning.

TECHNOLOGY

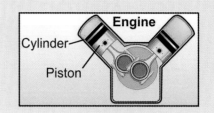

One of the ways to make car engines more efficient is to let them reach higher temperatures. Unfortunately, steel melts at about 1,500°C. Steel gets soft before it melts, so engines typically can't operate at temperatures even close to the melting point. Some new engine technologies use cylinders and pistons made of ceramic. Ceramic stays hard and strong at a much higher temperature than steel.

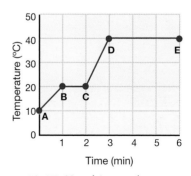

Figure 10.17: *Use this graph to answer question 12.*

The Fourth State of
Matter

The Sun is a giant ball of hot gases churned by magnetic fields and gravity. It produces and belches out enormous bubbles of plasma into space. What is plasma? Plasma is not a solid, a liquid, or a gas. It is a fourth state of matter, in which matter is heated to such a high temperature that the atoms begin to break apart.

The Sun is a plasma generator

The core of the Sun is subjected to intense pressure and heat because of its strong force of gravity. Gravity squeezes the matter at the core so tightly that atoms are ripped apart. Nuclear fusion occurs, releasing large amounts of energy. Gases, mostly hydrogen and helium, are then transformed into plasma, a super-heated gas in which electrons break loose from atoms. Astrophysicists believe that the fourth state of matter, plasma, makes up 99.9 percent of the universe.

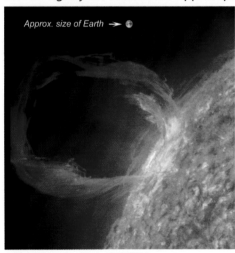

Approx. size of Earth ➜

Solar prominence   Photo: NASA/SDO

Where can we see plasma?

You have seen plasma on Earth in neon lights and in pictures of the Aurora Borealis, often called the Northern Lights. In the above photo from NASA, we can see a huge loop of plasma, called a solar prominence, extending from the Sun's surface. Scientists are still researching how solar prominences form. We do know from observation that they can form in a day and last for months. Extending into space for hundreds of thousands of miles, they are held in place by the Sun's magnetic fields. Most, however, will erupt at some point during their lifetime, sending glowing red loops of plasma into space.

Monitoring space weather

The Sun's outermost layer of its atmosphere, called the Corona, is about a million degrees Celsius, which is hot enough to host plasma. The Corona acts like a pot of boiling water by sending bubbles of plasma into space. The plasma blown off the Corona is known as solar wind. The solar wind travels at about 400 km per second or 1 million miles per hour. Solar wind travels to Earth and beyond. Made up of plasma guided by magnetic fields, solar wind pushes a million tons of matter into space every second.

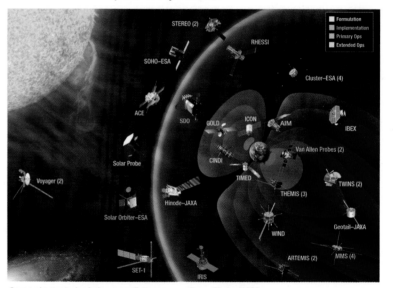

Space observatories help us predict space weather.   Photo: NASA

When the ejections of plasma and magnetic fields from the Corona are larger and more explosive, they are called coronal mass ejections. The largest coronal mass ejections contain billions of tons of matter travelling at several million miles per hour. These violent explosions can affect many things on Earth.

How does this plasma spewing from the Sun impact us? Solar winds and coronal mass ejections can interfere with technological systems in both space and here on Earth. Strong electrical currents from large solar weather events can affect GPS systems, the power grid, and high-frequency radio communications on Earth, as well as satellites and astronauts in space. NASA has a fleet of space observatories to study and predict space weather.

Aurora Borealis

Here on Earth, we can see some dramatic displays of light created by the solar winds and coronal mass ejections from the Sun. An aurora is a natural display of beautiful ribbons of colored light in the night sky that can be seen with the naked eye. In the Northern hemisphere, they are called Aurora Borealis, or the Northern Lights. In the Southern hemisphere, they are called Aurora Australis. They range from 50 to 200 miles above Earth's surface and can extend for thousands of miles across.

Auroras are caused by collisions between electrically charged particles, mostly electrons, from the solar wind, and with gas atoms in the part of Earth's atmosphere called the ionosphere. The violent collision gives the gas atoms energy and causes them to give off light. The color of light emitted depends on the type of atom being excited. Typical colors are blue, green, red, and purple.

Aurora Borealis in Alaska   Photo: NASA

Where can we see auroras? Earth's magnetic field is the key. Plasma can behave like a gas in some ways. It is affected by magnetic fields because it contains charged particles. Earth is like a giant magnet aligned along the geographic poles. The solar wind particles are guided by Earth's magnetic field toward the polar regions. So, the best place to see auroras are near Earth's magnetic poles: northern Greenland, the Scandinavian coast, Siberia, Alaska, and Antarctica. At the peak of the 11-year solar activity cycle, the auroras can sometimes be seen in places like Washington, D.C.; London; or Beijing.

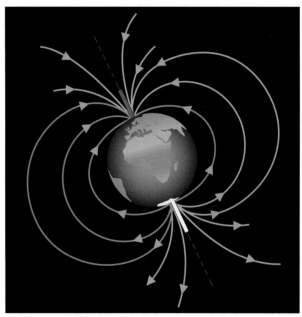

Earth's magnetic field lines

Questions:

1. What reaction occurring in the Sun creates plasma?

2. Why are we interested in forecasting space weather?

3. Where are you most likely to see an aurora? Why?

Chapter 10 *Assessment*

Vocabulary

Select the correct term to complete each sentence.

absolute zero	gas	mixture
atom	heterogeneous mixture	molecule
boiling point	homogeneous mixture	plasma
Celsius	intermolecular forces	pure substance
compound	Kelvin scale	solid
element	liquid	thermal energy
Fahrenheit	melting point	thermometer

Section 10.1

1. A pure substance that cannot be broken down into simpler substances by physical or chemical means is a(n) _____ .

2. The smallest particle of an element is a(n) _____ .

3. A(n) _____ is a substance that contains two or more elements that are chemically joined.

4. A(n) _____ is a group of two or more atoms joined together by chemical bonds.

5. A(n) _____ cannot be separated into other types of matter by physical means.

6. Matter that contains a combination of different elements and/or compounds and can be separated by physical means is called a(n) _____ .

7. A(n) _____ is a mixture that is the same throughout.

8. A(n) _____ is a mixture that is not the same throughout.

Section 10.2

9. _____ is a temperature scale in which water freezes at 32 degrees.

10. _____ is a scale in which water freezes at 0 degrees.

11. Energy due to temperature is called _____ .

12. You measure temperature with a(n) _____ .

13. The lowest possible temperature is called _____ .

14. The _____ is a temperature scale that starts with absolute zero.

Section 10.3

15. A(n) _____ holds its shape.

16. A(n) _____ does not hold its shape but has a volume.

17. A(n) _____ does not hold its shape and takes on the volume of its container.

18. The forces that determine the phase of matter are known as _____ .

19. The temperature at which a substance changes from solid to liquid is called _____ .

20. The temperature at which a substance changes from liquid to gas is called _____ .

21. _____ is a phase of matter in which some of the atoms begin to break apart.

Concepts

Section 10.1

1. What is Brownian motion? How does it provide evidence that matter is made of atoms and molecules?

2. Explain the differences between elements and compounds.

3. What are the two major categories of matter?

4. Name three foods that would be classified as heterogeneous mixtures, and three foods that are homogeneous mixtures.

5. Explain the difference between elements and compounds.

6. Explain the difference between an atom and a molecule.

Section 10.2

7. Compare the Celsius temperature scale with the Fahrenheit scale by answering the following questions:
   a. Which is the larger change in temperature, 1°C or 1°F?
   b. What are the freezing points and boiling points of water on each scale?
   c. Why are two different scales used?

8. How can the Fahrenheit and Celsius scales be converted from one to another?

9. Since it is fairly easy to tell when the temperature is high or low, why do we need thermometers, thermistors, and other devices for measuring temperature?

10. Compare the Celsius temperature scale with the Kelvin scale by answering the following questions:
   a. Which is the larger change in temperature, 1K or 1°C?
   b. What are the freezing points and boiling points of water on each scale?
   c. Why are two different scales used?

11. What is the difference between 0° on the Celsius scale and absolute zero?

12. Absolute zero is considered the lowest possible temperature. What is the highest possible temperature?

Section 10.3

13. A liquid takes the shape of its container, but why doesn't a liquid expand to fill the container completely?

14. Why doesn't a solid flow?

15. Name one similarity between gases and liquids.

16. Identify the phase represented by each diagram below and describe its basic properties.

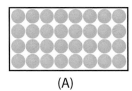 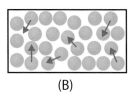 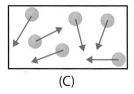

(A)　　　　　　　　(B)　　　　　　　　(C)

17. What is sublimation?

18. Explain how a liquid can enter the gas phase without reaching its boiling point.

19. Which has more thermal energy: gas, plasma, or liquid?

20. What is the most common phase of matter in the universe?

Problems

Section 10.1

1. Describe a method you would use to separate chicken soup into other forms of matter from which it is made.

2. Describe a method you would use to separate a mixture of sugar and water.

Section 10.2

3. Calculate the average human body temperature, 98.6°F, on the Celsius scale.

4. Convert –20°C to the Kelvin scale.

5. What is the Celsius equivalent of 100K?

6. A pizza box says to bake the pizza at 450°F, but your oven measures temperature in Celsius. At what temperature should you set the oven?

Section 10.3

7. The diagram to the right shows a graph of temperature vs. time for a material that starts as a solid. Heat is added at a constant rate. Using the diagram, answer the following questions:

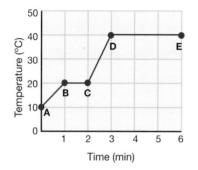

a. During which time interval does the solid melt?

b. During which time interval is the material all liquid?

c. What is the boiling point of the substance?

d. Does it take more heat energy to melt the solid or boil the liquid?

8. About 70 percent of the Earth's surface is covered by water. There is water underground, and even in the atmosphere. What is water's state at each of the following temperatures?

a. temperatures below 0 degrees Celsius

b. temperatures between 0 and 100 degrees Celsius

c. temperatures above 100 degrees Celsius

Applying Your Knowledge

Section 10.1

1. Identify each of the following in your classroom, school cafeteria, or home:

a. five homogeneous mixtures

b. five heterogeneous mixtures

c. three elements

d. five compounds

2. Design a poster to illustrate the classification of matter. Provide examples of everyday materials that belong in each category.

3. Air is a homogeneous mixture. Conduct research to find out the gases found in air and the percentage of each. Make a pie chart illustrating your findings.

Section 10.2

4. If you keep lowering the temperature of a material, the molecules vibrate less and less. If you could eventually reach a low enough temperature, the molecules might not vibrate at all. Is this possible, and what does it mean for the temperature scale? Is it possible to keep lowering the temperature indefinitely?

5. In the 1860s, English physicist James Clerk Maxwell (1831–1879) and Austrian physicist Ludwig Boltzmann (1844–1906) first gave a rigorous analysis of temperature in terms of the average kinetic energy of the molecules of a substance. Explore their lives and their contributions to the development of the theory of temperature.

Section 10.3

6. Design a poster or model to summarize for your classmates the differences between a solid, liquid, gas, and plasma.

7. Create a chart that illustrates the following phase changes: melting, boiling, freezing, evaporation, condensation, and sublimation.

8. Plasmas, or ionized gases as they are sometimes called, are of great interest both physically and technologically. Do some research to find out why plasmas are of great interest to scientists and manufacturers. Describe at least two current uses of plasmas, and describe one way scientists and engineers hope to use plasmas in the future.

There is a new kind of farm that is unlike any other. It doesn't produce food; it produces energy from wind. These farms can help solve the energy crisis by generating electricity from the powerful forces in wind. Most of Earth's energy comes from thermal radiation from the Sun, called solar radiation. A small fraction of that energy is used to drive Earth's winds. Huge turbines that collect wind energy are becoming a familiar sight, silhouetted against the skies. Wind power in Texas, for example, has more than quadrupled in recent years. Currently there are more than 10,000 wind turbines in Texas, most of them on land leased from farmers and ranchers.

Not that long ago, most farms in the United States had a windmill. It was used to pump water from a well. These days an electric motor pumps the water, and the old windmill is gone or just admired as an antique. New windmills, however, are going strong. Tower-mounted wind turbines that are far larger and more efficient have replaced the old models. When these big turbines are grouped, they form a wind farm. They are being built on land that is still used for farming. With support from industry and the government, wind farms are sprouting across the country. In this chapter, you will learn how winds are produced through solar radiation and the transfer of heat.

▌▌▌▌ CHAPTER 11 INVESTIGATIONS ▌▌▌▌

11A: Temperature and Heat
How are temperature and heat related?

11B: The Specific Heat of a Metal
How can you use specific heat to identify an unknown metal sample?

11.1 Heat and Thermal Energy

To change the temperature, you usually need to add or subtract energy. For example, when it's cold outside, you turn up the heat in your house or apartment and the temperature goes up. You know that adding heat increases the temperature, but have you ever thought about exactly what "heat" is? What does "heat" have to do with temperature?

Heat, temperature, and thermal energy

What is heat?

What makes chocolate melt in your hand?

What happens when you hold a chocolate bar in your hand? Thermal energy flows from your hand to the chocolate and it begins to melt. We call this flow of thermal energy **heat**. Heat is really just another word for thermal energy that is moving. In the scientific sense, heat flows any time there is a difference in temperature. Heat flows naturally from the warmer object (higher energy) to the cooler one (lower energy). In the case of the melting chocolate bar, the thermal energy lost by your hand is equal to the thermal energy gained by the chocolate bar.

> ### *Heat flows naturally from the warmer object (higher energy) to the cooler one (lower energy).*

Thermal energy depends on mass and temperature

Heat and temperature are related, but they are not the same thing. The amount of thermal energy depends on the temperature, but it also depends on the *amount* of matter you have. Think about heating up two pots of water. One pot contains 1,000 grams of water and the other contains 2,000 grams of water. Both pots are heated to the same final temperature (Figure 11.1). Which takes more energy? Or do both require the same amount of energy? The pot holding 2,000 grams of water takes twice as much energy as the pot with 1,000 grams, even though both start and finish at the same temperature. The two pots illustrate the difference between temperature and thermal energy. The pot of water with more mass has more energy, even though both are at the same temperature.

Both pots of water boil at 100°C

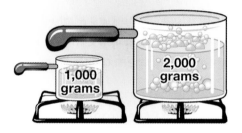

Figure 11.1: *It takes twice as much energy to heat a 2,000-gram mass of water compared to a 1,000-gram mass.*

Units of heat and thermal energy

Unit	Is equal to
1 calorie	4.186 joules
1 kilocalorie	1,000 calories
1 Btu	1,055 joules
1 Btu	252 calories

Figure 11.2: *Conversion table for units of heat.*

The joule   The metric unit for measuring heat is the *joule*. This is the same joule used to measure all forms of energy, not just heat. A joule is a small amount of heat. The average hair dryer puts out 1,200 joules of heat every second!

The calorie   One *calorie* is the amount of energy (heat) needed to increase the temperature of 1 gram of water by 1 degree Celsius. One calorie is a little more than 4 joules (Figure 11.2). You may have noticed that most food packages list "Calories per serving." The unit used for measuring the energy content of the food we eat is the *kilocalorie*, which equals 1,000 calories. The kilocalorie is often written as Calorie (with a capital *c*). If a candy bar contains 210 Calories, it contains 210,000 calories, or 879,060 joules!

The British thermal unit   Still another unit of heat energy you may encounter is the *British thermal unit*, or Btu. The Btu is often used to measure the heat produced by heating systems or heat removed by air-conditioning systems. A Btu is the quantity of heat it takes to increase the temperature of 1 pound of water by 1 degree Fahrenheit. One Btu is a little more than 1,000 joules.

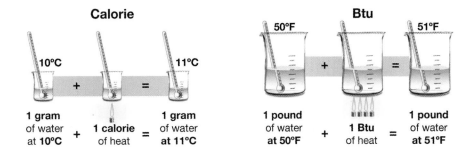

Why so many units?   The calorie and the Btu were being used to measure heat well before scientists knew that heat was really energy. The calorie and Btu are still used, even 100 years after heat was shown to be energy, because people give up familiar ways very slowly!

TECHNOLOGY

Heat and work

Work can be done whenever heat flows from a higher temperature to a lower temperature. Since heat flows from hot to cold, to get output work you need to maintain a temperature difference. Many inventions use heat to do work. The engine in your car uses the heat released by the burning of gasoline. In a car engine, the high temperature is inside the engine and comes from the burning gasoline. The low temperature is the air around the car. The output work produced by the engine is extracted from the flow of heat. Only a fraction of the heat is converted to work, and that is why a running car gives off so much heat through the radiator and exhaust.

Specific heat

Temperature, mass, and material

If you add heat to an object, how much will its temperature increase? It depends in part on the mass of the object. If you double the mass of the object, you need twice as much energy to get the same increase in temperature. The temperature increase also depends on what material you are heating up. It takes different amounts of energy to raise the temperature of different materials.

The temperature increase of an object depends on its mass and the material from which it is made.

Temperature and type of material

You need to add 4,184 joules of heat to one kilogram of water to raise the temperature by 1°C (Figure 11.3). You only need to add 470 joules to raise the temperature of one kilogram of steel by 1°C. It takes nine times more energy to raise the temperature of water by 1°C than it does to raise the temperature of the same mass of steel by 1°C.

Specific heat

Specific heat is a property of a material that tells us how much heat is needed to raise the temperature of one kilogram by one degree Celsius. Specific heat is measured in joules per kilogram per degree Celsius (J/kg°C). A large specific heat means you have to put in a lot of energy for each degree of increase in temperature.

Uses for specific heat

Knowing the specific heat tells you how quickly the temperature of a material will change as it gains or loses energy. If the specific heat is *low* (like steel), then temperature will change relatively quickly because each degree of temperature change takes less energy. If the specific heat is *high* (like water), then the temperature will change relatively slowly because each degree of temperature change takes more energy. Hot apple pie filling stays hot for a long time because it is mostly water, and therefore has a large specific heat. Pie crust has a much lower specific heat and cools much more rapidly. The table in Figure 11.4 lists the specific heat for some common materials.

VOCABULARY

specific heat – the amount of heat needed to raise the temperature of one kilogram of a material by 1 degree Celsius

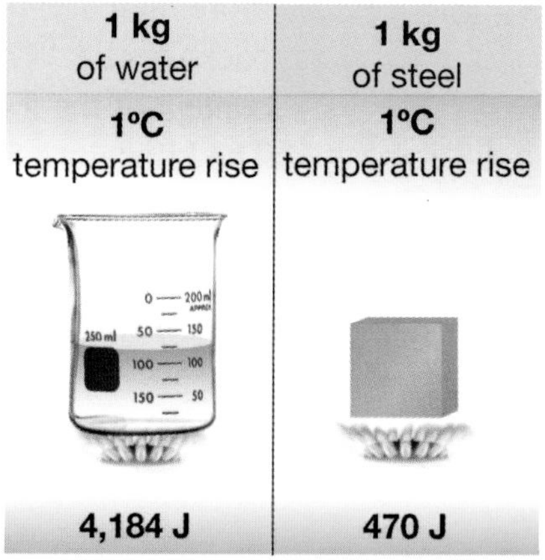

Figure 11.3: *Water and steel have different specific heats.*

Material	Specific heat (J/kg°C)
Water	4,184
Wood	1,800
Aluminum	900
Concrete	880
Glass	800
Steel	470

Figure 11.4: *Specific heat values of some common materials.*

Why is specific heat different for different materials?

Why specific heat varies In general, materials made up of heavy particles (atoms or molecules) have low specific heat compared with materials made up of lighter ones. This is because temperature measures the average kinetic energy *per particle*. Heavy particles mean fewer per kilogram. Energy that is divided between fewer particles means more energy per particle, and therefore more temperature change.

1 kilogram

Silver
Specific heat: 235 J/kg°C
Heavier atoms mean fewer atoms per kilogram

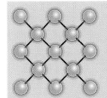

- Energy is spread over **fewer** atoms
- **More** energy per atom
- **Higher** temperature gain per joule (lower specific heat)

1 kilogram

Aluminum
Specific heat: 900 J/kg°C
Lighter atoms mean more atoms per kilogram

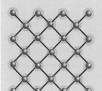

- Energy is spread over **more** atoms
- **Less** energy per atom
- **Lower** temperature gain per joule (higher specific heat)

An example: silver and aluminum Suppose you add 4,000 joules of energy to a kilogram of silver and 4,000 joules to a kilogram of aluminum. Silver's specific heat is 235 J/kg°C, and 4,000 joules is enough to raise the temperature of the silver by 17°C. Aluminum's specific heat is 900 J/kg°C, and 4,000 joules only raises the temperature of the aluminum by 4.4°C. The silver has fewer atoms than the aluminum because silver atoms are heavier than aluminum atoms. When energy is added, each atom of silver gets more energy than each atom of aluminum because there are fewer silver atoms in a kilogram. Because the energy per atom is greater, the temperature increase in the silver is also greater.

Calculating energy changes from heat

How could you figure out how much energy it would take to heat a swimming pool or boil a pot of water? The heat equation below tells you how much energy (E) it takes to change the temperature (T) of a mass (m) of a substance with a specific heat value (C_p). Figure 11.5 shows the specific heat values for some common materials.

HEAT EQUATION

Specific heat ($\frac{J}{kg°C}$)

Heat energy (J) — $E = mC_p\,(T_2 - T_1)$

Mass (kg)   Change in temperature (°C)

Material	Specific heat (J/kg°C)
Water	4,184
Wood	1,800
Aluminum	900
Concrete	880
Glass	800
Steel	470
Silver	235
Gold	129

Figure 11.5: *Use these specific heat values to solve the problems on this page.*

 **Solving Problems: Heat Equation**

How much heat is needed to raise the temperature of a 250-liter hot tub from 20°C to 40°C? (*Hint*: 1 liter of water has a mass of 1 kilogram.)

1. **Looking for:** You are looking for the amount of heat energy needed in joules.

2. **Given:** You are given the volume in liters, temperature change in °C, and specific heat of water in J/kg°C. You are also given a conversion factor for volume to mass of water.

3. **Relationships:** $E = mC_p(T_2 - T_1)$

4. **Solution:** $E = (250 \text{ L} \times 1 \text{ kg/L}) \times 4{,}184 \text{ J/kg°C } (40°C - 20°C) = 20{,}920{,}000 \text{ J}$

Your turn...

a. How much heat energy is needed to raise the temperature of 2.0 kg of concrete from 10°C to 30°C?

b. How much heat energy is needed to raise the temperature of 5.0 g of gold from 20°C to 200°C?

Solve First/Look Later

a. 35,200 J

b. 116.1 J

Section 11.1 *Review*

1. When you hold a piece of chocolate in your hand, why does the chocolate melt?

2. Which is a larger unit of heat: calorie, kilocalorie, Btu, or joule?

3. Which of the following would require more energy to heat it from 10°C to 20°C?

   a. 200 kg of water
   b. 200 kg of aluminum
   c. 100 kg of steel

4. What is the difference between temperature and heat?

5. What conditions are necessary for heat to flow?

6. How much heat energy is required to raise the temperature of 20 kilograms of water from 0°C to 35°C?

7. The temperature increase of an object depends on:

   a. its mass
   b. its velocity
   c. the material from which it is made
   d. answers a and c
   e. none of the above

8. On a night at the beach, which would you expect to cool faster: the ocean water or the beach sand? Explain your answer.

9. Why is the high specific heat of water important to our planet?

10. Which material would have a higher specific heat?

    a. a material made of heavier particles
    b. a material made of lighter particles
    c. the mass of the particles does not affect specific heat

CHALLENGE

A fast-food hamburger contains 870 kilocalories. Calculate the quantity of energy in calories, Btus, and joules.

STUDY SKILLS

You have learned many terms associated with heat and temperature. It is important to be able to distinguish the meanings of each term. Make a set of flash cards with the terms below. Write the term on one side and the definition on the other. Also use the term in a sentence. Write the sentence under the definition.

Temperature

Thermometer

Heat

Thermal energy

Specific heat

11.2 **Heat Transfer**

Thermal energy flows from higher temperature to lower temperature. This process is called **heat transfer**. How is heat transferred from material to material, or from place to place? It turns out there are three ways heat flows: *heat conduction*, *convection*, and *thermal radiation*.

Heat conduction

What is conduction?   **Heat conduction** is the transfer of heat by the direct contact of particles of matter. If you have ever held a warm mug of hot cocoa, you have experienced heat conduction. Heat is transferred from the mug to your hand. Heat conduction only occurs between two materials at different temperatures and when they are touching each other. In conduction, heat can also be transferred *through* materials. If you stir hot cocoa with a metal spoon, heat is transferred *from* the cocoa, *through* the spoon, and *to* your hand.

> ### *Heat conduction is the transfer of heat by the direct contact of particles of matter.*

How does conduction work?   Imagine placing a cold spoon into a mug of hot cocoa (Figure 11.6). The molecules in the cocoa have a higher average kinetic energy than those of the spoon. The molecules in the spoon exchange energy with the molecules in the cocoa through collisions. The molecules within the spoon itself spread the energy up the stem of the spoon through the intermolecular forces between them. Heat conduction works both through collisions and also through intermolecular forces between molecules.

Thermal equilibrium   As collisions continue, the molecules of the hotter material (the cocoa) lose energy and the molecules of the cooler material (the spoon) gain energy. The kinetic energy of the hotter material is transferred, one collision at a time, to the cooler material. Eventually, both materials are at the same temperature. When this happens, they are in **thermal equilibrium**. Thermal equilibrium occurs when two objects have the same temperature. No heat flows in thermal equilibrium because the temperatures are the same.

Cold spoon

Hot cocoa

➡ **Flow of heat energy**

Figure 11.6: *Heat flows by conduction from the hot cocoa into, and up, the spoon.*

Thermal conductors and insulators

Which state of matter conducts best? Heat conduction can happen in solids, liquids, and gases. Solids make the best conductors of heat because their particles are packed closely together. Because the particles in a gas are spread so far apart, relatively few collisions occur, making air a poor conductor of heat. This explains why many materials used to keep things warm, such as fiberglass insulation and down jackets, contain air pockets (Figure 11.7).

Thermal conductors and insulators Materials that conduct heat easily are called *thermal conductors*, and those that conduct heat poorly are called *thermal insulators*. For example, metal is a thermal conductor, and a foam cup is a thermal insulator. The words *conductor* and *insulator* are also used to describe a material's ability to conduct electrical current. In general, good electrical conductors like silver, copper, gold, and aluminum are also good thermal conductors.

Figure 11.7: *Because air is a poor conductor of heat, a down jacket keeps you warm in the cold of winter.*

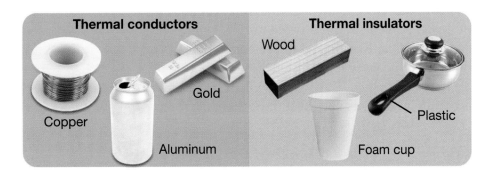

Heat conduction cannot occur through a vacuum Heat conduction happens only if there are particles available to collide with one another. Heat conduction does not occur in the vacuum of space. One way to create an excellent thermal insulator on Earth is to make a *vacuum*. A vacuum is empty of everything, including air. A thermos bottle keeps liquids hot for hours using a vacuum. A thermos is a container consisting of a bottle surrounded by a slightly larger bottle. Air molecules have been removed from the space between the bottles to create a vacuum (Figure 11.8).

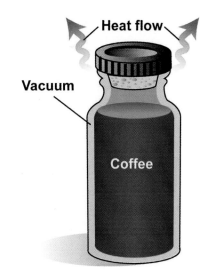

Figure 11.8: *A thermos bottle uses a vacuum to prevent heat transfer by conduction and convection.*

Convection

VOCABULARY

convection – the transfer of heat by the motion of matter, such as by moving air or water

What is convection? Have you ever watched water boil in a pot? Bubbles form on the bottom and rise to the top. Hot water near the bottom of the pan circulates up, forcing cooler water near the surface to sink. This circulation carries heat through the water (Figure 11.9). This heat transfer process is called **convection**. Convection is the transfer of heat through the motion of matter such as air and water.

Natural convection Fluids expand when they heat up. Because expansion increases the volume but not the mass, a warm fluid has a lower mass-to-volume ratio (called *density*) than the surrounding cooler fluid. In a container, warmer fluid floats to the top and cooler fluid sinks to the bottom. This is called *natural convection*.

Forced convection In many houses, a boiler heats water and then pumps circulate the water to rooms. Since the heat is being carried by a moving fluid, this is another example of convection. However, sicne the fluid is forced to flow by the pumps, this is called *forced convection*. Natural and forced convection often occur at the same time. Forced convection transfers heat to a hot radiator. The heat from the hot radiator then warms the room air by natural convection. Convection is mainly what distributes heat throughout the room.

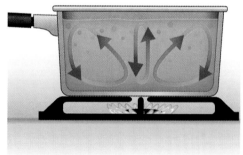

Figure 11.9: *Convection currents in water. The hot water at the bottom of the pot rises to the top and replaces the cold water.*

Thermal radiation

Definition of thermal radiation
If you stand in a sunny area on a cold, calm day, you will feel warmth from the Sun. Heat from the Sun is transferred to Earth by thermal radiation. **Thermal radiation** is electromagnetic waves (including light) produced by objects because of their temperature. All objects with a temperature above absolute zero (–273°C or –459°F) emit thermal radiation. To *emit* means to give off.

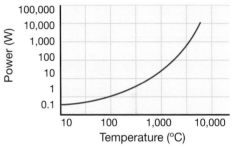

Thermal radiation is heat transfer by electromagnetic waves, including light.

Thermal radiation comes from atoms
Thermal radiation comes from the thermal energy of atoms. The power in thermal radiation increases with higher temperatures because the thermal energy of atoms increases with temperature (Figure 11.10). Because the Sun is extremely hot, its atoms emit lots of thermal radiation. Unlike conduction or convection, thermal radiation can travel through the vacuum of space. *All the energy the Earth receives from the Sun comes from thermal radiation.*

Figure 11.10: *The higher the temperature of an object, the more thermal radiation it emits.*

Objects emit and absorb radiation
Thermal radiation is also *absorbed* by objects. Otherwise all objects would eventually cool down to absolute zero by radiating their energy away. The temperature of an object rises if more radiation is absorbed. The temperature falls if more radiation is emitted. The temperature adjusts until there is a balance between radiation absorbed and radiation emitted.

Some surfaces absorb more energy than others
The amount of thermal radiation absorbed depends on the surface of a material. Black surfaces absorb almost all the thermal radiation that falls on them. For example, black asphalt pavement gets very hot in the summer sun because it effectively absorbs thermal radiation. A silver mirror surface reflects most thermal radiation, absorbing very little (Figure 11.11). You may have seen someone put a silver screen across their windshield after parking their car on a sunny day. This silver screen can reflect the Sun's heat back out the car window, helping the parked car stay cooler on a hot day.

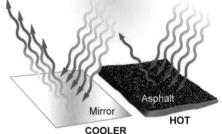

Figure 11.11: *Dark surfaces absorb most of the thermal radiation they receive. Silver or mirrored surfaces reflect thermal radiation.*

VOCABULARY

thermal radiation – electromagnetic waves produced by objects because of their temperature

Heat transfer, winds, and currents

Thermals are small convection currents in the atmosphere
Have you ever seen a hawk soaring above a highway and wondered how it could fly upward without flapping its wings? The hawk is riding a *thermal*—a convection current in the atmosphere (Figure 11.12). A thermal forms when a surface like a blacktop highway absorbs solar radiation and emits energy as heat. That heat warms the air near the surface. The warmed air molecules gain kinetic energy and spread out. As a result, the heated air near the highway becomes less dense than the colder air above it. The heated air rises, forcing the colder air to move aside and sink toward the ground. Then this colder air is warmed by the heat from the blacktop, and it rises. A convection current is created.

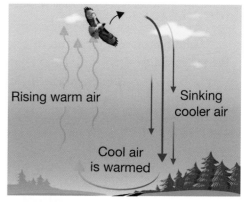

Figure 11.12: *Hawks ride convection currents called thermals.*

Giant convection currents
There are also giant convection currents in the atmosphere. These form as a result of the temperature difference between the equator and the poles. Warm air at the equator tends to rise and flows toward the poles. Cooler, denser air from the poles sinks and flows back toward the equator. When air flows horizontally from an area of high density and pressure into an area of low density and pressure, we call the flowing air *wind*.

Global wind cells
While it might seem logical that air would flow in giant circles from the equator to the poles and back, the reality is more complicated than that. The warm air from the equator doesn't make it all the way to the poles because of Earth's rotation. In fact, the combination of global convection and Earth's rotation sets up a series of wind patterns called *global wind cells* in each hemisphere (Figure 11.13). These cells play a large role in shaping weather patterns on Earth.

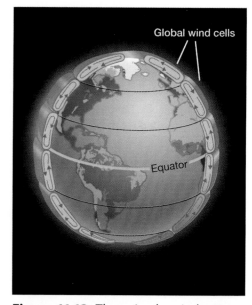

Figure 11.13: *These circular wind patterns exist in both the northern and southern hemispheres. We call them global wind cells.*

Ocean currents
The global wind patterns and Earth's rotation cause surface ocean currents to move in large circular patterns. Ocean currents can also occur deep in the ocean. These currents move slower than surface currents and are driven by temperature and density differences in the ocean. Surface and deep currents work together to move huge masses of water around the globe. Ocean currents play a big role in heating and cooling some parts of Earth.

Section 11.2 *Review*

1. What is thermal equilibrium?

2. Which state of matter—solid, liquid, or gas—is the best at conducting heat? Why?

3. Cooking pots are made of metal, but often the handle of a cooking pot is made of a type of plastic or rubber. Explain why this design makes sense.

4. A down jacket keeps your body warm mostly by reducing which form of heat transfer?
    a. conduction
    b. convection
    c. thermal radiation

5. What is the advantage of designing a thermos so that it has a vacuum layer surrounding the area where hot liquids are stored?

6. What is the difference between forced and natural convection?

7. Examine the scene below. Explain what types of heat transfer are occurring in the scene and where each is occurring:

8. How does heat from the Sun get to Earth?
    a. conduction
    b. convection
    c. thermal radiation

9. Explain the roles of density and temperature in convection.

10. A sailor on a sailboat depends on the process of convection. Explain why this is so.

Needed:
Efficient Buildings

The purpose of a building is to provide shelter from the weather and a climate control system that keeps us comfortable. However, climate control takes energy. In the United States, buildings account for 39 percent of total energy use. Most of this energy comes from burning fossil fuels—a process that many scientists believe contributes to global warming, pollution, and health problems. The need for energy-efficient building design is crucial.

What makes a building green?

"Green" building design is the term used to describe architecture that is energy efficient and environmentally friendly. The LEED (Leadership in Energy and Environmental Design) Green Building Rating System is nationally accepted by federal agencies and state governments as a means of evaluating green buildings.

To earn an LEED rating, a building project must satisfy a list of requirements in five key areas. Buildings are rated on their indoor environmental quality, energy efficiency, water savings, materials selection, and sustainable site development. A project is awarded a Certified, Silver, Gold, or Platinum rating depending on the credits it has earned in these five categories.

The CBF Merrill Environmental Center: A Platinum rating

The Chesapeake Bay Foundation Merrill Environmental Center in Annapolis, Maryland, received the first Platinum rating— the highest LEED rating. It is recognized as a pioneer in the green building field, showing others it could be done without losing comfort or beauty.

View of the Chesapeake Bay Foundation Merrill Environmental Center from the bay

The CBF Merrill Environmental Center uses both passive and active solar design to reduce energy consumption. Passive solar design features include south-facing windows equipped with overhangs or trellises.

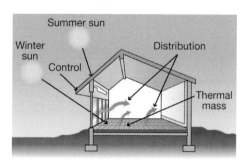
Elements of passive solar design

The windows capture the Sun's rays in the winter when the Sun is low in the sky, and the overhangs and trellises shade the building in the summer when the Sun is high in the sky. In addition, the CBF Center has a mechanical system to open windows and allow for natural ventilation during the year to keep the building cooler.

The center also uses active solar heating, which involves the use of a solar collector positioned on the roof. Photovoltaic panels produce a portion of the building's energy needs by converting the Sun's rays to electricity. Solar hot-water heating further reduces electricity demand.

Thermal mass: Material matters

When the school district of North Clackamas, Oregon, decided to focus on energy and resource efficiency, environmental responsibility, and high-quality indoor living, they chose a green building design for their new high school. Their energy study model shows that when the building is operated as the model outlines (using a nine-month school year and utilizing daylight and natural ventilation), the building is expected to achieve a 44 percent energy savings over conventional high schools of similar size.

The design team integrated a total site and building energy analysis in the design process. They established specific requirements for orientation; the building envelope; heating, ventilating, and air-conditioning (HVAC) systems; lighting; materials; and landscaping. The architects used extensive modeling of daylight and ventilation strategies. High school students assisted by building a full-scale mock-up of a classroom for physical testing.

Clackamas High School uses concrete slabs and concrete masonry walls as a thermal mass as well as a mechanically controlled system operating louvers and air stacks to control air flow by convection.
—Photo by Michael Mathers

Daylighting decreases the amount of required electric lighting, which in turn decreases the heat load and energy costs of electric lights. Windows, skylights, and light shelves provide natural light and great views of the outside.

Designers incorporated natural ventilation and cooling in all classrooms, common areas, and the gymnasium using mechanically controlled dampers, louvers, and air stacks. For instance, if the monitoring system senses that interior temperatures are getting too warm, cooler air from outside is allowed to enter the rooms through louver openings. Airflow is increased through ventilation stacks on the roof, so that heat moves up and out of the building by natural convection.

To help stabilize the building's internal temperature, the designers chose to use concrete slabs and concrete masonry walls. These dense materials have a high specific heat. During the day, thermal energy from the Sun streams through the large windows and is absorbed by the concrete. The concrete slabs and walls are known as a "thermal mass" because they can store a great deal of thermal energy without a significant rise in temperature. At night, the walls slowly release this thermal energy into the building, heating the air. As a result, the furnace doesn't have to run as often.

Not only does the school's design achieve significant energy savings, its light-filled spaces also provide an exciting atmosphere and great spaces for learning.

Greening your own home

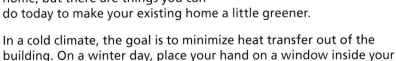

You may not be able to design and build your own brand-new green home, but there are things you can do today to make your existing home a little greener.

In a cold climate, the goal is to minimize heat transfer out of the building. On a winter day, place your hand on a window inside your home. If you have single-pane windows, the glass will feel very cold due to rapid heat conduction from your hand to the glass to the outside air. To minimize this heat transfer, you can purchase plastic wrap at a hardware store and tape or shrink-wrap it to the window frame, creating an insulating layer of air.

Have you ever felt a cold draft flowing under a door? Convection is to blame. Heated air from your radiators or vents is less dense than the surrounding air, so it rises. As the warm air moves upward, denser cold air from the outside rushes under your door to take its place. Seal a leaky door with weatherstripping to eliminate this problem.

Even in winter, you can maximize the benefit of the Sun's radiation in your home. Keep south-facing windows clean, and open curtains

or shades during daylight hours. Radiation passes easily through clean glass windows, and the energy is absorbed by interior surfaces. Just be sure to close the shades again at night!

In a warm climate, the goal is to prevent heat transfer into your home. Radiation passing through your windows is one of the main culprits here. White window shades will help reflect heat away from your house. Keep them closed on south- and west-facing windows during the day.

Understanding heat transfer can help you make simple changes that reduce your home's energy needs, your family's heating or cooling bills, and your impact on the environment.

Questions:

1. How are radiation, convection, and conduction used in green building design?

2. How does a thermal mass store energy?

3. What is one practical thing you can do to make your own home greener?

Source: Energy Efficiency and Renewable Energy Program, U.S. Department of Energy. CBF Merrill Environmental Center photos courtesy of The Chesapeake Bay Foundation/cbf.org

Chapter 11 *Assessment*

Vocabulary

Select the correct term to complete each sentence.

convection	heat transfer	thermal radiation
heat	specific heat	
conduction	thermal equilibrium	

Section 11.1

1. Thermal energy that is moving or capable of moving is called _____ .

2. The amount of heat needed to raise the temperature of one kilogram of a material by one degree Celsius is called its _____ .

Section 11.2

3. _____ is the flow of thermal energy from higher temperature to lower temperature.

4. Heat stops flowing when _____ is reached.

5. The transfer of heat by the direct contact of particles of matter is called _____ .

6. When heat is transferred by the motion of matter such as by moving air or water, it is called _____ .

7. Heat is transferred from the Sun to Earth by _____ .

Concepts

Section 11.1

1. Distinguish between heat and thermal energy.

2. When you hold a cold glass of water in your warm hand, which way does the heat flow?

3. Thermal energy depends on what two factors?

4. Name three units for measuring heat.

5. What is the relationship between the calorie used by scientists and the Calorie used by nutritionists?

6. Compare the size of a calorie to a joule.

7. Why does specific heat vary for different substances?

Section 11.2

8. Name the three methods by which heat can be transferred and give an example of each.

9. A metal cup containing water at 100°F is placed in an aquarium containing water at 80°F.
   a. Which way will heat flow? Why?
   b. When will the flow of heat stop?
   c. What is it called when heat no longer flows?

10. Why do thermos bottles keep cold beverages contained inside them from getting warm?

11. Name three good thermal insulators.

12. How do we know that we receive heat from the Sun by thermal radiation and not by conduction or convection?

13. Explain the difference between natural and forced convection. Give an example of each.

Problems

Section 11.1

1. How much energy, in joules, does it take to raise the temperature of 1.5 kg of aluminum from 20°C to 40°C?

2. Relative to 0°C, the amount of thermal energy in a quantity of water is its mass × specific heat × temperature. The specific heat of water is 4,184 J/kg°C.

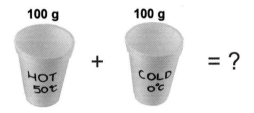

a. How much thermal energy is in 100 grams of water at 50°C?
b. How much thermal energy is in 100 grams of water at 0°C?
c. How much energy is there when both quantities of water are mixed together?
d. How much mass is this energy spread out over (in the mixture)?
e. What do you think the temperature of the mixture should be?

3. How much energy will it take to increase the temperature of 200 milliliters of water by 12°C? (*Hint*: 1L of water = 1 kg.)

4. Two beakers each contain 1 kilogram of water at 0°C. One kilogram of gold (specific heat = 129 J/kg°C) at 100°C is dropped into one beaker. One kilogram of aluminum (specific heat = 900 J/kg°C) at 100°C is dropped into the other beaker.

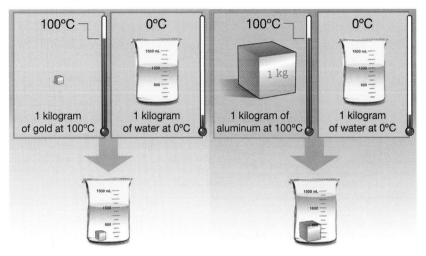

a. Compare the amount of thermal energy contained in the aluminum and gold.
b. After each beaker has reached thermal equilibrium, describe whether the temperatures are the same or different. If they are different, describe which is warmer and which is colder.
c. Explain your answer to part b. Use the concept of specific heat in your explanation.

Section 11.2

5. You pour some hot water into a metal cup. After a minute, you notice that the handle of the cup has become hot. Explain, using your knowledge of heat transfer, why the handle of the cup heats up. How would you design the cup so that the handle does not heat up?

6. A computer CPU chip creates heat because of the electric current it uses. The heat must be carried away, or the chip will melt. To keep the chip cool, a finned heat sink is used to transfer heat from the chip to the air. Which of the materials below would make the BEST heat sink (transfer the most heat)? Which would be the WORST material to use? *Note*: Thermal conductivity is a measure of a material's ability to conduct heat and is measured in units of watts per meter Kelvin.

Thermal conductivities of materials (W/m · K)		
Concrete = 1.7	Aluminum = 240	Asbestos = 0.1
Glass = 0.8	Copper = 400	Gold = 310
Wood = 0.1	Rubber = 0.2	Silver = 430

Applying Your Knowledge

Section 11.1

1. The first settlers in Colorado were very concerned about fruits and vegetables freezing in their root cellars overnight. They soon realized that if they placed a large tub of water in the cellar, the food would not freeze. Explain why the food would not freeze.

2. Scottish chemist Joseph Black developed the theories of specific and latent heat. Research his life and how he made these discoveries. Cite all sources using the required format.

Section 11.2

3. In an automobile, water and antifreeze are pumped through the engine block as a coolant. The mixture is pumped back to the radiator, where a fan blows air through the radiator. Explain, using conduction, convection, and radiation, how this system works to transfer heat from the engine to the air.

4. In a home aquarium, regulating the temperature of the water is critical for the survival of the fish. To keep a fish tank warm, a heating element with a thermostat is often placed on the bottom of the tank. Why is a heating element placed on the bottom of the tank instead of at the top?

5. A thermostat controls the switch on a furnace or air conditioner by sensing the temperature of the room. Explain, using conduction, convection, and radiation, where you would place the thermostat in your science classroom. Consider windows, inside and outside walls, and where the heating and cooling ducts are located. You can also sketch your answer—draw your classroom, showing room features and placement of the thermostat. (STEM)

6. Building materials such as plywood, insulation, and windows are rated with a number called the "R value." The R value has to do with the thermal conductivity of the material. Higher R values mean lower conductivity and better insulation properties. Design a window with a high R value. Sketch your window, and label its features and the materials it is made from. Explain the reasons for each of your design choices. (STEM)

Properties of Matter

Would you believe that someone has made a solid material that has about the same density as air? If someone put a chunk of it your hand, you might not even notice. Silica aerogel is a foam that's like solidified smoke. Aerogel is mostly air and has remarkable thermal, optical, and acoustical properties.

Aerogels are fantastic insulators. You could hold a flame under a chunk of the material and touch the top without being burned. Aerogels have the potential to replace a variety of materials in everyday life. If researchers could make a transparent version of an aerogel, it would almost certainly be used in double pane windows to keep heat inside your house in the winter and outside in the summer. Opaque aerogels are already being used as insulators. Aerogels have been put to use by NASA in several projects, including the Mars Pathfinder *Soujourner* Rover and the *Stardust* missions.

Read this chapter to find out more about matter and its properties.

CHAPTER 12 INVESTIGATIONS

12A: Mystery Material
How do solids and liquids differ?

12B: Buoyancy
Can you make a clay boat?

Photo courtesy of NASA

12.1 Properties of Solids

All matter is made up of tiny atoms and molecules. In a solid, the atoms or molecules are closely packed, and they stay in place. This is why solids hold their shape. In this section, you will learn how the properties of solids are a result of the behavior of atoms and molecules.

Matter has physical and chemical properties

Characteristics of matter Different kinds of matter have different characteristics. They melt and boil at a wide range of temperatures. They might be different colors or have different odors. Some can stretch without breaking, while others shatter easily. These and other properties help us distinguish one substance from another. These properties also help us choose which kind of material to use for a specific purpose.

Physical properties Characteristics that you can observe directly are called **physical properties**. Physical properties include color, texture, density, brittleness, and state (solid, liquid, or gas). Substances can often be identified by their physical properties. For example, water is a colorless, odorless substance that exists as a liquid at room temperature. Gold is shiny, exists as a solid at room temperature, and can be pounded into very thin sheets.

Physical changes A *physical change* is any change in the size, shape, or phase of matter in which the identity of a substance does not change. For example, when water is frozen, it changes from a liquid to a solid. This does not change the water into a new substance. It is still water, only in solid form, which we call ice. The change can easily be reversed by melting the solid water. Bending a steel bar is another example of a physical change. The bar is still steel, even after it is bent.

Chemical properties Properties that can only be observed when one substance changes into a different substance are called **chemical properties**. For example, if you leave an iron nail outside, it will eventually rust. A chemical property of iron is that it reacts with oxygen in the air to form iron oxide (rust). Any change that transforms one substance into a different substance is called a *chemical change* (Figure 12.1). Chemical changes are not easily reversible. Rusted iron will not turn shiny again, even if you remove it from oxygen in the air.

physical properties – characteristics that you can observe directly

chemical properties – characteristics that can only be observed when one substance changes into a different substance

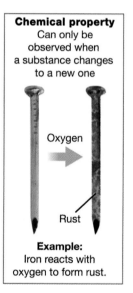

Physical property Can be observed directly	Chemical property Can only be observed when a substance changes to a new one
Example: Iron is a solid at room temperature.	**Example:** Iron reacts with oxygen to form rust.

Figure 12.1: *Physical and chemical properties of iron.*

Reviewing density

Reviewing the definition of density
In Chapter 2, you learned that density is the ratio of mass to volume. Physicists and engineers use units of kilograms per cubic meter (kg/m^3) for density. In classroom experiments, it is more convenient to use units of grams per cubic centimeter (g/cm^3). Earlier, you measured volume in milliliters (mL). One milliliter is exactly equal to one cubic centimeter (1 mL = 1 cm^3).

One milliliter (mL) is equal to one cubic centimeter (cm³).

Solids have a wide range of density
Solids have a wide range of densities. One of the densest metals is platinum, with a density of 21.5 g/cm^3. Platinum is twice as dense as lead and almost three times as dense as steel. A ring made of platinum has three times as much mass as a ring of the exact same size made of steel. Rock has a lower density than metals, between 2.2 and 2.7 g/cm^3. As you might expect, the density of wood is less than rock, ranging from 0.4 to 0.6 g/cm^3.

Density of liquids and gases
The density of water is 1.0 g/cm^3 and many common liquids have densities between 0.5 and 1.5 g/cm^3. The density of air and other gases is much lower. The air in your room has a density near 0.0001 g/cm^3. Gases have low density because the molecules in a gas are far away from each other.

Atoms have different masses
The density of a solid material depends on two things. One is the individual mass of each atom or molecule. The other is how closely the atoms or molecules are packed together. Solid lead is a very dense metal compared to solid aluminum. One atom of lead has 7.7 times more mass than one atom of aluminum. Solid lead is denser than solid aluminum mostly because a single lead atom has more mass than a single aluminum atom.

Atoms may be "packed" loosely or tightly
Density also depends on how tightly the atoms and molecules are "packed." Diamond is made of carbon atoms and has a density of 3.50 g/cm^3. The carbon atoms in diamond are relatively closely packed together. Paraffin wax is also mostly carbon, but the density of paraffin is only 0.87 g/cm^3. The density of paraffin is low because the carbon atoms are mixed with hydrogen atoms in long molecules that take up a lot of space (Figure 12.2).

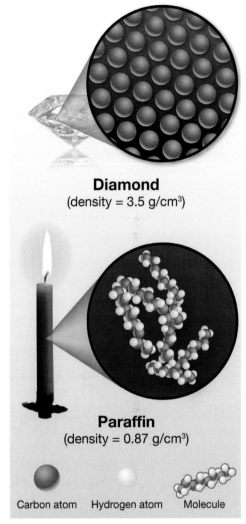

Diamond
(density = 3.5 g/cm^3)

Paraffin
(density = 0.87 g/cm^3)

Carbon atom   Hydrogen atom   Molecule

Figure 12.2: *The carbon atoms in diamond are packed relatively tightly, while the carbon atoms in paraffin are part of long molecules that take up a lot of space.*

The arrangement of atoms and molecules in solids

Crystalline and amorphous solids
The atoms or molecules in a solid are arranged in two ways. If the particles are arranged in an orderly, repeating pattern, the solid is called **crystalline**. Examples of crystalline solids include salts, minerals, and metals. If the particles are arranged in a random way, the solid is **amorphous**. Examples of amorphous solids include rubber, wax, and glass.

■■■■ VOCABULARY ■■■■

crystalline – an orderly, repeating arrangement of atoms or molecules in a solid

amorphous – a random arrangement of atoms or molecules in a solid

Crystalline solids

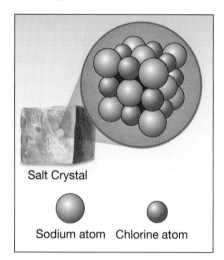

Salt Crystal

Sodium atom   Chlorine atom

Most solids on Earth are crystalline. Some materials, like salt, exist as single crystals, and you can see the arrangement of atoms reflected in the shape of the crystal. If you look at a crystal of table salt under a microscope, you'll see that it's cubic in shape. If you could examine the arrangement of atoms, you would see that the shape of the crystal comes from the cubic arrangement of sodium and chlorine atoms. Metals are also crystalline. They don't look like "crystals" because solid metal is made from very tiny crystals fused together in a jumble of different orientations (Figure 12.3).

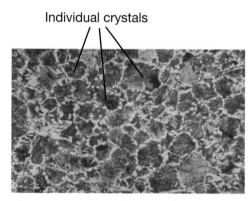

Individual crystals

Figure 12.3: *Metallic crystals in steel. Single crystals are very small. This image was taken with an electron microscope at very high magnification.*

Amorphous solids
The word *amorphous* comes from the Greek for "without shape." Unlike crystals, amorphous solids do not have a repetitive pattern in the arrangement of molecules or atoms. The atoms or molecules are randomly arranged. While amorphous solids also hold their shape, they are often softer and more elastic than crystalline solids. This is because a molecule in an amorphous solid is not tightly connected to as many neighboring molecules as it would be in a crystalline solid. Glass is a common amorphous solid. Glass is hard and brittle because it is made from molten silica crystals that are cooled quickly, before they have time to recrystallize. The rapid cooling leaves the silica molecules in a random arrangement. Plastic is another useful amorphous solid.

Mechanical properties of solids

The meaning of strength

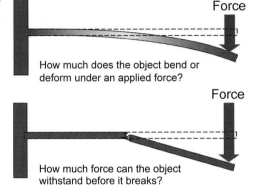

Force

How much does the object bend or deform under an applied force?

Force

How much force can the object withstand before it breaks?

When you apply a force to an object, the object may change its size, shape, or both. The concept of **strength** describes the ability of a solid object to maintain its shape even when force is applied. The strength of an object can be determined based on the answers to the two questions in the illustration at the left.

Tensile strength   **Tensile strength** is a measure of how much stress from pulling, or tension, a material can withstand before breaking (Figure 12.4). Strong materials like steel have high tensile strength. Weak materials like wax and rubber have low tensile strength. Brittle materials also have low tensile strength.

Hardness   **Hardness** measures a solid's resistance to scratching. Diamond is the hardest natural substance found on Earth. Geologists sometimes classify rocks based on hardness. Given six different kinds of rock, how could you line them up in order of increasing hardness?

Elasticity   If you pull on a rubber band, its shape changes. If you let it go, the rubber band returns to its original shape. Rubber bands can stretch many times their original length before breaking, a property called elasticity. **Elasticity** describes a solid's ability to be stretched and then return to its original size. This property also gives objects the ability to bounce and to withstand impact without breaking.

Brittleness   **Brittleness** is defined as the tendency of a solid to crack or break before stretching very much. Glass is a good example of a brittle material. You cannot stretch glass even one-tenth of a percent (0.001) before it breaks. To stretch or shape glass, you need to heat the glass until it is almost melted. Heating causes molecules to move faster, temporarily breaking the forces that hold them together.

Tensile Strength

Figure 12.4: *Tensile strength measures how much pulling or tension a material can withstand before breaking.*

Ductility   One of the most useful properties of metals is that they are ductile. A ductile material can be bent a relatively large amount without breaking. For example, a steel fork can be bent in half and the steel does not break. A plastic fork cracks when it is bent only a small amount. Steel's high **ductility** means steel can be formed into useful shapes by pulling, rolling, and bending. These processes would destroy a brittle material like glass. The ductility of many metals, like copper, allow them to be formed into wire, like the copper wire shown below.

What is malleability?   **Malleability** measures a solid's ability to be pounded into thin sheets. Aluminum is a highly malleable metal. Aluminum foil and beverage cans are two good examples of how manufacturers take advantage of the malleability of aluminum.

VOCABULARY

ductility – the ability to bend without breaking

malleability – the ability of a solid to be pounded into thin sheets

thermal expansion – the tendency of the atoms or molecules in a substance (solid, liquid, or gas) to take up more space as the temperature increases

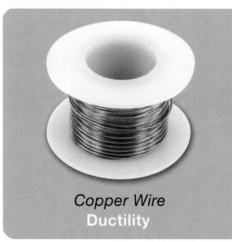

Copper Wire
Ductility

Aluminum Foil
Malleability

Expansion Joint

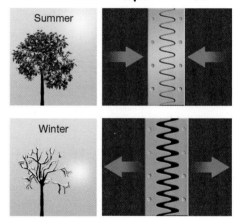

Summer

Winter

Figure 12.5: *Bridges have expansion joints to allow for thermal expansion of concrete.*

Thermal expansion   As the temperature increases, the kinetic energy in the vibration of atoms and molecules also increases. The increased vibration makes each particle take up a little more space, causing **thermal expansion**. Almost all solid materials expand as the temperature increases. Some materials (like plastic) expand a great deal. Other materials (like glass) expand only a little. All bridges longer than a certain size have special joints that allow the bridge surface to expand and contract with changes in temperature (Figure 12.5). The bridge surface would crack without these expansion joints.

Section 12.1 *Review*

1. Name one example of a physical change and one example of a chemical change.

2. Name one example of a material for each set of properties.
   a. high elasticity and high tensile strength
   b. amorphous and brittle
   c. crystalline and brittle
   d. amorphous and elastic
   e. ductile and crystalline

3. The strength of a material determines
   a. how dense the material is
   b. how much force it can withstand before breaking
   c. how good a thermal or electrical conductor it is

4. Latex is a soft, stretchy, rubber-like material. Would you expect latex to be crystalline or amorphous?

5. Explain, from an atomic-level perspective, why expansion joints are used in bridges.

6. Which property of a metal describes why it can be formed into wire?

7. When installing wood floors, it is often recommended that you leave a half-inch of space between the flooring and the wall (Figure 12.6). Why do you think this space would be recommended?

8. Aluminum can be made into foil because aluminum has high _____ .

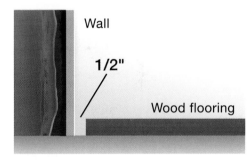

Figure 12.6: *Question 7.*

12.2 Properties of Fluids

A **fluid** is defined as any matter that flows when force is applied. Liquids like water are one kind of fluid. Gases, like air, are also fluids. You may notice cool air flowing into a room when a window is open, or the smell of someone's perfume drifting your way. These examples provide evidence that gases flow. What are some other properties of fluids?

Density of fluids

How could you find the density of *liquid* silver? A piece of pure silver in the shape of a candle holder has the same density as a pure silver ring (Figure 12.7). Size and shape do not change a material's density. But what if you heated a silver ring until it completely melted?

Atoms in liquid form tend to take up more space

Solid silver

Liquid silver

The density of a liquid is the ratio of mass to volume, just like the density of a solid. The mass of the silver does not change when the ring is melted. The volume of the liquid silver, however, is greater than the volume of the solid silver. The particles in a solid, as you remember, are fixed in position. Although the silver atoms in a ring are constantly vibrating, they cannot switch places with other atoms. They are neatly stacked in a repeating pattern. The atoms in the liquid silver are less rigidly organized. They can slide over and around each other and they take up a little more space.

Temperature and solid density The density of solids usually decreases slightly as temperature increases because solids expand when heated. As the temperature of the solid silver increases, the volume increases slightly, even before the silver melts. This is due to the increased vibration of the silver molecules.

Water is less dense in solid form Most materials are more dense in their solid phase than in their liquid phase. Water is a notable exception. Ice is less dense than liquid water! When water molecules freeze into ice crystals, they form a pattern that has an unusually large amount of empty space (Figure 12.8). The molecules are more tightly packed in water's liquid form!

Figure 12.7: *The density of solid silver is the same no matter what shape it is formed into.*

Figure 12.8: *Because of the spacing of molecules, ice forms hexagonal crystals, which give us the beautiful six-pointed shapes of snowflakes.*

Pressure

Forces in fluids When you push down on a bowling ball, what happens? Because the bowling ball is a solid, the force is transmitted down in the same direction as the applied force. Think about what happens when you push down on an inflated balloon. The downward force you apply creates forces that act sideways as well as down. Because fluids change shape, forces in fluids are more complicated than forces in solids.

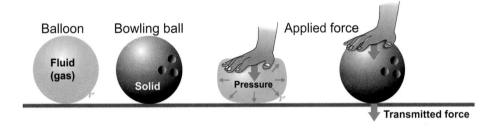

Pressure A force applied to a fluid creates **pressure**. Pressure acts in all directions, not just the direction of the applied force. When you inflate a car tire, you are increasing the pressure in the tire. This force acts up, down, and sideways in all directions inside the tire.

Units of pressure

35 pounds acts on every square inch

The units of pressure are force divided by area (Figure 12.9). If your car tires are inflated to 35 pounds per square inch (35 psi), then a force of 35 pounds acts on every square inch of area inside the tire (left). The pressure on the bottom of the tire is what holds up the car! The metric unit of pressure is the pascal (Pa). One pascal is one newton of force per square meter of area (N/m^2).

Units of pressure
Pressure is force per unit of area.

psi

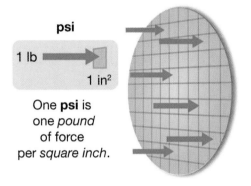

1 lb

1 in^2

One **psi** is one *pound* of force per *square inch*.

Pascal

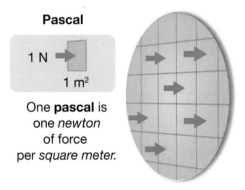

1 N

1 m^2

One **pascal** is one *newton* of force per *square meter*.

Figure 12.9: *Comparing units of pressure.*

Pressure, energy, and force

The atomic level explanation   What causes pressure? On the atomic level, pressure comes from collisions between atoms and molecules. Look at Figure 12.10. Molecules move around and bounce off each other and off the walls of the pitcher. It takes force to make a molecule reverse its direction and bounce the other way. The bouncing force is applied *to* the molecule *by* the inside surface of the pitcher. According to Newton's third law, an equal and opposite reaction force is exerted *by* the molecule *on* the pitcher. The reaction force is what creates the pressure acting on the inside surface of the pitcher. Trillions of molecules per second are constantly bouncing against every square millimeter of the inner surface of the pitcher. Pressure comes from the collisions of those many, many molecules.

Pressure is potential energy   Differences in pressure create potential energy in fluids just like differences in height create potential energy from gravity. A pressure difference of one newton per m^2 is equivalent to a potential energy of one joule per m^3. We get useful work when we allow a fluid under pressure to expand. In a car engine, high pressure is created by an exploding gasoline-air mixture. This pressure pushes the cylinders of the engine down, doing work that moves the car.

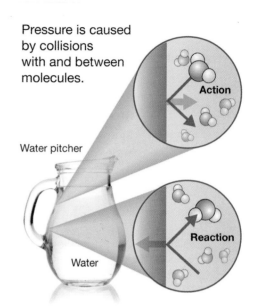

Pressure is caused by collisions with and between molecules.

Water pitcher

Action

Reaction

Water

Figure 12.10: *Pressure comes from constant collisions of trillions of molecules.*

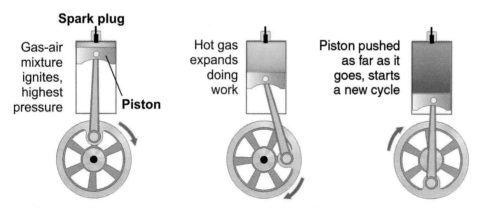

Spark plug

Gas-air mixture ignites, highest pressure

Piston

Hot gas expands doing work

Piston pushed as far as it goes, starts a new cycle

An engine uses pressure in an expanding gas to do work.

CHALLENGE

Car tires are usually inflated to a pressure of 32–40 pounds per square inch (psi). Racing bicycle tires are inflated to much higher pressure, 100–110 psi. A bicycle and rider are much lighter than a car. Why is the pressure in a bicycle tire higher than the pressure in a car tire?

Bernoulli's principle

Bernoulli's principle – a relationship that describes energy conservation in a fluid

Bernoulli's principle Everything obeys the law of energy conservation. But this law is more difficult to explain in a flowing fluid such as water coming out of a hole in a bucket. In addition to potential and kinetic energy, the fluid also has *pressure energy*. If friction is neglected, the total energy stays constant for any particular sample of fluid. This relationship is known as **Bernoulli's principle**.

Streamlines *Streamlines* are imaginary lines drawn to show the flow of fluid. We draw streamlines so that they are always parallel to the direction of flow. If water is coming out of a hole in a bucket, the streamlines look like the one shown in Figure 12.11. Bernoulli's principle tells us that the energy of any sample of fluid moving along a streamline is constant.

Figure 12.11: *Streamlines are imaginary lines drawn to show the flow of a fluid.*

Bernoulli's principle

Form of energy	Potential energy	Kinetic energy	Pressure energy	=	Constant along any streamline in a fluid
Variable	height	speed	pressure		

The three variables Bernoulli's principle says the three variables of height, pressure, and speed are related by energy conservation. Height is associated with potential energy, speed with kinetic energy, and pressure with pressure energy. If one variable increases along a streamline, *at least one of the other two must decrease*. For example, if speed goes up, pressure goes down.

The airfoil An important application of Bernoulli's principle is the airfoil shape of wings on a plane (Figure 12.12). The shape of an airfoil causes air flowing along the top (A) to move faster than air flowing along the bottom (B). According to Bernoulli's principle, if the speed goes up, the pressure goes down. When a plane is moving, the pressure on the top surface of the wings is lower than the pressure beneath the wings. The difference in pressure is what creates the lift force that supports the plane in the air.

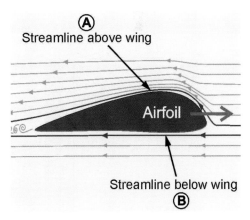

Figure 12.12: *Streamlines showing air moving from right to left around an airfoil (wing).*

Hydraulics and Pascal's principle

Hydraulic devices use pressure to multiply forces
Hydraulic lifts and other hydraulic devices use pressure to multiply forces and do work. The word *hydraulic* refers to anything that is operated by a fluid under pressure. Hydraulic devices operate on the basis of Pascal's principle, named after Blaise Pascal. **Pascal's principle** states that the pressure applied to an incompressible fluid in a closed container is transmitted equally in all parts of the fluid. An *incompressible fluid* does not decrease in volume when pressure is increased.

Input and output forces
Pressure is force divided by area (Figure 12.13). Suppose you have a small and large cylinder connected by a tube. If you apply a small force (the input force) to a piston at the small cylinder, you generate a given pressure. According to Pascal's principle, the pressure would be the same in the larger cylinder. Because the larger cylinder has more area, the output force exerted by the piston at the larger cylinder would be greater.

How a hydraulic lift multiplies force
To show this mathematically, rearrange $P = F/A$ to solve for force: $F = P \times A$ (Figure 12.14). The pressure stays the same in the larger cylinder, but area increased, resulting in a larger output force exerted by the piston. The greater the differences in the areas of the cylinders, the greater the output force exerted by the piston at the larger cylinder.

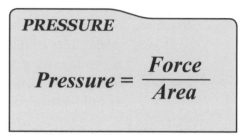

PRESSURE

$$Pressure = \frac{Force}{Area}$$

Figure 12.13: *Pressure is described mathematically as force divided by area.*

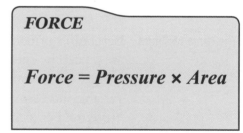

FORCE

$$Force = Pressure \times Area$$

Figure 12.14: *You can calculate the force exerted if you know the pressure and area.*

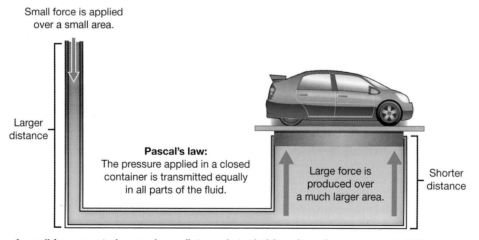

Small force is applied over a small area.

Larger distance

Pascal's law:
The pressure applied in a closed container is transmitted equally in all parts of the fluid.

Large force is produced over a much larger area.

Shorter distance

A small force exerted over a large distance is traded for a large force over a small distance.

You can't get something for nothing   In a hydraulic lift, a small input force produced a much larger output force. Does this mean we get more work for less? Unfortunately, no. Like a lever, the hydraulic lift converts work (force × distance) at the smaller piston for the same work at the larger one. In the example shown in Figure 12.15, a smaller piston in a hydraulic lift moves a distance of 5 m and displaces 500 cm³ of fluid. The amount displaced moves the piston in the larger cylinder only 1 m. This means a smaller force and larger distance have been exchanged for a large force through a smaller distance.

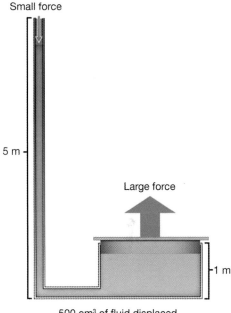

Figure caption:
Figure 12.15: *Like all simple machines, in a hydraulic lift, you trade force for distance.*

   ⊕ **Solving Problems: Pressure and Force**

On a hydraulic lift, 5 N of force are applied over an area of 0.125 m². What is the output force if the area of the larger cylinder is 5.0 m²?

1. **Looking for:**   You are looking for output force.

2. **Given:**   You are given the input force, input area, and output area.

3. **Relationships:** $P = F_{input} \div A$ and $F_{output} = PA$

4. **Solution:**   First, calculate the pressure exerted: $P = 5N \div 0.125$ m² $= 40$ N/m².

   Next, calculate the output force from the pressure:

   $$F_{output} = (40 \text{ N/m}^2) \times (5.0 \text{ m}^2) = 200 \text{ N}$$

Your turn...

a. A force of 35 N is exerted over a cylinder with an area of 5 cm². What pressure, in pascals, will be transmitted in the hydraulic system?

b. A force of 500 N is applied to a 5 cm² cylinder. What output force will be produced in the larger cylinder (500 cm²)?

SOLVE FIRST LOOK LATER

a. 70,000 Pa

b. 50,000 N

Viscosity

What is viscosity?   **Viscosity** is the measure of a fluid's resistance to flow. High-viscosity fluids take longer to pour from their containers than low-viscosity fluids. Ketchup, for example, has a high viscosity, and water has a low viscosity.

Viscosity is an important property of motor oils. If an oil is too thick, it may not flow quickly enough to parts of an engine. However, if an oil is too thin, it may not provide enough "cushion" to protect the engine from the effects of friction. A motor oil must function properly when the engine is started on a bitterly cold day, and when the engine is operating at high temperatures (see Science Fact on the next page).

Viscosity and particles   Viscosity is determined in large part by the shape and size of the particles in a liquid. If the particles are large and have bumpy surfaces, a great deal of friction will be created as they slide past each other. For instance, corn oil is made of large, chain-like molecules. Water is made of much smaller molecules. As a result, corn oil has greater viscosity than water.

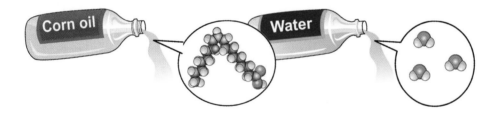

As a liquid gets warmer, its viscosity decreases   As the temperature of a liquid is raised, the viscosity of the liquid decreases. In other words, warm liquids have less viscosity than cold liquids. Fudge topping, for example, is much easier to pour when it's warm than when it's chilled. Why is this? When temperature rises, the vibration of molecules increases. This allows molecules to slide past each other with greater ease. As a result, the viscosity decreases (Figure 12.16).

Fudge 20°C   Fudge 60°C

Figure 12.16: *Heating fudge topping decreases viscosity so it is much easier to pour.*

Section 12.2 *Review*

1. Explain why liquid silver is less dense than solid silver.

2. The pressure at the bottom of Earth's atmosphere is about 100,000 N/m^2. This means there is a force of 100,000 N acting on every square meter of area! Your body has about 1.5 square meters of surface. Why aren't you crushed by the atmosphere?

3. The pressure at the bottom of the ocean is great enough to crush submarines with steel walls that are 10 centimeters thick. Suppose a submarine is at a depth of 1,000 meters. The weight of water above each square meter of the submarine is 9,800,000 newtons.

PRESSURE

$$Pressure = \frac{Force}{Area}$$

   a. What is the pressure?

   b. How does this pressure compare with the air pressure we experience every day on Earth's surface (100,000 N/m^2)?

4. What does pressure have to do with how a car engine works?

5. Bernoulli's principle relates the speed, height, and pressure in a fluid. Suppose speed goes up and height stays the same. What happens to the pressure?

6. At the atomic level, what causes fudge topping to pour faster when it is heated?

12.3 Buoyancy

If you drop a steel marble into a glass of water, it will sink to the bottom. The steel does not float because it has a greater density than the water. And yet many ships are made of steel. How does a steel ship float when a steel marble sinks? The answer has to do with gravity and weight.

VOCABULARY

buoyancy – the measure of the upward force that a fluid exerts on an object that is submerged

Weight and buoyancy

Weight and mass are not the same We all tend to use the terms *weight* and *mass* interchangeably. In science, however, *weight and mass are not the same thing.* Mass is a fundamental property of matter. Weight is a force caused by gravity. It is easy to confuse mass and weight because often heavy objects (more weight) have lots of mass and light objects (less weight) have little mass.

Buoyancy is a force It is much easier to lift yourself in a swimming pool than to lift yourself on land. This is because the water in the pool exerts an upward force on you that acts in a direction opposite to your weight (Figure 12.17). We call this force **buoyancy**. Buoyancy is a measure of the upward force that a fluid exerts on an object that is submerged.

Figure 12.17: *The water in the pool exerts an upward force on your body, so the net force on you is lessened.*

Pushing a ball under water

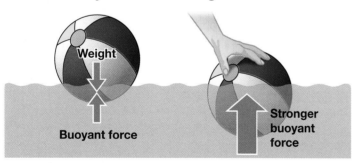

The strength of the buoyant force on an object in water depends on the volume of the object that is under water.

Suppose you have a large beach ball that you want to submerge in a pool. As you keep pushing downward on the ball, you notice the buoyant force getting stronger and stronger. The greater the part of the ball you manage to push under water, the stronger the force trying to push it back up. The strength of the buoyant force is proportional to the volume of the part of the ball that is submerged.

Archimedes' principle

What is Archimedes' principle? In the third century BCE, a Greek mathematician named Archimedes realized that buoyant force is equal to the weight of the fluid displaced by an object. We call this relationship **Archimedes' principle**. For example, suppose a rock with a volume of 1,000 cubic centimeters is dropped into water (Figure 12.18). The rock displaces 1,000 cm³ of water, which has a mass of 1 kilogram. The buoyant force on the rock is the weight of 1 kilogram of water, or 9.8 newtons.

VOCABULARY

Archimedes' principle – states that the buoyant force is equal to the weight of the fluid displaced by an object

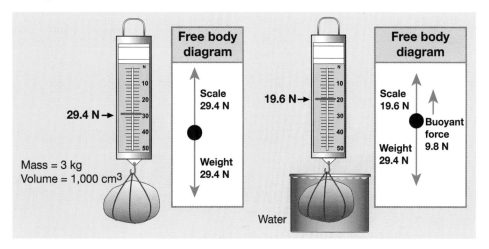

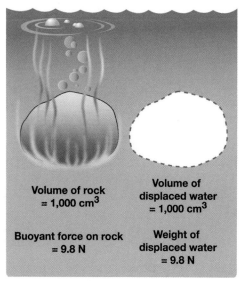

Figure 12.18: *A rock with a volume of 1,000 cm³ experiences a buoyant force of 9.8 newtons.*

A simple buoyancy experiment Look at the illustration above. A simple experiment can be done to measure the buoyant force on a rock (or any object) using a spring scale. Suppose you have a rock with a volume of 1,000 cubic centimeters and a mass of 3 kilograms. In air, the scale shows the rock's weight as 29.4 newtons. The rock is then gradually immersed in a container of water, but not allowed to touch the bottom or sides of the container. As the rock enters the water, the reading on the scale decreases. When the rock is completely submerged, the scale reads 19.6 newtons.

Calculating the buoyant force Subtracting the two scale readings, 29.4 newtons and 19.6 newtons results in a difference of 9.8 newtons. This is the buoyant force exerted on the rock, and it is the same as the weight of the 1,000 cubic centimeters of water the rock displaced.

Sinking and floating

Comparing buoyant force and weight   Buoyancy explains why some objects sink and others float. A submerged object floats to the surface if the buoyant force is greater than the object's weight (Figure 12.19). If the buoyant force is less than its weight, then the object sinks.

Equilibrium   Suppose you place a block of foam in a tub of water. The block sinks partially below the surface. Then it floats without sinking any farther. The upward buoyant force perfectly balances the downward force of gravity (the block's weight). But how does the buoyant force "know" how strong it needs to be to balance the weight?

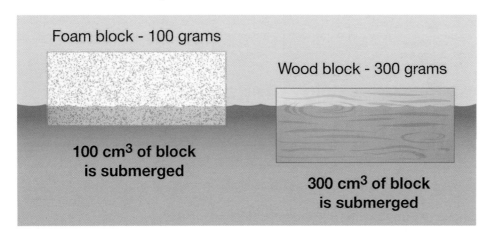

Foam block - 100 grams

Wood block - 300 grams

100 cm³ of block
is submerged

300 cm³ of block
is submerged

Denser objects float lower in the water   You can find the answer to this question in the illustration above. If a foam block and a wood block of the same size are both floating, the wood block sinks farther into the water. Wood has a greater density, so the wood block weighs more. A greater buoyant force is needed to balance the wood block's weight, *so the wood block displaces more water*. The foam block has to sink only slightly to displace water with a weight equal to the block's weight. A floating object displaces just enough water to make the buoyant force equal to the object's weight.

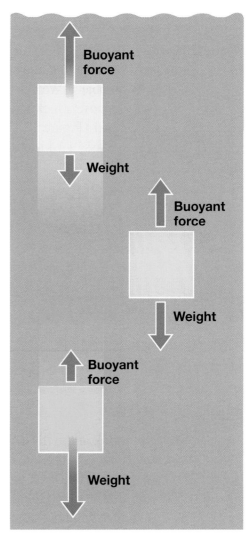

Buoyant force

Weight

Buoyant force

Weight

Buoyant force

Weight

Figure 12.19: *Whether an object sinks or floats depends on how the buoyant force compares with the object's weight.*

Density and buoyancy

Comparing densities   If you know an object's density, you can immediately predict whether it will sink or float—without measuring its weight. An object sinks if its density is greater than that of the liquid it is submerged into. It floats if its density is less than that of the liquid.

Two balls with the same volume but different densities   To see why, picture dropping two balls into a pool of water. The balls have the same size and volume but have different densities. The steel ball has a density of 7.80 g/cm^3, which is greater than the density of water (1.0 g/cm^3). The wood ball has a density of 0.75 g/cm^3, which is less than the density of water.

5 cm steel ball
Density = 7.8 g/cm^3

5 cm wood ball
Density = 0.75 g/cm^3

Water Density = 1.0 g/cm^3

Why one sinks and the other floats   When they are completely under water, both balls have the same buoyant force because they displace the same volume of water. However, the steel ball has more weight because it has a higher density. The steel ball sinks because steel's higher density makes the ball heavier than the same volume of water. The wood ball floats because wood's lower density makes the wood ball lighter than the same volume of displaced water.

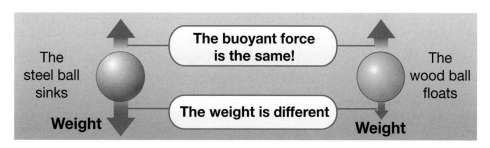

The steel ball sinks

The buoyant force is the same!

The wood ball floats

Weight

The weight is different

Weight

TECHNOLOGY

Buoyancy and submarines

STEM

Beneath the ocean surface, there are many clues to the past and present conditions of our planet. Exploring the deep ocean requires sophisticated engineering. *Alvin*, a U.S. Navy research submarine, can dive to 4,500 meters below the ocean surface. Scientists aboard *Alvin* have discovered strange life-forms near deep hot spots where there is no light and pressures are 400 times greater than at Earth's surface.

Alvin's depth is controlled by changing its average density. A tank aboard the submarine can be filled with air or water. To dive, water is pumped into the tank and air is released. The tank's average density becomes greater than the density of water, and the submarine sinks.

When the sub reaches the proper depth, the amount of air and water is adjusted with pumps until the average density of the submarine is the same as the density of water. This is called neutral buoyancy. When it is time for the sub to head back to the surface, water is pumped out of the tank and replaced with air. The submarine's average density decreases, and the submarine rises.

Boats and average density

How do steel boats float? If you place a solid chunk of steel in water, it immediately sinks because the density of steel (7.8 g/cm^3) is much greater than the density of water (1.0 g/cm^3). So how is it that thousands of huge ships made of steel are floating around the world? The answer is that *average density* determines whether an object sinks or floats (Figure 12.20).

Solid steel sinks because it is denser than water To make steel float, you have to reduce the *average* density somehow. Making the steel hollow does exactly that. Making a boat hollow expands its volume a tremendous amount without changing its mass. Steel is so strong that it is quite easy to reduce the average density of a boat to 10 percent of the density of water by making the shell of the boat relatively thin.

An object with an average density GREATER than the density of water will sink.

An object with an average density LESS than the density of water will float.

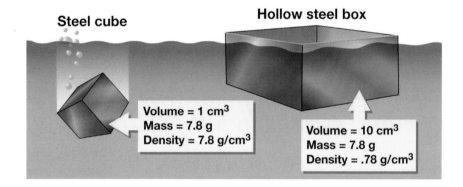

Steel cube

Volume = 1 cm³
Mass = 7.8 g
Density = 7.8 g/cm³

Hollow steel box

Volume = 10 cm³
Mass = 7.8 g
Density = .78 g/cm³

Average density
Average density is the total mass divided by the total volume.

Solid steel ball
volume = 25 mL
mass = 195 g

Hollow steel ball
volume = 25 mL
mass = 20 g

Avg. density = $\dfrac{195 \text{ g}}{25 \text{ mL}}$ Avg. density = $\dfrac{20 \text{ g}}{25 \text{ mL}}$

| Avg. Density = 7.8 g/mL | Avg. Density = 0.8 g/mL |
| **SINKS!** | **FLOATS!** |

Figure 12.20: *The meaning of* average density. *Note: 1 mL = 1 cm³.*

Increasing volume decreases density

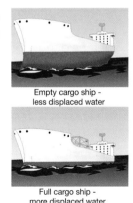

Empty cargo ship -
less displaced water

Full cargo ship -
more displaced water

Ah, you say, but that's an *empty ship*. True, so the density of a new ship must be designed to be less than 1.0 g/cm^3 to allow for cargo. When objects are placed in a boat, the boat's average density increases. The boat must sink deeper to displace more water and increase the buoyant force. If you have seen a loaded cargo ship, you might have noticed that it sat lower in the water than an unloaded ship nearby. In fact, the limit to how much a ship can carry is set by how low in the water the ship can get before rough seas cause waves to break over the sides of the ship.

Section 12.3 *Review*

1. The buoyant force on an object depends on the _____ of the object that is under water.

2. What happens to the buoyant force on an object as it is lowered into water? Why?

3. The buoyant force on an object is equal to the weight of the water it _____ .

4. When the buoyant force on an object is greater than its weight, the object _____ .

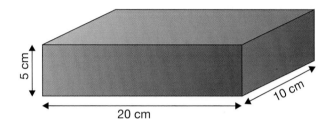

5. A rectangular object is 10 centimeters long, 5 centimeters high, and 20 centimeters wide. Its mass is 800 grams.
   a. Calculate the object's volume in cubic centimeters.
   b. Calculate the object's density in g/cm^3.
   c. Will the object float or sink in water? Explain.

6. Solid iron has a density of 7.9 g/cm^3. Liquid mercury has a density of 13.5 g/cm^3. Will iron float or sink in mercury? Explain.

7. Why is it incorrect to say that heavy objects sink in water?

8. Steel is denser than water, yet steel ships float. Explain.

9. A rock sinks in water, but suppose a rock was placed in a pool of mercury. Mercury is a liquid metal with a density of 13.5 g/cm^3. The density of rock is 2.6 g/cm^3. Will the rock sink or not? Explain how you got your answer.

The Hull

What Makes a Boat Stay Afloat?

There are many different types of boats, but all have one thing in common—the hull. The hull is the main body of the boat. It displaces the water that provides the upward buoyant force. It also provides stability. All hulls must have the ability to displace an amount of water equal to the weight of the boat in order to float. However, hulls come in a variety of shapes, each serving a different purpose. A sailboat needs a different type of hull than a speedboat or an oil tanker.

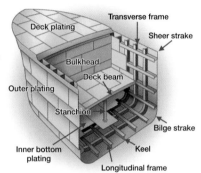

Deck plating
Transverse frame
Sheer strake
Bulkhead
Deck beam
Outer plating
Stanchion
Bilge strake
Inner bottom plating
Keel
Longitudinal frame

Two main types of hulls

The displacement hull: Boats with displacement hulls have cross-sections that are rounded or v-shaped. The hull displaces the water, and the water completely supports the weight of the boat. As the boat moves, a displacement hull pushes water out of the way, both sideways and down. The curves of the hull allow the water to be pushed away smoothly with a minimum of turbulence, making these types of hulls more energy efficient. The top speed of boats with displacement hulls is limited by the amount of the hull that is under water and by the hull's shape.

The planing hull: This type of hull has a flatter bottom designed to ride primarily on "plane," or on top of the water. When a boat with a planing hull is at rest or moving at low speeds, the entire weight of the boat

is supported by the water and the hull acts like a displacement hull. But at high speeds, the boat rides up on top of the water. The force of the water being thrown out of the way supports most of the weight of the boat. Boats with planing hulls are capable of moving very fast since they displace very little water. The limiting factor for this type of hull is usually rough water because planing hulls tend to pound or slam into waves, making the ride uncomfortable for passengers.

Variations on main hull types

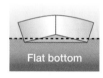
Flat bottom

Flat bottom: This design enables the boat to ride on top of the water at high speeds (planing). These boats are used on calmer waters like ponds, small lakes, and slow-moving rivers. Flat bottom boats have less depth, allowing them to be used in shallower waters. These boats tend to be very stable, but can be damaged by pounding in rough water.

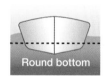
Round bottom

Round bottom: Boats with this displacement hull move easily and smoothly through the water, even at slow speeds. Round bottom hulls have a tendency to roll over. To fix this problem on sailboats, a deep keel is added. The deep keel of a sailboat sticks down in the water like a long wing and is very heavy. It helps the boat sail into the wind and provides stability.

Deep-V hull

Deep-V hull: This design comes nearly to a point at the very bottom of the boat. A deep-V hull moves more efficiently and with greater stability than a flat bottom boat, but it is less stable than a round bottom boat. A deep-V hull is used on high-powered pleasure boats. A boat with this hull can go very fast and can plane, but the hull slices somewhat through the water for a more comfortable ride in rough water.

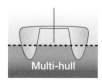
Multi-hull

Multi-hull: In this design, two or more hulls are connected together. The hulls are set wide apart with a platform on top. The multiple hulls allow the boat to move very fast through the water, using a displacement hull design, while remaining very stable.

Boat stability:
Center of gravity vs. center of buoyancy

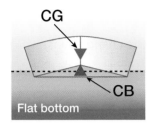

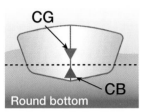

Have you ever tried to stand up in a rowboat? It's a lot harder than standing up on a sailboat or powerboat. One reason is that the rowboat has less mass and is easily influenced by your weight. When you're standing in a rowboat, the center of gravity is raised, making the boat less stable. The lower the center of gravity, the more stable the boat. A boat's stability is also dependent on the relationship between its center of gravity (CG) and its center of buoyancy (CB).

The CG is the exact center of the boat according to its weight distribution. If you had a toy boat and you put your finger directly under the CG, it would balance on your finger. The center of buoyancy, or CB, is the exact center of gravity of the volume of water displaced by the boat's hull. When weight is distributed evenly in a boat, the CB is directly below the CG. The CG is a downward force while the CB is an upward force, so together they create balance.

The stability of the boat depends on how the CB shifts as the boat tilts or "heels." For example, if you move from the center to one side of a small boat, the CG will move to that side. The boat tilts as the CB shifts toward that side to balance the new CG. That ability of the CB to shift is the measure of the boat's stability. The more easily the CB shifts, the more stable the boat is.

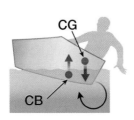

If you were to shift your weight all the way to the side of a small rowboat, you could shift the CG so far out that no matter how far the boat tilts, the CB won't be able to shift under the CG. If the CB can't get underneath the CG, the boat will capsize, and you'll be going for a swim.

With a sailboat, the boat is larger, has more mass, and is less influenced by your weight. Moreover, the CG is generally made lower in a big sailboat by the weight of the keel. The combination of the CG and the CB of the boat create a righting moment opposite to the direction the boat is heeling (tilting). This will keep it from capsizing. This is how a sailboat stays upright when the wind is blowing to the left or right—the coupled CG and CB work in opposite directions to counteract the tilt.

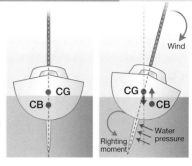

Hydrofoil boats: Boats that can fly

A hydrofoil boat is a boat built for speed. It uses wings, or foils, to lift the boat up and the hull out of the water. When a normal boat moves forward through the water, most of the energy expended by the engine goes into moving the water in the front of the boat out of the way by pushing the hull through it. This action creates the opposing force of fluid friction, slowing the boat down.

Hydrofoils engage in lift, the same phenomenon airplanes use to fly. A hydrofoil boat has foils beneath the hull. As a hydrofoil boat increases in speed, water flows over and under the foils like air flows over and under an airplane wing. The shape of the foil causes the water to move down, creating an upward force on the boat called lift. When this lifting force of the hydrofoils is balanced by the weight of the boat, the boat is in equilibrium. At this point, the displacement part of the hull is out of the water, greatly reducing the force of fluid friction.

Questions:
1. What are the two main types of hulls?
2. What is the advantage of a planing hull?
3. What two forces determine the stability of a boat?
4. How does a hydrofoil boat work?

Chapter 12 *Assessment*

Vocabulary

Select the correct term to complete each sentence.

amorphous	crystalline	physical properties
Archimedes' principle	ductility	pressure
Bernoulli's principle	elasticity	strength
brittleness	fluid	tensile strength
buoyancy	hardness	thermal expansion
chemical properties	malleability	viscosity
	Pascal's principle	

Section 12.1

1. _____ are properties that can be observed directly.

2. _____ can only be observed when one substance is changed to another substance.

3. A solid having randomly arranged atoms or molecules is called _____ .

4. The tendency to crack or break is called _____ .

5. A(n) _____ solid has an orderly, repeating arrangement of particles.

6. _____ is the ability to bend without breaking.

7. A solid that can be bent and stretched and then return to its original size has high _____ .

8. A solid's ability to resist being scratched is called _____ .

9. Gold has high _____ because it can be pounded into very thin sheets.

10. The ability to maintain shape under the application of forces is called _____ .

11. A measure of how much pulling a material can withstand before breaking is called _____ .

12. When a material changes size as temperature changes, it is called _____ .

Section 12.2

13. _____ is a measure of a fluid's resistance to flow.

14. Any matter that flows when force is applied is referred to as a(n) _____ .

15. _____ is the measure of force per unit of area.

16. _____ is a relationship that describes energy conservation in a fluid.

17. According to _____, an increase in pressure at one point in a closed container of fluid results in an equal increase of pressure everywhere in the fluid.

Section 12.3

18. _____ is a measure of the upward force a fluid exerts on an object that is submerged.

19. _____ states that the buoyant force is equal to the weight of the fluid displaced by an object.

Concepts

Section 12.1

1. In general, how do the densities of a material in solid, liquid, and gas phases compare? Name a common exception to the general rule.

2. Explain the difference between physical and chemical properties. Use an example in your explanation.

3. The density of a solid material depends on two things. Name those two things.

4. Compare the arrangement of atoms or molecules in an amorphous solid to the arrangement of atoms or molecules in a crystalline solid.

5. Classify the following as a physical property (P) or a chemical property (C).
   a. _____ ice melts at room temperature
   b. _____ an apple turns brown when it is peeled
   c. _____ mercury is a metal that is liquid at room temperature
   d. _____ rust is orange
   e. _____ copper is shiny
   f. _____ copper forms a blue-green patina after being exposed to the air for a long period of time

6. Use the words *amorphous* or *crystalline* to describe each of the materials listed below.
   a. metal   e. taffy candy
   b. glass   f. plastic
   c. rubber   g. sugar
   d. diamond   h. ice

7. Match each material below with the mechanical property associated with it.
   a. _____ gold   1. brittleness
   b. _____ rubber   2. ductility
   c. _____ glass   3. elasticity

Section 12.2

8. Compare the terms *liquid* and *fluid*.

9. Describe how Newton's third law is related to fluid pressure.

10. Explain how Bernoulli's principle helps to explain the lift that airplane wings experience.

11. Tamara sprays a garden hose at her brother, who is 20 feet off the ground in a tree. How would the speed of the water as it comes from the faucet compare to the speed of the water as it hits her brother? Explain.

12. What is Pascal's principle, and how does it apply to hydraulic lifts?

Section 12.3

13. Why does a glass marble sink in water?

14. What happens to the weight of a rock when it is placed under water? Why?

15. Compare the buoyant force to the weight of a floating block of foam.

16. Explain why a solid steel ball sinks in water but a steel ship floats in water.

17. A solid steel ball and a hollow steel ball of the same size are dropped into a bucket of water. Both sink. Compare the buoyant force on each.

18. What is the maximum density that a fully loaded cargo ship can have without sinking?

19. Why does ice float in a glass of water? Explain in terms of density and buoyancy.

Problems

Section 12.1

1. Your teacher gives you two stainless steel ball bearings. The larger has a mass of 25 g and a volume of 3.2 cm³. The smaller has a mass of 10 g. Calculate the volume of the smaller ball bearing.

2. At 20°C, the density of copper is 8.9 g/cm³. The density of platinum is 21.4 g/cm³. What does this tell you about how the atoms are "packed" in each material?

Section 12.2

3. What is the pressure if 810 N of force are applied on an area of 9 m²?

4. If the air pressure is 100,000 N/m², how much force is acting on a dog with a surface area of 0.5 m²?

5. A 4,000-pound car's tires are inflated to 35 pounds per square inch (psi). How much tire area must be in contact with the road to support the car?

6. A force of 15 N is exerted over an area of 0.1 m² in a hydraulic lift. What output force will be generated by the output cylinder if it has an area of 5 m²?

Section 12.3

7. What buoyant force is exerted on a toy balloon with a volume of 6,000 mL by the air surrounding the balloon? The density of air is 0.001 g/cm³.

8. An object weighing 45 newtons in air is suspended from a spring scale. The spring scale reads 22 newtons when the object is fully submerged in water. Calculate the buoyant force on the object.

9. A bucket is filled to the top with water and set into a large pan. When a 200-gram wooden block is carefully lowered into the water it floats, but some water overflows into the pan. What is the weight of the water that spills into the pan?

10. A stone that weighs 6.5 newtons in air weighs only 5.0 newtons when submerged in water. What is the buoyant force exerted on the rock by the water?

11. A 100-mL oak object is placed in water. Will the object sink or float? The density of oak is 0.60 g/cm³.

12. If an object has a buoyant force of 320 newtons acting on it, would the weight of the object have to be more or less than 320 newtons in order to float?

13. Neutral buoyancy is when an object stays in one position under water. It doesn't sink or float. An object weighs 135 newtons. What would the buoyant force have to be for the object to have neutral buoyancy?

Applying Your Knowledge

Section 12.1

1. You are an engineer who must choose a type of plastic to use for the infant car seat that you are designing. Name two properties of solids that would help you decide, and explain why each is important. STEM

Section 12.2

2. Many studies have been done about the viscosity of lava from various volcanic eruptions around the world. Do some research to find out how scientists determine the viscosity of lava, and find out if there is much variation in the viscosity of different lava flows.

Section 12.3

3. Scuba divers use weights and a buoyancy control device (BCD) to help them maintain neutral buoyancy. Explain the advantages of being neutrally buoyant when you are a scuba diver.

4. The Dead Sea is a body of water that lies between Israel and Jordan. It is so salty that almost no organisms other than a few types of bacteria can survive in it. The density of its surface water is 1.166 g/mL. Would you find it easier to float in the Dead Sea or in a freshwater lake? Give a reason for your answer.

The Behavior of Gases

Do you know what the oldest form of aircraft is? You may think it's the airplane flown by the Wright brothers in 1903, but the hot air balloon dates back much earlier than the Wright brothers' plane. In 1783, the first passengers to ride in a hot air balloon were a duck, a rooster, and a sheep. Several months later, the Montgolfier brothers of France made a balloon of paper and silk. This flight carried 2 men for 25 minutes across 5½ miles. Ballooning has come a long way since that historic flight. Balloons are used to forecast weather, explore space, and perform experiments, and for recreation.

The National Scientific Balloon Facility in Palestine, Texas, is a National Aeronautics and Space Administration (NASA) facility. NASA launches about 25 science balloons each year. These balloons do not carry people, but they carry a "payload." The payload includes equipment for experiments and may weigh up to 8,000 pounds. These experiments help scientists study Earth and space. Airplanes usually fly 5 to 6 miles above the ground. Science balloons fly up to 26 miles high!

Imagine you are floating over your community in a hot air balloon. How is it possible to perform such a feat? It has to do with the gases inside and outside the balloon. In this chapter, you will learn about the behavior of gases and also about Earth's atmosphere, which consists of gases. Once you've read this chapter, you will be able to explain why and how a hot air balloon works.

CHAPTER 13 INVESTIGATIONS

13A: Boyle's Law
How are pressure and volume of a gas related?

13B: Pressure and Temperature Relationship
How are temperature and pressure of a gas related?

13.1 Gases, Pressure, and the Atmosphere

Earth's is a layer of gases surrounding the planet, protecting and sustaining life. It insulates Earth so we don't freeze at night. Earth's atmosphere also contains the carbon dioxide needed by plants for photosynthesis and the oxygen we need to breathe.

What's in Earth's atmosphere?

Earth's atmosphere is 78% nitrogen   Nitrogen (N_2) gas makes up about 78 percent of Earth's atmosphere. Nitrogen is released into the air by volcanoes and decaying organisms and is a vital element for living things. Protein, a substance in body tissues, contains nitrogen. However, this nitrogen is not absorbed directly from the air. Instead, the nitrogen is changed into nitrates (NO_3) by nitrogen-fixing bacteria in the soil. Plants absorb nitrates from the soil and use them to make proteins. We eat plants (especially their seeds) or meat to obtain these proteins (Figure 13.1).

Earth's atmosphere is 21% oxygen   The second most abundant gas is oxygen (O_2), which makes up 21 percent of Earth's atmosphere. Atmospheric oxygen enables us to process the fuel we need for life. The remaining 1 percent of Earth's atmosphere is made up of 0.93 percent argon and 0.04 percent carbon dioxide. There are also tiny amounts of neon, helium, methane, krypton, and hydrogen, which we call trace gases.

Why Earth's atmosphere exists   The atmosphere exists around Earth because our planet has just the right balance of mass and distance from the Sun. Scientists explain that at the time of Earth's formation, the heat from the Sun drove off most of the lightweight elements such as hydrogen and helium. Earth would have remained a rocky, airless world except that, as it cooled, earthquakes and volcanoes spewed out heavier gases like nitrogen and carbon dioxide. Earth's mass gave it enough gravitational pull that these gases stayed around. Although the planet Mercury was formed in a similar way, its mass is too small and it is too close to the Sun to have retained much of a layer of gas surrounding it. Venus, Earth, and Mars, however, retained their atmospheres.

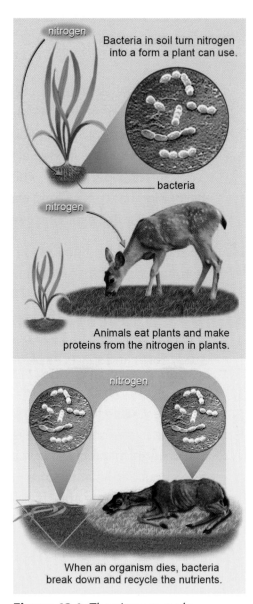

Bacteria in soil turn nitrogen into a form a plant can use.

nitrogen

bacteria

nitrogen

Animals eat plants and make proteins from the nitrogen in plants.

nitrogen

When an organism dies, bacteria break down and recycle the nutrients.

Figure 13.1: *The nitrogen cycle.*

Atmospheres of Earth, Venus, and Mars

Comparing atmospheres An **atmosphere** is a layer of gases surrounding a planet or other body in space. The atmospheres of Venus, Earth, and Mars were formed in similar ways, so we might expect them to contain similar elements. Figure 13.2 compares the atmospheres of these three planets. As you can see, Earth's atmosphere is much different from those of Venus and Mars.

Similarities between Venus and Mars Venus and Mars show striking similarities in the makeup of their atmospheres. They are mostly carbon dioxide, with a small amount of nitrogen. Earth, on the other hand, is very different. Ours is the only planet with a large amount of nitrogen and oxygen, and just a tiny amount of carbon dioxide in its atmosphere. Why is Earth so different?

Life changed Earth's atmosphere Through photosynthesis, life on Earth has actually changed the planet's atmosphere. Many forms of life use photosynthesis to obtain energy from the Sun. This process breaks down carbon dioxide, uses carbon to build the organism, and releases oxygen into the air (Figure 13.3).

Where does the carbon go? When organisms die and decompose, some of the carbon from their bodies is released as carbon dioxide back into the air. However, if all of the carbon used by life processes returned to Earth's atmosphere, it would still be like that of Venus and Mars. Instead, some of the carbon used to build living organisms ends up staying in the ground. Earth stores carbon in several ways.

How Earth stores carbon  Many marine organisms, such as microscopic phytoplankton, use the carbon dioxide dissolved in seawater to form shells of calcium carbonate. A greatly magnified picture of a phytoplankton is shown at the left. When these organisms die, their shells sink to the bottom of the water and stay there. The carbon doesn't return to the atmosphere. Huge piles of calcium carbonate have built up over the years, creating some of our land forms. Fossil fuels (oil, coal, and natural gas) also store carbon from decaying plants and animals in the ground. Another process stores carbon in a type of rock called limestone.

Planet	Major gases in atmosphere			
Venus	96% CO_2	3% N_2	0.1% H_2O	
Earth	0.04% CO_2	78% N_2	21% O_2	0.93% Ar
Mars	95% CO_2	3% N_2	1.6% Ar	

Figure 13.2: *Comparing the atmospheres of Venus, Earth, and Mars.*

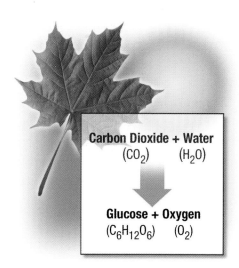

Carbon Dioxide + Water
(CO_2)   (H_2O)

Glucose + Oxygen
($C_6H_{12}O_6$)   (O_2)

Figure 13.3: *Photosynthesis is a process that uses carbon dioxide and produces oxygen.*

What is atmospheric pressure?

Air molecules exert pressure

The pressure of air molecules in the atmosphere is a result of the weight of a column of air pressing down on an area. **Atmospheric pressure** is a measurement of the force of air molecules per unit of area in the atmosphere at a given altitude.

atmospheric pressure – a measurement of the force of air molecules per unit of area in the atmosphere at a given altitude

How we withstand air pressure

At sea level, the weight of the column of air above a person is about 9,800 newtons (2,200 pounds)! This is equal to the weight of a small car (Figure 13.4). Why aren't we crushed by this pressure? First, there is air inside our bodies that is pushing out with the same amount of pressure, so the forces are balanced. Second, our skeletons are designed to withstand the pressure of our environment.

Air Pressure at Sea Level

Deep-sea animals are adapted to live under great pressure

Contrast these systems with those used by deep-sea animals, like the deep-sea angler shown to the right. Fish that live at a depth of 10,000 feet are under pressure 300 times greater than we withstand. Instead of thick, strong bones, deep-sea creatures have cell membranes that contain a material that would be liquid at Earth's surface. The intense water pressure makes the material more rigid, so that the organism's body tissues hold their shape and function properly. Each organism on Earth is uniquely adapted to thrive in the pressure of its particular environment.

Photo by E. Widder

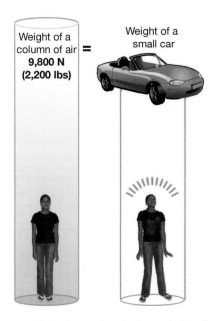

Figure 13.4: *At sea level, the weight of the column of air above a person is equal to the weight of a small car.*

Barometers and units of pressure

Barometers measure air pressure   Atmospheric pressure is measured with an instrument called a **barometer**. The oldest type of barometer is the mercury barometer (Figure 13.5). It consists of a tube sealed at one end and partially filled with mercury. The open end of the tube stands in a dish of mercury. As air presses down on the mercury in the dish, it forces the liquid in the tube to rise. When the air pressure is greater, the mercury travels farther up the tube. The air pressure at sea level generally causes the mercury in a barometer to rise 29.92 inches (760 millimeters). Table 13.1 compares units of pressure.

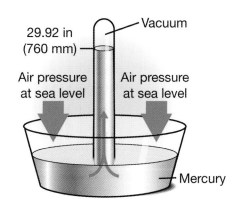

Figure 13.5: *A mercury barometer.*

Table 13.1: Units of pressure

Unit	Description	Relationship
Pascal (pa)	Metric unit commonly used to measure pressure of air in a container.	1 pa = 1 N/m²
Atmosphere (atm)	One atmosphere is the standard air pressure at sea level. Used by divers to compare pressure under water with surface pressure.	1 atm = 101,325 pa
Millimeter of mercury (mm Hg)	Unit describing the height of a column of mercury in a barometer.	760 mm Hg = 1 atm
Pounds per square inch (psi)	English unit commonly used to measure pressure of air in a container, like a tire or ball.	14.7 psi = 1 atm
Millibar (mb)	Metric unit used to measure atmospheric pressure.	1,013.25 mb = 1 atm

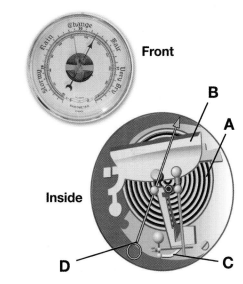

Aneroid barometers   Mercury barometers have a downside: Mercury is a poisonous liquid, and it creates unhealthy vapors. You would not want to have a mercury barometer in your living room! Most barometers in use today are aneroid barometers (Figure 13.6). They have an airtight cylinder made of thin metal. The walls of the cylinder are squeezed inward when the atmospheric pressure is high. At lower pressures, the walls bulge out. A dial attached to the cylinder moves as the cylinder changes shape, indicating the change in air pressure.

Figure 13.6: *Inside an aneroid barometer. Letter A shows the airtight cylinder, to which a spring, B, is attached. C is a series of levers that amplify the spring's movement. A small chain transfers the movement to the pointer, D.*

Atmospheric pressure and altitude

(from previous page)

barometer – an instrument that measures atmospheric pressure

A giant pile of cotton balls Atmospheric pressure decreases as altitude increases. Why? Imagine that the molecules of the atmosphere are like a giant pile of cotton balls. At the top of the pile, the cotton balls are loosely spread out. But the weight of the cotton balls at the top presses down on the ones underneath, and those cotton balls press down on the ones below them. As a result, the cotton balls at the bottom of the pile are packed together much more tightly than the ones at the top.

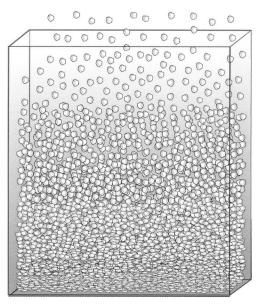

Air pressure is greatest at sea level A similar thing happens in the atmosphere. The molecules at the bottom are packed together more tightly, because the weight of the molecules above presses down on them. The air pressure is greatest at sea level (the bottom of the atmosphere). As you get farther and farther from sea level, the molecules are more and more spread out, so there are fewer molecules above you pushing down.

Figure 13.7 shows that as altitude increases, atmospheric pressure decreases rapidly. At sea level, atmospheric pressure averages about 1,013 millibars. At the top of Mt. Washington, New Hampshire (the highest point in the northeastern United States, at 1.917 kilometers), the average atmospheric pressure is 800.3 millibars. At the top of Mt. Everest, a height of 8.85 kilometers, atmospheric pressure averages only one-third of the pressure found at sea level.

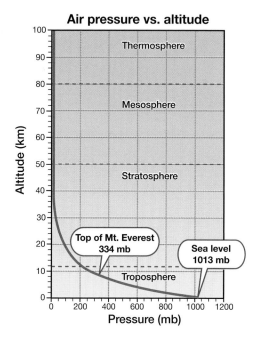

Figure 13.7: *As altitude increases, atmospheric pressure decreases.*

Section 13.1 *Review*

1. Earth's atmosphere is 78 percent nitrogen.
   a. Where does nitrogen in the atmosphere come from?
   b. How does nitrogen from the air get into the bodies of plants and animals?

2. What is the second most abundant gas in Earth's atmosphere? Where does it come from?

3. The atmospheres of Mars and Venus are mostly carbon dioxide. Earth's atmosphere contains only a fraction of a percent of carbon dioxide. Why is Earth's atmosphere so different from those of Mars and Venus?

4. What is atmospheric pressure?

5. Why doesn't Earth's atmospheric pressure crush our bodies?

6. What is a barometer? Explain how a mercury barometer works.

7. Which statement is true about atmospheric pressure?
   a. As altitude increases, atmospheric pressure increases.
   b. Atmospheric pressure does not change with altitude.
   c. As altitude increases, atmospheric pressure decreases.
   d. None of the above are true.

8. Use the graph in Figure 13.7 on page 300 to estimate the following:
   a. What is the atmospheric pressure at an altitude of 22 kilometers?
   b. If the atmospheric pressure measures 500 millibars, what is the altitude?
   c. Mt. Rainier (Figure 13.8) in Washington state has an elevation of about 14,000 feet. What is the atmospheric pressure at its summit?

▓▓▓▓▓▓ SOLVE IT! ▓▓▓▓▓▓

Converting units of pressure

Convert the following units of pressure. Use Table 13.1 on page 299 to find the conversion factors.

1. 2 atm = _____ mm Hg
2. 4.5 atm = _____ pa
3. 35 psi = _____ atm
4. 1,850 mm Hg = _____ atm
5. 45 psi = _____ pa

Figure 13.8: *Question 8c.*

13.2 The Gas Laws

Gases are fluids, but they are different from liquids because the molecules in a gas are completely separated from each other. Because gas molecules act independently, gases are free to expand or contract. Unlike liquids, a gas will expand to completely fill its container. In this section, you will learn about some laws that describe the behavior of gases.

Pressure, volume, and density

Pressure and volume When you squeeze a fixed quantity of gas into a smaller volume, the pressure goes up (Figure 13.9). This rule is known as **Boyle's law**. The pressure increases because the same number of molecules are now squeezed into a smaller space. The molecules hit the walls more often because there are more of them per unit of area. The formula for Boyle's law relates the pressure and volume of gas. If the mass and temperature are kept constant, the product of the pressure multiplied by the volume stays the same.

BOYLE'S LAW

$$\text{Initial pressure} - \underset{\text{Initial volume}}{P_1 V_1} = \underset{\text{New volume}}{P_2 V_2} - \text{New volume}$$

Initial volume   New pressure

Mass and temperature remain constant

Pressure and density The density of a gas increases when the pressure increases. (We say "usually" because density and pressure are also affected by temperature.) By increasing the pressure, you are doing one of two things: squeezing the same amount of mass into a smaller volume, or squeezing more mass into the same volume. Either way, the density goes up. For example, air has a density of $0.0009 \ g/cm^3$ at atmospheric pressure. When compressed in a diving tank to 150 times higher pressure, the density is about $0.135 \ g/cm^3$. The density of a gas can vary from near zero (in outer space) to greater than the density of some solids. This is very different from the behavior of liquids or solids.

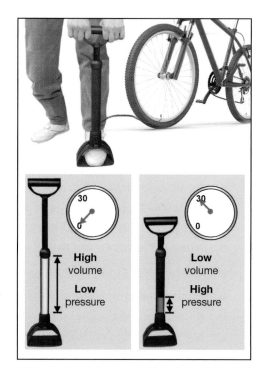

Figure 13.9: *Compressing the volume of air to increase the pressure.*

Boyle's law graph   The relationship between pressure and volume for a gas, when temperature remains constant, is evident in the graph in Figure 13.10. The example below shows you how to solve problems using Boyle's law.

   Solving Problems: Boyle's Law

A kit used to fix flat tires consists of an aerosol can containing compressed air and a patch to seal the hole in the tire. Suppose 5 liters of air at atmospheric pressure (1 atm) is compressed into a 0.5-liter aerosol can. What is the pressure of the compressed air in the can? Assume no change in temperature or mass.

1.	**Looking for:**	The pressure inside an aerosol can.
2.	**Given:**	You are given the initial volume (in liters), initial pressure (in atmospheres), and final volume.
3.	**Relationships:**	Use Boyle's law, $P_1V_1 = P_2V_2$; rearrange variables to solve for P_2: $$P_2 = \frac{P_1 \times V_1}{V_2}$$
4.	**Solution:**	Plug in the numbers and solve: $$P_2 = \frac{1\ \text{atm} \times 5.0\ \text{L}}{0.5\ \text{L}} = 10\ \text{atm}$$

Your turn...

a. A total of 0.50 L of O_2 is collected at a pressure of 300 mm Hg. What volume will this gas occupy at sea level (760 mm Hg) at constant temperature and mass?

b. A total of 1.0 L of helium is stored at sea level (1,013 mb). If the gas is carried to the top of Mt. Washington (pressure = 800 mb), what volume will it occupy at constant temperature and mass?

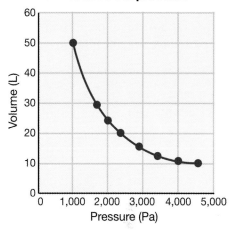

Volume vs. pressure

Figure 13.10: *This graph shows the relationship between the pressure and volume of a gas when the temperature does not change.*

SOLVE FIRST LOOK LATER

a. 0.20 L

b. 1.3 L

Pressure and temperature

Pressure and temperature The pressure of a gas is affected by temperature changes. If the mass and volume are kept constant, the pressure goes up when the temperature goes up, and the pressure goes down when the temperature goes down.

Why temperature affects pressure Pressure changes with temperature because the average kinetic energy of moving molecules is proportional to temperature. Hot molecules move faster than cold molecules. Faster molecules exert more force when they bounce off each other and off the walls of their container (Figure 13.11).

Gay-Lussac's law The mathematical relationship between the temperature and pressure of a gas at constant volume and mass was discovered by Joseph Gay-Lussac in 1802.

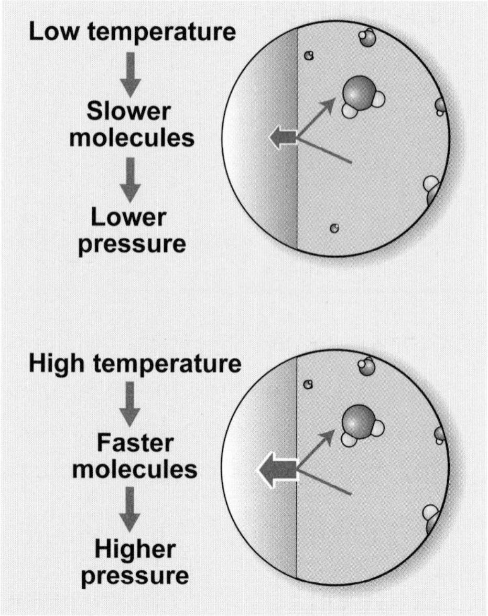

Figure 13.11: *Faster molecules create higher pressure because they exert larger forces as they collide with the sides of the container.*

PRESSURE-TEMPERATURE RELATIONSHIP

Initial pressure P_1 P_2 New pressure

$$\frac{P_1}{T_1} = \frac{P_2}{T_2}$$

Initial temperature (K) T_1 T_2 New temperature

Volume and mass remain constant

Use Kelvins for problems related to gas Any time you see a temperature in a formula in this section about gases, the temperature must be in Kelvins. This is because only the Kelvin scale starts from absolute zero. A temperature in Kelvins expresses the true thermal energy of the gas above zero thermal energy. A temperature in Celsius measures only the energy, relative to zero Celsius. Remember, temperature must be in Kelvins for problems related to the law of gases!

Kelvin temperature scale The Kelvin temperature scale starts at absolute zero. Add 273 to the temperature in Celsius to get the temperature in Kelvins (Figure 13.12). For example, a temperature of 21°C is equal to 294 K (21 + 273).

CONVERTING CELSIUS TO KELVIN

$$T_{Kelvin} = T_{Celsius} + 273$$

Figure 13.12: *To convert degrees Celsius to Kelvins, simply add 273 to the Celsius temperature.*

Buoyancy, volume, and temperature

Sinking in a gas   Like water, gases can create buoyancy forces. Because gas can flow and has a very low density, objects of higher density sink quickly. For example, if you drop a penny, it drops through the air quite easily. This is because the density of a penny is 9,000 times greater than the density of air.

Floating in a gas   Objects of lower density can float on gas of higher density. A hot air balloon floats because it is less dense than the surrounding air. What makes the air inside the balloon less dense? The word "hot" is an important clue. To get their balloons to fly, balloonists use a torch to heat the air inside the balloon. The heated air in the balloon expands and lowers the overall density of the balloon to less than the density of the surrounding cooler air (Figure 13.13).

Charles's law   The balloon example illustrates an important relationship, known as **Charles's law**, discovered by Jacques Charles in 1787. According to Charles's law, the volume of a gas increases with increasing temperature. The volume decreases with decreasing temperature.

CHARLES'S LAW

Initial volume V_1

$$\frac{V_1}{T_1} = \frac{V_2}{T_2}$$

Initial temperature (K) T_1 New volume V_2 New temperature (K) T_2

Pressure and mass remain constant

The buoyancy of hot air   Charles's law explains why the air inside the balloon becomes less dense than the air outside the balloon. The volume increases as the temperature increases. Since there is the same total mass of air inside, the density decreases and the balloon floats. Stated another way, the weight of the air displaced by the balloon provides buoyant force to keep the balloon in flight.

VOCABULARY

Charles's law – at constant pressure and mass, the volume of a gas increases with increasing temperature and decreases with decreasing temperature

15°C outside = denser

50°C inside = less dense

Figure 13.13: *A hot air balloon floats because the air inside is less dense than the air outside.*

Solving Problems: More Gas Laws

A can of hair spray has a pressure of 300 psi at room temperature (21°C). The can is accidentally moved too close to a fire, and its temperature increases to 295°C. What is the final pressure in the can (rounded to the nearest whole number)?

1. **Looking for:**	You are asked for final pressure in psi.
2. **Given:**	You are given initial pressure in psi, and initial and final temperatures in °C.
3. **Relationships:**	Convert temperatures to K: °C + 273
	Apply the pressure–temperature relationship: $P_1 \div T_1 = P_2 \div T_2$
4. **Solution:**	Convert °C to K: 21°C + 273 = 294 K and 295°C + 273 = 568 K
	Rearrange variables and solve:
	P2 = (P1 × T2) ÷ T1 = (300 psi × 568 K) ÷ 294 K = 580 psi.
	Note: This is why you should NEVER put spray cans near heat (Figure 13.14). The pressure can increase so much that the can explodes!

Figure 13.14: *NEVER put spray cans near heat!*

SOLVE FIRST LOOK LATER

a. 0.46 m³

b. 232 in³

c. 2.8 atm

Your turn...

a. A balloon filled with helium has a volume of 0.50 m³ at 21°C. Assuming the pressure and mass remain constant, what volume will the balloon occupy at 0°C?

b. A tire contains 255 in³ of air at a temperature of 28°C. If the temperature drops to 1°C, what volume will the air in the tire occupy? Assume no change in pressure or mass.

c. A gas in a container has a pressure of 3 atm at 21°C. What will the pressure be if the temperature drops to 5°C? Assume constant volume and mass.

Section 13.2 *Review*

1. Boyle's law states that if you squeeze a fixed amount of a gas into a smaller volume, the pressure will increase. Explain why in your own words.

2. Which statement is true?
   a. When the pressure of a gas increases, its density decreases.
   b. Generally, the density of a gas increases with an increase in pressure.
   c. The density of a gas is not related to pressure.

3. Atmospheric pressure at the top of Mt. Everest is about 150 mm Hg. That is why climbers always bring oxygen tanks. If a climber carries a 12.0-liter tank with a pressure of 35,000 mm Hg, what volume will the gas occupy if it is released at the top of Mt. Everest?

4. If the mass and volume are kept constant, what happens to the pressure of a gas if the temperature goes up?

5. Explain why pressure changes with temperature.

6. Average human body temperature is 98.6°F. What is average human body temperature in Kelvins?

7. A helium balloon has a pressure of 40 psi at 20°C. What will the pressure be at 40°C? Assume constant volume and mass.

8. If mass and pressure are constant, what is the relationship between temperature and volume?

9. Charles's law deals with what quantities?
   a. pressure and temperature
   b. pressure and volume
   c. volume and temperature
   d. volume, temperature, and pressure

10. A balloon has a volume of 0.5 L at 20°C. What will the volume be if the balloon is heated to 150°C? Assume constant pressure and mass.

CHALLENGE

Divers get "the bends" if they come up too fast during a dive. This is because gas in their blood expands, forming bubbles in their blood. If a diver has 0.03 L of gas in her blood under a pressure of 190 atm, then rises fast to a depth where the pressure is 50 atm, what will the volume of gas in her blood be? Do you think this will harm the diver? Explain your answer.

Up, Up, and Away

On the "Wings" of Air Pressure

Imagine you're traveling by airplane across the country. You're settled comfortably in your seat, a small pillow tucked behind your head, a cold drink in your hand, ready to relax. Before the movie begins, you take the opportunity to peer out the window. Suddenly you realize: I am hurtling through the air at more than 500 miles per hour, 30,000 feet above the ground, with only the structure of this plane around me!

Just how does an airplane stay in the air? Would you be surprised to know that something as simple as air itself makes flight possible? In fact, it is the behavior of air, specifically when it acts as a force called air pressure, that enables airplanes to do many different things.

Airplanes depend on air pressure for flight, technology, and comfort. Air pressure is used to keep an aircraft in the air as well as to propel it forward. Flight instruments use air pressure to measure altitude, air speed, and vertical speed. Finally, a pressurized airplane uses air pressure to maintain appropriate pressure for human survival in its cabin.

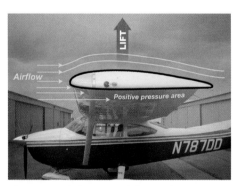

Airfoils, lift, and thrust

An airplane has shapes, called airfoils, that are designed specifically to perform in certain ways as the plane moves through the air. Airplane wings, tail surfaces, and propellers are all airfoils. The essential upward force that acts on an airfoil moving through the air is called lift. Lift is created by changes in the air pressure surrounding an airfoil.

Bernoulli's principle explains that the energy of any sample of fluid, like air, must remain constant while it is moving. If one variable, such as speed, is increased, then another variable, such as pressure, must decrease in order for the energy to remain constant. Due to the curved shape of an airplane wing, the air that flows over it during flight will have a longer distance to travel than the air that flows beneath it. The speed of the air flowing above the wing will be higher than the air flowing below it. Therefore, the air pressure above the wing will be lower than the pressure below the wing. This differential in air pressure is what generates lift and holds the wing aloft.

Flight instruments and air pressure

Air pressure is also essential to the function of several flight instruments in an airplane. These instruments are all applications of Boyle's law. Boyle's law states that for a fixed quantity of a gas, the pressure and volume are inversely related. If the volume of a gas decreases, its molecules will collide more rapidly, causing the pressure to increase. Conversely, when the volume of a gas expands, its pressure decreases, due to less frequent molecular collisions.

Altitude is the measure of the height of an airplane above a given level. An altimeter is an instrument that measures altitude. Essentially, an altimeter is an aneroid barometer linked to an indicator that points to measurements on a scale in feet. The air at Earth's surface is more dense, so the air pressure decreases as altitude increases. The partial vacuum inside a sealed case in the altimeter expands or contracts with changes in atmospheric pressure, mechanically moving the indicator to show altitude on the dial. As an airplane ascends, decreased pressure causes the altimeter to indicate increased altitude. As a plane descends, increased pressure causes the altimeter to indicate decreased altitude.

The altimeter shows the plane's cruising altitude of 8,000 feet.

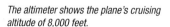

Why do your ears "pop"?

When your ears pop in a descending airplane, they are acting just like an altimeter! As the air pressure outside your body increases, the volume of air behind your eardrums decreases and you feel as though your ears are blocked. Your body then works to open an airway to equalize the pressure on both sides of your eardrums. When this happens, you feel a "pop" and relief from the blockage.

An airplane also uses air pressure to determine how fast it is moving, for both its forward and vertical speed. A vertical speed indicator is an instrument that shows whether an airplane is climbing, descending, or in level flight. It is made up of a diaphragm in a sealed case that is linked to an indicator. As the airplane moves upward or downward, the pressure in the diaphragm changes more quickly than the pressure in the case, causing the diaphragm to expand or contract, moving the indicator to show a climb, descent, or level flight.

The vertical speed indicator shows that this plane is ascending at a rate of 500 feet per minute.

According to the airspeed indicator, this plane is traveling at 100 knots.

The airspeed indicator is yet another sensitive pressure gauge used in an airplane. It is part of a larger system called the pitot-static system. This system uses two types of pressure, pitot (impact) and static, to take measurements. When an airplane is moving through the air, the impact pressure is greater than the static pressure. The airspeed indicator measures this difference and moves the pointer on the instrument to show the speed of the airplane in miles per hour or nautical miles per hour (knots).

Cabin pressure

One other use of air pressure in aviation is in a pressurized airplane. At altitudes above 10,000 feet, the decreased air pressure makes it difficult for the human body to process oxygen. This can lead to loss of consciousness or even death. To solve this problem, air is pumped into the cabin of the airplane. This compresses the air in the cabin, thus increasing the air pressure, which allows pilots and passengers to fly safely at higher altitudes.

It's amazing to think how the behavior of just one gas, air, plays such a significant role in aviation! Without the understanding of air pressure and its applications, inventors could not have made airplane flight possible for us. Thanks to Bernoulli, Boyle, and other scientists, we now can soar over Earth on the "wings" of air pressure.

Questions:

1. How does Bernoulli's principle explain how lift is generated?

2. Explain how an altimeter uses air pressure to measure altitude.

3. Prepare a report (written or oral) on the contributions of an aviation pioneer. Research this individual's life and work, specifically looking for examples of how he or she applied an understanding of science to aviation. Cite your sources using the required format.

Flight instrument photos and airplane photo courtesy of Michail Sheen.

Chapter 13 *Assessment*

Vocabulary

Select the correct term to complete each sentence.

atmosphere	barometer	Charles's law
atmospheric pressure	Boyle's law	

Section 13.1

1. The layer of gases surrounding a planet is called its _____ .

2. _____ is a measurement of the force of air molecules in the atmosphere at a given altitude.

3. An instrument used to measure atmospheric pressure is called a(n) _____ .

Section 13.2

4. _____ describes the relationship between the pressure and volume of a gas when temperature and mass are constant.

5. _____ describes the relationship between the volume and temperature of a gas when pressure and mass are constant.

Concepts

Section 13.1

1. What gases are found in Earth's atmosphere? How has life on Earth changed Earth's atmosphere?

2. Explain how Earth's atmosphere formed.

3. What is atmospheric pressure? Explain how our bodies are adapted to survive Earth's atmospheric pressure.

4. How does a mercury barometer measure atmospheric pressure? What is atmospheric pressure at sea level on a mercury barometer?

5. Which statement is true?
   a. The atmospheric pressure at sea level is greater than the atmospheric pressure at the top of Mt. Everest.
   b. Atmospheric pressure increases with altitude.
   c. Atmospheric pressure is not related to altitude.
   d. Atmospheric pressure decreases with altitude.
   e. Statements a and d are true.

Section 13.2

6. What does Boyle's law say about the relationship between the pressure and volume of a gas at constant mass and temperature?

7. What does Charles's law say about the relationship between the volume and temperature of a gas at constant pressure and mass?

8. How is the pressure of a gas affected by temperature changes? Assume no change in volume or mass in your explanation.

9. Explain how the density of a gas is different from the density of a liquid.

10. Explain, using what you have learned in this chapter, why it is not wise to place an aerosol can near a heat source.

11. A hot air balloon floats because:
    a. The heated air inside the balloon becomes less dense than the air outside the balloon.
    b. The weight of the air displaced by the balloon provides buoyant force to keep the balloon in flight.
    c. Both statements are true.
    d. Neither statement is true.

Problems

Section 13.1

1. Would you expect a barometer to have a higher reading in Alaska's Denali National Park or in Florida's Everglades National Park? (*Hint*: An atlas may help you.)

2. Convert the following barometric readings to atmospheres (atm):
   a. 1,890 mm Hg
   b. 306,000 pa
   c. 100 psi
   d. 5,000 mb

Section 13.2

3. A total of 1.00 L of helium at 1 atm is compressed to 350 mL. What is the new pressure of the gas? Assume that temperature and mass are constant.

4. A total of 5.00 L of oxygen is pumped from a tank with a pressure of 20 atm into another tank. The new pressure is 80 atm. What is the new volume of the oxygen? Assume that temperature and mass are constant.

5. A tank of helium is stored at 273 K and 10 atm of pressure. The tank is moved into a room at 293 K. What is the new pressure of the helium? Assume no change in volume or mass.

6. At 225°C, a gas has a volume of 350 mL. What is the volume of this gas at 120°C? Assume constant pressure and mass.

7. At 210°C, a gas has a volume of 7.5 L. What is the volume of this gas at –20.0°C? Assume constant pressure and mass.

8. A 7.25 L sample of nitrogen is heated from 80.5°C to 86.0°C. Find its new volume if the pressure and mass remain constant.

Applying Your Knowledge

Section 13.1

1. Find out the atmospheric pressure for today. You can find this value by listening to a local TV weather report or by going to a weather website on the Internet. Convert this pressure reading so that you have the value in inches of mercury, atmospheres, and millibars.

2. Earth's atmosphere is composed of layers. Find out why there are layers and the names of each layer.

Section 13.2

3. Describe how your body makes use of Boyle's law in order to breathe.

4. SCUBA stands for self-contained underwater breathing apparatus. A number of inventors have contributed to developing the technology for scuba diving. The invention of the aqualung by Jacques-Yves Cousteau and Emile Gagnan  in 1943 made scuba diving available to anyone who wanted to do underwater exploring. A standard-sized scuba tank is filled with the equivalent of 80 cubic feet of air at 1 atm compressed into a 0.39-cubic-foot space. What is the pressure in the tank, in psi?

5. Decompression sickness, also known as the bends, can happen when a diver surfaces too quickly. Research decompression sickness. What are some other situations where it can occur? What happens inside the body? How is it treated? Prepare a pamphlet explaining the condition.

Unit 5

Atoms, Elements, and Compounds

CHAPTER 14 **Atoms**

CHAPTER 15 **Elements and the Periodic Table**

CHAPTER 16 **Compounds**

Try this at home → You know that your food contains "nutrients," but did you ever stop to think about what those "nutrients" are? Take some packages of food and the periodic table of elements in your book and compare them. Do you see the names of any elements in the list of food ingredients? Do you see any elements listed on the nutrition label? What about parts of words that look or sound similar to the names of any elements? Are you surprised at what you find?

Atoms

Have you ever seen fireflies on a warm summer night? These amazing creatures use a process called bioluminescence (bio means "living" and luminesce means "to glow") to create light signals to attract a mate. A firefly has special light-emitting organs in its abdomen where a chemical reaction takes place, causing the emission of light. Each species of firefly has a unique flashing pattern that they use to locate other members of the same species. Bioluminescence is a very efficient process. About 90 percent of the energy used by a firefly to create light is transformed into visible light. Contrast that with an incandescent light bulb that converts only 10 percent of its energy into visible light.

There are many other examples of bioluminescent creatures on land and in the sea. How does bioluminescence work? It has to do with atoms! After reading this chapter on atoms, you can read about how living things produce light in the chapter's Biology Connection.

CHAPTER 14 INVESTIGATIONS

14A: The Atom
What is inside an atom?

14B: Atomic Challenge!
How were the elements created?

14.1 The Structure of the Atom

Scientists once believed atoms were the smallest particles of matter. With the advancement of technology, it became clear that atoms themselves are made of simpler particles. Today, we believe atoms are made of three basic particles: protons, electrons, and neutrons. It's amazing that the incredible variety of matter around us can all be built from just three subatomic particles!

Electric charge

| VOCABULARY |

electric charge – a fundamental property of matter that can be either positive or negative

elementary charge – the smallest unit of electric charge that is possible in ordinary matter; represented by the lowercase letter e

Electric charge is a property of matter In order to understand atoms, we need to understand the idea of *electric charge*. **Electric charge** is a fundamental property of matter that can be either positive or negative. One of the two forces that hold atoms together comes from electric charge.

Positive and negative There are two different kinds of electric charge—*positive* and *negative*. Because there are two kinds of charge, the force between electric charges can be either *attractive* or *repulsive*.

- A positive and a negative charge will attract each other.

- Two positive charges will repel each other.

- Two negative charges will also repel each other.

Attract

Repel

The elementary charge Scientists use the letter e to represent the **elementary charge**. At the size of atoms, electric charge always comes in units of $+e$ or $-e$. It is *only* possible to have charges that are multiples of e, such as $+e$, $+2e$, $-e$, $-2e$, $-3e$, and so on. Scientists believe it is *impossible* for ordinary matter to have charges that are fractions of e. For example, a charge of $+0.5e$ is impossible in ordinary matter. Electric charge only appears in whole units of the elementary charge (Figure 14.1).

Electric charge only appears in multiples of the elementary charge, e.

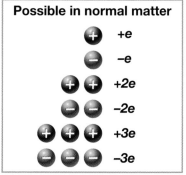

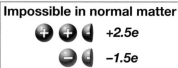

Figure 14.1: *Just as normal matter is divided into atoms, electric charge appears only in whole units of the elementary charge, e.*

Inside an atom: Solving the puzzle

The electron identified The first strong evidence that something smaller than an atom existed was found in 1897. English physicist J. J. Thomson discovered that electricity passing through a gas caused the gas to give off particles that were too small to be atoms. The new particles had negative electric charge. Atoms have zero charge. Thomson's particles are now known as **electrons**. Electrons were the first particles discovered that are smaller than atoms.

An early model of an atom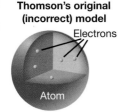

Thomson's original (incorrect) model
Electrons
Atom

Thomson proposed that negative electrons were sprinkled around inside atoms like raisins in a loaf of raisin bread. The "bread" was positively charged and the electrons were negatively charged. This was the first real model for the inside of an atom. As it soon turned out, it was not the *right* model, but it was a good place to start.

Testing the model with an experiment In 1911, Ernest Rutherford, Hans Geiger, and Ernest Marsden did an experiment to test Thomson's model of the atom. They launched positively charged helium ions (a charged atom is an *ion*) at a very thin gold foil (Figure 14.2). They expected most of the helium ions to be deflected a little as they plowed through the gold atoms.

An unexpected result! They found something quite unexpected. Most of the helium ions passed right through the foil with no deflection at all. Even more surprising—a few bounced back in the direction they came! This unexpected result prompted Rutherford to remark, "It was as if you fired a fifteen-inch (artillery) shell at a piece of tissue paper and it came back and hit you!"

The nuclear model of the atom The best way to explain the pass-through result was if a gold atom was mostly empty space. If most of the helium ions hit nothing, they wouldn't be deflected. The best way to explain the bounce-back result was if nearly all the mass of a gold atom were concentrated in a tiny, dense core at the center. Further experiments confirmed Rutherford's idea about this dense core. We now know that every atom has a tiny **nucleus**, which contains more than 99 percent of the atom's mass.

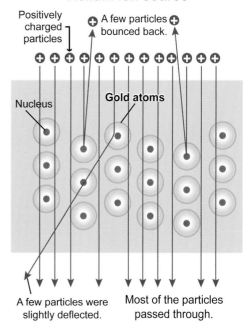

Helium ion source

Positively charged particles
A few particles bounced back.
Nucleus
Gold atoms
A few particles were slightly deflected.
Most of the particles passed through.

Figure 14.2: *Rutherford's famous experiment led to the discovery of the nucleus.*

Three particles make up all atoms

VOCABULARY

proton – a particle found in the nucleus with a positive charge exactly equal and opposite to the electron

neutron – a particle found in the nucleus with mass similar to the proton but with zero electric charge

Protons and neutrons
Today we know that the nucleus of an atom contains *protons* and *neutrons*. **Protons** have positive charge, opposite of electrons. The charge on a proton (+e) and an electron (–e) are exactly equal and opposite. **Neutrons** have zero electric charge.

The nucleus contains most of the mass

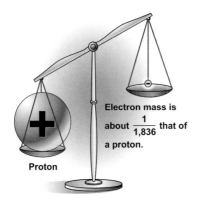

Electron mass is about $\frac{1}{1,836}$ that of a proton.

Proton

Protons and neutrons are *much* more massive than electrons. A proton has 1,836 times as much mass as an electron. A neutron has about the same mass as a proton. The chart below compares electrons, protons, and neutrons in terms of charge and mass. Because protons and neutrons have so much more mass, more than 99 percent of an atom's mass is in the nucleus.

	Occurrence	Relative charge	Mass (g)	Relative mass
⊖ **Electron**	Found outside nuclei in energy levels	−1	9.109×10^{-28}	1
⊕ **Proton**	Found in all nuclei	+1	1.673×10^{-24}	1,836
⬤ **Neutron**	Found in almost all nuclei (*exception*: most H nuclei)	0	1.675×10^{-24}	1,839

Electrons define the volume of an atom
Electrons occupy the space *outside* the nucleus in a region called the *electron cloud*. The diameter of an atom is really the diameter of the electron cloud (Figure 14.3). Compared to the tiny nucleus, the electron cloud is enormous, more than 10,000 times larger than the nucleus. As a comparison, if an atom were the size of a football stadium, the nucleus would be the size of a pea, and the electrons would be equivalent to a small swarm of gnats buzzing around the stadium at an extremely high speed. Can you imagine how much empty space there would be in the stadium? An atom is mostly empty space!

Size and structure of the atom

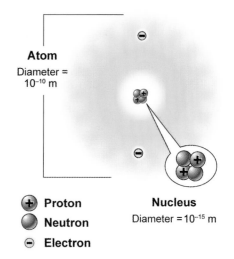

Atom
Diameter = 10^{-10} m

⊕ Proton

⬤ Neutron

⊖ Electron

Nucleus
Diameter = 10^{-15} m

Figure 14.3: *The overall size of an atom is the size of its electron cloud. The nucleus is much, much smaller.*

Forces inside atoms

Electromagnetic forces

Electrons are bound to the nucleus by the attractive force between electrons (–) and protons (+). The electrons don't fall into the nucleus because they have kinetic energy, or momentum. The energy of an electron causes it to move around the nucleus instead of falling in (Figure 14.4). A good analogy is Earth orbiting the Sun. Gravity creates a force that pulls Earth toward the Sun. Earth's kinetic energy causes it to orbit the Sun rather than fall straight in. While electrons don't really move in orbits, the energy analogy is approximately right.

Strong nuclear force

Because of electric force, all the positively charged protons in the nucleus *repel* each other. So what holds the nucleus together? There is another force that is even stronger than the electric force. We call it the *strong nuclear force*. The strong nuclear force is the strongest force known to science (Figure 14.5). This force attracts neutrons and protons to each other and works only at the extremely small distances inside the nucleus. If there are enough neutrons, the attraction from the strong nuclear force wins out over repulsion from the electromagnetic force and the nucleus stays together. In every atom heavier than helium, there is at least one neutron for every proton in the nucleus.

Weak force

There is another nuclear force called the *weak force*. The weak force is weaker than both the electric force and the strong nuclear force. If you leave a single neutron outside the nucleus, the weak force eventually causes it to break down into a proton and an electron. The weak force does not play an important role in a stable atom, but comes into action in certain special cases when atoms break apart.

Gravity

The force of gravity inside the atom is much weaker than even the weak force. It takes a relatively large mass to create enough gravity to make a significant force. We know that particles inside an atom do not have enough mass for gravity to be an important force on the scale of atoms. But there are many unanswered questions. Understanding how gravity works inside atoms is an unsolved scientific mystery.

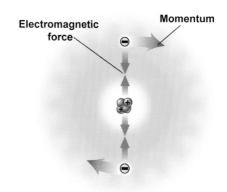

Figure 14.4: *The negative electrons are attracted to the positive protons in the nucleus, but their momentum keeps them from falling in.*

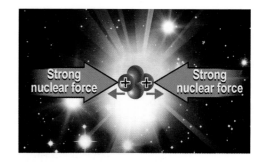

Figure 14.5: *When enough neutrons are present, the strong nuclear force wins out over the repulsion between positively charged protons and pulls the nucleus together tightly. The strong nuclear force is the strongest known force in the universe.*

How atoms of various elements are different

The atomic number is the number of protons How is an atom of one element different from an atom of another element? The atoms of different elements contain varying numbers of protons in the nucleus. For example, all atoms of carbon have six protons in the nucleus, and all atoms of hydrogen have one proton in the nucleus (Figure 14.6). Because the number of protons is so important, it is called the **atomic number**. The atomic number of an element is the number of protons in the nucleus of every atom of that element.

Elements have unique atomic numbers

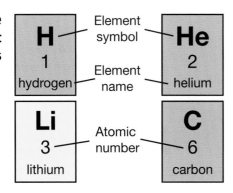

Each element has a unique atomic number. On a periodic table of elements, the atomic number is usually written above or below the atomic symbol. An atom with only one proton in its nucleus is the element hydrogen, atomic number 1. An atom with six protons is the element carbon, atomic number 6. Atoms with seven protons are nitrogen, atoms with eight protons are oxygen, and so on.

Complete atoms are electrically neutral Because protons and electrons attract each other with very large forces, the number of protons and electrons in a *complete* atom is always equal. For example, hydrogen has one proton in its nucleus and one electron outside the nucleus. The net electric charge of a hydrogen atom is zero because the negative charge of the electron cancels the positive charge of the proton. Each carbon atom has six electrons, one for each of carbon's six protons. Like hydrogen, a complete carbon atom is electrically neutral.

Ions *Ions* are atoms that have a different number of protons than electrons and so have a net electric charge. Positively charged ions have more protons than electrons. Negatively charged ions have more electrons than protons. You will read more about ions in Chapter 16.

All carbon atoms have 6 protons.

All hydrogen atoms have 1 proton.

Figure 14.6: *Atoms of the same element always have the same number of protons in the nucleus.*

Isotopes

Isotopes All atoms of the same element have the same number of protons in the nucleus. However, atoms of the same element may have different numbers of neutrons in the nucleus. **Isotopes** are atoms of the *same* element that have different numbers of neutrons.

The isotopes of carbon Figure 14.7 shows three isotopes of carbon that exist in nature. Most carbon atoms have six protons and six neutrons in the nucleus. However, some carbon atoms have seven or eight neutrons. They are all carbon atoms because they all contain six protons, but they are different *isotopes* of carbon. The isotopes of carbon are called carbon-12, carbon-13, and carbon-14. The number after the name is called the mass number. The **mass number** of an isotope tells you the number of protons plus the number of neutrons.

 Solving Problems: Isotopes

How many neutrons are present in an aluminum atom that has an atomic number of 13 and a mass number of 27?

1. **Looking for:**	You are asked to find the number of neutrons.
2. **Given:**	You are given the atomic number and the mass number.
3. **Relationships:**	Use the relationship: protons + neutrons = mass number.
4. **Solution:**	Plug in and solve: $13 + x = 27$; $x = 14$ The aluminum atom has 14 neutrons.

Your turn...

a. How many neutrons are present in a magnesium atom with a mass number of 24?

b. Find the number of neutrons in a calcium atom that has a mass number of 40.

VOCABULARY

isotopes – atoms of the same element that have different numbers of neutrons in the nucleus

mass number – the number of protons plus the number of neutrons in the nucleus

Carbon-12
$\oplus$ = 6
$\bullet$ = 6
$\ominus$ = 6

Carbon-13
$\oplus$ = 6
$\bullet$ = 7
$\ominus$ = 6

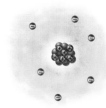

Carbon-14
$\oplus$ = 6
$\bullet$ = 8
$\ominus$ = 6

Figure 14.7: *The isotopes of carbon.*

SOLVE FIRST LOOK LATER

a. 12

b. 20

Radioactivity

What if there are too many neutrons? Almost all elements have one or more isotopes that are **stable**. Stable means the nucleus stays together. For complex reasons, the nucleus of an atom becomes unstable if it contains too many or too few neutrons relative to the number of protons. If the nucleus is unstable, it breaks apart. Carbon has two stable isotopes, carbon-12 and carbon-13. Carbon-14 is **radioactive** because it has an unstable nucleus. An atom of carbon-14 eventually changes into an atom of nitrogen-14.

Radioactivity If an atomic nucleus is unstable for any reason, the atom eventually changes into a more stable form. Radioactivity is a process in which the nucleus spontaneously emits particles or energy as it changes into a more stable isotope. Radioactivity can change one element into a completely different element.

Alpha decay When *alpha decay* occurs, the nucleus ejects two protons and two neutrons (Figure 14.8). Check the periodic table and you can quickly find that two protons and two neutrons are the nucleus of a helium-4 (He-4) atom. Alpha radiation is actually fast-moving He-4 nuclei. When alpha decay occurs, the atomic number is reduced by two because two protons are removed. The atomic mass is reduced by four because two neutrons go along with the two protons. For example, uranium-238 undergoes alpha decay to become thorium-234.

Beta decay *Beta decay* occurs when a neutron in the nucleus splits into a proton and an electron. The proton stays in the nucleus, but the high-energy electron is ejected (this is called beta radiation). During beta decay, the atomic number increases by one because one new proton is created. The mass number stays the same because the atom lost a neutron but gained a proton.

Gamma decay *Gamma decay* is how the nucleus gets rid of excess energy. In gamma decay, the nucleus emits pure energy in the form of gamma rays. The number of protons and neutrons stays the same.

VOCABULARY

stable – a nucleus is stable if it stays together

radioactive – a nucleus is radioactive if it spontaneously breaks up, emitting particles or energy in the process

Alpha decay
Nucleus ejects a helium-4 nucleus

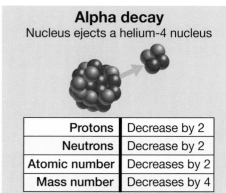

Protons	Decrease by 2
Neutrons	Decrease by 2
Atomic number	Decreases by 2
Mass number	Decreases by 4

Beta decay
Nucleus converts a neutron to a proton and an electron, ejecting the electron.

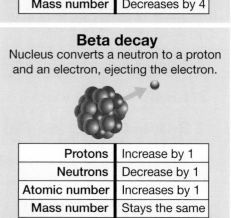

Protons	Increase by 1
Neutrons	Decrease by 1
Atomic number	Increases by 1
Mass number	Stays the same

Figure 14.8: *Two common radioactive decay reactions.*

Section 14.1 *Review*

1. Which of the following statements regarding electric charge is TRUE?
   a. A positive charge repels a negative charge and attracts other positive charges.
   b. A positive charge attracts a negative charge and repels other positive charges.

2. Is electric charge a property of just electricity, or is charge a property of all atoms?

3. Which of the drawings in Figure 14.9 is the most accurate model of the interior of an atom?

4. There are four forces in nature. Name the four forces and rank them from strongest to weakest.

5. There are three particles inside an atom. One of them has zero electric charge. Which one is it?

6. All atoms of the same element have (choose one):
   a. the same number of neutrons
   b. the same number of protons
   c. the same mass

7. The atomic number is:
   a. the number of protons in the nucleus
   b. the number of neutrons in the nucleus
   c. the number of neutrons plus protons

8. The diagram in Figure 14.10 shows three isotopes of the element carbon. Which one is radioactive?

9. *Radioactive* means:
   a. an atom gives off radio waves
   b. the nucleus of an atom is unstable and will eventually change
   c. the electrons in an atom have too much energy

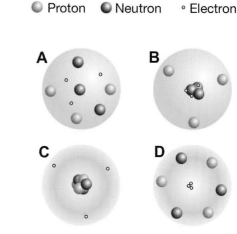

○ Proton ● Neutron ∘ Electron

Which best illustrates the interior of an atom?

Figure 14.9: *Question 3.*

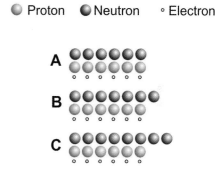

○ Proton ● Neutron ∘ Electron

Which one of these shows the particles in a radioactive isotope of carbon?

Figure 14.10: *Question 8.*

14.2 **Electrons**

Atoms interact with each other through their electrons. This is why almost all the properties of the elements (except mass) are due to electrons. Chemical bonds involve only electrons, so electrons determine how atoms combine into compounds. We find a rich variety of matter because electrons inside atoms are organized in unusual and complex patterns. Exactly how electrons create the properties of matter was a puzzle that took bright scientists a long time to figure out!

The spectrum

The spectrum is a pattern of colors

Almost all the light you see comes from atoms. For example, light is given off when electricity passes through the gas in a fluorescent bulb or a neon sign. When scientists look carefully at the light given off by a pure element, they find that the light does not include all colors. Instead, they see a few very specific colors, and the colors are different for different elements (Figure 14.11). Hydrogen has a red line, a green line, a blue line, and a violet line in a characteristic pattern. Helium and lithium have different colors and patterns. Each different element has its own characteristic pattern of colors called a **spectrum**. The colors of clothes, paint, and everything else around you come from this property of elements that allows them to emit or absorb light of only certain colors.

Spectroscopes and spectral lines

Each individual color in a spectrum is called a **spectral line** because each color appears as a line in a **spectroscope**. A spectroscope is a device that separates light into its different colors. The illustration below shows a spectroscope made with a prism. The spectral lines appear on the screen at the far right.

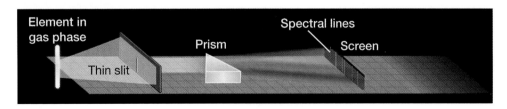

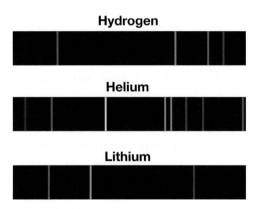

Figure 14.11: *When light from energized atoms is directed through a prism, spectral lines are observed. Each element has its own distinct pattern of spectral lines.*

VOCABULARY

spectrum – the characteristic colors of light given off or absorbed by an element

spectral line – a bright, colored line in a spectroscope

spectroscope – an instrument that separates light into a spectrum

The Bohr model of the atom

Energy and color Light is a form of pure energy that comes in tiny bundles called *photons*. A photon is the smallest possible quantity of light energy. The amount of energy in a photon determines the color of the light. Red light has lower energy, and blue light has higher energy. Green and yellow light have energy between red and blue. The fact that atoms only emit certain colors of light tells us that something inside an atom can only have certain values of energy.

Neils Bohr Danish physicist Neils Bohr (1885–1962) proposed the concept of **energy levels** to explain the spectrum of hydrogen. In Bohr's model, the electron in a hydrogen atom must be in a specific energy level. You can think of energy levels like steps on a staircase. You can be on one step or another, but you cannot be between steps except in passing. Electrons must be in one energy level or another and cannot remain in between energy levels. Electrons change energy levels by absorbing or emitting light (Figure 14.12).

Explaining the spectrum When an electron moves from a higher energy level to a lower one, the atom gives up the energy difference between the two levels. The energy comes out as different colors of light. The specific colors of the spectral lines correspond to the differences in energy between the energy levels. The diagram below shows how the spectral lines of hydrogen come from electrons falling from the 3rd, 4th, 5th, and 6th energy levels down to the 2nd energy level.

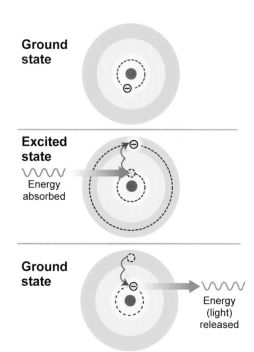

VOCABULARY

energy level – one of the discrete allowed energies for electrons in an atom

Figure 14.12: *When the right amount of energy is absorbed, an electron in a hydrogen atom jumps to a higher energy level. When the electron falls back to the lower energy level, the atom releases the same amount of energy it absorbed. The energy comes out as light of a specific color.*

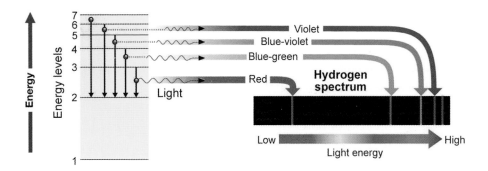

The quantum theory

VOCABULARY

quantum theory – the theory that describes matter and energy at very small (atomic) sizes

uncertainty principle – it is impossible to know variables precisely in the quantum world

Quantum versus classical

Quantum theory says that when things get very small, like the size of an atom, matter and energy do *not* obey Newton's laws or other laws of *classical* physics. That is, the classical laws are not obeyed in the same way as with a larger object, like a baseball. According to quantum theory, when a particle (such as an electron) is confined to a small space (such as inside an atom) then the energy, momentum, and other variables of the particle become restricted to certain specific values.

Everything is fuzzy in the quantum world

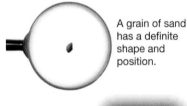

A grain of sand has a definite shape and position.

On a much smaller scale, an electron has no definite shape or position.

A particle like a grain of sand is small, but you can easily imagine it has a definite shape, size, position, and speed. According to quantum theory, particles the size of electrons are fundamentally different. When you look closely, an electron is "smeared out" into a wave-like "cloud."

The uncertainty principle

The work of German physicist Werner Heisenberg (1901–1976) led to Heisenberg's **uncertainty principle**. According to the uncertainty principle, in the quantum world, a particle's position, momentum, energy, and time can never be precisely known at the same time. For example, if you choose to measure the location of the electron, its momentum cannot be determined.

Understanding the uncertainty principle

The uncertainty principle arises because the quantum world is so small. To "see" an electron, you have to bounce a photon of light off it, or interact with the electron in some way (Figure 14.13). Because the electron is so small, even a single photon moves it and changes its motion. That means the moment you use a photon to locate an electron, you push it, so you no longer know precisely how fast it was going. However, you know its position at that moment in time. In fact, any process of observing in the quantum world changes the very system you are trying to observe. The uncertainty principle exists because measuring any variable disturbs the others in an unpredictable way.

An electron is moving.

To see the electron, you must bounce a photon of light off it.

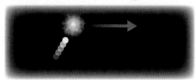

When you receive the photon, you know where the electron *was*, but the photon disturbed it, so you don't know its speed and direction anymore.

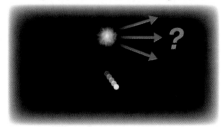

Figure 14.13: *The act of observing anything in the quantum world means disturbing in unpredictable ways the very thing you are trying to observe.*

Electrons and energy levels

The energy levels are at different distances from the nucleus

The positive nucleus attracts negative electrons like gravity attracts a ball down a hill. The farther down the "hill" an electron slides, the less energy it has. Conversely, electrons have more energy farther up the hill, and away from the nucleus. The higher energy levels are farther from the nucleus and the lower energy levels are closer.

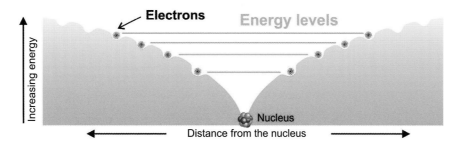

The electron cloud

While Bohr's model of electron energy levels explained atomic spectra and the periodic behavior of the elements, it was incomplete. Electrons are so fast and light that their exact position in an atom cannot be defined. Remember, in the current model of the atom, we think of the electrons as moving around the nucleus in an area called an electron cloud. The energy levels occur because electrons in the cloud are at different average distances from the nucleus.

Rules for energy levels

Inside an atom, electrons always obey these rules:

- The energy of an electron must match one of the energy levels in the atom.
- Each energy level can hold only a certain number of electrons, and no more.
- As electrons are added to an atom, they settle into the lowest unfilled energy level.

Quantum mechanics

Energy levels are predicted by *quantum mechanics*, the branch of physics that deals with the microscopic world of atoms. While quantum mechanics is outside the scope of this book, you should know that it is a very accurate theory and it explains the characteristics of the energy levels.

Orbitals

The energy levels in an atom are grouped into different shapes called *orbitals*.

The s-orbital

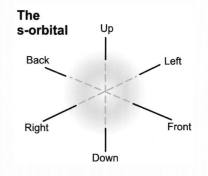

The s-orbital is spherical and holds two electrons. The first two electrons in each energy level are in the s-orbital.

The p-orbitals

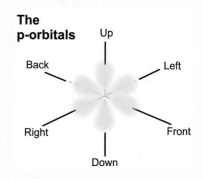

The p-orbitals hold 6 electrons and are aligned along the three directions on a 3-D graph.

The energy levels in an atom

How electrons fill in the energy levels   In the Bohr model of the atom, the first energy level can accept up to 2 electrons. The second and third energy levels hold up to 8 electrons each. The fourth and fifth energy levels hold up to 18 electrons each (Figure 14.14). A good analogy is to think of the electron cloud like a parking garage. The first level of the garage only has spaces for 2 cars, just as the first energy level only has spaces for 2 electrons. The second and third levels of the garage can hold 8 cars each, and the fourth and fifth levels can each hold 18 cars. Each new car that enters the garage parks in the lowest level with an unfilled space, just as each additional electron occupies the lowest unfilled energy level in the atom.

How the energy levels fill   The number of electrons in an atom depends on the atomic number because the number of electrons equals the number of protons. That means each element has a different number of electrons and therefore fills the energy levels to a different point. For example, a helium atom (He) has two electrons (Figure 14.15). The two electrons completely fill up the first energy level (diagram below). The next element is lithium (Li) with three electrons. Because the first energy level only holds two electrons, the third electron must go into the second energy level. The diagram shows the first 10 elements that fill the first and second energy levels.

Energy levels (First four)

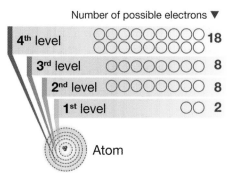

Figure 14.14: *Electrons occupy energy levels around the nucleus. The farther away an electron is from the nucleus, the higher the energy it possesses.*

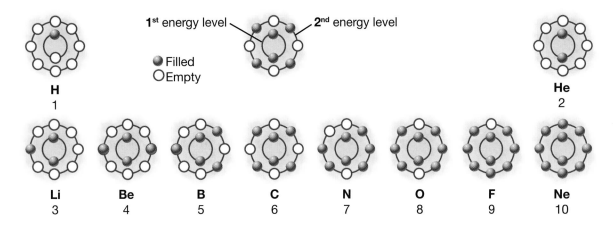

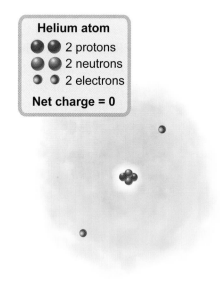

Figure 14.15: *A helium-4 atom has two protons in its nucleus and two electrons.*

Section 14.2 *Review*

1. The pattern of colors given off by a particular atom is called:

   a. an orbital

   b. an energy level

   c. a spectrum

2. Which of the diagrams in Figure 14.16 corresponds to the element lithium?

3. When an electron moves from a lower energy level to a higher energy level, the atom:

   a. absorbs light

   b. gives off light

   c. becomes a new isotope

4. Two of the energy levels can hold eight electrons each. Which energy levels are they?

5. How many electrons can fit in the fourth energy level?

6. The element beryllium has four electrons. Which diagram in Figure 14.17 shows how beryllium's electrons are arranged in the first four energy levels?

7. Which two elements have electrons only in the first energy level?

   a. hydrogen and lithium

   b. helium and neon

   c. hydrogen and helium

   d. carbon and oxygen

8. On average, electrons in the fourth energy level are:

   a. farther away from the nucleus than electrons in the second energy level

   b. closer to the nucleus than electrons in the second energy level

   c. about the same distance from the nucleus as electrons in the second energy level

Which belongs to lithium?

Figure 14.16: *Question 2.*

Which is correct for normal beryllium?

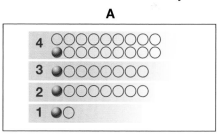

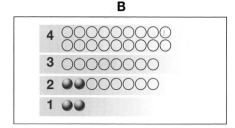

Figure 14.17: *Question 6.*

Bioluminescence
Glow Live!

Imagine you could make your hands glow like living flashlights. No more fumbling around for candles when the power goes out! You could read in bed all night, or get a job directing airplanes to their runways.

Although a glowing hand might sound like something from a science fiction movie, many living things can make their own light. On warm summer evenings, fireflies flash signals to attract a mate. A fungus known as foxfire glows in decaying wood. While there are only a few kinds of glowing creatures that live on land, about 90 percent of the animals that live in the deep parts of the ocean make their own light!

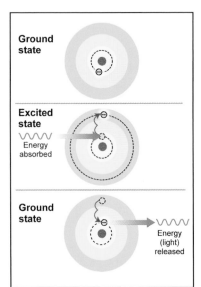

How do they do that?

Almost everything that creates light is made of atoms. If an atom absorbs energy, an electron can move to a higher energy level. When the electron moves back down to its original energy level, the atom may give off visible light.

Atoms can absorb energy from a number of sources. Electrical energy is used to light ordinary light bulbs. Mechanical energy can be used, too. Hit two quartz rocks together in a dark room, and you'll see flashes of light as the energized electrons fall back down to lower energy levels.

Atoms can also use the energy from a chemical reaction. When you bend a glow stick, you break a vial inside so that two chemicals can combine. When they react, energy is released and used to make light.

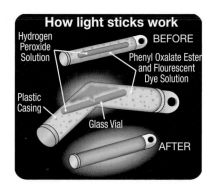

Bioluminescence

Like a glow stick, living things produce their own light using a chemical reaction. We call this process *bioluminescence* (*bio-* means "living" and *luminesce* means "to glow"). Bioluminescence is "cold light" because it doesn't produce a lot of heat. While it takes a lot of energy for a living thing to produce light, almost 100 percent of the energy becomes visible light. In contrast, only 10 percent of the energy used by an incandescent electric light bulb is converted to visible light. Ninety percent of the energy is lost as heat.

The chemical reaction

Three ingredients are usually needed for a bioluminescent reaction to occur: an organic chemical known as luciferin, a source of oxygen, and an enzyme called luciferase. Luciferin in a firefly is not exactly the same as the luciferin in foxfire fungus. However, both luciferin chemicals are carbon based and have the ability to give off light under certain conditions.

Firefly light

In a firefly, luciferin and luciferase are stored in special cells in the abdomen called *photocytes*. To create light, fireflies push oxygen into the photocytes. When the luciferin and luciferase are exposed to oxygen, they combine with ATP (a chemical source of energy) and magnesium. This chemical reaction drives some of the luciferin electrons into a higher energy state. As they fall back down to their "ground state," energy is given off in the form of visible light.

Why make light?

Living creatures don't have an endless supply of energy. Since it takes a lot of energy to make light, there must be good reasons for doing it.

Fireflies flash their lights in patterns to attract a mate. The lights also warn predators to stay away, because the light-producing chemicals taste bitter. Light can also be used as a distress signal, warning others of their species that there is danger nearby. The female of

one firefly species has learned to mimic the signal of other types of fireflies. She uses her light to attract males of other species and then she eats them!

It's a little harder to figure out why foxfire fungus glows. Some scientists think that the glow attracts insects that help spread the fungus spores.

Bioluminescent ocean creatures use their lights in amazing ways. The deep-sea angler fish looks like it has a glowing lure attached to its head. It is actually a modified spine with a fleshy bulb (called an esca) at the tip. Bioluminescent bacteria grow in the esca, causing it to glow. When a smaller fish comes to munch on the "lure," it is gobbled up by the angler fish instead.

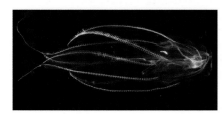

Comb jellies are some of the ocean's most beautiful glowing creatures. Comb jellies are mostly colorless, but they have iridescent plates that reflect sunlight. This picture (left) shows a comb jelly under reflected light. You are seeing iridescence, not bioluminescence.

Comb jellies can produce bright flashes of light to startle a predator. This photo set (right) shows the comb jelly's bioluminescence. When threatened, some comb jellies release a cloud of bioluminescent particles into the water, temporarily blinding the attacker.

So far, we know that living creatures use bioluminescence to attract mates, to communicate, to find food, and to ward off attackers. Perhaps someday you will be part of a research team that discovers even more uses for bioluminescence.

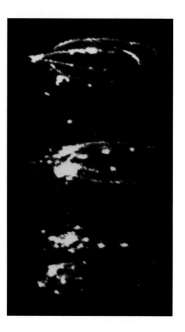

Questions:

1. Name three sources of energy that can be absorbed by atoms to produce light. Which source is used in bioluminescent organisms?

2. Bioluminescence is found in a wide range of living organisms, including bacteria, fungi, insects, crustaceans, and fish. However, no examples have been found among flowering plants, birds, reptiles, amphibians, or mammals. Why do you think this is so?

3. Use the Internet or a library to find out more about bioluminescent sea creatures. Here are some questions to pursue: What is the most common color of light produced? What other colors of bioluminescence have been found?

Angler fish and comb jelly photos by Edith Widder. Fungus photos by Garth Fletcher.

Chapter 14 *Assessment*

Vocabulary

Select the correct term to complete each sentence.

atomic number	isotopes	spectral line
electron	mass number	quantum theory
elementary charge	neutron	radioactive
energy level	nucleus	spectroscope

Section 14.1

1. The sum of protons plus neutrons in the nucleus of an atom is known as the _____ .

2. The smallest unit of electric charge in matter is called _____ .

3. The core of the atom containing most of the atom's mass and all of its positive charge is called the _____ .

4. A light particle with a negative charge, found in atoms, is called a(n) _____ .

5. A neutral particle with nearly the same mass as the proton is the _____ .

6. The number of protons in an atom, unique to each element, is known as the _____ .

7. A nucleus that spontaneously breaks apart, emitting particles of energy, is referred to as _____ .

8. Atoms of the same element containing different numbers of neutrons are called _____ .

Section 14.2

9. One of the allowed energies for electrons in an atom is known as a(n) _____ .

10. The theory that describes matter and energy at atomic sizes is the _____ .

11. A bright-colored line produced by a spectroscope is a(n) _____ .

12. An instrument that is used to separate light into spectral lines is a(n) _____ .

Concepts

Section 14.1

1. Explain why Rutherford assumed most of the atom to be empty space.

2. Explain how Rutherford concluded that positive charge was concentrated in a small area.

3. How did Rutherford's model of the atom differ from Thomson's model?

4. Summarize the characteristics of the electron, proton, and neutron, comparing their relative mass, charge, and location within the atom by completing the table below.

Particle	Location in atom	Charge	Relative mass
Electron	?	?	1
Proton	?	+1	?
Neutron	?	?	?

5. Name the four forces of nature and compare their relative strengths.

6. Explain the effect of the electromagnetic and strong forces on the structure of the atom.

7. What do the atomic number and mass number tell you about an atom?

8. Compare the number of protons and electrons in a neutral atom.

9. Compare the mass number and atomic number for isotopes of an element. Explain your answer.

10. Describe the radioactive disintegrations known as alpha, beta, and gamma decay.

Section 14.2

11. Which particle in an atom is most responsible for its chemical properties?

12. What is the source of the light you see?

13. How can a spectroscope be used to identify an element heated to incandescence?

14. Cite evidence that electrons are restricted to having only certain amounts of energy.

15. How did Neils Bohr explain spectral lines?

16. What is the difference between an electron in ground state and one in an excited state?

17. What would occur if an electron were to move from a certain energy level to a lower energy level?

18. Summarize the uncertainty principle.

19. Why can't the position of an electron be determined with certainty?

20. How is the location of an electron described?

Problems

Section 14.1

1. Which of the following charges do *not* appear in normal matter?

  a. $+2e$ **d.** $-5.4e$

  b. $+1/4e$ **e.** $+3/4e$

  c. $-4e$ **f.** $-1e$

2. What charge would an atom have if it lost one electron?

3. For each of the nuclei shown below, do the following.

Atom A	Atom B	Atom C	Atom D
17 protons 18 neutrons	20 protons 20 neutrons	29 protons 34 neutrons	35 protons 45 neutrons

  a. Name the element.
  b. Give the atomic number.
  c. Give the mass number.

4. A neutral atom has 7 protons and 8 neutrons. Determine its:

  a. mass number
  b. atomic number
  c. number of electrons

5. A carbon atom contains 6 protons in the nucleus. If an atom of carbon-14 were to undergo alpha decay, determine each of the following for the new element.

  a. mass number
  b. atomic number
  c. number of protons
  d. number of neutrons

6. A uranium atom contains 92 protons in the nucleus. If an atom of uranium-238 were to undergo alpha decay, determine each of the following for the new element.

   a. mass number
   b. atomic number
   c. number of protons
   d. number of neutrons

Section 14.2

7. If electrons in the hydrogen atom become excited and then fall back to the 2nd energy level from levels 3, 4, 5, and 6, four colors of light are emitted: violet, red, blue-violet, and blue-green.

   a. Which transition is responsible for the blue-violet light: 6 to 2, 5 to 2, 4 to 2, or 3 to 2?

   b. If an electron on the 2nd level were struck by a photon, then it could be excited to the 6th energy level. What color photon would be absorbed by the electron?

8. An atom has an atomic number of 6. Sketch a diagram that correctly represents the electron arrangement in energy levels around the nucleus. What is the name of this atom?

Applying Your Knowledge

Section 14.1

1. Make a poster illustrating the different models of the atom that scientists have proposed since the 1800s. Explain how each model reflects the new knowledge that scientists gained through their experiments. When possible, comment on what scientists learned about charge, mass, and location of subatomic particles.

2. Radioactive isotopes emit particles that can cause harm to our cells. However, scientists have figured out ways to use radioisotopes in ways that are beneficial to our health. Nuclear medicine is a branch of medicine that uses medical radioisotopes to diagnose and treat diseases. Research a disease that is either diagnosed or treated with medical radioisotopes. Create a pamphlet that provides information about the disease and how medical radioisotopes are used to diagnose and/or treat it.

Section 14.2

3. Research how helium was discovered and how it got its name. The answer may surprise you.

4. Choose an atom and make a three-dimensional model of its structure, using the Bohr model. Choose different materials to represent protons, neutrons, and electrons. Attach a key to your model that explains what each material represents.

Elements and the Periodic Table

What are metals like? Think of things that are made with metals like aluminum, copper, iron, and gold. What do they have in common? They are usually shiny, and they can often be bent into different shapes without breaking. Did you know there is a metal that is shiny, but is so soft it can be cut with a knife? This metal is very reactive. If you place a piece of this metal in water, it will race around the surface, and the heat given off is often enough to melt the metal and ignite the hydrogen gas that is produced! This strange metal is called sodium. You can look at the periodic table of elements to find other metals that behave like sodium. In this chapter, you will become familiar with how you can predict the properties of different elements by their location on the periodic table.

CHAPTER 15 INVESTIGATIONS

15A: The Periodic Table
How is the periodic table organized?

15B: Periodic Table Challenge
What information can you get from the periodic table?

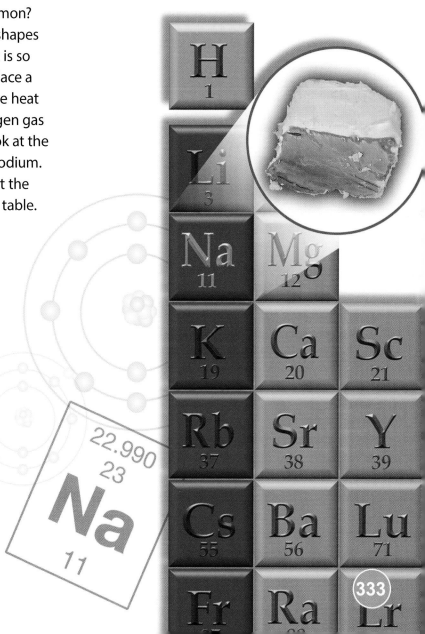

15.1 The Periodic Table of the Elements

Long before scientists understood atoms, they grouped elements by their chemical properties. Chemical properties can only be observed when a chemical change occurs. In this section, you will learn how the *periodic table* gives us a way to organize all the known elements. The periodic table also shows how chemical properties are related to the arrangement of electrons inside the atom.

Chemical changes

Physical properties Recall that the physical properties of water include that it is colorless, odorless, and exists as a liquid at room temperature. Physical properties include color, texture, density, brittleness, and state (solid, liquid, or gas). Melting point, boiling point, and specific heat are also physical properties.

Physical changes are reversible A **physical change** in matter, such as melting, boiling, or bending, is usually *reversible* and does not result in the formation of a new substance. When water freezes, it undergoes a physical change from a liquid to a solid. This does not change the water into a new substance. It is still water, only in solid form. The change can easily be reversed by melting the solid water (ice). Bending a steel bar is another physical change. Bending changes the shape of the bar, but it is still steel.

Chemical properties As you have read, properties that can only be observed when one substance changes into a different substance are called chemical properties. For example, if you leave an iron nail outside, it will eventually rust (Figure 15.1). A chemical property of iron is that it reacts with oxygen in the air to form iron oxide (rust).

Chemical changes are hard to reverse Any change that transforms one substance into a different substance is called a **chemical change**. The transformation of iron into rust is a chemical change. Chemical changes are not easily reversible. Rusted iron will not turn shiny again even if you remove it from the oxygen in the air.

VOCABULARY

physical change – a change that does not result in a new substance being formed

chemical change – a change that transforms one substance into a different substance

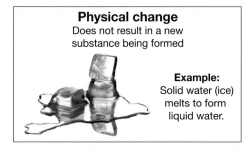

Physical change
Does not result in a new substance being formed

Example: Solid water (ice) melts to form liquid water.

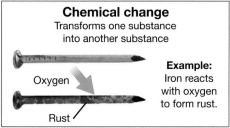

Chemical change
Transforms one substance into another substance

Oxygen

Rust

Example: Iron reacts with oxygen to form rust.

Figure 15.1: *Rusting is an example of a chemical change.*

The periodic table

How many elements are there?

How many elements make up the universe? The only way to tell if a substance is an element is to try to break it down into other substances by any possible means. A substance that can be chemically broken apart cannot be an element. As of this writing, scientists have identified 118 confirmed elements. Only about 90 of these elements occur naturally. The others are made in laboratories.

The modern periodic table

As chemists worked on identifying the true elements, they noticed that some elements acted like other elements. For example, the soft metals lithium, sodium, and potassium always combine with oxygen in a ratio of two atoms of metal to one atom of oxygen (Figure 15.2). By keeping track of how each element combined with other elements, scientists began to recognize repeating patterns. From this data, they developed the first periodic table of the elements. The **periodic table** organizes the elements according to how they combine with other elements due to their chemical properties.

Organization of the periodic table

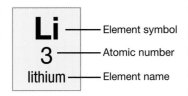

The periodic table is organized in order of increasing atomic number. The lightest element (hydrogen) is at the upper left. The heaviest (#118) is on the lower right. Each element corresponds to one box on the periodic table, identified with the element symbol.

The periodic table is further divided into *periods* and *groups*. Each horizontal row is called a **period**. Across any period, the properties of the elements gradually change. Each vertical column is called a **group**. Groups of elements have similar properties. The *main group elements* are Groups 1 and 2 and Groups 13 through 18 (the tall columns of the periodic table). Elements in Groups 3 through 12 are called the *transition elements*. The inner transition elements, called lanthanides and actinides, are often shown below the bottom row of the chart in order for the chart to fit on a page.

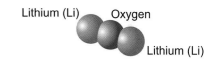

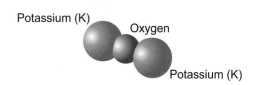

Figure 15.2: *The metals lithium, sodium, and potassium all form compounds with a ratio of two atoms of the metal to one atom of oxygen. All the elements in Group 1 of the periodic table form similar compounds.*

VOCABULARY

periodic table – a chart that organizes the elements by their chemical properties and increasing atomic number

period – a row of the periodic table

group – a column of the periodic table

Reading the periodic table

Metals, nonmetals, and metalloids Most of the elements are metals. A **metal** is typically shiny, opaque, and a good conductor of heat and electricity as a pure element. Metals are also ductile, which means they can be bent into different shapes without breaking. **Nonmetals** are poor conductors of heat and electricity. Solid nonmetals are brittle and appear dull. With the exception of hydrogen, the nonmetals are on the right side of the periodic table. The elements on the border between metals and nonmetals are called *metalloids*. Silicon is an example of a metalloid element with properties in between those of metals and nonmetals.

■■■■ **VOCABULARY** ■■■■

metals – elements that are typically shiny and good conductors of heat and electricity

nonmetals – elements that are poor conductors of heat and electricity

Periodic Table of the Elements

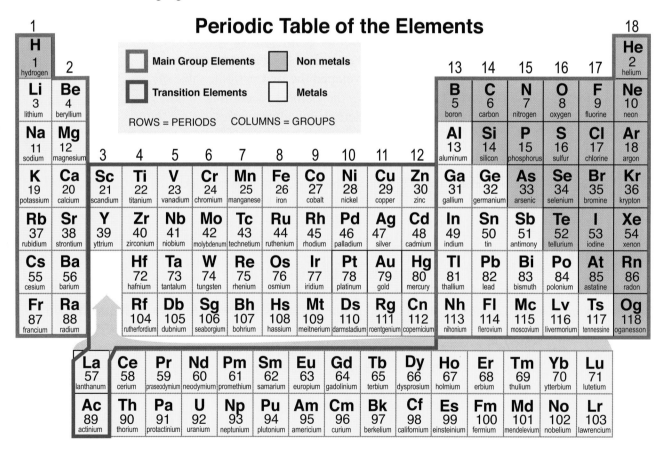

Atomic mass

Atomic mass units   The mass of individual atoms is so small that the numbers are difficult to work with. To make calculations easier, scientists came up with the **atomic mass unit** (amu). One atomic mass unit is about the mass of a single proton (or neutron). In laboratory units, 1 amu is 1.66×10^{-24} grams. That's 0.00000000000000000000000166 grams!

Atomic mass and isotopes   The **atomic mass** is the *average* mass (in amu) of an atom of each element. Atomic masses differ from mass numbers because most elements in nature contain more than one isotope (see chart below). For example, the atomic mass of lithium is 6.94 amu. That does *not* mean there are 3 protons and 3.94 neutrons in a lithium atom! On average, out of every 100 atoms of lithium, 6 atoms are Li-6 and 94 atoms are Li-7 (Figure 15.3). The *average* atomic mass of lithium is 6.94, because of the mixture of isotopes.

Atomic number review   As you learned earlier, the atomic number is the number of protons that all atoms of that element have in their nuclei. If the atom is neutral, it will have the same number of electrons as well.

VOCABULARY

atomic mass unit – a unit of mass equal to 1.66×10^{-24} grams

atomic mass – the average mass of all the known isotopes of an element, expressed in amu

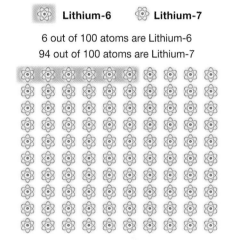

Lithium-6   Lithium-7

6 out of 100 atoms are Lithium-6
94 out of 100 atoms are Lithium-7

Figure 15.3: *Naturally occurring elements have a mixture of isotopes.*

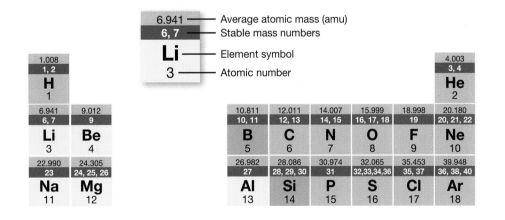

6.941 ——— Average atomic mass (amu)
6, 7 ——— Stable mass numbers
Li ——— Element symbol
3 ——— Atomic number

Groups of the periodic table

Alkali metals

All of the elements in the different groups of the periodic table have similar chemical properties. The first group is known as the **alkali metals**. This group includes the elements lithium (Li), sodium (Na), and potassium (K). The alkali metals are soft and silvery in their pure form and are highly reactive. Each of them combines in a ratio of two to one with oxygen. For example, lithium oxide has two atoms of lithium per atom of oxygen.

Oxides

BeO   MgO   CaO

Li_2O Na_2O K_2O

Group 2 metals

The Group 2 metals include beryllium (Be), magnesium (Mg), and calcium (Ca). These metals also form oxides; however, they combine one-to-one with oxygen. For example, beryllium oxide has one beryllium atom per each oxygen atom.

Salts

LiF   NaCl   KCl

Halogens

The **halogens** are on the right-hand side of the periodic table. These elements tend to be toxic in their pure form. Some examples are fluorine (F), chlorine (Cl), and bromine (Br). The halogens are also very reactive and are rarely found in pure form. When combined with alkali metals, they form *salts* such as sodium chloride (NaCl) and potassium chloride (KCl).

Noble gases

On the far right of the periodic table are the **noble gases**, including the elements helium (He), neon (Ne), and argon (Ar). These elements do not naturally form chemical bonds with other atoms and are almost always found in their pure state. They are sometimes called *inert gases* for this reason.

Transition metals

In the middle of the periodic table are the transition metals, including titanium (Ti), iron (Fe), and copper (Cu). These elements are usually good conductors of heat and electricity. For example, the wires that carry electricity in your school are made of copper. Figure 15.4 shows the location of the groups of elements on the periodic table.

alkali metals – elements in the first group of the periodic table

halogens – elements in the group containing fluorine, chlorine, and bromine, among others

noble gases – elements in the group containing helium, neon, and argon, among others

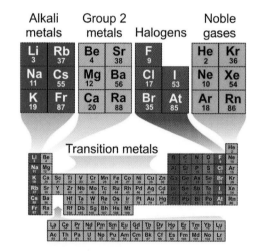

Figure 15.4: *Groups of the periodic table.*

Energy levels and the periodic table

Period 1 is the first energy level — The periods (rows) of the periodic table correspond to the energy levels in the Bohr model of the atom (Figure 15.5). The first energy level can accept up to two electrons. Hydrogen (H) has one electron and helium (He) has two. These two elements complete the first period.

Period 2 is the second energy level — The next element, lithium (Li), has three electrons. Lithium begins the second period because the third electron goes into the second energy level. The second energy level can hold eight electrons, so there are eight elements in the second row of the periodic table, ending with neon (Ne). Neon has 10 electrons, which completely fill the second energy level.

Period 3 is the third energy level — Sodium (Na) has 11 electrons and starts the third period because the 11th electron goes into the third energy level. We know of elements with up to 118 electrons. These elements have their outermost electrons in the seventh energy level.

Outer electrons — As we will see in the next chapter, the outermost electrons in an atom are the ones that interact with other atoms. The outer electrons are the ones in the highest energy level. Electrons in the completely filled inner energy levels do not participate in forming chemical bonds.

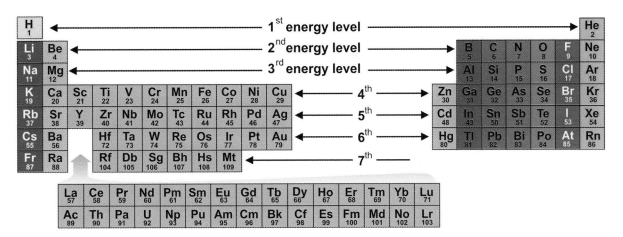

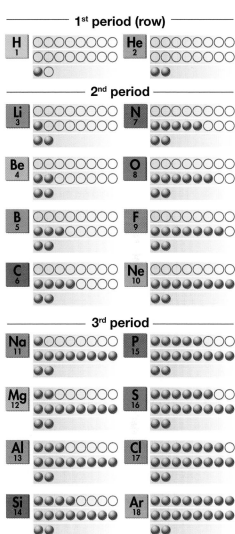

Figure 15.5: *The rows (periods) of the periodic table correspond to the energy levels for the electrons in an atom.*

Section 15.1 *Review*

1. Which two of the following are physical properties of matter and *not* chemical properties?

    a.   melts at 650°C

    b.   density of 1.0 g/mL

    c.   forms molecules with two oxygen atoms

2. Groups of the periodic table correspond to elements with:

    a.   the same color

    b.   the same atomic number

    c.   similar chemical properties

    d.   similar numbers of neutrons

3. Which element is the atom shown in Figure 15.6?

4. Name three elements that have similar chemical properties to oxygen.

5. The atomic mass unit (amu) is:

    a.   the mass of a single atom of carbon

    b.   one millionth of a gram

    c.   approximately the mass of a proton

    d.   approximately the mass of an electron

6. Which element belongs in the empty space in Figure 15.7?

7. The outermost electrons of the element vanadium (atomic #23) are in which energy level of the atom? How do you know?

8. The elements fluorine, chlorine, and bromine are in which group of the periodic table?

    a.   the alkali metals

    b.   the oxygen-like elements

    c.   the halogens

    d.   the noble gases

9. Which three metals are in the third period (row) of the periodic table?

Energy levels

What element is this?

Figure 15.6: *Question 3.*

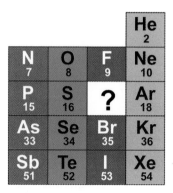

Figure 15.7: *Question 6.*

15.2 **Properties of the Elements**

The elements have a wide variety of chemical and physical properties. Some are solid at room temperature, like copper. Others are liquid (like bromine) or gas (like oxygen). Some solid elements (like zinc) melt at very low temperatures and some (like titanium) melt at very high temperatures. Chemically, there is an equally wide variety of properties. Some elements, like sodium, form salts that dissolve easily in water. Other elements, like neon, do not form compounds with any other elements.

State of matter at room temperature

Most elements are solid at room temperature   Most of the pure elements are solid at room temperature. Only 11 of the 92 naturally occurring elements are a gas, and 10 of these 11 are found on the far right of the periodic table. Only two elements (Br and Hg) are liquid at room temperature.

What this tells us about intermolecular forces   An element is solid when intermolecular forces are strong enough to overcome the thermal motion of atoms. At room temperature, this is true for most of the elements. The noble gases and elements to the far right of the periodic table are the exceptions. *These elements have completely filled or nearly filled energy levels* (Figure 15.8). When an energy level is completely filled, the electrons do not interact strongly with electrons in other atoms, reducing intermolecular forces.

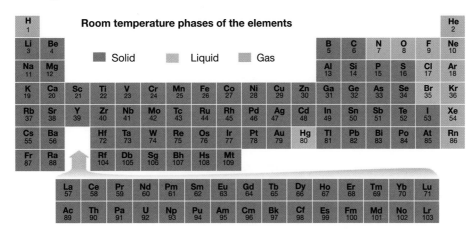

Room temperature phases of the elements

■ Solid   ■ Liquid   ■ Gas

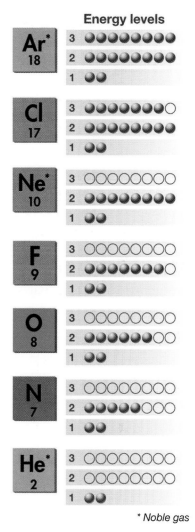

Energy levels

* *Noble gas*

Figure 15.8: *The noble gases have completely filled energy levels. All of the elements that are gas at room temperature have filled or nearly filled energy levels.*

Periodic properties of the elements

The pattern in melting and boiling points We said earlier that the periodic table arranges elements with common properties into groups (columns). The diagram below shows the melting and boiling points for the first 36 elements. The first element in each row (Li, Na, K) always has a low melting point. The melting (and boiling) points rise toward the center of each row and then decrease again.

Periodicity The pattern of melting and boiling points is an example of **periodicity**. Periodicity means properties repeat each period (row) of the periodic table (Figure 15.9). Periodicity tells us a property is strongly related to the filling of electron energy levels. Melting points reflect the strength of intermolecular forces. The diagram below shows that intermolecular forces are strongest when energy levels are about half full (or half empty). Elements with half-filled energy levels have the greatest number of electrons that can participate in bonding.

VOCABULARY

periodicity – the repeating pattern of chemical and physical properties of the elements

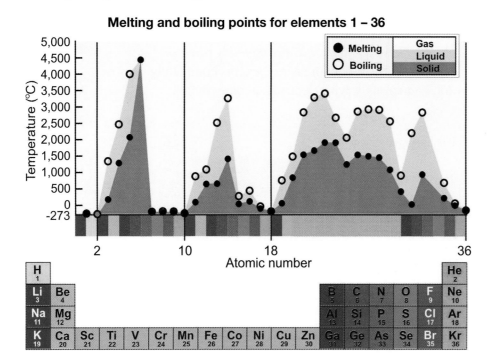

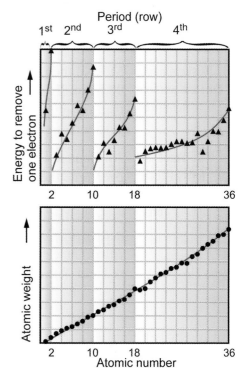

Figure 15.9: *One of these graphs shows periodicity and the other does not. Can you tell which one is periodic? The top graph shows the energy it takes to remove an electron. The bottom graph shows the atomic weight.*

Thermal and electrical conductivity

Metals are good electrical conductors

Copper wire

Electricity is something we often take for granted because we use it every day. Fundamentally, electricity is the movement of electric charge, usually electrons. Some materials allow electrons to flow easily through them. If you connected a battery and a light bulb through one of these materials, the bulb would light. We call these materials **electrical conductors**. Copper and aluminum are excellent electrical conductors. Both belong to the family of metals, which are elements in the center and left-hand side of the periodic table (Figure 15.10). Copper and aluminum are used for almost all electrical wiring.

electrical conductor – a material that allows electricity to flow through easily

thermal conductor – a material that allows heat to flow easily

insulator – a material that slows down or stops the flow of either heat or electricity

Metals are good conductors of heat

If you hold one end of a copper pipe with your hand and heat the other end with a torch, your hand will quickly get hot. That is because copper is a good conductor of heat and electricity. Like copper, most metals are good **thermal conductors**. That is one reason pots and pans are made of metal. Heat from a stove can pass easily through the metal walls of a pot to transfer energy (heat) to the food inside.

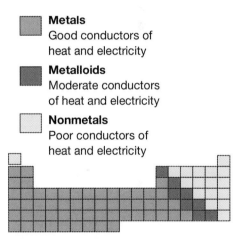

■ **Metals**
Good conductors of heat and electricity

■ **Metalloids**
Moderate conductors of heat and electricity

☐ **Nonmetals**
Poor conductors of heat and electricity

Figure 15.10: *Dividing the periodic table into metals, metalloids, and nonmetals.*

Nonmetals are typically insulators

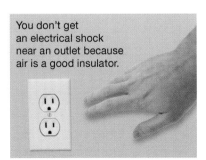

You don't get an electrical shock near an outlet because air is a good insulator.

Elements to the far right on the periodic table are not good conductors of electricity or heat, especially since many are gases. Because they are so different from metals, these elements are called *nonmetals*. Nonmetals make good insulators. An **insulator** is a material that slows down or stops the flow of either heat or electricity. Air is a good insulator. Air is made of oxygen, nitrogen, and argon.

Metals and metal alloys

Steel is an alloy of iron and carbon When asked for an example of a metal, many people immediately think of **steel**. Steel is made from iron, which is the fourth most abundant element in Earth's crust. However, steel is not pure iron. Steel is an *alloy*. An alloy is a solid mixture of one or more elements. Most metals are used as alloys and not in their pure elemental form. Common steel contains mostly iron with a small percentage of carbon. Stainless steel and high-strength steel alloys also contain small percentages of other elements such as chromium, manganese, and vanadium. More than 500 different types of steel are in everyday use (Figure 15.11).

Aluminum is light Aluminum is a metal widely used for structural applications. Aluminum alloys are not quite as strong as steel, but aluminum has one-third the density of steel. Aluminum alloys are used when the product needs to be lightweight, such as an airplane. The frames and skins of airplanes are built of aluminum alloys (Figure 15.12).

Titanium is both strong and light

Titanium combines the strength and hardness of steel with the light weight of aluminum. Titanium alloys are used for military aircraft, racing bicycles, and other high-performance machines. Titanium is expensive because it is somewhat rare and difficult to work with.

Brass

Brass is a hard, gold-colored metal alloy. Ordinary (yellow) brass is an alloy of 72 percent copper, 24 percent zinc, 3 percent lead, and 1 percent tin. Hinges, door knobs, keys, and decorative objects are made of brass because brass is easy to work with. Because it contains lead, however, you should never eat or drink from anything made of ordinary (yellow) brass.

Stainless steel kitchen knife (will not rust)

Ordinary steel nails (will rust)

Figure 15.11: *Nails are made of steel that contains 95 percent iron and 5 percent carbon. Kitchen knives are made of stainless steel that is an alloy containing vanadium and other metals.*

Figure 15.12: *This aircraft is made mostly from aluminum alloys. Aluminum combines high strength and light weight.*

Carbon and carbon-like elements

Carbon is an important element for life

Carbon represents less than one percent of Earth's crust by mass, yet it is the element most essential for life on our planet. Virtually all the molecules that make up plants and animals are constructed around carbon. The chemistry of carbon is so important that it has its own name, *organic chemistry* (Figure 15.13).

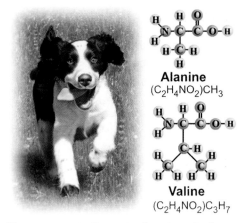

Alanine
($C_2H_4NO_2$)CH_3

Valine
($C_2H_4NO_2$)C_3H_7

Figure 15.13: *Organic chemistry is the chemistry of living organisms and is based on the element carbon.*

Diamond and graphite

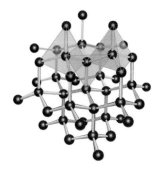

Pure carbon is found in nature in two very different forms. Graphite is a black solid made of carbon that becomes a slippery powder when ground up. Graphite is used for lubricating locks and keys. Diamond is also pure carbon. Diamond is the hardest natural substance known and also has the highest thermal conductivity of any material. Diamond is so strong because every carbon atom in diamond is bonded to four neighboring atoms in a tetrahedral crystal (shown left).

Silicon

Directly under carbon on the periodic table is the element silicon. Silicon is the second most abundant element in Earth's crust, second only to oxygen. Like carbon, silicon has four electrons in its outermost energy level. This means silicon can also make bonds with four other atoms. Sand, rocks, and minerals are predominantly made of silicon and oxygen (Figure 15.14). Most gemstones, such as rubies and emeralds, are compounds of silicon and oxygen with traces of other elements. In fact, when you see a glass window, you are looking at (or through) pure silica (SiO_2).

Examples of silica (SiO_2)

Glass

Sand

Figure 15.14: *Sand and glass are two common materials made of silicon.*

Silicon and semiconductors

Perhaps silicon's most famous application today is for making semiconductors. Virtually every computer chip and electronic device uses crystals of very pure silicon. The area around San Jose, California, is known as Silicon Valley because of the electronics companies located there. Germanium, the element just below silicon on the periodic table, is also used to make semiconductors.

Nitrogen, oxygen, and phosphorus

Nitrogen and oxygen make up most of the atmosphere
Nitrogen is a colorless, tasteless, and odorless gas that makes up 77 percent of Earth's atmosphere. Oxygen makes up another 21 percent of the atmosphere (Figure 15.15). Both oxygen and nitrogen gas consist of molecules with two atoms (N_2, O_2).

Oxygen in rocks and minerals
Oxygen makes up only 21 percent of the atmosphere; however, oxygen is by far the most abundant element in Earth's crust. Almost 46 percent of Earth's crust is oxygen (Figure 15.16). Because it is so reactive, all of this oxygen is bonded to other elements in rocks and minerals in the form of oxides. Silicon dioxide (SiO_2), calcium oxide (CaO), aluminum oxide (Al_2O_3), and magnesium oxide (MgO) are common mineral compounds. Hematite (Fe_2O_3), an oxide of iron, is a common ore from which iron is extracted.

Liquid nitrogen
With a boiling point of $-196°C$, liquid nitrogen is used for rapid freezing in medical and industrial applications. A common treatment for skin warts is to freeze them with liquid nitrogen.

Oxygen and nitrogen in living organisms
Oxygen and nitrogen are crucial to living animals and plants. For example, proteins and DNA both contain nitrogen. Nitrogen is part of a key ecological cycle. Bacteria in soil convert nitrogen dioxide (NO_2) in the soil into complex proteins and amino acids. These nutrients are taken up by the roots of plants, and later eaten by animals. Waste and dead tissue from animals are recycled by the soil bacteria that return the nitrogen to begin a new cycle.

Phosphorus
Directly below nitrogen on the periodic table is phosphorus. Phosphorus is a key ingredient of DNA, the molecule responsible for carrying the genetic code in all living creatures. Like nitrogen and oxygen, phosphorus is vital to plant nutrition, and its primary use by industry is in agricultural fertilizer.

Composition of Earth's atmosphere

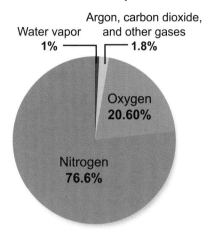

Percentages by weight at 70% relative humidity

Figure 15.15: *Earth's atmosphere is predominantly made up of nitrogen and oxygen.*

Oxygen is a major component of rocks and minerals.

Figure 15.16: *Oxygen makes up 46 percent of the mass of Earth's crust. This enormous quantity of oxygen is bound up in rocks and minerals.*

Section 15.2 *Review*

1. Name two elements that are liquid at room temperature.

2. Which of the following is *not* true about the noble gases?
   a. They have completely filled energy levels.
   b. They have weak intermolecular forces.
   c. They do not bond with other elements in nature.
   d. They have boiling points above room temperature.

3. Describe what it means if a chemical or physical property is periodic.

4. Name three elements that are good conductors of electricity.

5. Name three elements that are good conductors of heat.

6. A metalloid is an element that:
   a. has properties between those of a metal and a nonmetal
   b. is a good thermal conductor but a poor electrical conductor
   c. is a good electrical conductor but a poor thermal conductor
   d. belongs to the same group as carbon on the periodic table

7. Steel is a metallic-like material but is not a pure element. What is steel?

8. Almost all of the oxygen on the planet Earth is found in the atmosphere. Is this statement true or false?

9. Name an element that is abundant in Earth's crust and combines with oxygen to form rocks and minerals.

10. An element that has strong intermolecular forces is most likely to have:
    a. a boiling point below room temperature
    b. a melting point below room temperature
    c. a boiling point very close to its melting point
    d. a very high melting point

11. Which element in Figure 15.17 is likely to be the best conductor of electricity?

12. Which element in Figure 15.17 is likely to be the best insulator?

CHALLENGE

One of the elements with an atomic number less than 54 has the honor of being the first man-made element. Which element is this, and how was it discovered?

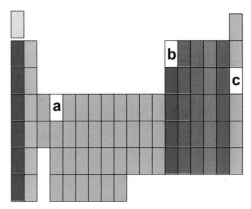

Figure 15.17: *Questions 11 and 12.*

STEM Silicon:

The Super Element of the Information Age

Silicon is one of the most useful elements on the planet, and it's easy to find—in fact, it's the second most common element in Earth's crust, making up over 28.2 percent by weight. However, you're not likely to find chunks of pure silicon in your backyard—or anywhere in nature. Silicon, like carbon, has four electrons in its outermost energy level. This means it can, and usually does, bond with other atoms.

Silicon compounds

Silicon often bonds with oxygen, in a compound called silicon dioxide (SiO_2), commonly known as silica. Sand, quartz, amethyst, and opal are all silicas found in nature.

Diatoms are a type of tiny photosynthetic plankton (one-celled organism) found in both salt water and fresh water. They have beautiful outer "shells" called frustules that are made of silica.

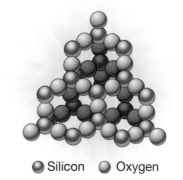

● Silicon ○ Oxygen

Silicon dioxide (silica) structure

Different species of diatoms

There are about 70,000 known species of diatoms, each with a unique frustule shape. Biologists estimate that diatoms make up one-quarter of all photosynthetic life and produce at least one-quarter of the oxygen we breathe!

Common glass is made by mixing silica (sand) with soda (Na_2CO_3), which lowers the melting point, and lime (CaO), which makes the glass more chemically stable. This mixture is heated to a very high temperature (over 1,700°C!) until it melts. The liquid glass is poured into a mold and allowed to cool.

Silicon also forms minerals called silicates when it bonds with oxygen and a metal ion, such as aluminum, iron, magnesium, calcium, sodium, or potassium. Most of the rocks on Earth's surface are silicates: granite, feldspar, mica, hornblende, and sandstone are just a few examples. Clay is made of silicates, too. Silicates are important natural resources for the construction industry since they are used to make concrete and brick.

Most of the rocks on Earth's surface are silicates.

Silicon is not the same as silicone, although the two are often confused. Silicone is a man-made compound created by repeating groups of silicon, oxygen, and a carbon compound. A simple silicone is shown below.

$$CH_3 - Si - O \left(- Si - O \right)_n - Si - CH_3$$

(with CH_3 groups above and below each Si)

Silicones are never found in nature—they have to be created in a lab. They're incredibly useful because they don't melt or break down easily, they're water repellent, and they don't conduct electricity. You'll find silicones in glue, caulk, paint, contact lenses, and many other objects.

Pure silicon: A high-tech marvel

Although silicon can form a myriad of useful compounds, it's best known for its impact on the modern world in the form of the microchip, a miniaturized electric circuit that fits into computers, cellular telephones, microwaves, and other digital appliances. Pure silicon enables the electronic transfer of data with remarkable efficiency.

Microchips, also known as computer chips or integrated circuits, are loaded with tiny transistors. Transistors are devices that act as switches, responsible for turning current on and off. By doing so, transistors relay electrical signals from one part of a circuit to another. Silicon has proven to be an excellent material for making transistors. Why? Silicon is a semiconductor. This means its ability to conduct electricity lies somewhere between a conductor, which conducts electricity well, and an insulator, which conducts electricity poorly. The conductivity of a semiconductor is suitable for turning the electric signal on and off without complications.

Creating the computer chip

How do you get from silicon dioxide, or common sand, to the microchip? The first step is a chemical reaction that separates the silicon from the oxygen. The silicon is then melted down at extremely high temperatures. Next, a seed crystal (a small sample of pure silicon) is positioned on a rod and dipped into the molten silicon. An example of a seed crystal is an ice cube; it's a solid sample of an element or compound where the atoms or molecules have a particular repeating pattern.

The liquid particles of the silicon attach themselves to the solid seed crystal in the same pattern and a larger monocrystal is formed. This large single crystal of silicon, known as a boule, is now a usable shape. Standard boules are typically 200 mm to 300 mm in diameter and 1 to 2 meters in length. A diamond blade or wire is then used to cut the boule into wafers that are 0.5 mm thick.

To fabricate a chip on top of a wafer, several processes must be performed, one after the other. These processes may include depositing a film, patterning the film, and etching or removing part of the film. In other words, a coating or film is applied to the wafer and the diagram of the circuit is etched into it, then other parts, such as tiny copper wires, are deposited in the etching.

Silicon wafers

Computer chips must be created in sterile environments known as clean rooms. Just one speck of dust can interfere with the circuitry, making the chip unusable. Chip makers wear gowns, hoods, goggles, and masks that industry insiders call "bunny suits." Pencils are forbidden in clean rooms, because bits of graphite could flake off and stick to a chip. Since graphite is a conductor, even a microscopic particle could short-circuit a chip!

Silicon: It's all around you

Look around the room you're in. How many devices can you find that contain these silicon-based computer chips? What else can you see that contains silicon? It's easy to see why silicon is known as "the super element of the information age."

Questions:

1. What is the difference between silicon, silica, silicates, and silicone?

2. Why is silicon used for computer chips?

3. **Research:** Where is Silicon Valley? How did this region of the United States acquire this nickname?

Diatoms photo by Randolph Femmer/NBII.

Chapter 15 *Assessment*

Vocabulary

Select the correct term to complete each sentence.

atomic mass	insulator	periodic table
chemical change	nonmetals	physical change
electrical conductor	period	steel
group	periodicity	thermal conductor

Section 15.1

1. A chart that organizes elements by their chemical properties and increasing atomic number is the _____ .

2. A change in matter that can be seen through direct observation is called _____ .

3. A row of the periodic table is referred to as a(n) _____ .

4. A change in matter that can be observed only as one substance changes to another is called _____ .

5. A column of the periodic table is known as a(n) _____ .

6. The average mass of all known isotopes of an element, expressed in amu, is that element's _____ .

7. Elements that are generally poor conductors, and in solid form, and are generally dull and brittle are called _____ .

Section 15.2

8. A repeating pattern of chemical and physical properties of the elements is called _____ .

9. A material that allows heat to flow easily is called a(n) _____ .

10. A material that slows or stops the flow of heat or electricity is called a(n) _____ .

11. An alloy of iron and carbon is called _____ .

12. A material that allows electricity to flow easily is called a(n) _____ .

Concepts

Section 15.1

1. Label each of the following changes or properties as being a physical (**P**) or chemical (**C**) change or property.
   a. One cm^3 of water has a mass of one gram.
   b. Burning hydrogen in the presence of oxygen produces water.
   c. Candle wax will melt when heated gently.
   d. An iron nail left outside for a year will rust.
   e. To raise the temperature of 1 kg of lead by 1°C requires 130 joules of energy.
   f. If ice is heated enough, the ice will change to steam.

2. Melting, boiling, and bending are considered physical changes, but burning is a chemical change. Explain why this is so.

3. How may a substance be tested to determine whether it is an element?

4. Supply the missing number for each of the following.
   a. The number of naturally occurring elements.
   b. The atomic number of the heaviest element.
   c. The atomic number of the lightest element.
   d. The total number of elements identified (as of the publication of this book).

5. Describe the difference between a period and a group on the periodic table.

6. What property of elements was used to organize the periodic table?

7. Describe the difference between the mass number and the atomic mass of an element.

8. Identify each of the following as a metal (**M**), nonmetal (**N**), or metalloid (**T**).
   a. includes most of the elements
   b. as solids they are dull, poor conductors, and brittle
   c. generally located on the right side of the periodic table
   d. ductile
   e. share properties in between metals and nonmetals

9. Briefly describe each group below and give an example of one element in each group.
   a. alkali metals
   b. halogens
   c. noble gases

10. How does the energy level of an element on the periodic table compare to its period number?

Section 15.2

11. Most elements occur as solids at room temperature.
    a. Name the two elements that are found as liquids at room temperature.
    b. Name 5 elements (out of 11) that are found as gases at room temperature.

12. Explain why the elements in Group 18 are all gases.

13. Name two properties that display periodicity across the periodic table.

14. Name three elements that are good conductors of both heat and electricity.

15. Name three elements that are poor conductors but are good insulators of both heat and electricity.

16. Carbon is not an exceptionally abundant element, but it is the most essential element for life on our planet. Why?

17. Name two reasons why silicon is an important element economically.

18. Name the following.
    a. The two most abundant gases and their approximate percentage of occurrence in Earth's atmosphere.
    b. The most abundant element in Earth's crust and its percentage of occurrence.

Problems

Section 15.1

1. How many electrons can be held in the first energy level? In the second energy level?

2. Aluminum has 13 electrons. How many electrons are found in the outermost energy level for this atom? Which energy level is the outermost?

3. Name the elements found at the following positions.
   a. Group 1, Period 2
   b. Period 4, Group 9

4. Give the symbol, name, and atomic number of the two elements in Period 4 that are most similar to cobalt (Co).

Section 15.2

5. State one property for each labeled element on this periodic table. Write your answers in a data table on your own paper. An example of a data table is shown below.

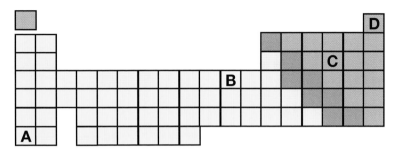

Element	Property
A	
B	
C	
D	

Section 15.2

3. Choose an element from the periodic table. Create a fact sheet about the element. Include the following on your fact sheet.

   a. symbol, name, atomic number, and atomic mass
   b. date of discovery and people responsible for discovery
   c. physical properties of the element including state of matter at room temperature, melting point, boiling point, appearance, etc.
   d. use(s) for the element, including useful compounds that contain the element
   e. any other important information about the element. If possible, find pictures of the element on the Internet and include them on your fact sheet

4. Suppose the periodic table arranged Periods 1 to 4 in order of increasing average atomic mass instead of increasing atomic number. Would this arrangement show periodicity? Explain your answer.

Applying Your Knowledge

Section 15.1

1. Create a pie graph showing the elements classified as nonmetals, metalloids, and metals. You can use The Periodic Table of the Elements on page 336 and Figure 15.10 to figure out the percentages.

2. List the elements of the periodic table for which the symbol does not match the name. For example, the symbol for lead is Pb. Choose any three of those elements and find out where the symbol comes from.

Compounds

What do sugar, aspirin, ethanol, and wood have in common?
They are all compounds made from different combinations of the same three elements: carbon, hydrogen, and oxygen. By themselves, these elements cannot sweeten your tea, relieve pain, fuel a car, or build a house. But when these elements are combined in different ways to form various compounds, they can be useful in many, many ways.

Most of the matter you use every day is in the form of compounds and mixtures, not elements. All compounds are made from combinations of less than 100 of the elements on the periodic table. The number of different compounds that exist is mind boggling! Study this chapter to learn how and why compounds form.

CHAPTER 16 INVESTIGATIONS

16A: Chemical Bonds
Why do atoms form chemical bonds?

16B: Chemical Formulas
Why do atoms combine in certain ratios?

16.1 Chemical Bonds and Electrons

Most matter exists as compounds, not as pure elements. That's because most pure elements are chemically unstable. They quickly form *chemical bonds* with other elements to make compounds. For example, water (H_2O) is a compound of hydrogen and oxygen. The salt used in food is a compound that contains two elements, sodium and chlorine, that are poisonous by themselves. In this section, you will learn why and how the atoms of elements form compounds.

Covalent bonds

Electrons form chemical bonds A **chemical bond** forms when atoms transfer or share electrons. Almost all elements form chemical bonds easily. This is why most of the matter you experience is in the form of compounds.

Covalent bonds A **covalent bond** is formed when atoms share electrons. A group of atoms held together by covalent bonds is called a *molecule*. The bonds between oxygen and hydrogen in a water molecule are covalent bonds (Figure 16.1). There are two covalent bonds in a water molecule, between the oxygen and each of the hydrogen atoms. Each bond represents a shared electron pair.

Chemical formulas A molecule's **chemical formula** tells you the ratio of atoms of each element in the compound. For example, the chemical formula for water is H_2O. The subscript 2 indicates there are two hydrogen atoms in a water molecule. No subscript after the O indicates there is only one oxygen atom for every two hydrogen atoms in the molecule.

VOCABULARY

chemical bond – a bond that forms when atoms transfer or share electrons

covalent bond – a chemical bond formed by atoms that are sharing one or more electrons

chemical formula – a representation of a compound that includes the symbols and ratios of atoms of each element in the compound

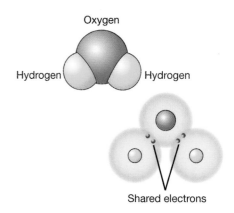

Figure 16.1: *In a covalent bond, electrons are shared between atoms.*

Reading a chemical formula

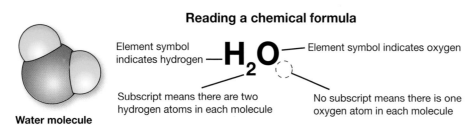

Element symbol indicates hydrogen — H_2O — Element symbol indicates oxygen

Subscript means there are two hydrogen atoms in each molecule

No subscript means there is one oxygen atom in each molecule

Water molecule

Ratio of two hydrogen atoms to one oxygen atom in the compound

Ionic bonds

An ion is a charged atom Not all compounds are made of molecules. For example, sodium chloride (NaCl) is a compound of sodium (Na) and chlorine (Cl) in a ratio of one sodium atom per chlorine atom. The difference is that in sodium chloride, the electron is transferred (instead of shared) from the sodium atom to the chlorine atom. When atoms gain or lose an electron, they become **ions**. An ion is a charged atom. By losing an electron, the sodium atom becomes a sodium ion with a charge of +1. By gaining an electron, the chlorine atom becomes a chloride ion with a charge of -1. (Note that when chlorine becomes an ion, the name changes to chlor*ide*.)

Ionic bonds Sodium and chlorine form an **ionic bond** because the positive sodium ion is attracted to the negative chloride ion. Ionic bonds are bonds in which one or more electrons are transferred from one atom to another.

Ionic compounds do not form molecules Unlike covalent bonds, ionic bonds are not limited to a single pair of atoms. In sodium chloride, each positive sodium ion is attracted to all of the neighboring chloride ions (Figure 16.2). Likewise, each chloride ion is attracted to all the neighboring sodium atoms. Because the bonds are not just between pairs of atoms, *ionic compounds do not form molecules!* In an ionic compound, each atom bonds with *all* of its neighbors through attraction between positive and negative charges.

The chemical formula for ionic compounds Like covalent compounds, ionic compounds have fixed ratios of elements. For example, there is one sodium ion per chloride ion in sodium chloride (NaCl). This means we can use the same type of chemical formula for ionic compounds and covalent compounds.

Ions may be multiply charged Sodium chloride involves the transfer of one electron. However, ionic compounds may also be formed by the transfer of two or more electrons. A good example is magnesium chloride ($MgCl_2$). The magnesium atom gives up two electrons to become a magnesium ion with a charge of +2 (Mg^{2+}). Each chlorine atom gains one electron to become a chloride ion with a charge of -1 (Cl^-). The ion charge is written as a superscript after the element symbol (Mg^{2+}, Fe^{3+}, Cl^-, etc.).

Sodium and Chlorine form an ionic compound

 Chloride ion   Sodium ion

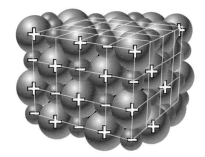

Figure 16.2: *Sodium chloride is an ionic compound in which each positive sodium ion is attracted to all of its negative chloride ion neighbors and vice versa.*

Why chemical bonds form

Atoms form bonds to reach a lower energy state Imagine pulling tape off a surface. It takes energy to pull the tape off. It also takes energy to separate atoms that are bonded together. If it takes energy to separate bonded atoms, then the same energy must be released when the bond forms. *Energy is released when chemical bonds form.* Energy is released because atoms bonded together have less total energy than the same atoms separately. Like a ball rolling downhill, atoms form compounds because the atoms have lower energy when they are together in compounds. For example, one carbon atom and four hydrogen atoms have more total energy apart than they do when combined in a methane molecule (Figure 16.3).

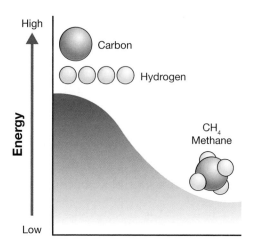

Figure 16.3: *The methane (CH_4) molecule has lower total energy than four separate hydrogen atoms and one separate carbon atom.*

Chemical reactivity All elements, except the noble gases, form chemical bonds. However, some elements are much more reactive than others. In chemistry, *reactive* means an element readily forms chemical bonds, often releasing energy. For example, sodium is a highly reactive metal. Chlorine is a highly reactive gas. If pure sodium and pure chlorine are placed together, a violent explosion occurs as the sodium and chlorine combine. The energy of the explosion is the energy given off by the formation of the chemical bonds.

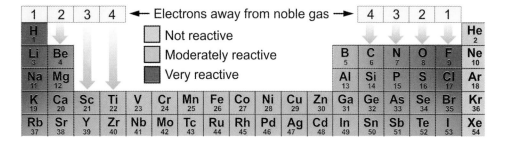

Some elements are more reactive than others The closer an element is to having the same number of electrons as a noble gas, the more reactive the element is. The alkali metals are very reactive because they are just one electron away from the noble gases. The halogens are also very reactive because they are also one electron away from the noble gases. The beryllium group and the oxygen group are less reactive because each element in these groups is two electrons away from a noble gas.

CHALLENGE

The noble gases (He, Ne, Ar, etc.) are called *inert* because they do not ordinarily react with anything. You can put sodium in an atmosphere of pure helium and nothing will happen. However, scientists have found that a few noble gases *do* form compounds, in very special circumstances. Research this topic and see if you can find a compound involving a noble gas.

Valence electrons

Compounds contain particular ratios of elements

The discovery of energy levels in the atom solved a 2,000-year-old mystery. Why do elements combine with other elements only in particular ratios (or not at all)? For example, why do two hydrogen atoms bond with one oxygen atom to make water? Why isn't there a molecule with three (H_3O) or even four (H_4O) hydrogen atoms? Why does sodium chloride have a precise ratio of one sodium ion to one chloride ion? Why don't helium, neon, and argon form compounds with any other elements? The answer has to do with the electrons in the outermost energy levels.

What are valence electrons?

Chemical bonds are formed only between the electrons in the highest unfilled energy level. These electrons are called **valence electrons**. You can think of valence electrons as the outer "skin" of an atom. Electrons in the inner (filled) energy levels do not interact with other atoms because they are shielded by the valence electrons. For example, chlorine has 7 valence electrons. The first 10 of chlorine's 17 electrons are in the inner (filled) energy levels (Figure 16.4).

Most elements bond to reach eight valence electrons

It turns out that *eight is the stable number for chemical bonding.* All elements heavier than boron form chemical bonds to become more stable, which means they have a configuration with eight valence electrons. When sodium and chlorine form an ionic bond, they each have a configuration of eight valence electrons (Figure 16.5). Eight is a stable number because eight electrons completely fill a part of the outermost energy level. The noble gases already have a stable number of eight valence electrons. They don't form chemical bonds because they are already stable!

Light elements bond to reach two valence electrons

For elements with atomic number of five (boron) or less, the stable number is two instead of eight. For these light elements, two valence electrons completely fill the *first* energy level. The elements H, He, Li, Be, and B form bonds with a stable number of two valence electrons.

Hydrogen is special

Because of its single electron, hydrogen can also have zero valence electrons! Zero is a stable number for hydrogen, as well as two. This flexibility makes hydrogen a very "friendly" element; hydrogen can bond with almost any other element.

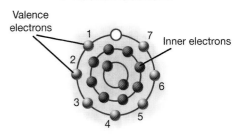

Chlorine
17 total electrons
7 valence electrons

Valence electrons

Inner electrons

Figure 16.4: *Chlorine has 7 valence electrons. The other 10 electrons are in filled (inner) energy levels.*

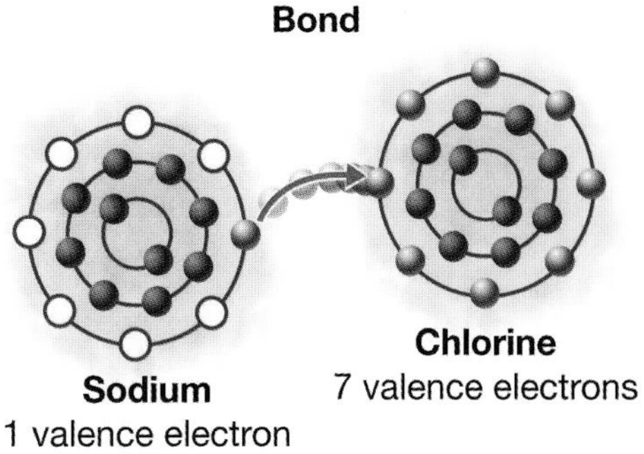

Bond

Sodium
1 valence electron

Chlorine
7 valence electrons

Figure 16.5: *Chlorine and sodium bond so each can reach a configuration with eight valence electrons.*

Valence electrons and the periodic table

Period 2 elements The illustration below shows how the electrons in the elements in the second period (lithium to neon) fill the energy levels. Two of lithium's three electrons go in the first energy level. Lithium has one valence electron because its third electron is the only one in the second energy level.

Each successive element has one more valence electron Going from left to right across a period, each successive element has one more valence electron. Beryllium has two valence electrons, boron has three, and carbon has four. Each element in the second period adds one more electron until all eight spots in the second energy level are full at atomic number 10, which is neon, a noble gas. Neon has eight valence electrons.

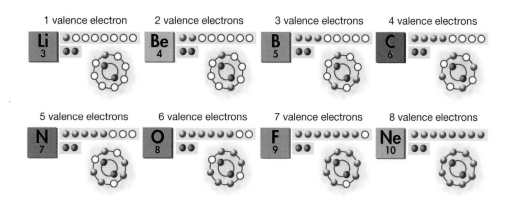

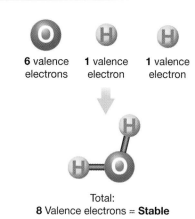

Figure 16.6: *Oxygen has 6 valence electrons and hydrogen has 2. In a water molecule, each hydrogen supplies one electron to make a total of 8 valence electrons.*

Bonding Oxygen has six valence electrons. To get to the magic number of eight, oxygen needs to add two electrons. *Oxygen forms chemical bonds that provide these two extra electrons.* For example, a single oxygen atom combines with two hydrogen atoms because each hydrogen can supply only one electron (Figure 16.6).

Double bonds share two electrons Carbon has four valence electrons. That means two oxygen atoms can bond with a single carbon atom, with each oxygen sharing two of carbon's four valence electrons. The bonds in carbon dioxide (CO_2) are *double bonds* because each bond involves four electrons (Figure 16.7).

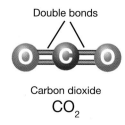

Figure 16.7: *Carbon forms two double bonds with oxygen to make carbon dioxide.*

Lewis dot diagrams

Dot diagrams of the elements
A **Lewis dot diagram** is a way to represent an atom's valence electrons. A dot diagram shows the element symbol surrounded by one to eight dots representing its valence electrons. Each dot represents one electron. Lithium has one dot, beryllium has two, boron has three, etc. Figure 16.8 shows dot diagrams for some of the elements.

Dot diagrams of molecules
Each element forms bonds to reach one of the stable numbers of valence electrons: two or eight. In dot diagrams of a complete molecule, each element symbol has either two or eight dots around it. Both configurations correspond to completely filled (or empty) energy levels.

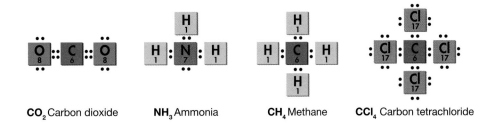

CO_2 Carbon dioxide   **NH_3** Ammonia   **CH_4** Methane   **CCl_4** Carbon tetrachloride

Example dot diagrams
Carbon has four dots and hydrogen has one. One carbon atom bonds with four hydrogen atoms, giving the carbon atoms eight valence electrons (eight dots)—four of its own and four shared with the hydrogen atoms. The picture above shows dot diagrams for carbon dioxide (CO_2), ammonia (NH_3), methane (CH_4), and carbon tetrachloride (CCl_4), a flammable solvent.

The formation of an ionic bond
A sodium atom is neutral with 11 positively charged protons and 11 negatively charged electrons. When sodium loses 1 electron, it has 11 protons (+) and 10 electrons (-) and becomes an ion with a net charge of +1. This is because it now has one more positive charge than its negative charges. A chlorine atom is neutral with 17 protons and 17 electrons. When chlorine gains 1 electron to have a stable 8 electrons, it has 17 protons (+) and 18 electrons (-) and becomes an ion with a charge of -1. This is because it has gained one negative charge. When sodium and chlorine form an ionic bond, the resulting compound is neutral (+1) + (-1) = 0.

Lewis dot diagram – a method for representing an atom's valence electrons using dots around the element symbol

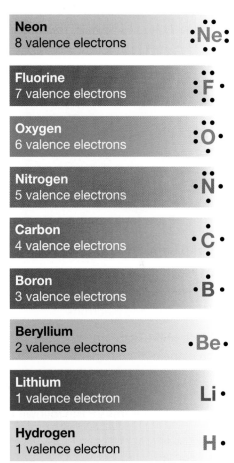

Neon
8 valence electrons

Fluorine
7 valence electrons

Oxygen
6 valence electrons

Nitrogen
5 valence electrons

Carbon
4 valence electrons

Boron
3 valence electrons

Beryllium
2 valence electrons

Lithium
1 valence electron

Hydrogen
1 valence electron

Figure 16.8: *Dot diagrams for some of the elements.*

Section 16.1 *Review*

1. Molecules are held together by:
   a. ionic bonds
   b. covalent bonds
   c. both a and b

2. How many atoms of chlorine (Cl) are in the carbon tetrachloride molecule (CCl_4)?

3. Which of the compounds below has a chemical formula of C_3H_8?

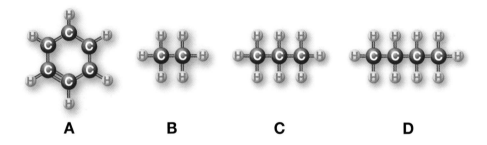

| A | B | C | D |

4. True or False: Ionic compounds do not form molecules.

5. Atoms form chemical bonds using:
   a. electrons in the innermost energy level
   b. electrons in the outermost energy level
   c. protons and electrons

6. Which of the diagrams in Figure 16.9 shows an element with three valence electrons? What is the name of this element?

7. Name two elements that have the Lewis dot diagram shown in Figure 16.10.

8. Draw dot diagrams for the following.
   a. silicon
   b. xenon
   c. calcium
   d. H_2O

Which of these diagrams shows three valence electrons?

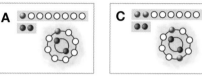
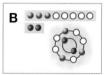

Figure 16.9: *Question 6.*

Name two elements that have this Lewis dot diagram.

Figure 16.10: *Question 7.*

16.2 **Chemical Formulas**

In the previous section, you learned how and why atoms form chemical bonds with one another. You also learned that atoms combine in certain ratios with other atoms. These ratios determine the chemical formula for a compound. In this section, you will learn how to write the chemical formulas for compounds. You will also learn how to name compounds based on their chemical formulas.

Chemical formulas and oxidation numbers

Ionic compounds Recall that the chemical formula for sodium chloride is NaCl. This formula indicates that every formula unit of sodium chloride contains one atom of sodium and one atom of chlorine—a 1:1 ratio. Why do sodium and chlorine combine in a 1:1 ratio? When sodium loses an electron, it becomes an ion with a charge of +1. When chlorine gains an electron, it becomes an ion with a charge of -1. When these two ions combine to form an ionic bond, the net electrical charge is zero (Figure 16.11). This is because (+1) + (-1) = 0.

All compounds have an electrical charge of zero; that is, they are neutral.

Oxidation numbers A sodium atom always ionizes to become Na⁺ (a charge of +1) when it combines with other atoms to make a compound. Therefore, we say that *sodium has an oxidation number of 1+.* An **oxidation number** indicates the electric charge on an atom when electrons are lost, gained, or shared during chemical bond formation. Notice that the convention for writing oxidation numbers is the opposite of the convention for writing the charge. When writing the oxidation number, the positive (or negative) symbol is written *after* the number, not *before* it.

What is chlorine's oxidation number? If you think it is 1-, you are right. This is because chlorine gains one electron, one negative charge, when it bonds with other atoms. Figure 16.12 shows the oxidation numbers for some of the elements.

VOCABULARY

oxidation number – a quantity that indicates the charge on an atom when it gains, loses, or shares electrons during bond formation

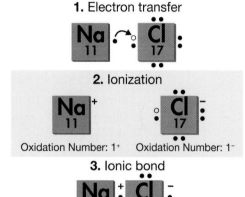

1. Electron transfer

2. Ionization

Oxidation Number: 1⁺   Oxidation Number: 1⁻

3. Ionic bond

Neutral compound: (1⁺) + (1⁻) = 0

Figure 16.11: *Sodium and chlorine combine in a 1:1 ratio.*

Atom	Electrons gained or lost	Oxidation number
K	loses 1	1+
Mg	loses 2	2+
Al	loses 3	3+
P	gains 3	3−
Se	gains 2	2−
Br	gains 1	1−

Figure 16.12: *Oxidation numbers of some common elements.*

Predicting oxidation numbers from the periodic table

Valence electrons and oxidation numbers In the last section, you learned that you can tell how many valence electrons an element has by its location on the periodic table. If you can determine how many valence electrons an element has, you can predict its oxidation number. An oxidation number corresponds to the need of an atom to gain or lose electrons (Figure 16.13).

Beryllium has an oxidation number of 2+ For example, locate beryllium (Be) on the periodic table below. It is in the second column, or Group 2, which means beryllium has two valence electrons. Will beryllium get rid of two electrons or gain six in order to obtain a stable number? Of course, it is easier to lose two electrons. When these two electrons are lost, beryllium becomes an ion with a charge of +2. Therefore, the most common oxidation number for beryllium is 2+. In fact, the most common oxidation number for all elements in Group 2 is 2+.

The periodic table The periodic table below shows the most common oxidation numbers of most of the elements. The elements known as transition metals (in the middle of the table) have variable oxidation numbers.

Oxidation number of 1+
(need to lose 1 electron)

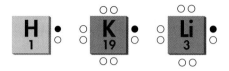

Oxidation number of 2+
(need to lose 2 electrons)

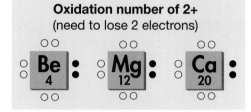

Oxidation number of 2–
(need to gain 2 electrons)

Oxidation number of 1–
(need to gain 1 electron)

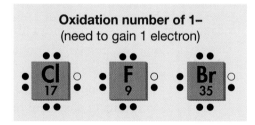

Figure 16.13: *Oxidation numbers correspond to the need to gain or lose electrons.*

1+	2+	Most common oxidation number			3+	4+	3-	2-	1-

NOTE: Many elements have more than one possible oxidation number.

																	He 2
Li 3	Be 4											B 5	C 6	N 7	O 8	F 9	Ne 10
Na 11	Mg 12											Al 13	Si 14	P 15	S 16	Cl 17	Ar 18
K 19	Ca 20	Sc 21	Ti 22	V 23	Cr 24	Mn 25	Fe 26	Co 27	Ni 28	Cu 29	Zn 30	Ga 31	Ge 32	As 33	Se 34	Br 35	Kr 36
Rb 37	Sr 38	Y 39	Zr 40	Nb 41	Mo 42	Tc 43	Ru 44	Rh 45	Pd 46	Ag 47	Cd 48	In 49	Sn 50	Sb 51	Te 52	I 53	Xe 54

Predicting ionic and covalent bonds

Why bonds are ionic or covalent
Whether or not a compound is ionic or covalently bonded depends on how much each element "needs" an electron to get to a magic number (2 or 8). Elements that are very close to the noble gases tend to give or take electrons rather than share them. These elements often form ionic bonds rather than covalent bonds.

Sodium chloride is ionic
As an example, sodium has one electron more than the noble gas neon. Sodium has a very strong tendency to give up that electron and become a positive ion. Chlorine has one electron less than argon. Therefore, chlorine has a very strong tendency to accept an electron and become a negative ion. Sodium chloride is an ionic compound because sodium has a strong tendency to give up an electron and chlorine has a strong tendency to accept an electron.

Widely separated elements form ionic compounds
On the periodic table, strong electron donors are the left side (alkali metals). Strong electron acceptors are on the right side (halogens). The farther separated two elements are on the periodic table, the more likely they are to form an ionic compound.

Nearby elements form covalent compounds
Covalent compounds form when elements have roughly equal tendency to accept electrons. Elements that are nonmetals and therefore close together on the periodic table tend to form covalent compounds with each other because they have approximately equal tendency to accept electrons. Compounds involving carbon, silicon, nitrogen, and oxygen are often covalent.

SOLVE IT!

You can use the periodic table to predict whether or not two elements will form ionic or covalent compounds. For example, potassium combines with bromine to make potassium bromide (KBr). Are the chemical bonds in this compound likely to be ionic or covalent? To solve this problem, look at the periodic table at the left.

K is a strong electron donor and Br is a strong electron acceptor. KBr is an ionic compound because K and Br are from opposite sides of the periodic table.

Now you try these:

1. Are the chemical bonds in silica (SiO_2) likely to be ionic or covalent?

2. Are the chemical bonds in calcium fluoride (CaF_2) likely to be ionic or covalent?

Oxidation numbers and chemical formulas

Oxidation numbers in a compound add up to zero
When elements combine in molecules and ionic compounds, the total electric charge is always zero. This is because any electron donated by one atom is accepted by another. The rule of zero charge is easiest to apply using oxidation numbers. The total of all the oxidation numbers for all the atoms in a compound must be zero. This important rule allows you to predict many chemical formulas.

The oxidation numbers for all the atoms in a compound must add up to zero.

Example: carbon tetrachloride
To see how this works, consider the compound carbon tetrachloride (CCl_4). Carbon has an oxidation number of 4+. Chlorine has an oxidation number of 1-. It takes four chlorine atoms to cancel carbon's 4+ oxidation number.

Element	Oxidation number
Copper (I)	Cu^+
Copper (II)	Cu^{2+}
Iron (II)	Fe^{2+}
Iron (III)	Fe^{3+}
Chromium (II)	Cr^{2+}
Chromium (III)	Cr^{3+}
Lead (II)	Pb^{2+}
Lead (IV)	Pb^{4+}

Figure 16.14: *In some cases, roman numerals are used to distinguish the oxidation number for an element with multiple numbers.*

Carbon Tetrachloride (CCl_4)

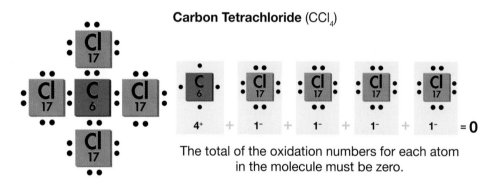

$4^+ \; + \; 1^- \; + \; 1^- \; + \; 1^- \; + \; 1^- \; = 0$

The total of the oxidation numbers for each atom in the molecule must be zero.

Most elements have more than one possible oxidation number
Some periodic tables list multiple oxidation numbers for most elements. This is because more complex bonding is possible. This course gives you the fundamental ideas, but there is much more! When multiple oxidation numbers are shown, the most common one is usually in bold type. For example, nitrogen has possible oxidation numbers of 5+, 4+, 3+, 2+, and 3-, even though 3- is the most common (shown right). In some reference materials, roman numerals are used to distinguish the oxidation number. Figure 16.14 shows a few of these elements.

5+, 4+, 3+
2+, **3-**

N

7
nitrogen

Predicting chemical formulas for binary compounds

Rules for predicting chemical formulas Once you know how to find the oxidation numbers of the elements, you can predict the chemical formulas of binary compounds (Figure 16.15). A **binary compound** is a compound that consists of two elements. Sodium chloride (NaCl) is a binary compound. To predict and write the chemical formula of a binary compound, use the following rules.

1. Write the symbol for the element that has a positive oxidation number first. Do not write the oxidation number.
2. Write the symbol for the element that has a negative oxidation number second. Do not write the oxidation number.
3. Find the least common multiple between the oxidation numbers to make the sum of their charges equal zero. Use the numbers you multiply the oxidation numbers by as subscripts.

 Solving Problems: Binary Compounds

Iron (III) (3+) and oxygen (2-) combine to form a compound. Predict the chemical formula of this compound.

1.	**Looking for:**	Chemical formula for a binary compound
2.	**Given:**	Elements and oxidation numbers: Fe (III) = 3+ and O = 2-
3.	**Relationships:**	Write the subscripts so that the sum of the oxidation numbers equals zero.
4.	**Solution:**	The least common multiple between 3 and 2 is 6. For iron (III): $2 \times (3+) = 6+$. For oxygen: $3 \times (2-) = 6-$ $(6+) + (6-) = 0$. The chemical formula is Fe_2O_3 because it took 2 Fe atoms and 3 O atoms to make a neutral compound.

Your turn...

a. Predict the chemical formula of the compound containing beryllium (2+) and fluorine (1-).

b. Predict the chemical formula of the compound containing lead (IV) and sulfur (2-).

Predict the chemical formula for a compound made from iron (oxidation number 3+) and oxygen (oxidation number 2–).

1. Write the symbol for the element that has a positive oxidation number first. Do not write the oxidation number.

$$Fe$$

2. Write the symbol for the element that has a negative oxidation number second. Do not write the oxidation number.

$$O$$

3. Add subscripts so that the sum of the oxidation numbers of all the atoms in the formula is zero.

$$2 \times Fe^{3+} = 6+$$
$$3 \times O^{2-} = 6-$$
$$(6+) + (6-) = 0$$
Chemical formula: Fe_2O_3

Figure 16.15: *The steps to predicting the chemical formula of a binary compound.*

 SOLVE FIRST/LOOK LATER

a. BeF_2

b. PbS_2

Compounds with more than two elements

Not all compounds are made of only two types of atoms Have you ever taken an antacid for an upset stomach? Many antacids contain calcium carbonate, or $CaCO_3$. How many types of atoms does this compound contain? You are right if you said three: calcium, carbon, and oxygen. Some compounds contain more than two elements. Some of these types of compounds contain polyatomic ions. A **polyatomic ion** contains more than one atom. The prefix *poly-* means "many." Figure 16.16 lists some common polyatomic ions. The example below illustrates how to write chemical formulas for these types of compounds.

polyatomic ion – an ion that contains more than one atom

Oxidation number	Name of ion	Formula
1+	Ammonium	NH_4^+
1–	Acetate	$C_2H_3O_2^-$
2–	Carbonate	CO_3^{2-}
2–	Chromate	CrO_4^{2-}
1–	Hydrogen carbonate	HCO_3^-
1+	Hydronium	H_3O^+
1–	Hydroxide	OH^-
1–	Nitrate	NO_3^-
2–	Peroxide	O_2^{2-}
3–	Phosphate	PO_4^{3-}
2–	Sulfate	SO_4^{2-}
2–	Sulfite	SO_3^{2-}

Figure 16.16: *Oxidation numbers of some common polyatomic ions.*

   Solving Problems: More Chemical Formulas

Aluminum (3+) combines with the sulfate (SO_4^{2-}) to make aluminum sulfate. Write the chemical formula for aluminum sulfate.

1. **Looking for:** Chemical formula for a compound containing more than two elements

2. **Given:** Al 3+ and SO_4^{2-}

3. **Relationships:** The oxidation numbers for all of the atoms in the compound must add up to zero.

4. **Solution:** Two aluminum ions have a charge of 6+. It takes three sulfate ions to get a charge of 6–. To write the chemical formula, parentheses must be placed around the polyatomic ion. The subscript is placed on the outside of the parentheses. The formula is: $Al_2(SO_4)_3$

Your turn...

a. Write the chemical formula for hydrogen (1+) peroxide (O_2^{2-}).

b. Write the chemical formula for calcium (2+) phosphate (PO_4^{3-}).

SOLVE FIRST LOOK LATER

a. H_2O_2

b. $Ca_3(PO_4)_2$

Naming compounds

Naming binary ionic compounds You can name a binary ionic compound if you are given its chemical formula by following these rules. A *binary ionic compound* is held together by ionic bonds. *Binary molecular compounds* consist of covalently bonded atoms. Naming binary molecular compounds is in discussed the Solve It! on the next page. To name a binary ionic compound:

1. Write the name of the first element.

2. Write the root name of the second element.

3. Add the suffix *-ide* to the root name.

What is the name of MgBr₂? $MgBr_2$ is *magnesium* (name of first element) + *brom* (root name of second element) + *ide* = magnesium bromide (Figure 16.17, top).

If the positive element has more than one oxidation number, you must first figure out that number. Then use a roman numeral to indicate the oxidation number. For example, $FeCl_3$ = iron (III) chloride because iron (III) has a charge of 3+, so it would take 3 chloride ions (oxidation number = 1-) to make the sum of the oxidation numbers equal zero.

Naming compounds with polyatomic ions Naming compounds with polyatomic ions is easy.

1. Write the name of the first element or polyatomic ion first. Use the periodic table or ion chart (Figure 16.16, previous page) to find its name.

2. Write the name of the second element or polyatomic ion second. Use the periodic table or ion chart (Figure 16.16, previous page) to find its name. If the second one is an element, use the root name of the element with the suffix *-ide*.

What is the name of NH₄Cl? NH_4Cl is *ammonium* (the name of the polyatomic ion from Figure 16.16) + *chlor* (root name of the second element) + *ide* = ammonium chloride (Figure 16.17, bottom).

Again, if an element has more than one oxidation number, you must figure out that number. For example, Cu_2SO_3 would be named copper (I) sulfite, and $CuSO_3$ would be copper (II) sulfite.

Naming a binary compound

$MgBr_2$

1. Write the name of the first element.

    Mg = magnesium

2. Write the root name of the second element.

    Br = bromine = brom-

3. Add the suffix *-ide* to the root name.

    brom + -ide = bromide

Name of the compound:

Magnesium bromide

Naming compounds with polyatomic ions

NH_4Cl

1. Write the name of the first element or polyatomic ion first. Use the periodic table or ion chart to find its name.

    NH₄ = ammonium

2. Write the name of the second element or polyatomic ion second. Use the periodic table or ion chart to find its name. If the second one is an element, use the root name of the element with the suffix *-ide*.

    Cl = chloride

Name of the compound:

ammonium chloride

Figure 16.17: *Naming compounds.*

Section 16.2 *Review*

1. The oxidation number is:

   a. the number of oxygen atoms an element bonds with

   b. the positive or negative charge acquired by an atom in a chemical bond

   c. the number of electrons involved in a chemical bond

2. Name three elements that have an oxidation number of 3+.

3. What is the oxidation number for the elements in Group 17?

4. When elements form a molecule, what is *true* about the oxidation numbers of the atoms in the molecule?

   a. The sum of the oxidation numbers must equal zero.

   b. All oxidation numbers from the same molecule must be positive.

5. True or False: All oxidation numbers from the same molecule must be negative.

6. Which of the following elements will bond with oxygen resulting in a 1:1 ratio of oxygen and the element?

   a. lithium

   b. boron

   c. beryllium

   d. nitrogen

7. Name the following compounds.

   a. $NaHCO_3$

   b. $BaCl_2$

   c. LiF

   d. $Al(OH)_3$

   e. SrI

8. Would a bond between potassium and iodine most likely be covalent or ionic? Explain your answer.

SOLVE IT!

Naming binary molecular compounds

Naming binary molecular compounds is similar to the methods used in naming binary ionic compounds described on the previous page. However, in this case, the *number* of each type of atom (the subscript) is also specified in the name of the compound. The Greek prefixes are, from 1 to 10: *mono, di, tri, tetra, penta, hexa, hepta, octa, nona, deca.*

To name a binary molecular compound, specify the number of each type of atom using the Greek prefix. As with binary ionic compounds, the ending of the name of the second element in the compound is modified by adding the suffix *-ide* as shown in this example:

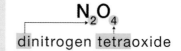

If the first element in the compound does not have a subscript, do not use a Greek prefix for that element, but use one for the second element. For example, CO_2 is carbon dioxide.

Name the following binary molecular compounds:

(a) CCl_4 (b) N_4O_6 (c) S_2F_{10}

16.3 **Molecules and Carbon Compounds**

Do you know which compounds you are made of? Excluding water, 91 percent of your body mass consists of compounds that are made up of only four elements: carbon, oxygen, nitrogen, and hydrogen. Of those four, carbon is the largest part at 53 percent. The molecules of some of those compounds are large and complex. In this section, you will learn more about molecules and why carbon is such an important element in the molecules of living things.

Structural diagrams of molecules

Molecules are represented using structural diagrams   In addition to the elements from which it is made, the shape of a molecule is also important to its function and properties. For this reason, we use *structural diagrams* to show the shape and arrangement of atoms in a molecule. Single bonds between atoms are shown with solid lines connecting the element symbols. Double and triple bonds are shown with double and triple lines. Figure 16.18 shows the chemical formula and structural diagram for some compounds.

Properties come from the molecule   *Both chemical formula and the structure of the molecules determine the properties of a compound.* For example, aspirin, a pain reducer, is a molecule made from carbon, hydrogen, and oxygen, according to the chemical formula $C_9H_8O_4$. The same 21 atoms in aspirin can be combined in other structures with the same chemical formula! But the resulting molecules do not have the pain-relieving properties of aspirin.

Structural diagram	Chemical formula
	Sodium bicarbonate (baking soda) $NaHCO_3$
	Benzene C_6H_6
	Ethane C_2H_6
	Acetic acid (in vinegar) $HC_2H_3O_2$

Figure 16.18: *Chemical formulas and structural diagrams.*

Three different molecules, same chemical formula

Acetylsalicylic acid
(aspirin)

$C_9H_8O_4$

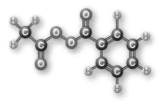

Benzodioxole-5 carboxylic acid
methyl ester

$C_9H_8O_4$

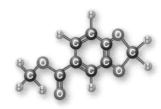

Acetyl benzoyl peroxide

$C_9H_8O_4$

Carbon molecules

VOCABULARY

organic chemistry – a branch of chemistry that specializes in the study of carbon compounds, also known as organic molecules

polymer – a compound that is composed of long chains of smaller molecules

Many compounds contain carbon Most of the compounds you are made of contain the element carbon. **Organic chemistry** is the branch of chemistry that specializes in carbon compounds, also known as organic molecules. But carbon compounds are not only found in living things. Plastic, rubber, and gasoline are carbon compounds. In fact, there are more than 12 million known carbon compounds! Carbon is unique among the elements because a carbon atom can form chemical bonds with other carbon atoms in long chains or rings. Some carbon compounds contain several thousand carbon atoms.

Carbon forms ring and chain molecules Carbon atoms have four valence electrons and can share one or more of these electrons to make covalent bonds with other carbon atoms or as many as four other elements. Carbon molecules come in three basic forms: straight chains, branching chains, and rings. The three basic shapes can be combined in the same molecule.

Ring

Benzene
C_6H_6

Chain

Butane
C_4H_{10}

Branched

Valine
$C_5H_{11}NO_2$

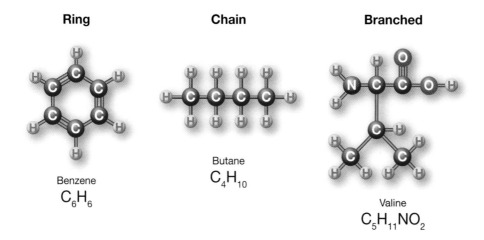

Ethylene

Plastic

Figure 16.19: *Polyethylene plastic is a polymer made from long chains of ethylene molecules.*

Polymers A **polymer** is a compound that is composed of long chains of smaller, repeating molecules. Polyethylene plastic is a polymer that is composed of long chains of a smaller molecule called ethylene. You can think of a polyethylene molecule as a chain of paperclips. Each paperclip represents an ethylene molecule (Figure 16.19).

Carbohydrates

The four types of biological molecules
Scientists classify the organic molecules in living things into four basic groups: carbohydrates, proteins, fats, and nucleic acids. All living things contain *all four types* of molecules. And each type of molecule includes thousands of different compounds, some specific to plants, some to animals. It is only in the past few decades that biotechnology has been able to reveal the rich chemistry of living things.

What are carbohydrates?
Carbohydrates are energy-rich compounds made from carbon, hydrogen, and oxygen. Carbohydrates are classified as sugars and starches. Sugars are smaller molecules. *Glucose* is a simple sugar made of 6 carbon, 12 hydrogen, and 6 oxygen atoms (Figure 16.20). The sugar you use to sweeten food is called *sucrose*. A sucrose molecule is made from two glucose molecules.

Starches are chains of sugar
Starches are long chains of glucose molecules joined together to make natural polymers. Because starches are larger molecules, they are slower to break down in the body and therefore can provide energy for a longer period than sugars. Corn, potatoes, and wheat contain substantial amounts of starches.

Cellulose
Cellulose is the primary molecule in plant fibers, including wood. The long-chain molecules of cellulose are what give wood its strength. Like starch, cellulose is a polymer made of thousands of glucose molecules. However, in starch, all the glucose units are the same orientation. In cellulose, alternate glucose units are inverted. This difference makes cellulose difficult for animals to digest. Although most animals cannot digest cellulose, some, like cows, have bacteria in their gut that break down the cellulose into starches that are digestible.

VOCABULARY

carbohydrates – a group of energy-rich compounds that are made from carbon, hydrogen, and oxygen and that include sugars and starches

Glucose molecule

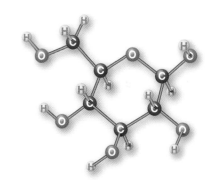

Figure 16.20: *A glucose molecule.*

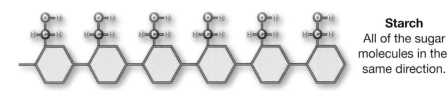

Starch
All of the sugar molecules in the same direction.

Cellulose
Direction of sugar molecules alternates.

16.3 Molecules and Carbon Compounds   **371**

Lipids

Lipids Like carbohydrates, **lipids** are energy-rich compounds made from carbon, hydrogen, and oxygen (Figure 16.21). Lipids include fats, oils, and waxes. Lipids are made by cells to store energy for long periods of time. Animals that *hibernate* (sleep through the winter) live off the fat stored in their cells. Polar bears have a layer of fat beneath their skin to insulate them from very cold temperatures. Can you name some foods that contain lipids?

Cholesterol is a lipid Like fat, cholesterol is listed on food labels. *Cholesterol* is a lipid that makes up part of the outer membrane of your cells. Your liver normally produces enough cholesterol for your cells to use. Too much cholesterol in some people's diets may cause fat deposits on their blood vessels. This may lead to coronary artery disease. Foods that come from animals are often high in cholesterol.

Saturated fats A lipid molecule has a two-part structure. The first part is called *glycerol*. Attached to the glycerol are 3 carbon chains. In a **saturated fat**, the carbon atoms are surrounded by as many hydrogens as possible. (See graphic below, left.)

Unsaturated fats An **unsaturated fat** has fewer hydrogen atoms than it could have, because double bonds exist between some of the carbon atoms. (See graphic below, right.) Chemical processing of food adds some hydrogen to unsaturated fats in a process called *hydrogenation*. While *partially hydrogenated* fats have a longer shelf life, research is showing that consuming them may be unhealthy.

Lipid molecule

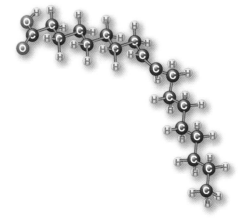

Figure 16.21: *A lipid molecule.*

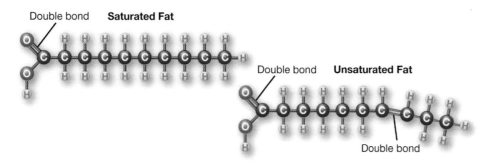

Proteins

Proteins **Proteins** are very large molecules made of carbon, hydrogen, oxygen, nitrogen, and sometimes sulfur. Many animal parts like hair, fingernails, muscle, and skin contain proteins. *Hemoglobin* is a protein in your blood that carries oxygen to your cells. Enzymes are also proteins. An **enzyme** is a type of protein that cells use to speed up chemical reactions in living things.

Proteins are made of amino acids Protein molecules are made of smaller molecules called **amino acids**. Your cells combine different amino acids in various ways to make different proteins. There are 20 amino acids used by cells to make proteins. You can compare amino acids to letters in the alphabet. Just as you can spell thousands of words with just 26 letters, you can make thousands of different proteins from just 20 amino acids (Figure 16.22).

Shape and function Only certain parts of a protein are chemically active. The shape of a protein determines which active sites are exposed. Many proteins work together by fitting into each other like a lock and key. This is one reason why proteins that perform a function in one organism cannot perform the same function in another organism. For example, a skin protein from an animal cannot replace a skin protein from a human.

Amino acids from food are used to build proteins Food supplies new proteins that a body needs to live and grow. However, proteins from one organism cannot be directly used by another. Fortunately, the same 20 amino acids are found in proteins from almost all living things. In your body, digestion breaks down food protein into its component amino acids. Cells reassemble the amino acids into new proteins suitable for your body's needs.

Protein A   **Protein B**

Active sites

Active sites

Proteins have complex shapes that fit other proteins or molecules in the body

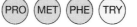

20 amino acids . . .

ALA VAL LEU ISO

PRO MET PHE TRY

GLY SER THR CYS

ASP GLM TYR ASA

GLA LYS ARG HIS

Skin proteins

Enzymes (proteins)

Hemoglobin equals protein in blood

. . . in different combinations make up all proteins.

Figure 16.22: *Proteins are made from smaller molecules called amino acids.*

DNA and nucleic acids

What are nucleic acids? **Nucleic acids** are compounds made of long, repeating chains called *nucleotides*. Nucleotides are made from carbon, hydrogen, oxygen, nitrogen, and phosphorus. Each nucleotide contains a sugar molecule, a phosphate molecule, and a base molecule, as shown in the graphic below.

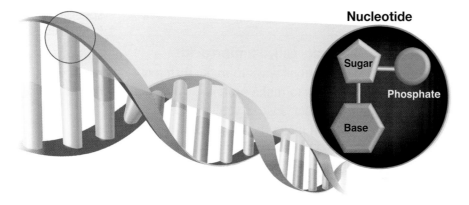

Nucleotide

DNA base-pairing
A pairs with **T** **G** pairs with **C**

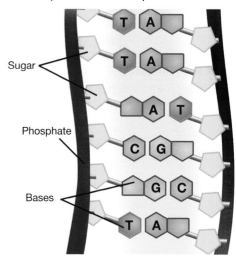

DNA **DNA** (deoxyribonucleic acid) is a nucleic acid that contains the information cells need to make all of their proteins. A DNA molecule is put together like a twisted ladder, or *double helix*. Each rung of the DNA ladder consists of a base pair. A base on one side of the molecule always matches up with a certain base on the other side (Figure 16.23). The base adenine (**A**) only pairs with thymine (**T**), and cytosine (**C**) only pairs with guanine (**G**). This base pairing is very important to the function of DNA. A single DNA molecule contains more than one million atoms!

Figure 16.23: *The DNA molecule.*

The four bases

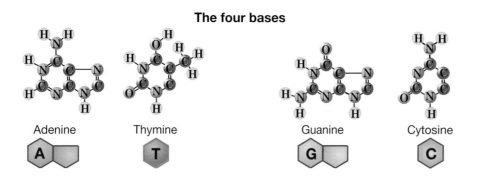

Adenine   Thymine   Guanine   Cytosine

Section 16.3 *Review*

1. Explain why life is often referred to as "carbon-based."

2. What are the four groups of carbon compounds found in living things?

3. You may have heard the saying, "You are what you eat." Use information learned in this section to explain what this statement means.

4. Classify each substance as either sugar, starch, protein, or nucleic acid.

   a. the major compound that makes up the skin

   b. glucose

   c. the major compound in potatoes

   d. DNA

5. Complete the table below.

Carbon compound	Elements it is made from	Importance to living things	Example
Carbohydrate			
Lipid			
Protein			
Nucleic acid			

6. What is the difference between saturated and unsaturated fat? Why are partially hydrogenated fats useful for making potato chips but not particularly healthy for humans to eat?

7. Simple sugars are the building blocks of carbohydrates. What are the simple units that make up proteins?

8. How many amino acids are used by cells to make the proteins? How many different kinds of proteins can be made by this number of amino acids?

9. What type of biological molecule is an enzyme, and why are enzymes so important to living things.

SOLVE IT!

Counting calories

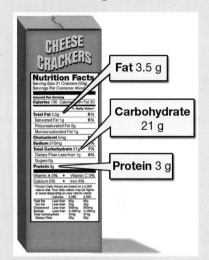

A food calorie (kilocalorie) tells you how much energy is in different foods. Each type of carbon compound has a certain number of food calories per gram. Fat contains 9 food calories per gram. Carbohydrate and protein each contain 4 food calories per gram. Based on this information, answer the following.

1. How many food calories in the product above come from fat?

2. How many food calories come from carbohydrate?

3. How many food calories come from protein?

4. How many food calories are in a serving of the product?

STEM The Spin on
Scrap Tires

The next time you travel in a car, think about the tires on which you are riding. Did you know that approximately one tire for every person in the United States is discarded each year? For New York State alone, that's 18–20 million tires. As the number of cars on the road increases each year, so does the number of scrap tires. For many years, the only disposal options were to throw scrap tires into landfills or burn them, which caused air pollution. Today, scientists and engineers are coming up with innovative ways to put a new spin on discarding old tires.

Tire composition

A typical automobile tire is made of rubber, steel and textile cables, and numerous other chemical agents. The rubber found in tires is vulcanized, or chemically treated, to increase the number of sulfur bonds. While this process produces a rigid, strong, and puncture-resistant substance, it also makes it harder to chemically break down the rubber into useful substances. Some experts have said that reclaiming the original components from scrap tires is like trying to recycle a cake back to its original ingredients. That's why it is so important to find innovative uses for scrap tires.

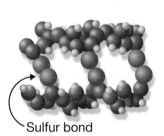

Sulfur bond

Keeping tires out of the dump

These days, a tire can have more than one life. After manufacture and use, tires are either recycled, reconditioned, used as fuel, or, in fewer and fewer cases, discarded. In 1990, nearly 3 million tons of tires ended up in a landfill. Today, we have reduced that number by more than 90 percent (*Source: U.S. Environmental Protection Agency, or EPA*).

Recycle and reuse

Whole-tire recycling involves using the old tire whole for other purposes, such as landscape borders, playground structures, bumpers, and highway crash barriers. Some scrap tires can be reconditioned and reused as tires again. It is more difficult to recycle tires for other purposes. Most tires are made from rubber, steel, and plastic fiber, bonded together in layers as shown below. To recycle tires for other uses, these materials must be separated. The process involves chopping up the scrap tires into pieces, and then separating the rubber and fiber from the steel.

Life Cycle of a Tire

Manufacture

Use

Reuse/retread

Disposal

Recycle

Tire-derived fuel

Tire cross-section

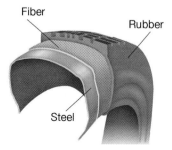

Fiber

Rubber

Steel

An expensive but very effective way to separate the rubber, fiber, and steel involves cooling the small pieces of tire with liquid nitrogen. This releases the bond between the steel, rubber, and fiber pieces. Next, magnets are used to take out the steel. The pieces of rubber can then be separated from the fiber using density techniques.

Uses for scrap rubber and steel

The small particles of rubber can be used immediately as a substitute for new rubber in products such as footwear, carpet underlay, and waterproofing compounds. The rubber crumbs can be added to artificial turf athletic fields and used as a soil additive. When

mixed with asphalt, ground tire rubber can make roads quieter and more durable. In fact, it has been found that adding scrap rubber to the asphalt used to pave roads can significantly decrease braking distances! Currently, 12 million scrap tires per year are used in paving highways (*Source: NSF*). Asphalt–scrap rubber mixtures are also used for track and field grounds, equestrian tracks, and playground surfaces.

The steel that is recovered from tires is used to make new steel. In fact, almost everything we make out of steel contains some percentage that is recycled. For nearly as long as steel has been made, recycling has been part of the process.

Scrap tires as fuel

Another use for scrap tires is to burn them as fuel. The tires are either incinerated whole or ground up into smaller pieces first. Many asphalt plants across the country use scrap tires as a fuel source. A total of 130 million scrap tires were used as tire-derived fuel (TDF) in 2003, up from 25 million in 1990. Tires produce the same amount of energy as oil and 25 percent more energy than coal. In fact, one passenger car tire is equivalent in energy to about seven gallons of oil! The amount of energy in each pound of scrap tire is about 15,800 kilojoules (*Source for all: EPA*). In addition, burning scrap tires produces fewer pollutants and less carbon dioxide than burning coal or fossil fuels!

Chemically changing rubber

Like plastic, rubber is a polymer—that is, a molecule that consists of long chains of repeating smaller molecules. Rubber is a polymer that is very difficult to break down—especially vulcanized rubber. But recent advances in technology have created an environmentally friendly process for breaking the carbon–carbon, carbon–sulfur, and sulfur–sulfur bonds in order to produce smaller molecules. These smaller molecules can be used to make liquid and gaseous fuels, ingredients for other polymers, lubricating oils, and a charcoal that can be used to decontaminate water or soil. However, this process, called pyrolysis, is difficult and expensive. Only a small fraction of the rubber from scrap tires is used in this way.

A shortage of discarded tires?

Currently, there are so many uses for discarded tires that a good question seems to be, Why not recycle all of our discarded tires? Perhaps in the near future, instead of an overabundance of discarded tires, there will be a shortage!

Questions:

1. What are the advantages and disadvantages to whole-tire recycling?

2. How does tire-derived fuel (TDF) compare to oil and coal as a source of energy?

3. What are the advantages and disadvantages to chemically changing scrap rubber?

4. Imagine that a manufacturer designs a new tire using completely different materials. How might this affect the tire recycling options? What should tire manufacturers take into consideration before changing their tire designs? STEM

Chapter 16 *Assessment*

Vocabulary

Select the correct term to complete each sentence.

amino acids	enzyme	oxidation number
binary compound	ion	polymer
carbohydrates	ionic bond	proteins
chemical bond	organic chemistry	saturated fat
chemical formula	Lewis dot diagram	unsaturated fat
covalent bond	lipids	valence electrons
DNA polyatomic ion	nucleic acids	

Section 16.1

1. H_2O is the _____ of water.

2. A(n) _____ is formed when atoms share one or more electrons.

3. A(n) _____ is formed when atoms transfer or share electrons.

4. A(n) _____ is formed when atoms transfer electrons.

5. You can use a(n) _____ to represent the valence electrons of an atom.

6. A charged atom is called a(n) _____ .

7. The electrons involved in chemical bonds are called _____ .

Section 16.2

8. A(n) _____ indicates the electric charge on an atom when it gains, loses, or shares electrons during chemical bond formation.

9. A compound consisting of two elements is called a(n) _____ .

10. The type of ion that contains more than one atom is called a(n) _____ .

Section 16.3

11. Fats, oils, and waxes are examples of _____ .

12. _____ are molecules composed of long chains of smaller, repeating molecules.

13. Sugars and starches are examples of _____ .

14. The building blocks of proteins are called _____ .

15. A branch of chemistry that specializes in the study of carbon compounds is _____ .

16. A fat that has fewer hydrogen atoms because double bonds exist among some of the carbon atoms is called a(n) _____ .

17. A fat in which the carbon atoms are surrounded by as many hydrogen atoms as possible is called a(n) _____ .

18. Very large molecules composed of carbon, hydrogen, oxygen, nitrogen, and sometimes sulfur are called _____ .

19. A type of protein that speeds up a chemical reaction in living things is called a(n) _____ .

20. Compounds made of many repeating nucleotides are known as _____ .

21. _____ is a nucleic acid that contains the genetic code for an organism.

Concepts

Section 16.1

1. What is the chemical formula for water? What atoms make up this compound?

2. Why do atoms form compounds instead of existing as single atoms?

3. What type of bond holds a water molecule together?

4. What do we call the particle that is a group of atoms held together by covalent bonds?

5. What does the subscript 2 in H_2O mean?

6. What do the subscripts in the formula for ethane represent?

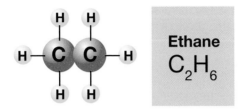

Ethane

C_2H_6

7. Name two ways that ionic compounds are different than molecules.

8. Summarize the differences between a covalent compound and an ionic compound.

9. Which group of elements usually doesn't form chemical bonds?

10. When atoms form chemical bonds, which of their electrons are involved in the bonds?

11. In a Lewis dot diagram, what is represented by the dots surrounding the element symbol?

12. Name a very reactive group of metals and a very reactive group of nonmetals. Why do they behave this way?

13. Noble gases usually don't form chemical bonds. Why?

14. What pattern do you see with the number of valence electrons as you go from left to right across a period of the periodic table?

Section 16.2

15. How does the oxidation number indicate if an electron will be lost or gained by the bonding atom?

16. Using the periodic table, what is the oxidation number of:
   a. calcium
   b. aluminum
   c. fluoride

17. What is the total electric charge on molecules and compounds?

18. Elements close to the noble gases tend to form what type of bond?

19. Elements that are widely separated on the periodic table tend to form _____ compounds.

20. Elements that are close together on the periodic table tend to form _____ compounds.

21. Strong electron donors are on the _____ side of the periodic table, while strong electron acceptors are on the _____ side.

Section 16.3

22. What do all organic molecules have in common?

23. What makes carbon uniquely suited to being the basis for biological molecules?

24. An organic compound contains carbon, hydrogen, oxygen, and nitrogen. Is this compound likely to be a lipid, carbohydrate, or protein? Explain.

25. Describe the four types of biological molecules. Give an example for each type.
   a. carbohydrate
   b. fat
   c. protein
   d. nucleic acid

26. What elements are carbohydrates made of?

27. Why do sugars break down so quickly in your body?

28. Identify each of the following as a carbohydrate, lipid, protein, or nucleic acid.

  a. glucose

  b. hemoglobin

  c. DNA

  d. digestive enzymes

  e. cholesterol

  f. cellulose

Problems

Section 16.1

1. For each of the molecule formulas listed below, name each element and tell how many of atoms of each element are in that molecule.

  a. $C_6H_{12}O_6$

  b. $CaCO_3$

  c. Al_2O_3

  d. $B(OH)_3$

2. Draw Lewis dot diagrams for the following.

  a. Bi

  b. Ge

  c. Ne

  d. SrI_2

Section 16.2

3. Predict the formula for a molecule containing carbon (C) with an oxidation number of 4+ and oxygen (O) with an oxidation number of 2-.

4. Which of the following would be a correct chemical formula for a molecule of N^{3-} and H^+?

  a. HNO_3

  b. H_3N_6

  c. NH_3

5. Refer to the diagram of the periodic table in Chapter 15 and determine which element in each pair is more active.

  a. Li or Be

  b. Ca or Sc

  c. P or S

  d. O or Ne

6. Nitrogen has 5 valence electrons and an oxidation number of 3-. In order for nitrogen to form a compound with other elements, how many electrons will it need to accept?

7. Using the periodic table:

  a. determine the oxidation number of Ca and Cl

  b. write the chemical formula for calcium chloride

8. Write the chemical formulas for the following compounds. Consult Figure 16.16 on page 366 if necessary.

  a. sodium iodide

  b. aluminum hydroxide

  c. magnesium sulfide

  d. ammonium nitrate

9. Name the following compounds.

  a. KI

  b. $SrCl_2$

  c. KNO_3

  d. Al_2O_3

Section 16.3

10. Classify each of the following carbohydrates as containing mostly sugar, starch, or cellulose.

    a. a stack of firewood
    b. rice
    c. jelly beans
    d. a shirt made of cotton
    e. an apple

11. The human body is made mostly of:

    a. carbon, oxygen, nitrogen, and hydrogen
    b. oxygen, calcium, carbon, and hydrogen
    c. hydrogen, iron, nitrogen, and oxygen

12. Which of the following compounds are organic?

    a. nucleic acid
    b. CH_4
    c. H_2O
    d. hydrochloric acid
    e. table salt
    f. sugar

13. The diagram below shows an enzyme and three different molecules. Which of the three molecules would this enzyme target for a reaction?

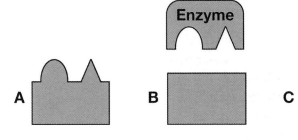

Applying Your Knowledge

Section 16.1

1. The noble gases used to be called "inert" gases until 1962, when scientists were able to cause them to react and form compounds. Using a search engine, and keywords "noble gas compound," conduct research on this topic. Find the names of some noble gas compounds, who discovered them, their chemical formulas, and how they are used.

Section 16.2

2. Answer each of the following questions about compounds.

    a. Ammonium sulfate is often used as a chemical fertilizer. What is its chemical formula?
    b. Calcium carbonate is a main ingredient in some antacids. What is its chemical formula?
    c. Kidney stones, a painful problem, are partially made from a compound whose chemical formula is $Ca_3(PO_4)_2$. What is the name of this compound?

Section 16.3

3. Suppose that there are only three amino acids called 1, 2, and 3. If all three are needed to make a protein, how many different proteins could be made? Each amino acid may only appear in each protein once. Also, the position of the amino acid is important—123 is not the same as 321. Show your number arrangements to support your answer.

4. You are entering a contest to design a new advertising campaign for National Nutrition Awareness Week. Create a slogan and a written advertisement that encourages teens to eat the right amounts of carbohydrates, lipids, and proteins. Use at least three facts to make your advertisement convincing.

Unit 6
Changes in Matter

acid

base

COOKING CHEMISTRY

YEAST

BREAD:
Ingredients:
eggs
flour
baking powder
yeast

respiration

$C_6H_{12}O_6 \rightarrow 2C_2H_5OH + 2CO_2$
glucose   ethanol   carbon dioxide

oxidation

silver - copper

vinegar →

oil

Endothermic reaction - Photosynthesis

white- 90% water, 10% protein

yolk- 50% water, 34% fat, 16% protein

shell - calcium carbonate

egg protein strand unravels

cooked egg forms bridge structure

Try this at home ➤ Make invisible ink! Mix equal parts of baking soda and water in a small cup. Use a paintbrush or cotton swab dipped in the baking soda mixture to create a message on a piece of paper. Let the message dry. Paint over the message with some concentrated purple grape juice. Watch the message appear! The purple grape juice changes color in the presence of certain substances like the baking soda; it's an acid/base reaction.

HELLO

AKING SODA

Chemical Change

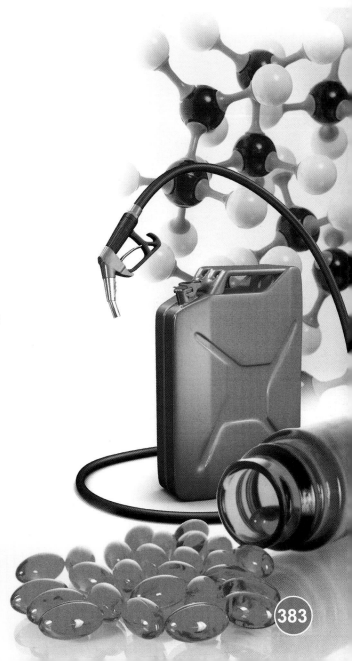

When studying science, it is common to be told "Look around!" However, in chemistry, the objects of study aren't cars or people in motion, which are easy to see. The objects of study are atoms and molecules, which are extremely small. Nonetheless, these tiny particles are *all* around you. In fact, some scientists describe the space around them as "chemical space." They think that the number of possible arrangements of atoms in the universe may be as many as up to 10^{60} compounds! That is a huge number! To date, "only" about 27 million compounds are known to be on Earth or have been made by scientists. For a new compound to be made, a chemical change has to occur. That means the atoms in the starting materials are rearranged to make different or even new compounds. What might be the motivation for making a new compound? Here are some ideas: New compounds can mean new medicines, new materials to make lighter cars or airplanes, or even new fuels to run a car or an airplane! For this reason, being able to predict the outcome of chemical changes is important. You are going to learn the basics of doing just that in this chapter.

CHAPTER 17 INVESTIGATIONS

17A: Chemical Equations

How are atoms conserved in a chemical reaction?

17B: Conservation of Mass

How do scientists describe what happens in a chemical reation?

17.1 **Chemical Reactions**

Atoms and molecules are all around us and so are *chemical reactions*. How do you know a chemical reaction is occurring? When you make pizza, for example, some of your work involves physical changes and some involves chemical changes (Figure 17.1). You know a chemical reaction has occurred if a chemical change has occurred as well. In this section, you will learn about chemical reactions.

Physical vs. chemical changes

A review of changes   In Chapter 15, you learned that matter undergoes chemical changes and physical changes. Recall that a *physical change* is a change that affects only the physical properties of a substance. Examples of physical changes include chopping pizza toppings (like vegetables) into smaller pieces and melting an ice cube into liquid water. Both of these changes involve a change in size, shape, or state of matter. A *chemical change* is a change in a substance that involves the breaking and reforming of chemical bonds to make one or more different substances.

Physical and chemical changes in making pizza   The process of making pizza involves some physical changes (like chopping vegetables) and chemical changes. Pizza dough is made of flour, oil, salt, and *yeast* (a type of fungus). As pizza dough is made, the yeast produces carbon dioxide gas in a process called *cellular respiration*. The carbon dioxide causes the dough to rise. This gas, the result of a chemical change, is responsible for the small holes you see in any kind of bread made with yeast. The action of the yeast and heat from an oven causes chemical changes that transform the sticky pizza dough into a tasty crust.

Energy and changes   Both physical and chemical changes involve energy. For example, you need energy to chop a green pepper into smaller pieces. Energy is also required for a substance to change its state from a solid to a liquid to a gas. Because chemical changes involve breaking and forming bonds, energy is also involved in these changes. Heat or light—forms of energy—are produced or used during a chemical change. The chemical changes in making pizza require the yeast to use and release energy and the heat of an oven to cook the pizza.

Figure 17.1: *This woman is making pizza from scratch. Here, she is preparing the dough. What part of making a pizza involves physical changes? What part of the process involves chemical changes?*

▨ JOURNAL ▨

Science in your mouth

Place a saltine cracker in your mouth. How does it taste? Hold it there for about 10 minutes. Now how does it taste? Is this evidence of a chemical or a physical change? Write down what you observe and think.

What is a chemical reaction?

Chemical reaction defined You have just learned something about the physical and chemical changes involved in making pizza. Any time there is a chemical change, a chemical reaction has occurred. A **chemical reaction** is the process of breaking chemical bonds in one or more substances, and the reforming of new bonds to create new substances. The process of cellular respiration performed by yeast in making pizza dough is a chemical reaction. The process used to generate heat in a gas stove to bake the pizza is also a chemical reaction and is illustrated below. When methane gas (a fuel) and oxygen react, the bonds in these molecules are broken to form the compounds carbon dioxide and water.

VOCABULARY

chemical reaction – the process of breaking chemical bonds in one or more substances and the reforming of new bonds to create new substances

precipitate – a solid that forms and is insoluble in a reaction mixture

Substances that change   Substances that are formed

$$CH_4 \quad + \quad O_2 \quad \longrightarrow \quad CO_2 \quad + \quad H_2O$$

Bubbling
A new gas is forming

Turns cloudy
A new solid is forming

Evidence of a chemical reaction When you combine two or more compounds, how do you know whether a chemical reaction has occurred? You can't see atoms and molecules actually breaking and forming bonds, but you can observe other events that indicate a chemical reaction. Figure 17.2 illustrates the type of evidence you can expect. For example, if you see a newly formed substance like a gas or a solid, you can suspect a chemical reaction. If a gas is a product in the reaction, you might see bubbles. If a new solid is produced, you might see powder forming in the reaction mixture so that it turns cloudy. A solid that forms and is insoluble in the reaction mixture is called a **precipitate**. Similarly, if you see a color change in the reaction mixture, a new substance may have been formed. Finally, evidence of a chemical reaction includes a temperature change. Keep in mind that any heat added to the reaction to get it started is not part of the evidence of a chemical reaction.

Color change
A new substance
is forming

Temperature change
Energy is
released or absorbed

Figure 17.2: *These are all different kinds of evidence that a chemical reaction is occurring.*

Reactants and products

Parts of a chemical reaction You can think of a chemical reaction as a kind of recipe. A recipe calls for specific amounts of ingredients to make a food—like a cake. The starting ingredients for a chemical reaction are called the reactants. A **reactant** is a substance present at the start of a chemical reaction. The chemical bonds of the reactants are broken during the chemical reaction. The resulting substances formed in a chemical reaction are called the products. A **product** is a compound that results from new chemical bonds formed when a chemical reaction occurs.

Reactants are chemically changed to form products On the previous page, you saw a reaction involving methane and oxygen. Below is that reaction presented again so that you can see what happens when reactants are chemically changed to become products. In the reaction, methane (a natural gas) is *burned* or *combusted*. Some energy is added to get the reaction started. Once this happens, a carbon atom from the methane molecule reacts with oxygen in the air to form carbon dioxide. Single oxygen atoms and hydrogen atoms also combine to form water. This reaction is particularly useful in making gas stoves work because a great deal of heat is released by this reaction.

States of matter in chemical reactions You know that the reactants in the reaction below are gases because of the symbol (g) listed next to the molecules (Figure 17.3). Likewise, you know that the products are gases—carbon dioxide gas and water vapor. In the next section, you will learn more about the ways that chemical reactions are written and the symbols that are used.

VOCABULARY

reactant – a starting ingredient in a chemical reaction

product – a new substance formed in a chemical reaction

Symbol	Meaning
(s)	Substance is a solid.
(l)	Substance is a liquid.
(g)	Substance is a gas.
(aq)	Substance is dissolved in water (aqueous).

Figure 17.3: *Symbols used for states of matter.*

SOLVE IT!

For the methane reaction, do the number of atoms of the reactants equal the number of atoms of the products? Count and see. How could you make the numbers match? (*Note:* In the next section, you'll find out!)

Chemical reaction

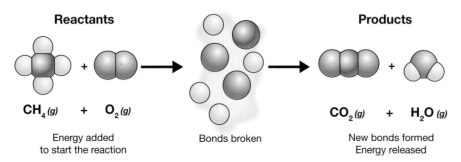

Reactants   **Products**

$CH_4 (g)$ + $O_2 (g)$ $CO_2 (g)$ + $H_2O (g)$

Energy added to start the reaction   Bonds broken   New bonds formed Energy released

Section 17.1 *Review*

1. Is the formation of rust on an iron nail a chemical change or a physical change? Explain your answer.

2. Which of the statements below is correct?

   a. In the process of cooking a frozen pizza, only chemical changes occur.

   b. Both physical and chemical changes occur when you make pizza from scratch.

   c. Only chemical changes occur when you cook a frozen pizza, and only physical changes occur when you make pizza from scratch.

   d. You cause chemical changes in vegetables by cutting up them into small pieces.

3. In your own words, explain how energy is involved in physical changes and in chemical changes.

4. For the examples below, explain whether a physical or a chemical change has occurred. Justify your answer.

   a. When you mix baking soda and vinegar, carbon dioxide is released.

   b. You build a tall sand castle at the beach. After a wave washes over it, the sand castle turns into a big pile of sand.

   c. Boiling water turns a raw egg into a hardboiled egg.

   d. Max divided cookie dough into small piles on a cookie sheet.

   e. A loaf of freshly baked bread tastes better and looks much different than a lump of bread dough.

   f. A glass of water is left in the sun. In time, the water evaporates, leaving the glass empty.

5. List the kinds of evidence that indicate that a chemical reaction has occurred.

6. Identify the reactants and products in this chemical reaction. For each compound, identify if it is a gas, solid, or liquid, or is in solution.

$$BaCO_3\ (s) + HBr\ (aq) \longrightarrow BaBr_2\ (aq) + H_2O\ (l) + CO_2\ (g)$$

17.2 **Balancing Equations**

Have you ever wondered what happens to wood in a fireplace or campfire as it is burned? The burning of wood is a chemical reaction. By writing this reaction as a *chemical equation*, you can figure out what happened to the wood. It doesn't just disappear! In this section, you will learn how reactants and products are related, and how to write and balance chemical equations.

The relationship between reactants and products

The law of conservation of mass In the 18th century, chemical reactions were still a bit of a mystery. A French scientist, Antoine Laurent Lavoisier, established an important principal based on his experiments with chemical reactions. He stated that the total mass of the products of a reaction is equal to the total mass of the reactants. This statement, which relates reactants and products, is known as the **law of conservation of mass**.

Investigating a reaction To understand the law of conservation of mass, let's look at the reaction of burning wood. It is easy to find the mass of a piece of wood you want to burn. But what happens to the mass of the wood after it burns (Figure 17.4)? To find out, look at the reaction below. The combined mass of the burning wood and oxygen is converted to carbon dioxide and water.

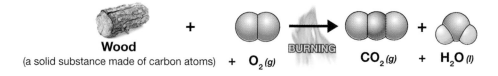

Wood
(a solid substance made of carbon atoms)   + O_2 (g)   BURNING → CO_2 (g)   + H_2O (l)

Using a closed system to study a reaction How can you prove that the mass of the reactants is equal to the mass of the products in the reaction of burning wood? Lavoisier showed that a *closed system* must be used when studying chemical reactions. When chemicals are reacted in a closed container, you can show that the mass before and after the reaction is the same (Figure 17.5).

For a chemical reaction, the total mass of reactants always equals the total mass of the products.

VOCABULARY

law of conservation of mass – a principle that states that the total mass of the reactants equals the total mass of the products in a chemical reaction

Figure 17.4: *What happens to wood when it is burned?*

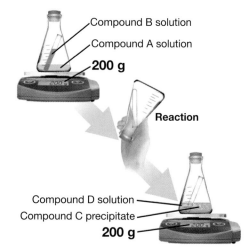

Compound B solution
Compound A solution
200 g
Reaction
Compound D solution
Compound C precipitate
200 g

Figure 17.5: *A closed system illustrates the law of conservation of mass.*

Formula mass and the mole

Formula mass The chemical formula of a reactant or product helps you calculate the mass of one unit of the compound relative to the mass of other compounds. The sum of the atomic mass values of the atoms in a chemical formula is called the **formula mass**. Formula mass is a way to compare the masses of units of different substances.

Determining formula mass The formula mass of a compound is determined by adding up the atomic mass values of all of the atoms in the compound. For example, the chemical formula for water is H_2O. This means that there are two hydrogen atoms for every one oxygen atom in a molecule of water. Using the periodic table, you can see that the atomic mass of hydrogen is 1.007 atomic mass unit (amu). We will round all atomic mass values to the hundredths place. Using 1.01 amu for hydrogen, we can multiply this number by the number of atoms present to determine atomic mass of hydrogen in a water molecule. The atomic mass of oxygen, rounded off, is 16.00. Using this information, the formula mass of a water molecule is calculated to be 18.02 amu (Figure 17.6).

Avogadro number In order to perform chemical reactions, it is helpful to be able to measure compounds in grams. Therefore, units of amu and grams have been related using a number called the **Avogadro number**. This number is equal to 6.02×10^{23} and it is a very, very large number! The Avogadro number refers to the number of molecules in the formula mass of a compound when this mass is expressed in grams. The Avogadro number is also the number of atoms in the atomic mass of an element when that value is expressed in grams. For example, 1.01 grams of hydrogen contains 6.02×10^{23} hydrogen atoms.

The mole The Avogadro number was named in honor of Amedeo Avogadro, an Italian physicist and mathematician who discovered that a *mole* of any gas under the same conditions has the same number of molecules. Johann Josef Loschmidt, a German physicist, discovered and named the Avogadro number. Loschmidt realized that a **mole** of any substance—be it a gas, liquid, or solid—contains 6.02×10^{23} atoms or molecules (Figure 17.7). The mass, in grams, of one mole of a compound is called its **molar mass**.

VOCABULARY

formula mass – the sum of the atomic mass values of the atoms in a chemical formula

Avogadro number – the number of atoms or molecules in a mole of any substance; the number equals 6.02×10^{23}

mole – a unit of any substance that contains the Avogadro number of atoms or molecules

molar mass – the mass, in grams, of one mole of a compound

1.01 amu 1.01 amu 16.00 amu

Formula mass of a water molecule =
(1.01 amu × 2) + 16.00 amu = 18.02 amu

amu = atomic mass unit

Figure 17.6: *The formula mass of a water molecule.*

The formula mass of H_2O is **18.02 amu**

18.02 amu = 1 molecule of H_2O

18.02 g = 6.02×10^{23} molecules of H_2O

Figure 17.7: *Comparing numbers for water.*

Solving Problems: Formula Mass and Moles

What is the molar mass of one mole of $CaCO_3$?

1. **Looking for:**	The molar mass of one mole of $CaCO_3$.
2. **Given:**	The chemical formula: $CaCO_3$.
3. **Relationships:**	One mole of $CaCO_3$ has a molar mass in grams equal to the formula mass in atomic mass units.
4. **Solution:**	Find the atomic mass units of each element in the chemical formula for $CaCO_3$ on a periodic table. These values are equivalent to the value for mass in grams of one mole of each element.

Atom	amu	Grams	Total mass of $CaCO_3$ (grams)
Ca	40.08	40.08	40.08
C	12.01	12.01	12.01
O	16.00	16.00	$16.00 \times 3 = 48.00$
Total			100.09

The formula mass of $CaCO_3$ is 100.09 amu and, therefore, the molar mass of one mole of $CaCO_3$ is 100.09 grams.

Your turn...

a. What is the molar mass of two moles of $CaCO_3$?

b. What is the formula mass of these compounds: CH_4, O_2, and CO_2?

c. You have 18.02 grams of each of the following states of water: water vapor (a gas), liquid water, and ice. How many molecules of each of these states of water do you have?

SOLVE IT!

Looking ahead to the next topic

You have just calculated the formula mass of the reactants and one product for the reaction below. You also know that the formula mass of H_2O is 18.02 amu.

$$CH_4 (g) + O_2 (g) \rightarrow CO_2 (g) + H_2O (g)$$

Convert the formula mass of each molecule to grams and determine if the mass of the reactants in this reaction equals the mass of the products.

As it is written, does this reaction satisfy the law of conservation of mass?

How are chemical reactions written?

The chemical equation So far you have seen how a chemical reaction—like the methane reaction below—is written. When a chemical reaction is written using chemical formulas and symbols, it is called a **chemical equation**. Chemical equations are a convenient way to describe chemical reactions. Here you see the methane reaction written as a chemical equation and as a sentence. What advantages do you see for writing the reaction as an equation?

$$CH_4\,(g) + O_2\,(g) \longrightarrow CO_2\,(g) + H_2O\,(g)$$

Methane gas reacts with oxygen gas to produce carbon dioxide gas and water vapor.

Parts of a chemical equation In Chapter 16, you learned how to write chemical formulas (Figure 17.8). Recall that the symbols for elements are used along with subscripts. Additional parts of a chemical equation are symbols that indicate the state of matter for each reactant and product. An arrow is always included between reactants and products. The arrow means "to produce" or "to yield."

Accounting for the atoms You know that a chemical reaction involves breaking and reforming chemical bonds. See if you can account for how atoms are distributed on the reactant side versus the product side in the methane reaction above. What's wrong? Notice that there are only two oxygen atoms on the reactant side, but there are three on the product side. You might also notice that you have four hydrogen atoms on the reactant side and only two on the product side (Figure 17.9). This means that the chemical equation above is not completely correct.

Numbers and types of atoms must balance The law of conservation of mass is always applied to chemical equations. The law is applied by *balancing* the number and type of atoms on either side of the equation. When balancing a chemical equation, you consider whole atoms rather than fractions of an atom because only whole atoms react. Also, you are not allowed to change the chemical composition of any of the compounds on the reactants or products side. To learn how to balance chemical equations, let's take another look at the methane reaction.

■■■■■ **VOCABULARY** ■■■■■

chemical equation – an expression of a chemical reaction using chemical formulas and symbols

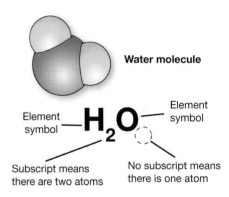

Water molecule

Element symbol — H_2O — Element symbol

Subscript means there are two atoms

No subscript means there is one atom

Figure 17.8: *The parts of a chemical formula.*

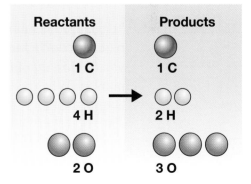

Reactants	Products
1 C	1 C
4 H	2 H
2 O	3 O

Figure 17.9: *This graphic illustrates that the number of oxygen and hydrogen atoms are not balanced for the methane reaction.*

Balancing a chemical equation

Begin by counting the number of atoms The first step of balancing a chemical equation involves counting the number of each type of atom on both sides of the reaction. Recall that the subscripts in a chemical formula tell you the number of each type of atom. The table below summarizes this information for the methane reaction (Figure 17.10).

Type of atom in methane reaction	Total on reactants side	Total on products side	Balanced?
C	1	1	yes
H	4	2	no
O	2	3	no

coefficient – a whole number placed in front of a chemical formula in a chemical equation

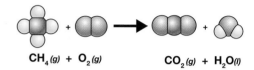

$$CH_4\,(g) + O_2\,(g) \longrightarrow CO_2\,(g) + H_2O\,(l)$$

Figure 17.10: *Graphic of the unbalanced methane reaction.*

When an equation is unbalanced As you can see, the chemical equation for the methane reaction is not balanced. The number of hydrogen and oxygen atoms are different on each side of the equation. To make them equal and balance the equation, you must figure out what number to multiply each compound by in order to make the numbers add up. Remember, you cannot change the number of individual atoms in a compound. That would change its chemical formula, and you would have a different compound.

Adding coefficients To change the number of molecules of a compound, you can write a whole number **coefficient** in front of the chemical formula (Figure 17.11). When you do this, *all* of the types of atoms in that formula are multiplied by that number. When there is no coefficient in front of a chemical formula, you assume that one molecule of that compound is sufficient.

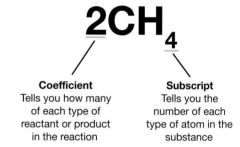

$$2CH_4$$

Coefficient
Tells you how many of each type of reactant or product in the reaction

Subscript
Tells you the number of each type of atom in the substance

Figure 17.11: *What do coefficients and subscripts mean?*

A coefficient of **2** in front of methane **CH₄** gives you:

$$2 \times CH_4$$

$2 \times 1C = 2C$
$2 \times 4H = 8H$

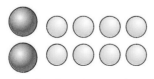

2 carbon atoms and 8 hydrogen atoms

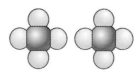

Enough carbon and hydrogen to make 2 molecules of methane

Checking your work Figuring out where to place coefficients to multiply the numbers of atoms in a chemical formula is largely a process of trial and error. Let's look at the methane reaction after the correct coefficients have been added:

$$CH_4 (g) + 2O_2 (g) \rightarrow CO_2 (g) + 2H_2O (g)$$

Counting the atoms on both sides again, we see that the equation is balanced.

Type of atom in methane reaction	Total on reactants side	Total on products side	Balanced?
C	1	1	yes
H	4	2(× 2) = 4	yes
O	2(× 2) = 4	2 + 1(× 2) = 4	yes

Reading a balanced equation Now that the equation is balanced, it can be read as follows: "One molecule of methane reacts with two molecules of oxygen to produce one molecule of carbon dioxide and two molecules of water." Figure 17.12 reviews key points to remember when balancing chemical equations.

When balancing a chemical equation...
1. Make sure you have written the correct chemical formula for each reactant and product.
2. The subscripts in the chemical formulas of the reactants and products cannot be changed during the process of balancing the equation. Changing the subscripts will change the chemical makeup of the compounds.
3. Numbers called coefficients are placed in front of the formulas to make the number of atoms on each side of the equation equal.

Figure 17.12: *Key points for balancing a chemical equation.*

 Your turn... balanced or unbalanced?

Identify which of the following equations are balanced.

a. $2H_2 + O_2 \rightarrow 2H_2O$

b. $MgO + H_2O \rightarrow Mg(OH)_2$

c. $Ca + O_2 \rightarrow CaO$

d. $Na_2O + H_2O \rightarrow NaOH$

e. $2HCl + Ca(OH)_2 \rightarrow CaCl_2 + 2H_2O$

SOLVE FIRST LOOK LATER

a. balanced

b. balanced

c. not balanced

d. not balanced

e. balanced

Example: Balancing a common reaction

What happens when you take an antacid? *Hydrochloric acid* (HCl) is a compound your stomach normally produces to help you break down food. Sometimes, if you eat spicy foods or worry excessively about studying chemistry, your stomach produces too much hydrochloric acid and you get acid indigestion. Most people take antacids to relieve this painful condition. Many antacids contain calcium carbonate ($CaCO_3$), which neutralizes the hydrochloric acid (Figure 17.13). The products formed by the chemical reaction are calcium chloride, carbon dioxide, and water. How might you write and balance the chemical equation for this reaction? The following steps outline the process for you.

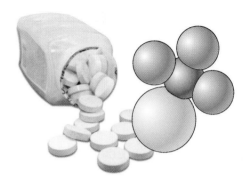

Figure 17.13: *Calcium carbonate is the main ingredient in antacid tablets used to fight acid indigestion.*

1. Write the word form of the equation.

Hydrochloric acid reacts with calcium carbonate to produce calcium chloride, carbon dioxide, and water.

2. Write the chemical equation.

Figure 17.14 gives you the chemical formulas for the compounds in the reaction. Additionally, you can use the ion charts in Chapter 16 to help you determine the chemical formulas for compounds. Using the chemical formulas, the chemical equation for this reaction is:

Some chemical formulas	
Calcium carbonate	$CaCO_3$
Hydrochloric acid	HCl
Calcium chloride	$CaCl_2$
Carbon dioxide	CO_2
Water	H_2O

Figure 17.14: *Chemical formulas for each compound in the reaction.*

Reactants		Products

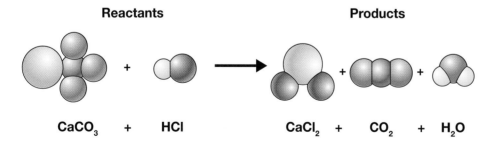

$$CaCO_3 + HCl \longrightarrow CaCl_2 + CO_2 + H_2O$$

3. Count the number of each type of atom on both sides.

The graphic below summarizes how many atoms of each type are on the reactants and products sides of the chemical equation. Notice that there is an extra hydrogen and an extra chlorine on the products side. These two extra atoms have to come from somewhere. We need to add something to the reactants that will give us an extra chlorine and hydrogen.

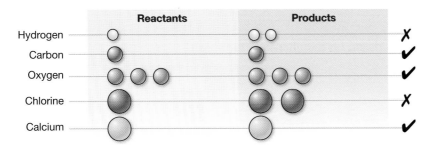

4. Add coefficients to balance the equation.

Fortunately, one of the reactants is HCl, so we can add one more molecule of HCl to the reactants side. In the equation, we put a 2 in front of the HCl to indicate that we need 2 molecules. Remember, you cannot change the subscripts. In this case, you just need to put a coefficient of 2 in front of HCl to balance the equation.

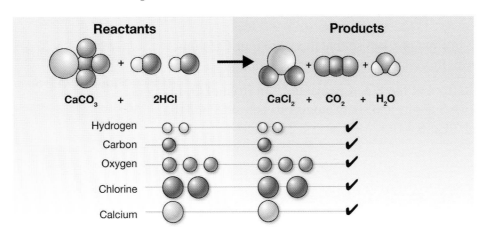

Solving Problems: **Balancing Equations**

In this reaction, chalcocite (a mineral) reacts with oxygen in the presence of heat. The products are a type of copper oxide and sulfur dioxide. Balance this equation: $Cu_2S + O_2 \rightarrow Cu_2O + SO_2$ (Figure 17.15).

1. **Looking for:**	Coefficients that will balance the chemical equation.

2. **Given:**   The following information is based on the chemical equation.

Type of atom	Reactants	Products	Balanced?
Cu	2	2	yes
S	1	1	yes
O	2	3	no

3. **Relationships:** Coefficients can be added in front of any chemical formula in a chemical equation. When a coefficient is added in front of a chemical formula, all atoms in that formula are multiplied by that number.

4. **Solution:**   First try: Add a 2 in front of O_2 and in front of Cu_2O so that there are four O atoms on each side. However, this changes the number of Cu atoms.

Second try: Add a 2 in front of Cu_2S so that there are four Cu atoms on each side. However, this changes the number of S atoms.

Third try: Add a 2 in front of the SO_2. Change the 2 in front of O_2 to a 3. Now there are two S atoms and six O atoms on each side, and the equation is balanced: $2Cu_2S + 3O_2 \rightarrow 2Cu_2O + 2SO_2$

Your turn...

a. $KClO_3 \rightarrow KCl + O_2$

b. $Al_2S_3 + H_2O \rightarrow Al(OH)_3 + H_2S$

First try:
$Cu_2S + \mathbf{2}O_2 \rightarrow \mathbf{2}Cu_2O + SO_2$

Atom	Reactants	Products
Cu	2	$2(\times 2) = 4$
S	1	1
O	$2(\times 2) = 4$	$1(\times 2) + 2 = 4$

Second try:
$\mathbf{2}Cu_2S + \mathbf{2}O_2 \rightarrow \mathbf{2}Cu_2O + SO_2$

Atom	Reactants	Products
Cu	$2(\times 2) = 4$	$2(\times 2) = 4$
S	$1(\times 2) = 2$	1
O	$2(\times 2) = 4$	$1(\times 2) + 2 = 4$

Third try:
$2Cu_2S + \mathbf{3}O_2 \rightarrow 2Cu_2O + \mathbf{2}SO_2$

Atom	Reactants	Products
Cu	$2(\times 2) = 4$	$2(\times 2) = 4$
S	$1(\times 2) = 2$	$1(\times 2) = 2$
O	$2(\times 3) = 6$	$1(\times 2) + 2(\times 2) = 6$

Figure 17.15: *Balancing the equation.*

SOLVE FIRST / LOOK LATER

a. $2KClO_3 \rightarrow 2KCl + 3O_2$

b. $Al_2S_3 + 6H_2O \rightarrow 2Al(OH)_3 + 3H_2S$

Section 17.2 *Review*

1. What is the law of conservation of mass? How is it related to balancing chemical equations?

2. Why is it important to study chemical reactions in closed containers?

3. In one of his experiments, Lavoisier placed 10.0 grams of mercury (II) oxide into a sealed container and heated it. The mercury (II) oxide then reacted in the presence of heat to produce 9.3 grams of mercury. Oxygen gas was another product in the reaction. According to the law of conservation of mass, how much oxygen gas would have been produced?

4. Which answer below is the formula mass of a water molecule?
   a. 18.02 amu
   b. 18.02 moles
   c. 18.02 grams
   d. 6.02×1023 molecules

5. What is the difference between the formula mass and the molecular mass of a compound?

6. What is the difference between a subscript and a coefficient in a chemical equation?

7. Are the following chemical equations balanced or unbalanced? Balance any unbalanced equations.
   a. $2KClO_3 \rightarrow KCl + 3O_2$
   b. $Fe + O_2 \rightarrow FeO$
   c. $2Li + Cl_2 \rightarrow 2LiCl$
   d. $NH_3 + HCl \rightarrow NH_4Cl$

8. $BaO_2 (s) \rightarrow BaO (s) + O_2 (g)$
   a. Balance the chemical equation above.
   b. Use the information in Figure 17.16 to write the equation in words. Be sure to describe the state of matter for each compound.
   c. Challenge: Read the Science Fact at the right. How many moles of barium peroxide would be needed to produce four moles of barium oxide and two moles of oxygen?

Some chemical formulas	
Barium peroxide	BaO_2
Barium oxide	BaO
Oxygen	O_2

Figure 17.16: *Question 8.*

▬▬▬ ■ SCIENCE FACT ■ ▬▬▬

More about coefficients

A coefficient also represents the number of moles of a compound that are involved in a reaction. For example, for the balanced methane reaction in the text, you can say:

One mole of methane reacts with two moles of oxygen to produce one mole of carbon dioxide and two moles of water.

$$CH_4 (g) + 2O_2 (g) \longrightarrow CO_2 (g) + 2H_2O (l)$$

17.3 Classifying Reactions

Most of the products you use every day are the result of one or more chemical reactions. As you might imagine, there are many possible chemical reactions. This section provides you with information on how to classify the different types of chemical reactions.

Addition reactions

Making compounds   In an **addition reaction**, two or more substances combine to form a new compound. A good example of an addition reaction is the formation of rust.

$$\text{Fe } (s) + \text{O}_2\,(g) \longrightarrow \text{Fe}_2\text{O}_3\,(s)$$

From this example, how might you describe the reaction in general terms? The answer to this question is below. In this general equation for an addition reaction and the other reactions in this section, A and B represent ions, atoms, or molecules.

$$\text{A} + \text{B} \longrightarrow \text{AB}$$

Polymerization   Recall that a *polymer* is a large molecule made up of repeating segments. **Polymerization**, or the formation of polymers, is a series of addition reactions taking place to produce a very large molecule. Polymers are made by joining smaller molecules called *monomers*.

monomers   polymer   monomer   larger polymer

Table 17.1: Polymers

Common polymers	Polymer products
Polystyrene	Foam containers
Polyethylene	Food packaging
Polyester	Clothing
Polyvinyl chloride	Plumbing (PVC pipes)
Polyvinyl acetate	Chewing gum

Decomposition reactions

Breaking down compounds As you might suspect, chemical reactions are used to make compounds. However, a chemical reaction is also used to break down compounds. A chemical reaction in which a single compound is broken down to produce two or more smaller compounds is called a **decomposition reaction**. The general equation for decomposition is:

$$AB \xrightarrow{\text{Energy}} A + B$$

Energy is required In most cases, energy is required to get a decomposition reaction going. The most common form of energy used in these chemical reactions is heat. For example, the reaction below was involved in the discovery of oxygen. Heat was used in the decomposition of mercury (II) oxide.

$$2HgO\,(s) \xrightarrow{\text{Heat}} 2Hg\,(l) + O_2\,(g)$$

For the decomposition of water into hydrogen and oxygen, the energy source is electricity. In fact, this particular reaction, illustrated in Figure 17.17, is called *electrolysis*.

$$2H_2O\,(l) \xrightarrow{\text{Electricity}} 2H_2\,(g) + O_2\,(g)$$

The number of products formed The simplest kind of decomposition is the breakdown of a binary compound into its elements. However, larger compounds can also decompose to produce other compounds. The number of compounds that form as products in a decomposition reaction depends on the number of elements in the reactant compound. For example, baking soda ($NaCO_3$) has four elements. When it undergoes a decomposition reaction with heat, three products form.

$$2NaHCO_3\,(s) \xrightarrow{\text{Heat}} 2CO_2\,(g) + H_2O\,(g) + Na_2O\,(s)$$

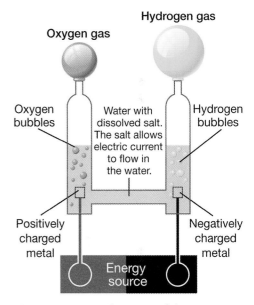

Figure 17.17: *A diagram of the experimental setup for performing the electrolysis of water. Why do you think the balloon for hydrogen gas is twice as big as the one for oxygen gas?*

Displacement reactions

VOCABULARY

single-displacement reaction – a chemical reaction in which one element replaces a similar element in a compound

double-displacement reaction – a chemical reaction in which ions from two compounds in solution exchange places to produce two new compounds

Single-displacement reactions

In a **single-displacement reaction**, one element replaces a similar element in a compound. For example, if you place an iron nail into a beaker of copper (II) chloride, you will begin to see reddish copper forming on the iron nail. In this reaction, iron *replaces* copper in the solution and the copper *falls out* of the solution onto the nail as a metal.

$$Fe\,(s) \;+\; CuCl_2\,(aq) \longrightarrow FeCl_2\,(aq) \;+\; Cu\,(s)$$

The general equation for a single-displacement reaction is:

$$A \;+\; BX \longrightarrow AX \;+\; B$$

In this equation, A and B are elements, and AX and BX are compounds.

Double-displacement reactions

In a **double-displacement reaction**, ions from two compounds in solution exchange places to produce two new compounds. One of the compounds formed is usually a precipitate that settles out of the solution, a gas that bubbles out of the solution, or a molecular compound such as water. The other compound formed often remains dissolved in the solution. Precipitates are first recognizable by the cloudy appearance they give to a solution. A precipitate is the result of one of the products in a double-displacement reaction being insoluble in water (Figure 17.18). Depending on the compound formed, the precipitate can be many different colors from white to fluorescent yellow, as in the reaction between lead (II) nitrate and potassium iodide.

$$Pb(NO_3)_2\,(aq) \;+\; 2KI\,(aq) \longrightarrow PbI_2\,(s) \;+\; 2KNO_3\,(aq)$$

The general formula for a double-displacement reaction is given below. Each pairing of letters—AB and CD, and AD and CB—are ionic compounds in a solution.

$$AB \;+\; CD \longrightarrow AD \;+\; CB$$

Solution A

Precipitate

Solution B

Figure 17.18: *The formation of a cloudy precipitate is evidence that a double-displacement reaction has occurred. If left undisturbed in a beaker, a precipitate will settle to the bottom. The precipitate in the image is potassium iodide.*

Combustion reactions

In a combustion reaction, energy is released A **combustion reaction**, also called *burning*, occurs when a substance, such as wood, natural gas, or propane, combines with oxygen and releases a large amount of energy in the form of light and heat. The products of this kind of combustion reaction are carbon dioxide and water. What do reactants like wood, natural gas, and propane have in common? The answer is that they are all carbon compounds. Following is the general equation for a combustion reaction.

$$\text{Carbon compound} + O_2\,(g) \longrightarrow CO_2\,(g) + H_2O\,(g)$$

Carbon compounds The methane reaction, which you have seen before, is a good example of a combustion reaction. Carbon compounds that are a mixture of only carbon and hydrogen atoms, like methane, are called hydrocarbons. The general formula for a hydrocarbon compound is C_xH_y where x and y represent different subscripts. Examples of hydrocarbon compounds can be found in Figure 17.19.

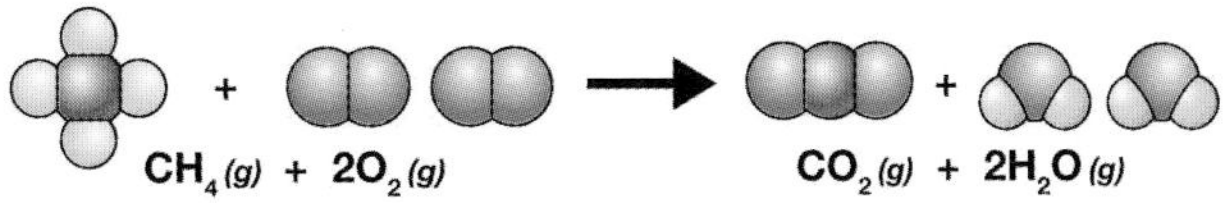

$$CH_4\,(g) + 2O_2\,(g) \longrightarrow CO_2\,(g) + 2H_2O\,(g)$$

Another kind of combustion reaction Not all combustion reactions use carbon compounds as a reactant. These types of combustion reactions do not produce carbon dioxide. For example, when hydrogen gas is burned in oxygen, water is the only product.

$$2H_2\,(g) + O_2\,(g) \longrightarrow 2H_2O\,(l)$$

The value of an alternative combustion reaction Perhaps, in the future, some of our cars will run by the reaction above. Instead of using gasoline, which is a mixture of carbon compounds, cars will run on hydrogen. Currently, automobile manufacturers are developing technologies that utilize hydrogen combustion in the internal combustion engines of cars. Another way hydrogen can be used to power cars is in an electrochemical process that uses a "fuel cell." In either case, the use of hydrogen fuel could help reduce the amount of carbon dioxide emissions related to transportation. However, it would still take energy, sometimes in the form of fossil fuels, to make the hydrogen fuel. What do you think? Should hydrogen technologies be developed for cars?

Hydrocarbon compound	Chemical formula
Methane	CH_4
Ethane	C_2H_6
Propane	C_3H_8
Butane	C_4H_{10}
Pentane	C_5H_{12}
Hexane	C_6H_{14}
Heptane	C_7H_{16}
Octane	C_8H_{18}

Figure 17.19: *Examples of hydrocarbon compounds.*

CHALLENGE

Hydrogen technology

In the text, you learned about two forms of hydrogen technology used for running an automobile. Find out more about each one. Is hydrogen fuel a viable alternative to fossil fuels?

Reviewing the types of reactions

Table 17.2 provides a summary of the types of reactions you have learned. Review the information in the table then practice identifying the different types of reactions by answering the problems below.

Table 17.2: Summary of the types of reactions

Type	General equation	Example
Addition	$A + B \rightarrow AB$	$2H_2 + O_2 \rightarrow 2H_2O$
Decomposition	$AB \rightarrow A + B$	$2NaHCO_3 \rightarrow 2CO_2 + H_2O + Na_2O$
Single-displacement	$A + BX \rightarrow AX + B$	$Fe + CuCl_2 \rightarrow FeCl_2 + Cu$
Double-displacement	$AB + CD \rightarrow AD + CB$	$Pb(NO_3)_2 + 2KI \rightarrow PbI_2 + 2KNO_3$
Combustion	carbon compound $+ O_2 \rightarrow CO_2 + H_2O$	$C_6H_{12}O_6 + 6O_2 \rightarrow 6CO_2 + 6H_2O$

 Your turn... **identifying types of reactions**

Identify each type of reaction.

a. $CaCO_3 \ (s) \rightarrow CaO \ (s) + CO_2 \ (g)$

b. $Na \ (g) + KCl \ (l) \rightarrow K \ (g) + NaCl \ (l)$

c. $C \ (s) + O_2 \ (g) \rightarrow CO_2 \ (g)$

Consumer chemistry: Preserving dried fruit

Have you ever opened a box of dried fruit such as golden raisins or apricots and smelled a slight "sulfur" odor, like a lit match? The odor is caused by sulfur dioxide, a gas that is used to preserve the color of dried fruits. This gas is produced in the reaction between sodium sulfite and hydrochloric acid:

$Na_2SO_3 \ (aq) + 2HCl \ (aq) \rightarrow$
$2NaCl \ (aq) + H_2O \ (l) + SO_2 \ (g)$

The fruit is exposed to the gas, which is absorbed into the skin of the fruit. When you open the box for the first time, some of the gas that has escaped the fruit may not escape your nose!

Which type of reaction produces this gas?

SOLVE FIRST / LOOK LATER

a. decomposition

b. single-displacement

c. addition

Section 17.3 *Review*

1. Why is polymerization a type of addition reaction?

2. You have learned about the different kinds of chemical reactions. Come up with a set of simple rules that you can use to help you identify each kind of chemical reaction. There are no right or wrong answers. Write down rules that make sense to you.

3. The graphic at the right shows the electrolysis of water.

    a. Come up with an explanation for why oxygen forms near the positively charged metal and hydrogen forms near the negatively charged metal.

    b. Why is a greater amount of hydrogen gas collected in this reaction?

    c. Is this reaction occurring in a closed container? Justify your answer.

4. How does the involvement of energy in a decomposition reaction compare to how energy is involved in a combustion reaction?

5. Compare and contrast single-displacement and double-displacement reactions.

6. Identify the following reactions as addition, decomposition, single- or double-displacement, or combustion.

    a. $N_2 (g) + 3H_2 (g) \rightarrow 2NH_3 (g)$

    b. $NH_4NO_3 (s) \rightarrow N_2O (g) + 2H_2O (g)$

    c. $AgNO_3 (aq) + NaCl (aq) \rightarrow AgCl (s) + NaNO_3 (aq)$

    d. $Fe (s) + H_2SO_4 (aq) \rightarrow H_2 (g) + FeSO_4 (aq)$

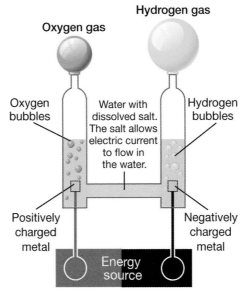

Oxygen gas

Hydrogen gas

Oxygen bubbles

Water with dissolved salt. The salt allows electric current to flow in the water.

Hydrogen bubbles

Positively charged metal

Negatively charged metal

Energy source

Hydrogen-Powered Cars

Imagine driving to school in a sleek, super-quiet sports car. You turn on the radio, and—surprise!—the news reporter isn't talking about the high price of oil or global climate change. You get stuck in a traffic jam, but roll down your windows anyway, and breathe in the fresh, clean air. The only emission coming from your car's tailpipe is water vapor. That's it! No carbon monoxide, no carbon dioxide, no nitrous oxides—just plain water. You notice your fuel gauge is getting close to empty, so you pull into a filling station and refuel your car with pressurized hydrogen.

Sound like a fairy tale? It's not! Today, we have cars, buses, trains, motorcycles, and even rockets that run on hydrogen fuel cell technology. Hydrogen-powered vehicles offer reduced vehicle greenhouse gas emissions and decrease American dependence on foreign oil supplies.

Scientists and engineers from government agencies, universities, and all of the major automobile manufacturers are designing, building, and testing hydrogen fuel cell vehicles, also known as FCVs.

Under the hood: How a fuel cell works

The most common kind of fuel cell placed in test vehicles is the Proton Exchange Membrane, or PEM, fuel cell. Basically, a fuel cell uses oxygen from the air and hydrogen gas to generate electricity to power the car's motor.

Like a battery, a fuel cell has two electrodes: an anode, to collect positive charges; and a cathode, to draw negative charges. First, hydrogen gas flows to the anode, where hydrogen atoms are separated into protons and electrons. The membrane is a solid organic polymer with the feel of plastic wrap and a thickness of two to seven sheets of paper; it allows only protons to pass through. Electrons are repelled by the anode and directed toward an external circuit. This flow of electrons is the electricity used to power a motor that moves the car.

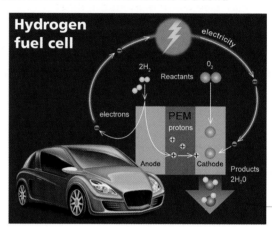

At the cathode side, oxygen taken from the outside air flows in and reacts with the hydrogen protons that have come through the membrane and the electrons from the circuit to produce water and heat. There are no other by-products of this reaction.

A typical gasoline-fueled car, in contrast, is powered by a combustion reaction that produces carbon dioxide, carbon monoxide, nitrous oxides, and sulfuric acids along with water and heat.

Not only is the FCV less polluting than a gasoline-powered vehicle, it's also remarkably more efficient. Internal combustion engines are at best 20 percent efficient. This means that only 20 percent of the gasoline's energy is converted into motion. Hydrogen fuel cell vehicles convert 40–60 percent of their fuel's energy into motion.

Fuel cell power is dependent upon the size, type of membrane, operating temperature, and gas pressure for the hydrogen and oxygen gases. A typical PEM fuel cell produces approximately one volt of electricity; this is only one-third the voltage used to operate a common flashlight. Therefore, fuel cells must be "stacked" to generate the power needed to propel a car. Some prototype vehicles use hundreds of stacked cells. These stacks can get expensive because each cathode and anode contains platinum, a costly metal.

Where would we get hydrogen fuel?

Hydrogen is the most abundant element in the universe, but it is most commonly found in compounds like water or in hydrocarbons, including fossil fuels and natural gas. These compounds' chemical bonds must be broken to obtain hydrogen gas.

Two of the main processes used to produce hydrogen are:

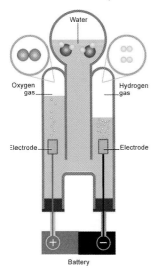

In electrolysis, an electric current splits water into oxygen and hydrogen.

Electrolysis: An electric current splits water into oxygen and hydrogen. If this electric current is produced by a "clean" energy source such as wind or solar energy, then the hydrogen fuel cell car is truly a zero-emissions vehicle. However, if the electricity comes from a coal or natural-gas-burning power plant, the total greenhouse gas and particulate emissions used to extract the hydrogen can actually cancel out the benefit of zero tailpipe emissions.

Gasification: Natural gas is combined with steam to extract hydrogen. Carbon monoxide and carbon dioxide are also produced. However, using hydrogen produced in this manner to power an automobile still reduces the total greenhouse gas emissions involved by about 50 percent, compared with the total emissions involved in operating a gasoline-powered automobile.

Once the hydrogen is produced, it has to be made available to drivers. Most FCVs store hydrogen on board in highly pressurized tanks. A tank at 5,000 psi (that's almost 100 times the pressure in most car tires) allows a car to travel about 250 miles without refueling.

Currently, there are very few places in the United States where you can fill a hydrogen tank. One of the biggest hurdles to switching to hydrogen-powered cars is getting the necessary infrastructure built.

Infrastructure is a term used to describe the support system needed for a new technology—such as hydrogen refineries, transport systems, and filling stations. After all, who is going to buy a vehicle if there's nowhere to fill its tank? And who is going to manufacture a vehicle if nobody can use it?

Some FCVs (called Hydrogen-Rich Fuel FCVs) are designed to take in gasoline or methanol and store it in a conventional gas tank, eliminating the infrastructure problem. These FCVs need a reformer to conduct gasification on board. The reformer breaks down the gasoline or methanol to extract hydrogen for the fuel cell. The by-products of the reformer are carbon dioxide and water, yet the total greenhouse gas emissions for this type of vehicle are less than half of a traditional gasoline-powered vehicle. Some scientists see this as an intermediate step that could get the technology on the road. Others point out that you can achieve nearly the same reduction in greenhouse gas emissions with hybrid gas–electric vehicles.

Challenges and breakthroughs

There is still a great deal of research and development ahead for FCVs. In order for this new type of automobile to take hold with the public, infrastructure must be developed, processes for producing hydrogen must be made cleaner and cheaper, and the vehicle designs must be proven safe and reliable. The United States Department of Energy is working with federal and international agencies and industry to see if these hurdles can be overcome.

Questions:

1. Compare and contrast hydrogen fuel cell vehicles with vehicles that have internal combustion engines.

2. Explain two ways hydrogen for fuel cell vehicles can be produced.

3. **Research:** Each of the major automobile manufacturers has a hydrogen fuel cell vehicle in production or development. Use a library or the Internet to learn about one of these projects. Prepare a report for your classmates. Include the timetable for the project, a description of challenges and breakthroughs, how much money is invested in the project, and, if possible, illustrations of the vehicle(s).

Chapter 17 *Assessment*

Vocabulary

Select the correct term to complete each sentence.

addition	combustion	mole
Avogadro number	decomposition	polymerization
chemical reaction	formula mass	product
chemical equation	law of conservation of mass	precipitate
coefficient	molar mass	reactant

Section 17.1

1. A(n) _____ is a process that involves reactants and products.

2. A starting ingredient in a chemical reaction is called a(n) _____ .

3. A substance that is the result of the forming of new bonds in a chemical reaction is called a(n) _____ .

4. An insoluble product in a double-displacement reaction is called a(n) _____ .

Section 17.2

5. A(n) _____ is the written form of a chemical reaction.

6. There are 6.02×10^{23} atoms in a(n) _____ of the element carbon.

7. The _____ of a compound is measured in atomic mass units.

8. The _____ states that the mass of reactants always equals the mass of the products.

9. You can change the number of atoms in a chemical equation by placing a(n) _____ in front of a chemical formula.

10. The number of molecules in a mole is equal to the _____ .

11. The _____ is the mass in grams of one mole of a substance.

Section 17.3

12. A large molecule of repeating units is made by the process of _____ .

13. A(n) _____ reaction is used to make a compound from other compounds or elements.

14. A carbon compound is usually one of the reactants in a(n) _____ .

15. Water can be broken down into hydrogen gas and oxygen gas by a(n) _____ .

Concepts

Section 17.1

1. Correct or incorrect? If the statement is incorrect, rewrite it so it is correct.
   a. Crushing ice and melting ice are examples of physical changes.
   b. The evaporation of water is an example of a chemical change.
   c. The action of enzymes on food during digestion results in chemical changes.
   d. Both physical and chemical changes involve chemical reactions and energy.
   e. Rolling out a lump of pizza dough into a pizza crust is an example of a physical change.
   f. Frying an egg in a pan causes a physical change in the egg.

2. Your body produces heat and maintains a stable, warm body temperature of about 98°F (37°C). Is this evidence that your body is undergoing chemical changes or physical changes?

3. Look at the ice cube in this picture. Is it undergoing a physical change or a chemical change? Justify your answer.

Section 17.2

4. In your chemistry lab, you mix baking soda and vinegar in a beaker. You carefully find the mass of the baking soda and vinegar you use in the reaction. However, after you are done with the reaction, you find that the product of the reaction has much less mass than the combined mass of the reactants. Evaluate your results. Are they correct? How might you perform this reaction again to make sure?

5. Identify each of the following as the formula mass, a mole, or the molar mass of a carbon monoxide (CO) molecule. One item does not make sense. Identify it.

  a. 28.01 grams
  b. 12.01 grams
  c. 28.01 amu
  d. 6.02×10^{23} molecules

6. Answer the following for the reaction below. You should be able to recognize three of the compounds. The compound $C_6H_{12}O_6$ is a molecule of glucose, a sugar.

$$C_6H_{12}O_6 \, (s) + 6O_2 \, (g) \rightarrow 6CO_2 \, (g) + 6H_2O \, (l)$$

  a. What are the reactants and products in this reaction? Give the state of matter for each.
  b. What does the arrow in a chemical equation mean?
  c. Is this equation balanced? Justify your answer.

Section 17.3

7. Write the general equations for each type of chemical reaction.

8. You perform a reaction with two compounds. One of the reactants contains oxygen. The products are oxygen gas and another compound. What kind of reaction is this? Justify your answer. Illustrate this reaction with symbols and a diagram.

Problems

Section 17.1

1. Identify whether a physical or chemical change is occurring in each situation. State your evidence.

  a. You place a beaker of water on a hot plate and heat it up. Eventually, it starts to boil and you can see bubbles forming. You also see steam rising from the water surface.
  b. You mix two ionic solutions. After you mix them in a beaker, the solution turns cloudy. After five minutes, the solution becomes clear and a white powder has settled on the bottom of the beaker.
  c. You heat up some pure sugar in a beaker. In time, you see a black substance and water droplets.
  d. By accident you drop an empty beaker on the floor and it breaks into pieces. You and your partner alert your teacher, who safely cleans up the mess.

Section 17.2

2. Determine the formula mass and the molar mass of each compound in this unbalanced reaction.

$$Cl_2 + KBr \rightarrow KCl + Br_2$$

3. How many molecules would be in two moles of bromine gas (Br_2)? Explain your answer.

4. A compound that contains both potassium and oxygen formed when potassium metal was burned in oxygen gas. The mass of the compound was 7.11 grams. The mass of the potassium metal was 3.91 grams. What mass of oxygen was involved in this reaction? Justify your answer.

5. Which of the following equations is balanced?

   a. $Al + Br_2 \rightarrow 2AlBr_3$
   b. $2Al + 2Br_2 \rightarrow 3AlBr_3$
   c. $2Al + 3Br_2 \rightarrow 2AlBr_3$
   d. $Al + Br_2 \rightarrow AlBr_3$

6. Balance the following equations. If an equation is already balanced, say so in your answer.

   a. $Cl_2 + Br \rightarrow Cl + Br_2$
   b. $CaO + H_2O \rightarrow Ca(OH)_2$
   c. $Na_2SO_4 + BaCl_2 \rightarrow BaSO_4 + NaCl$
   d. $ZnS + O_2 \rightarrow ZnO + SO_2$
   e. $Cl_2 + KBr \rightarrow KCl + Br_2$
   f. $H_2SO_4 + NaOH \rightarrow Na_2SO_4 + H_2O$

Section 17.3

7. When your body "burns" food for energy, carbon dioxide and water are released. This process is called respiration, and it is exactly like the respiration process performed by yeast in making pizza dough. Oxygen is needed for respiration to occur. Answer these questions:

   $+ O_2 \rightarrow CO_2 + H_2O$

   a. Where do the carbon dioxide and water come from in this reaction?
   b. Classify this reaction. Justify your answer.
   c. What kind of substance is an apple (or any food, for that matter)?

Applying Your Knowledge

Section 17.1

1. The process of digestion of food begins in your mouth and involves many other internal parts of your body, such as your stomach and intestines. Find out how physical changes and chemical changes are part of the human digestion process. Write up your findings as a descriptive essay or make a detailed poster of the digestion process.

Section 17.2

2. Balance this equation and then answer the questions.

   $Cl_2 + KBr \rightarrow KCl + Br_2$

   a. You found the formula mass and the molecular mass of the compounds in this reaction in question 2 in the problems set. Using your balanced equation, does the reaction follow the law of conservation of mass?
   b. Write this equation as a sentence in two ways. First, describe the reaction in terms of molecules, then describe the reaction in terms of moles.

Section 17.3

3. Balance this equation and then answer the questions.

   $SiI_4 (g) + heat \rightarrow Si (s) + I_2 (g)$

   a. What kind of reaction is this?
   b. Pure silicon is very useful in the electronics industry. How is it used?
   c. Oxygen is the most abundant element in Earth's crust. How does silicon compare in abundance?

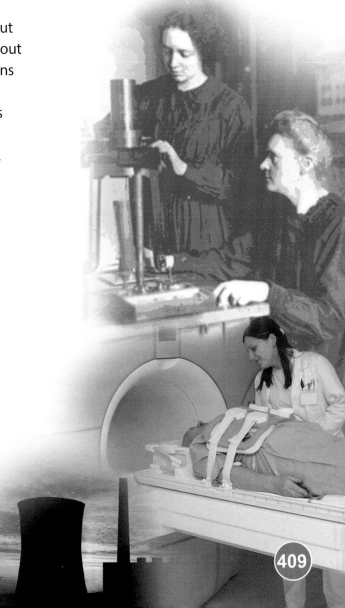

Energy and Reactions

***What do you think of when you hear the word* chemistry?**
You might think about making new substances from other substances or you might think of mixing things. These ideas do apply to the topic of chemistry, but chemistry is also all about energy. In this chapter, you will learn more about chemical reactions and how energy is involved in them. You will also learn about nuclear reactions—reactions that involve the nuclei of atoms. Nuclear reactions can produce much more energy than chemical reactions. The field of nuclear science—which ranges from energy production to medical technology—was pioneered by some fascinating people. For example, Marie Curie and her husband were early pioneers, as were their daughter, Irene, and her husband. Marie and Irene were the first and second women, respectively, to win Nobel Prizes. To learn more about the work of these women and more about the workings of chemical and nuclear reactions, read on!

CHAPTER 18 INVESTIGATIONS

18A: Energy and Chemical Changes
How do chemical changes involve energy?

18B: Thermodynamics of Hot/Cold Packs
Can we measure the heat released/energy absorbed by instant hot and cold packs?

18.1 Energy and Chemical Reactions

All chemical reactions involve energy. If you have ever sat near a campfire, you have experienced this energy as heat and light. In addition to producing energy, chemical reactions also *use* energy. For example, plants perform photosynthesis, which is a chemical reaction that uses energy from sunlight.

The two types of reactions

Energy is involved in two ways Energy is involved in chemical reactions in two ways: (1) At the start of a chemical reaction, energy is used to break some (or all) bonds between atoms in the reactants so that the atoms are available to form new bonds; and (2) Energy is released when new bonds form as the atoms recombine into the new compounds of the products. We classify chemical reactions based on how the energy used in (1) compares to the energy released in (2).

Exothermic reactions If forming new bonds releases *more* energy than it takes to break the old bonds, the reaction is **exothermic** (Figure 18.1, top). Once started, exothermic reactions tend to keep going because each reaction releases enough energy to start the reaction in neighboring molecules. A good example is the reaction of hydrogen with oxygen. If we include energy, the balanced reaction looks like this:

$$O_2\,(g) \ + \ 2H_2\,(g) \longrightarrow 2H_2O\,(l) \ + \ \textit{Energy}$$

Oxygen   Hydrogen   Water

Endothermic reactions If forming new bonds in the products releases *less* energy than it took to break the original bonds in the reactants, the reaction is **endothermic** (Figure 18.1, bottom). Endothermic reactions absorb energy. In fact, endothermic reactions need more energy to keep going. An example of an important endothermic reaction is *photosynthesis*. In photosynthesis, plants use energy from sunlight to make glucose and oxygen from carbon dioxide and water.

$$6CO_2 \ + \ 6H_2O \ + \ \textit{Energy} \longrightarrow C_6H_{12}O_6 \ + \ 6O_2$$

Carbon dioxide   Water   Glucose   Oxygen

Exothermic
Energy released > Energy used

Endothermic
Energy used > Energy released

Figure 18.1: *Exothermic and endothermic reactions.*

Activation energy

An interesting question
Exothermic reactions occur because the atoms arranged as compounds of the products have lower energy and are more stable than they are when arranged as compounds of the reactants. Chemical reactions—like other systems—move toward more stable circumstances. If this is true, why don't all the elements combine into the molecules that have the lowest possible energy?

Activation energy
The answer has to do with **activation energy**, which is the energy needed to begin a reaction and break chemical bonds in the reactants. Without enough activation energy, a reaction will not happen even if it is exothermic. That is why a flammable material like gasoline does not burn without a spark or flame. The spark supplies the activation energy to start the reaction.

VOCABULARY

activation energy – energy needed to break chemical bonds in the reactants to start a reaction

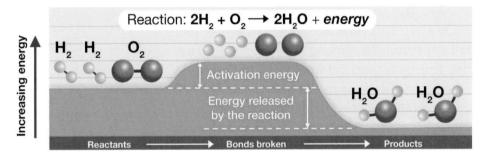

Reaction: $2H_2 + O_2 \longrightarrow 2H_2O + energy$

Increasing energy

H_2 H_2 O_2

Activation energy

Energy released by the reaction

H_2O H_2O

Reactants → Bonds broken → Products

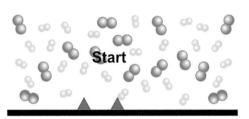

Start

Energy from a spark splits a few nearby molecules.

An example of a reaction
The diagram above shows how the energy flows in the reaction of hydrogen and oxygen. The activation energy must be supplied to break apart the molecules of hydrogen and oxygen. Energy is then released when the four free hydrogen and two free oxygen atoms combine to form two water molecules. The reaction is exothermic because the energy released by forming water is greater than the activation energy. Once the reaction starts, it supplies its own activation energy and quickly grows (Figure 18.2).

Reactions only occur when conditions are right
A reaction begins by itself when thermal energy is greater than the activation energy. However, any reaction that could start by itself probably already has! The compounds and molecules in substances around you need more activation energy to change into anything else. For example, table salt in a dish will remain table salt for a long time unless the conditions change to cause a chemical reaction between the salt and another compound.

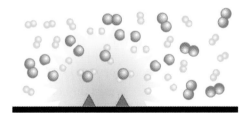

Released energy splits more molecules and the reaction becomes an explosion.

Figure 18.2: *Because energy released by one reaction supplies activation energy for new reactions, exothermic reactions can grow quickly once activation energy has been supplied.*

18.1 Energy and Chemical Reactions **411**

Carbon reactions and energy

Respiration and photosynthesis

At the beginning of this section, you learned that photosynthesis is an endothermic reaction. Through photosynthesis, plants capture energy from the Sun. The way humans and other life forms obtain this energy is by a chemical reaction called *respiration*, which is the reverse of photosynthesis. Glucose is metabolically "burned" in the presence of oxygen and carbon dioxide, and water is released. Evidence that your body is performing respiration includes the fact that you are breathing in oxygen and breathing out carbon dioxide right now as you read this paragraph! What kind of reaction is respiration—exothermic or endothermic? How do you know?

$$\underset{\text{Glucose}}{C_6H_{12}O_6} + \underset{\text{Oxygen}}{6O_2} \longrightarrow \underset{\text{Carbon dioxide}}{6CO_2} + \underset{\text{Water}}{6H_2O} + \textit{Energy}$$

Fossil fuels and combustion

Fossil fuels are mainly ancient, decayed plant and animal material. Today, we extract this carbon-based fuel from the ground and process it so that it is useful for many things. Energy is obtained from fossil fuels (like natural gas, kerosene, and coal) by using combustion reactions, which are exothermic. In addition to producing energy, combustion reactions yield carbon dioxide as a product. Currently, there is worldwide concern about how much carbon dioxide is being added to our atmosphere due to combustion reactions, mainly from sources of transportation such as cars.

$$\text{Fossil Fuel} + \underset{\text{Oxygen}}{O_2\,(g)} \overset{\text{⚡}}{\longrightarrow} \underset{\text{Carbon dioxide}}{CO_2\,(g)} + \underset{\text{Water}}{H_2O\,(l)} + \textit{Energy}$$

Global warming

Carbon dioxide in our atmosphere traps heat from the Sun to make the Earth warm and comfortable. This is why carbon dioxide and other atmospheric gases are called "greenhouse gases." But as more carbon dioxide is added, Earth is experiencing a global climate change. Climate change refers to the average increase in Earth's surface temperature since the mid-1900s, due to increased greenhouse gas concentrations in the atmosphere. The temperature increase has not been much—$1.0°F$ to $1.7°F$ from 1906 to 2005—but scientists believe this rise and the rise in the amount of greenhouse gases will continue. Scientists also believe that the consequences of global climate change will include continually rising sea levels and changes in weather.

SOLVE IT!

Chemistry and global climate change

Following is a balanced combustion reaction

$$CH_4\,(g) + 2O_2\,(g) \longrightarrow CO_2\,(g) + 2H_2O\,(l)$$

You read about this reaction in Chapter 17. It's the combustion of methane (natural gas), a fossil fuel.

Answer the following questions.

1. Fill in the blank. For every _____ mole(s) of methane gas, _____ mole(s) of carbon dioxide are produced.

2. What is the molecular mass of methane and of carbon dioxide?

3. Is the statement below correct or incorrect? Explain your answer.

The mass of carbon dioxide released is greater than the mass of fossil fuel burned.

JOURNAL

Global climate change is a worldwide concern. Find out what is being done to address this climate issue. Write a list of 10 actions you can take to make a difference.

Examples of endothermic reactions

Endothermic reactions in industry — Certainly, it's useful for chemical reactions to *produce* more energy than they use. But how do we benefit from reactions that *use* more energy than they produce? It turns out that most of the reactions used in industry to produce useful materials require more energy than they produce. This is one of the reasons sources of energy are so important to industry. In other words, exothermic reactions are needed to cause endothermic reactions to run! One example of an industry process that frequently uses endothermic reactions is the refining of ores to produce useful metals. Here is a specific example, the refinement of aluminum ore from aluminum oxide.

$$2Al_2O_3\,(s) + \textit{Energy} \longrightarrow 4Al\,(s) + 3O_2\,(g)$$

This reaction requires the input of energy because it takes more energy to break the bonds in the aluminum oxide than is released when the products are formed.

Cold packs — Have you ever used an "instant cold pack" as a treatment for a twisted ankle or a bruise? These products, found in your local drugstore, work by using an endothermic chemical reaction. The fact that more energy is used than produced is what makes the cold pack "cold." The reaction, shown below, works as follows. The product usually comes in a plastic bag. Inside the bag is a sealed packet of water surrounded by crystals of ammonium nitrate. To activate the cold pack, you squeeze the plastic bag to break the packet of water. When the water contacts the ammonium nitrate crystals, a reaction occurs and the pack becomes icy cold (Figure 18.3).

$$NH_4NO_3\,(s) + H_2O\,(l) + \textit{Energy} \longrightarrow NH_4^+\,(aq) + NO_3^-\,(aq)$$

Dissolution reactions — The ice pack gets very cold because it takes energy to dissolve the ionic bonds in the ammonium nitrate. Besides being endothermic, this reaction is also a dissolution reaction. A **dissolution reaction** occurs when an ionic compound (like ammonium nitrate) dissolves in water to make an ionic solution. In the cold pack reaction, the ions are an ammonium ion (NH_4^+) and a nitrate ion (NO_3^-).

VOCABULARY

dissolution reaction – an endothermic reaction that occurs when an ionic compound dissolves in water to make an ionic solution

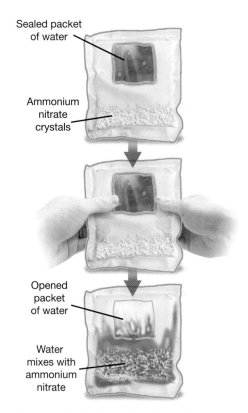

Sealed packet of water

Ammonium nitrate crystals

Opened packet of water

Water mixes with ammonium nitrate

Figure 18.3: *A cold pack works because of an endothermic reaction.*

Section 18.1 *Review*

1. List the two ways that energy is involved in chemical reactions.

2. Identify the following as describing either an exothermic or an endothermic reaction.

    a. More energy is released than is used by the reaction.
    b. The chemical reaction involved in burning wood
    c. Less energy is released than is used by the reaction.
    d. Photosynthesis

3. Once an exothermic reaction begins, the reaction tends to keep going. Why?

4. Why is a "spark" of energy required to begin the chemical reaction of burning a fossil fuel? What is another name for this spark of energy?

5. The reaction below is an exothermic reaction.

    $K_2O\ (s) + CO_2\ (g) \rightarrow K_2CO_3\ (s)$

    a. Rewrite this reaction and add "+ energy" in the correct location.
    b. Describe how the energy level of the reactants compares to the energy level of the products.

6. The reaction below is an endothermic reaction.

    $2HgO\ (s) \rightarrow 2Hg\ (l) + O_2\ (g)$

    a. Rewrite this reaction and add "+ energy" in the correct location.
    b. Describe how the energy level of the reactants compares to the energy level of the products.

7. How is a combustion reaction like the respiration reaction? List as many similarities as you can.

8. What is the main cause of global climate change? What are the consequences of global climate change?

9. How are exothermic and endothermic reactions linked in the process of refining metal ore?

10. Describe how you could perform a dissolution reaction with some table salt (NaCl) and a beaker of water. What are the products of this reaction?

Marie Curie

Later in this chapter, you will learn about nuclear chemistry. This biography introduces you to an incredible woman who was a pioneer in this field—Marie Curie.

The field of nuclear chemistry began when Marie Curie and her husband, Pierre Curie, discovered radioactivity. In 1898, Marie Curie, a Polish-born chemist, coined the word "radio-activity" to describe the peculiar behavior of elements she and her husband had discovered. They shared a Nobel Prize in 1903 for their discovery of radioactivity. Marie Curie was awarded a second Nobel Prize in 1911 for her discovery of the elements radium and polonium.

Marie Curie was the first woman to graduate with a degree in physics (1893) from the University of Sorbonne in Paris. She was the first woman to receive a Nobel Prize and the first person to receive two Nobel Prizes. She was also the first woman professor at the University of Sorbonne.

18.2 Chemical Reaction Systems

In Chapter 17, you learned that it's important to have chemical reactions occur in a closed system. When you study chemical reactions as systems, you are able to find out what types of products and how much of each are produced. You are also able to measure the amount of energy used or produced. In this section, you will look closer at chemical reaction systems.

Chemical equations are like recipes

Recipes and balanced chemical equations
Have you ever tried to make something from a recipe—say, a chocolate cake—and it didn't turn out quite the way you hoped? A recipe requires you to follow directions and add the correct amounts and types of ingredients (Figure 18.4). If you left out the eggs, for example, your cake wouldn't turn out well at all. A balanced chemical equation is like a recipe. It tells you the ingredients needed and the amount of each. It also tells you how much of each product will result if the precise amounts of reactants are combined.

Ingredients and quantities
Figure 18.5 illustrates the chemical equation for making water. This is the correct way to write a "recipe" for making a compound in chemistry. However, if you write the equation for making water as a recipe, you will see it is similar to a real recipe for making chocolate cake. Both recipes give you the ingredients needed, the instructions, and the product that will be produced. Both recipes also give you the quantities of ingredients (reactants) needed and the quantities of products that are produced.

Figure 18.4: *To successfully make a cake, you need to follow a recipe.*

Recipe #1: Chocolate cake		Recipe #2: Water
1 cup flour 1/2 cup cocoa powder 1/2 cup butter 1 teaspoon vanilla	1 cup sugar 1 teaspoon baking powder 1/2 cup milk 1 egg	2 molecules of hydrogen gas 1 molecule of oxygen gas
In a bowl, combine flour, sugar, cocoa powder, and baking power. Add butter, milk, vanilla, and egg. Mix until smooth. Bake in a 350°F oven for 35 minutes. *Makes 8 servings.*		Combine the molecules in a closed container. Add a spark of electricity. *Makes two molecules of water.*

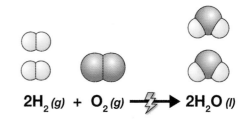

$$2H_2 \, (g) + O_2 \, (g) \longrightarrow 2H_2O \, (l)$$

Figure 18.5: *A chemical equation, such as this one for the formation of water, is like a recipe.*

Information from balanced chemical equations

1 Cup Flour   8 Servings

1/2 Cup Flour   4 Servings

The ratios of ingredients   The recipe for chocolate cake on the previous page gives you ratios among the ingredients needed to make eight servings. For example, if you double the ingredients, you can make twice as many servings. How many servings are possible if you only have half a cup of flour? To make a good-tasting chocolate cake with half as much flour, you would have to use half as much of the other ingredients, too. By halving the recipe, you can make four servings of chocolate cake (Figure 18.6).

Figure 18.6: *Using half the amount of each ingredient will make half as many servings of cake.*

Proportional relationships   Like a recipe, a balanced chemical equation shows the ratios of the number of molecules of reactants needed to make a certain number of molecules of products. The ratios are determined by the coefficients of the balanced equation. In the formation of water, two molecules of H_2 react with one molecule of O_2 to produce two molecules of H_2O. If you reacted four molecules of H_2 with two molecules of O_2, you would produce four molecules of H_2O instead of two (Figure 18.7). Doubling the reactants doubles the amount of product formed. What happens if you only double the number of H_2 molecules and not the number of O_2 molecules? How many H_2O molecules could you make? Would you have anything left over?

Balanced equations also show the ratio of relative masses   If you had a beaker with 6.02×10^{23} molecules of water, the water would have a molar mass of 18.02 grams. In the balanced chemical equation for water, you have a coefficient of 2 in front of the water molecule. This means you have two times the amount of water molecules. Therefore, the formula mass of water would be 36.04 amu and $2 \times (6.02 \times 10^{23})$ molecules would have an actual mass of 36.04 grams.

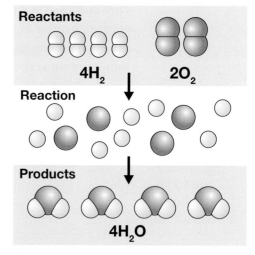

Reactants

$4H_2$ $2O_2$

Reaction

Products

$4H_2O$

What doesn't a balanced equation tell you?   A balanced equation does not describe the exact conditions under which a reaction will occur. For example, simply putting hydrogen and oxygen molecules together does not produce water. A spark of energy is needed for this reaction. In fact, many reactions require specific conditions to occur and these are not shown in the balanced equation. Additionally, some reactions may not occur at all, even though you can write a balanced equation for them. The right conditions for most of the reactions that are used in science and industry are the result of careful research and experimentation.

Figure 18.7: *If you react four hydrogen gas molecules with two oxygen molecules, you can make four water molecules.*

Limiting and excess reactants

When a reactant is in short supply

If a cookie recipe calls for two eggs and you only have one, you cannot make a whole batch of cookies. Having only one egg limits the number of cookies you can make. The same is true of chemical reactions. When a chemical reaction occurs, the reactants are not always present in the exact ratio indicated by the balanced equation. In fact, this is rarely the case unless the reaction is performed in a laboratory. What usually happens is that a chemical reaction will run until the reactant that is in short supply is used up.

Limiting reactants and excess reactants

The reactant that is used up first in a chemical reaction is called the **limiting reactant**. The limiting reactant limits the amount of product that can be formed. A reactant that is not completely used up is called an **excess reactant** because some of it will be left over when the reaction is complete.

Do reactions always turn out as expected?

Not all reactions turn out exactly as planned. If you use a specific amount of a limiting reactant and expect the exact amount of product to be formed, you will usually be disappointed. Often the amount of product you are able to collect and measure is less than the amount you would expect. This is because there are various factors that affect reactions as they are performed (Figure 18.8). For example, *experimental error* often affects how much product is produced. Additionally, just as it is hard to get all the cookie dough out of your bowl when you are ready to bake the cookies, some product is hard to collect and measure.

Percent yield

The amount of product you expect from a chemical reaction is called the *predicted yield*. The predicted yield is determined from the balanced equation for the reaction. The amount of product that you are able to measure is called the *actual yield*. The actual yield is determined by measuring the mass of the product produced in the actual reaction. The **percent yield** is the *actual yield* divided by the *predicted yield* and then multiplied by one hundred.

$$\textbf{\textit{PERCENT YIELD}}$$
$$\textit{Percent yield} = \left(\frac{\textit{Actual yield}}{\textit{Predicted yield}} \right) \times 100$$

VOCABULARY

limiting reactant – a reactant that is used up first in a chemical reaction

excess reactant – a reactant that is not completely used up in a chemical reaction

percent yield – the actual yield of a product in a chemical reaction divided by the predicted yield and multiplied by 100

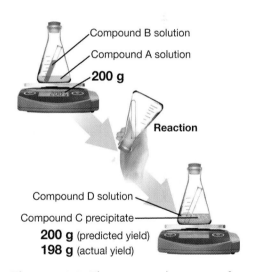

Compound B solution
Compound A solution
200 g
Reaction
Compound D solution
Compound C precipitate
200 g (predicted yield)
198 g (actual yield)

Figure 18.8: *The measured amount of reactants for a reaction yield less than the predicted amount of product in most cases. This is due to experimental error or other factors.*

Solving Problems: Percent Yield

Aspirin can be made in the laboratory through a series of reactions. If the actual yield for aspirin was 461.5 grams when the reactions were performed, and the predicted yield was 500 grams, what was the percent yield?

1. **Looking for:**	The percent yield for the reaction.
2. **Given:**	The predicted yield was 500 grams. The actual yield is 461.5 grams.
3. **Relationships:**	percent yield = (actual yield ÷ predicted yield) × 100
4. **Solution:**	percent yield = (461.5 g + 500 g) × 100 = 92.3% The percent yield for this reaction is 92.3 percent.

Your turn...

a. In another set of reactions for making aspirin, the predicted yield was 32.6 grams and the actual yield was 24.3 grams. What was the percent yield in this case?

b. For the equation below, the percent yield for NH3 is 78.0 percent. If 2 grams of NH4Cl is used, then the predicted yield is 0.636 gram. What is the actual yield?
NH4Cl (s) → NH3 (g) + HCl (g)

c. If one mole of NH4Cl was used in the reaction above, what would be the predicted yield of NH3?

d. Because the percent yield for the reaction above is 78 percent, what would be the actual yield of NH3, if one mole of NH4Cl was used in the reaction?

e. You make a cookie recipe that is designed to make 24 cookies. However, when you finish with your baking, you find that you have made only 22 cookies. List the predicted and actual yields for the recipe. What is the percent yield?

SOLVE FIRST LOOK LATER

a. 74.5 percent

b. 0.496 gram

c. When one mole of NH_4Cl is used in this reaction, you can expect one mole of NH_3 to be produced. The molar mass or the predicted yield of NH_3 would be 17.04 grams.

d. 13.29 grams

e. The predicted yield was 24 cookies. The actual yield was 22 cookies. The percent yield is 92 percent.

Kinetic molecular theory In all phases of matter, atoms and molecules exhibit random motion. This concept is part of the **kinetic molecular theory**. The speed at which atoms or molecules move depends on the state of matter and temperature. As you know, gas molecules move faster than molecules in a solid, and warmer substances have greater molecular motion than cold ones.

What is reaction rate? The **reaction rate** for a chemical reaction is the change in concentration of reactants or products over time. For a reaction involving two or more reactants, the reaction only works if the molecules collide. If we want the reaction to go faster, what kinds of things could we do to increase the motion and number of collisions among the reactants?

Increasing collisions For starters, you can add heat to a reaction to increase molecular motion. For example, to dissolve salt faster in water in a dissolution reaction, you increase the temperature of the water. Other ways to increase collisions include stirring the reaction mixture and using powdered reactants. Fine particles in powders have more available surface area for reacting.

Increasing concentration of reactants Another way to increase collisions among atoms or molecules is to increase the concentration of the reactants. When you increase the concentration of a reactant, it is like adding an extra team member to complete a project. If the project involved many calculations, the team could complete them more quickly with six people than with five. As you know, doing calculations by hand takes a while. What if the team had a computer or calculator?

Catalysts and inhibitors A **catalyst** is a molecule that can be added to a reaction to speed it up, but it doesn't get used up. A catalyst is a little like using a computer or a calculator to help you speed up the job of making calculations. Catalysts work by increasing the chances that two molecules will be positioned in the right way for a reaction to occur. Because a catalyst ensures the correct orientation of colliding molecules, less energy is needed in the collision for the reaction to occur. In effect, a catalyst provides a "shortcut" because a lower activation energy is needed for a reaction to proceed (Figure 18.9). Reactions can also be slowed down by molecules call **inhibitors**. Inhibitors bind with reactant molecules and effectively block them from combining to form products.

VOCABULARY

kinetic molecular theory – the concept that all atoms and molecules exhibit random motion

reaction rate – the change in concentration of reactants or products in a chemical reaction over time

catalyst – a molecule added to a chemical reaction that increases the reaction rate without getting used up in the process

inhibitor – a molecule that slows down a chemical reaction

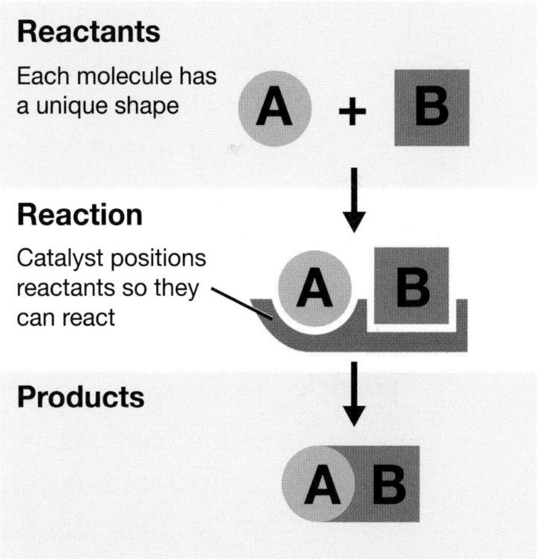

Reactants

Each molecule has a unique shape

A + B

Reaction

Catalyst positions reactants so they can react

A B

Products

A B

Figure 18.9: *By bringing together reactants, a catalyst lowers the activation energy needed.*

Chemical equilibrium

The direction of a chemical reaction

Up until now, we have thought about chemical reactions as going in only one direction. Reactants react to make products. This has been shown in chemical equations with a right-pointing arrow that points toward the products of the reaction. Therefore, chemical reactions are commonly described as proceeding "to the right." In some cases, once a reaction goes "to the right," the reaction reverses and goes "to the left." The products become reactants and the reactants become products (Figure 18.10).

Chemical equilibrium

Eventually, a reaction may reach **chemical equilibrium**, the state in which the rate of the forward reaction equals the rate of the reverse reaction. When we talk about chemical equilibrium, we acknowledge that the reaction can go left and right simultaneously. Chemical equilibrium is represented by arrows going both ways, or a double-headed arrow (Figure 18.10).

Characteristics of chemical equilibrium

Because chemical reactions are often open systems, the reactants and products can easily react with other compounds. If this happens, the products cannot revert back to reactants because they are unavailable. A gas that is a product, for example, easily leaves the reaction system. Therefore, for chemical equilibrium to be established, the chemical reaction has to occur in a closed system. When a chemical reaction occurs in a closed system at constant temperature, the forward and reverse reactions occur at the same rate, and the amounts of reactants and products are constant.

An advanced topic: Le Chatelier's principle

Let's say you have a chemical reaction at chemical equilibrium in a closed container in your laboratory. You leave the system alone, but someone turns up the heat by accident and the room you are in gets hotter and hotter. What happens to the chemical reaction in the container? A change in temperature is considered to be a "stress" on the system. In response to this stress, the system reacts until chemical equilibrium is re-established. This phenomenon is called *Le Chatelier's principle*. This principle states that a chemical reaction at chemical equilibrium reacts to any stress on the system until equilibrium is re-established. A "stress" could include increasing the concentration of a reactant or product, or a change in the temperature or pressure conditions of the reaction.

chemical equilibrium – the state at which the rate of the forward reaction equals the rate of the reverse reaction for a chemical reaction

A reaction going to the right

Reactants ⟶ Products

A + B ⟶ AB

A reaction going to the left

Products ⟵ Reactants

A + B ⟵ AB

Chemical equilibrium

Products ⟷ Reactants

A + B ⟷ AB

Figure 18.10: *The direction of reactions are indicated with arrows. When a reaction is in chemical equilibrium, a double-headed arrow is used.*

Section 18.2 *Review*

1. Is a recipe a good analogy for describing a chemical equation? Why or why not?

2. A recipe that you have for making lasagna says that it will make 10 servings. However, you have 20 people coming over for dinner. What do you need to do to feed everyone a full serving?

3. Here is the balanced chemical equation for making water:

$$2H_2 (g) + O_2 (g) \rightarrow 2H_2O (l).$$

   a. What is the predicted yield for water based on this equation?
   b. What would you have to do to triple the predicted yield?

4. Find the following for this reaction:

$$MgO + H_2O \rightarrow Mg(OH)_2$$

   a. This reaction is carried out with one mole of MgO and three moles of H_2O. What is the predicted yield for $Mg(OH)_2$?
   b. The percent yield for this reaction is 83.2 percent. What is the actual yield of $Mg(OH)_2$?
   c. In the reaction, which compound is the limiting reactant and which is the excess reactant?

Table 18.1: A review of the factors affecting reaction rate

	Do collisions increase?	Does the energy of the collisions increase?
Stirring	Yes	No
Increasing temperature	Yes	Yes
Increasing surface area	Yes	No
Increasing concentration	Yes	No
Adding a catalyst	Improves the effectiveness of collisions so less energy is needed for a reaction	
Adding an inhibitor	Prevents or diminishes the effectiveness of collisions so more energy is needed for a reaction	

SCIENCE FACT

Many reactions in the human body require enzymes, a kind of catalyst, to get reactions going. Not surprisingly, the temperature of the human body, 37°C or 98°F, is ideal for enzymes to work well.

Getting a mild fever indicates that you might be sick. However, you are dangerously ill if you have a high fever for too long. Based on the information above, what might be a consequence of having a high fever in terms of how your body works?

CHALLENGE

Table 18.1 organizes information related to the factors affecting reaction rate.

1. List the two most effective factors in increasing reaction rate. Explain your choices.

2. What is the most effective way to slow down a reaction rate? Explain your choice.

3. Would stirring affect a chemical reaction that has chemical equilibrium? Explain your answer.

18.3 Nuclear Reactions

What do you think of when you hear the terms *nuclear reactions* or *nuclear science*? You may think that anything involving nuclear reactions is considered controversial. Why might that be? For starters, a great deal of energy can be produced by nuclear reactions, and humans need energy constantly. You were introduced to the basics of nuclear science in Chapter 14. Earlier in this chapter, you learned that Marie Curie and her husband were pioneers in nuclear science. This section will provide you with more information about this fascinating topic.

Chemical vs. nuclear reactions

Chemical reactions When you mix two compounds like calcium carbonate ($CaCO_3$) and hydrochloric acid (HCl), something happens (Figure 18.11). In this case, you get calcium chloride ($CaCl_2$), carbon dioxide (CO_2), and water (H_2O). In the transition between the reactants and the products of a chemical reaction, either energy is mostly released (as in an exothermic reaction) or used (as in an endothermic reaction). The involvement of energy in chemical reactions has to do with the breaking and forming of chemical bonds. As you have learned, these bonds involve the outermost electrons of atoms.

Introducing nuclear reactions In the case of nuclear reactions, the main events and source of energy occur in the nuclei of the atoms involved. A **nuclear reaction** involves altering the number of protons and/or neutrons in an atom. Recall from Chapter 14 that protons have positive charge, opposite of electrons. The charge on a proton (+*e*) and an electron (-*e*) are exactly equal and opposite. Neutrons have zero electric charge (Figure 18.12).

Energy and reactions A great deal of energy is needed to begin a nuclear reaction. However, a great deal of energy is also released by this kind of reaction. Although they can produce a lot of energy, chemical reactions fall short of producing as much energy as nuclear reactions. For example, a coal power plant uses chemical reactions to produce energy, and a nuclear plant uses nuclear reactions. The fuel for a nuclear power plant, uranium-235, can produce *3.7 million times* as much energy as an equivalent amount of coal!

VOCABULARY

nuclear reaction – a reaction in which the number of protons and/or neutrons is altered in one or more atoms

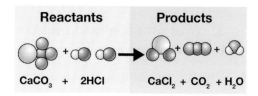

Figure 18.11: *A chemical reaction.*

Size and structure of the atom

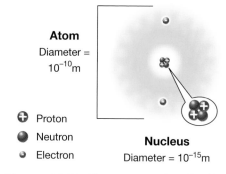

Figure 18.12: *Electrons are involved in chemical reactions. Protons and neutrons are involved in nuclear reactions.*

A review of radioactivity

Why is so much energy generated by nuclear reactions? *Strong nuclear force*, which you learned about in Chapter 14, is a special force inside the nuclei of atoms that holds them together (Figure 18.13). The strong nuclear force is the strongest force known to science, and it is crucial to maintain the structure of atoms. Why? Remember, the positively charged protons repel each other in the nucleus. Keeping a nucleus together, the strong nuclear force is an attractive force between neutrons and protons that works at extremely small distances. If there are enough neutrons in a nucleus, the attraction from the strong nuclear force wins out over the repulsion between protons (an electromagnetic force).

Why are isotopes radioactive? In every atom heavier than helium, there is at least one neutron for every proton in the nucleus. For complex reasons, the nucleus of an atom becomes unstable if it contains too many or too few neutrons relative to the number of protons. If the nucleus is unstable, it breaks apart and releases particles or energy. This process is described as *radioactivity* or *radioactive decay*. For example, radioactive decay results in an unstable, radioactive isotope like carbon-14 becoming the more stable isotope nitrogen-14.

Reviewing the types of radioactivity In Chapter 14, you learned about three types of radioactive decay (Figure 18.14). In *alpha decay*, the nucleus of an unstable atom ejects two protons and two neutrons, essentially the nucleus of a helium-4 (^{4}He) atom. The atomic number is reduced by two and the atomic mass is reduced by four when a nucleus undergoes alpha decay. For example, uranium-238 undergoes alpha decay to become thorium-234. *Beta decay* occurs when a neutron in the nucleus splits into a proton and an electron. The proton stays in the nucleus, but the high-energy electron is ejected and is called beta radiation. What is the consequence of beta decay? You are right if you said that the atomic number increases by one because one new proton is created. However, the mass number stays the same because the atom lost a neutron but gained a proton. Excess energy in an unstable isotope is released by *gamma decay*. In gamma decay, the nucleus emits pure energy in the form of gamma rays. Does atomic number or atomic mass change after gamma decay? No, because energy is released and not particles, the number of protons and neutrons stays the same.

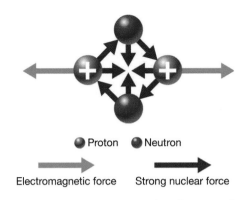

● Proton ● Neutron

→ Electromagnetic force   → Strong nuclear force

Figure 18.13: *Strong nuclear force works at very small distances between protons and neutrons.*

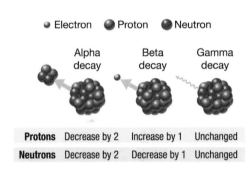

● Electron ● Proton ● Neutron

	Alpha decay	Beta decay	Gamma decay
Protons	Decrease by 2	Increase by 1	Unchanged
Neutrons	Decrease by 2	Decrease by 1	Unchanged

Figure 18.14: *A review of the three types of radioactive decay.*

Two types of nuclear reactions

Getting to the nucleus of the matter   As you have just learned, the nucleus of an atom can change. All by itself, an unstable isotope can experience radioactive decay and become a new, more stable isotope. Atoms that are unstable and prone to radioactive decay are useful in nuclear reactions. There are two kinds of these reactions: fusion and fission.

Fusion   **Nuclear fusion** is the process of combining the nuclei of lighter atoms to make heavier atoms. This process is occurring all the time in a very familiar object—the Sun. What exactly happens in nuclear fusion? The process that occurs matches its name. Two nuclei are "fused" together, a particle is emitted, and a lot of energy is released. The reaction below shows the fusion of hydrogen-3 (1 proton + 2 neutrons) with hydrogen-2 (1 proton + 1 neutron) to produce a helium nucleus, a neutron, and energy. This process occurs in the Sun and the resulting energy released provides Earth with heat and light.

nuclear fusion – a nuclear reaction in which the nuclei of lighter atoms are combined to make heavier atoms

nuclear fission – a nuclear reaction in which the nuclei of heavier atoms are split to make lighter atoms

Nuclear fusion

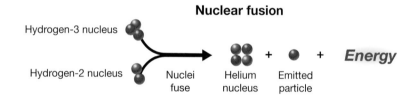

Fission   **Nuclear fission** is the process of splitting the nucleus of an atom. A fission reaction can be started when a neutron bombards a nucleus. A chain reaction results. A free neutron bombards a nucleus and the nucleus splits, releasing more neutrons. These neutrons then bombard other nuclei (Figure 18.15).

Performing fusion and fission reactions   Both fusion and fission reactions can be performed in a special machine called a particle accelerator. The particle accelerator bombards particles and atoms in order to achieve these reactions. However, only a very small number of atoms can be made in this way at one time. Fission, and the resulting energy production in nuclear reactors, is controlled by releasing neutrons to start a chain reaction or by capturing neutrons to slow or stop a chain reaction. As you have just learned, the largest nuclear reactor in our solar system is the Sun.

Nuclear fission

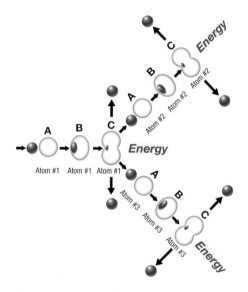

Figure 18.15: *Nuclear fission can be started when a free neutron (blue ball, step A) bombards a nucleus (step B). A chain reaction results as the nucleus splits, releasing more neutrons, which bombard other nuclei (step C).*

Using nuclear reactions for our energy needs

Sun power is nuclear power
You have just learned that the Sun is a giant nuclear reactor. All life on Earth depends on the energy produced by the fusion reactions that occur in the Sun. Plants rely on sunlight to photosynthesize and make sugars. Animals and people, in turn, eat plant products like fruits, vegetables, and grains. By eating these plant products, we are eating the Sun's energy! Additionally, fossil fuels derived from the fossil remains of plants and animals can be attributed to the Sun's energy. All of this energy is related to a nuclear reaction in which hydrogen isotopes are fused together to make helium.

Can we produce energy like the Sun?
The interior of the Sun, where fusion takes place, is about 15 million degrees Celsius. On Earth, we would need to generate about 100 million degrees Celsius to create fusion of hydrogen for producing energy. This high temperature is necessary to overcome the difficulty of forcing positively charged protons together. Given this information, is fusion a possible energy source for humans? Not currently. It would take too much energy to actually produce the energy humans need. However, because fusion reactions do not produce any waste, scientists are studying this kind of nuclear reaction to see if there are economical ways to produce energy using fusion.

How do nuclear reactors work?
A nuclear reactor is a power plant that uses fission to produce heat. Like the heat that is produced in a coal power plant by the burning of coal, the heat in a nuclear reactor is used to generate steam for running turbines. In turn, the turbines generate electricity for homes and businesses.

Nuclear reactors produce hazardous nuclear waste
Almost all of our energy technologies produce some harmful waste products. Burning coal and oil creates waste gases that contribute to global climate change and acid rain. Although nuclear reactors do not normally produce harmful emissions related to global climate change, they do produce nuclear waste.

What is nuclear waste?
The radioactive element in nuclear reactor fuel is uranium. When a uranium atom breaks up (fission), giving off energy, the resulting lighter atoms (the nuclear waste) are also radioactive and remain radioactive for a long time. Substances that are radioactive are extremely harmful. The particles and energy emitted from radioactive elements can cause diseases like cancer.

BIOGRAPHY

Chien-Shiung Wu

During World War II, Chinese-American physicist Chien-Shiung Wu played an important role in the Manhattan Project, the army's secret work to develop the atomic bomb, which worked using fission reactions. In 1957, Wu overthrew what was considered to be an indisputable law of physics, changing the way we understand the weak nuclear force.

Chien-Shiung Wu was born in 1912 near Shanghai, China. After graduating first in her class from high school, she was invited to attend a prestigious university in Nanjing. After earning a physics degree, Wu emigrated to the United States. She earned a Ph.D. from the University of California-Berkeley in 1940.

After working on the Manhattan Project, Wu continued research in nuclear physics at Columbia University. She won the National Medal of Science, the nation's highest award for scientific achievement, in 1975.

A plan for storing nuclear waste

In 1974, the U.S. Congress established the Nuclear Regulatory Commission (NRC) as a monitoring organization for nuclear fuel use and the storage of nuclear waste. Although construction for a permanent storage facility for radioactive nuclear waste was started in Yucca Mountain, Nevada, the plans were put on hold in 2011. Presently, nuclear waste is stored in cooling pools or dry casks at nuclear power plants around the country. Storing nuclear waste is a very controversial issue because the waste is radioactive, usable for making nuclear weapons, and very costly to store.

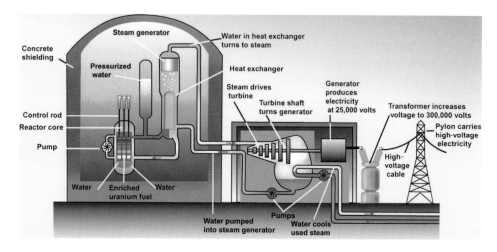

How much energy comes from nuclear reactors?

The United States gets about 19 percent of its energy from nuclear fission reactors. The remaining energy comes from coal, natural gas, oil, and hydroelectric dams. Many foreign countries get more of their electricity from nuclear fission reactors. France is the most dependent on nuclear power. About 76 percent of electricity generated in France comes from nuclear fission. If you look at the European Union as a whole, almost 30 percent of its energy comes from nuclear fission.

SCIENCE FACT

Blocking radiation

You don't always need thick walls of lead or concrete to block radiation (the product of radioactive decay). For example, a simple sheet of paper or your skin can block alpha particles. Your clothing or wood can block beta particles. However, gamma rays and high-speed neutrons will pass right through you, potentially causing damage along the way. Nuclear reactor facilities and the dry casks used to store nuclear waste use thick walls of concrete to block neutrons and gamma rays.

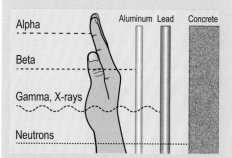

Find out more about protecting yourself from radiation at the U.S. Environmental Protection Agency website.

Using nuclear reactions in medicine and science

Half-life On this page, you will learn more about why radioactive waste is harmful, but also how radioactivity is useful. Let's begin by looking again at nuclear waste. The atoms that are a part of nuclear waste from nuclear reactors have long *half-lives* ranging from thousands (as in the case of plutonium-239) to millions of years. A **half-life** is a certain length of time after which *half* of the amount of radioactive element has decayed. For example, the half-life of carbon-14 (one of the radioactive isotopes of carbon) is 5,730 years. This means that if you start out with 100 atoms of carbon-14, 5,730 years from now, only 50 atoms will still be carbon-14. The rest of the carbon will have decayed to nitrogen-14 (a stable isotope). As a radioactive element decays, it emits harmful radiation such as alpha and beta particles and gamma rays. By breaking chemical bonds, radiation can damage cells and DNA. Exposure to radiation is particularly harmful if it is intense or for a long time period.

Radioactive dating *Radioactive dating* is a process that is used to figure out the age of objects by measuring the amount of radioactive material in a substance and by knowing the half-life of that substance. For example, *carbon dating* is used to date material made from plants or animals. Much of the carbon absorbed by plants and animals is carbon-12 and carbon-13 because these are the most abundant carbon isotopes. However, some carbon-14 is also absorbed. By measuring the amount of carbon-14 remaining in a plant or animal fossil, the age of the fossils that are between 50,000 and a few thousand years old can be estimated. For older material, the amount of uranium-238 can be measured. It takes 4.5 billion years for one half of the uranium-238 atoms in a sample to turn into lead (Figure 18.16). If a rock contains uranium-238, scientists can determine the rock's age by the ratio of uranium-238 to lead atoms in the sample. Understanding radioactive decay of uranium-238 has allowed scientists to determine that the Earth is 4.6 billion years old.

Radioisotopes detect problems in systems *Radioisotopes* (also called radioactive isotopes) are commonly used as tracers in medicine and science. By adding a radioactive isotope to a system (such as the human body or an underground water supply), problems can be detected. The tracer's radiation allows it to be detected using a Geiger counter or other machine and followed as it travels through the system.

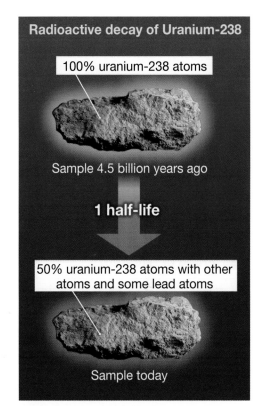

Radioactive decay of Uranium-238

100% uranium-238 atoms

Sample 4.5 billion years ago

1 half-life

50% uranium-238 atoms with other atoms and some lead atoms

Sample today

Figure 18.16: *The radioactive decay of uranium to lead. Radioactive decay is measured in half-lives. After one half-life, 50 percent of the uranium-238 atoms have decayed.*

Nuclear vs. chemical reactions

A summary of the differences between chemical and nuclear reactions is listed in the table below.

	Chemical reactions	Nuclear reactions
What part of the atom is involved?	Outermost electrons	Protons and neutrons in the nucleus
How is the reaction started?	Atoms are brought close together with high temperature or pressure, or catalysts, or by increasing concentrations of reactants	High temperature is required or atoms are bombarded with high-speed particles
What is the outcome of the reaction?	Atoms form ionic or covalent bonds	The number of protons and neutrons in an atom usually changes and/or energy is released
How much energy is absorbed or released?	A small amount	A huge amount
What are some examples?	Burning fossil fuels, digesting food, making medicines and commercial products	Generating nuclear energy, taking X-rays, treating cancer, irradiating food to sterilize it, the Sun generating heat and light

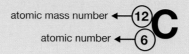

Section 18.3 *Review*

1. Why is so much energy required and released in a nuclear reaction?

2. Explain the interaction and roles of electromagnetic force and strong nuclear force in the nucleus of an atom.

3. Gold-185 decays to iridium-181. Is this an example of alpha or beta decay?

4. What has to happen, in terms of radioactive decay, for carbon-14 to decay to nitrogen-14?

5. How is gamma decay different from alpha or beta decay?

6. In your own words, describe the difference between fusion and fission. Why do certain elements undergo fusion or fission?

7. Which type of nuclear reaction is used in modern-day nuclear reactors? Why is the other type of nuclear reaction *not* used in modern-day energy production?

8. When an atom of beryllium-9 is bombarded by an alpha particle, an atom of carbon-12 is produced and a neutron is emitted. What kind of nuclear reaction has just occurred?

9. What is the half-life of each of these radioactive isotopes?
   a. A radioactive isotope decreased to one-half its original amount in 18 months.
   b. A radioactive isotope decreased to one-fourth its original amount in 100 years.

10. For each scenario below, indicate whether a chemical reaction or a nuclear reaction is occurring.
   a. When two compounds are combined, heat is released.
   b. A sample of galium-68 is reduced to one-half its original amount in 68.3 minutes.
   c. Radium-226 decays to radon-222.
   d. A spark of energy is used to begin the combustion of methane gas.
   e. Hydrogen nuclei are fused in the Sun to make helium atoms.

BIOGRAPHY

Irene Joliot-Curie

Irene Joliot-Curie was a remarkable woman. She was the oldest daughter of Marie and Pierre Curie. Irene studied both mathematics (her forte) and physics at the University of Sorbonne in Paris. However, her education was interrupted by World War I. Irene joined her mother in military hospitals and on the battlefield. Marie had developed portable X-ray machines, which she set up and used to treat wounded soldiers. For her service, Irene was awarded France's Military Medal. By 1925, Irene had earned her Ph.D., studying alpha rays of the element polonium, which was discovered by her parents. Around this time, she also met her future husband and scientific collaborator, Frederic Joliot. Frederic and Irene were both passionate about science and also shared interests in politics, art, and sports. The couple married in 1926 and had two children. They earned the Nobel Prize in 1935 for discovering that nonradioactive elements could be turned into radioactive isotopes using alpha particles.

Your Footprint
Matters

According to the U.S. Department of Energy, the concentration of carbon dioxide in Earth's atmosphere in 2016 was 402 ppm (parts per million). This means if you had a million kilograms of atmospheric gases, 402 kilograms of that million would be carbon dioxide. That doesn't seem like much, does it? But around 150 years ago, before the industrial revolution, the concentration was only 280 ppm. The concentration has increased 44 percent over 150 years!

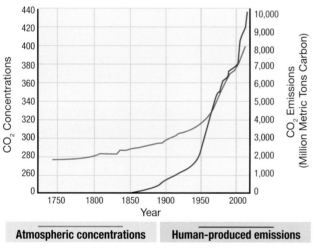

Atmospheric concentrations   Human-produced emissions

Source: Oak Ridge National Laboratory, Carbon Dioxide Information Analysis Center

So what, you ask. Here's the issue: Carbon dioxide is a greenhouse gas, which means that it helps trap heat from the Sun near Earth's surface. Without naturally occurring carbon dioxide, Earth's average surface temperature would be about 30°C colder. But too much carbon dioxide can cause too much solar energy to be trapped as heat. Since the industrial revolution, Earth's average surface temperature has risen about 0.99°C. This is not a huge increase, but it's enough to have caused the sea level to rise 10 to 25 centimeters. Further increases may impact global climate patterns.

Your personal carbon footprint

Every person on the planet is responsible for some of the atmosphere's carbon dioxide. Your contribution is known as your personal "carbon footprint." Most U.S. residents have a large carbon footprint, an average of 25 tons per year per person, far more than individuals in most other countries. In 2016, the U.S. Department of Energy estimated that people in the United States are responsible for about 15 percent of global carbon dioxide emissions.

Why? The average American's lifestyle choices (eating habits, driving habits, and leisure activities, for example) play a significant role. The good news is that there are many ways we can change our habits to reduce our individual carbon footprints.

Making choices, making a difference

The problem of carbon dioxide emissions and global climate change is huge. What can one person do? This is just what students at a Rhode Island high school asked themselves. To help find ways to reduce their individual carbon

footprints, physical science students researched different carbon dioxide reduction strategies and then shared their research with one another and members of their community at "Energy Night."

One student decided to teach people about the impact of their food choices. She explained, "Most of the food that we eat has traveled well over a thousand miles to get to our plates. The burning of all that fuel for transportation puts a lot of carbon dioxide into the air. The solution is to eat more locally grown foods."

"I also compared unprocessed foods versus highly processed foods. The unprocessed foods were less expensive, healthier, and required less energy to produce."

Three students focused on how Americans choose to spend leisure time. According to the United States Bureau of Labor Statistics, in 2015, Americans over the age of 15 spent an average of 2.8 hours per day watching television, an activity that increases a person's carbon footprint. To reduce this leisure time carbon footprint, one student suggested that families institute a game night. Pick a time every week when everyone in the house must turn off their televisions and computers and get together as a family to play a game, share stories, or make music.

Two students wanted to relay a similar message. To get their message out to lots of people, they hosted a community dance. "If kids spend time with people instead of computers or televisions, they can save energy and form better relationships," they explained.

To get the message out to lots of people, students hosted a community dance.

"Other things like reading, exercising, or playing games that don't use electricity can help make us smarter, stronger, and better connected to the people around us. So we organized a dance so that people in our community could spend time together. It was contra dance, not the usual kind of high school dance. This was fun for all ages." More than 100 people in the community came to the dance, and it was written up in three local newspapers.

Another student taught people about bioplastics. "Unlike regular plastics, which are made from petroleum, bioplastics can be made from plants or algae. So they are made from renewable resources that take carbon dioxide out of the atmosphere during part of their production. Plus, many bioplastics are compostable, which means we don't have to send them to the landfill."

Bioplastics can be made from plants or algae.

Doing your part

Every person can help address the problem of increasing carbon emissions by making changes in personal habits that reduce his or her individual carbon footprint. You can find an online personal carbon footprint calculator using an internet keyword search: "carbon footprint + calculator".

Solving this problem will take the combined efforts of all people. What will you do to help make the planet healthier for everyone?

Questions:

1. How much has the amount of carbon dioxide in the atmosphere changed since the industrial revolution?

2. What is meant by "carbon footprint"?

3. What are three things you can do to reduce your carbon footprint?

Energy statistics courtesy of Energy Information Administration, U.S. Department of Energy.

Chapter 18 *Assessment*

Vocabulary

Select the correct term to complete each sentence.

activation energy	excess reactant	nuclear fission
catalyst	exothermic	nuclear fusion
chemical equilibrium	half-life	nuclear reaction
dissolution reaction	inhibitor	percent yield
endothermic	limiting reactant	reaction rate

Section 18.1

1. A catalyst can lower the _____ of a reaction.

2. Once started, a(n) _____ reaction tends to keep going.

3. A(n) _____ reaction uses more energy than it releases.

4. A(n) _____ has occurred when an ionic compound dissolves in water.

Section 18.2

5. If the actual yield of a chemical reaction was 5 grams and the predicted yield was 10 grams, the _____ is 50 percent.

6. A(n) _____ slows down the reaction rate whereas a(n) _____ can speed up the reaction rate.

7. A reactant that is left over after a reaction is complete is called the _____ .

8. A reaction in an open system is less likely to achieve _____ .

9. A reactant that is used up first in a reaction is called the _____ .

10. One way to increase _____ is by increasing the temperature of the reaction.

Section 18.3

11. _____ occurs in the Sun, the largest nuclear reactor in our solar system.

12. The _____ of uranium-238 is 4.5 billion years.

13. Fusion is an example of a(n) _____ .

14. A(n) _____ reaction can be started when a neutron bombards a nucleus.

Concepts

Section 18.1

1. Write a general equation that illustrates the difference between an exothermic reaction and an endothermic reaction. You only need to use the following items in your general equation: reactants, products, and energy. Be sure to include an arrow in writing your equation.

2. Your teacher asks you to mix two compounds to find out if the reaction is endothermic or exothermic. What will you do to determine which type of reaction is occurring?

3. Compare a dissolution reaction with a combustion reaction.

Section 18.2

4. Your lab partner did not take careful notes during today's chemistry experiment. You see these numbers written without labels: 37.3 grams, 40 grams, 93.3 percent. Before asking your partner, you decide to predict which number is the predicted yield, the actual yield, and the percent yield. Justify your answer. Make a list of all the factors that will increase the reaction rate of a chemical reaction.

5. You are investigating the reaction rate of a chemical reaction. How might the required activation energy be affected if you add a catalyst to the reaction? If you add an inhibitor to the reaction?

6. Imagine you are a molecular-sized reporter and are able to witness a chemical reaction achieve chemical equilibrium. Write down what you observe. Be sure to describe the actions of the reactants and products, and the setting!

Section 18.3

7. Explain why nuclear reactions produce so much energy.

8. Compare and contrast a nuclear power plant with a coal-based power plant.

9. Say you know a certain fossil is more than a million years old. Can you use carbon dating to date it? Why or why not?

10. Why are radioisotopes useful in detecting problems in the human body? Why might a technique involving a radioisotope be a better option than surgery?

Problems

Section 18.1

1. The graph below illustrates the change in energy for an exothermic reaction.

a. Add labels to A and B on the graph.

b. Make a sketch that would show the change of energy that occurs for an endothermic reaction.

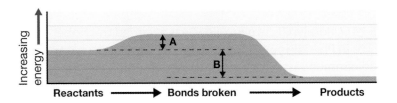

2. The U.S. Army developed an MRE (Meal Ready to Eat) for the 1991 Gulf War. These meals have a special sleeve placed around the food, which is wrapped in aluminum foil. When water is added to the sleeve, a chemical reaction produces enough heat to cook the food. The sleeve contains a pad with suspended particles of magnesium metal. When water is added, the magnesium reacts with it to produce magnesium hydroxide and heat. The heat is conducted through the aluminum foil to heat the food. What kind of reaction is this?

$$\text{Mg } (s) + 2\text{H}_2\text{O } (l) \rightarrow \text{Mg(OH)}_2 \ (aq) + \text{H}_2 \ (g)$$

Section 18.2

3. In the equation below, the actual yield of bromide (Br_2) was 19.8 grams when the reaction was performed with 10.0 grams of chlorine (Cl_2). The molecular mass of Br_2 is 159.8 grams, and the molecular mass of Cl_2 is 70.9 grams.

$$\text{Cl}_2 + 2\text{KBr} \rightarrow 2\text{KCl} + \text{Br}_2$$

a. What is the predicted yield for Br_2?

b. What is the percent yield for Br_2?

4. For the reaction in question 2 above, determine the predicted yield for $Mg(OH)_2$ and H_2. What is the actual yield for these products if the percent yield is 80 percent?

Section 18.3

5. The half-life of cesium-137 is 30 years. Make a graph that shows its radioactive decay over a period of 300 years. Show Time on the *x*-axis of the graph and Number of Atoms on the *y*-axis. The starting amount of cesium-137 is 100 atoms. Be sure to title the graph and label the axes.

6. Write the isotope notation for the following isotopes:

a. carbon-14
b. nitrogen-14
c. hydrogen-3
d. beryllium-9
e. gold-195
f. polonium-209

7. The decay series for uranium-238 and plutonium-240 are listed below. Above each arrow, write *a* for alpha decay or *b* for beta decay to indicate which type of decay took place at each step.

a. $^{238}_{92}\text{U} \rightarrow ^{234}_{90}\text{Th} \rightarrow ^{234}_{91}\text{Pa} \rightarrow ^{234}_{92}\text{U} \rightarrow ^{230}_{90}\text{Th} \rightarrow$

$^{226}_{88}\text{Ra} \rightarrow ^{222}_{86}\text{Rn} \rightarrow ^{218}_{84}\text{Po} \rightarrow ^{214}_{82}\text{Pb} \rightarrow ^{214}_{83}\text{Bi} \rightarrow$

$^{214}_{84}\text{Po} \rightarrow ^{210}_{82}\text{Pb} \rightarrow ^{210}_{83}\text{Bi} \rightarrow ^{210}_{84}\text{Po} \rightarrow ^{206}_{82}\text{Pb}$

b. $^{240}_{94}\text{Pu} \rightarrow ^{240}_{95}\text{Am} \rightarrow ^{236}_{93}\text{Np} \rightarrow ^{232}_{91}\text{Pa} \rightarrow ^{232}_{92}\text{U} \rightarrow$

$^{228}_{90}\text{Bi} \rightarrow ^{224}_{88}\text{Ra} \rightarrow ^{224}_{89}\text{Ac} \rightarrow ^{220}_{87}\text{Fr} \rightarrow ^{216}_{85}\text{At} \rightarrow$

$^{212}_{83}\text{Bi} \rightarrow ^{212}_{84}\text{Po} \rightarrow ^{208}_{82}\text{Pb} \rightarrow ^{208}_{83}\text{Bi}$

Applying Your Knowledge

Section 18.1

1. Many drain cleaners are a mixture of sodium hydroxide and aluminum filings. When these two substances mix in water, they react to produce enough heat to melt the fat in your clogged drain. The bubbles produced are hydrogen gas. From this description, do your best to write the chemical formulas for the reactants and products and their state of matter. What kind of reaction is this? How do you know?

Section 18.2

2. One chocolate sundae includes $^1/_2$ cup of ice cream, 2 ounces of chocolate sauce, and 1 cherry. You have 10 cups of ice cream, 25 ounces of chocolate sauce, and 10 cherries.

a. What is the maximum number of sundaes you can make with each sundae having exactly the same amount of each ingredient?
b. Which ingredient is the limiting component of the system?
c. What quantities of the other two "reactants" will be left over when you are finished?

Section 18.3

3. For every atom heavier than helium, there needs to be at least as many neutrons as protons to hold the nucleus together. For example, calcium-40 has 20 protons and 20 neutrons. For heavier atoms, more neutrons are needed than protons. For atoms with more than 83 protons, even the added strong nuclear force from neutrons is not enough to hold the nucleus together. How would you describe the elements that have more than 83 protons?

4. Theoretical physicist Hideki Yukawa was the first person from Japan to receive a Nobel Prize. For what accomplishment did he receive the Nobel Prize?

5. In 1948, Roscoe L. Koontz, a young, African American chemist from Missouri, was invited to participate in the Atomic Energy Health Physics Fellowship Training Program at the University of Rochester. Eventually, he became one of the world's first health physicists. Find out more about the field of health physics. What does a "health physicist" do? Also, find out more about Koontz and his accomplishments.

6. In 1977, Rosalyn Sussman Yalow was awarded the Nobel Prize in Physiology or Medicine for her work in developing radioimmunoassay, or RIA. Find out about this important medical technique and how it involves the use of radioisotopes.

Solutions

Let's say you're really thirsty. You go to a store and see a wide range of beverages. You might pick a plain bottle of water or, if you want something more interesting, you might buy carbonated water. Other types of drinks include sports drinks, sodas, fruit juice, dairy beverages, and even tea or coffee drinks. What do these drinks have in common? For starters, they all contain a very precious compound—water! And here's another thing they have in common: They are all types of solutions. In this chapter, you will learn about solutions. You will discover what makes plain water different from carbonated water and that water is a special substance because it has a neutral pH. What does that mean? Here are some clues: It's not acidic like orange juice and it's not basic like liquid soap. You definitely don't want to drink liquid soap. Yuck!

CHAPTER 19 INVESTIGATIONS

19A: Solubility Curve of KNO₃
What is a solubility curve?

19B: Acids, Bases, and pH
What is pH?

19.1 **Water**

We live on a watery planet. All life on Earth depends on this combination of hydrogen and oxygen atoms. Fortunately, Earth has a lot of water—75 percent of our planet's surface is covered with it! Interestingly, our bodies are mostly water, too—about 60 to 75 percent (Figure 19.1). With these facts in mind, let's find out about the properties of water that make it so valuable.

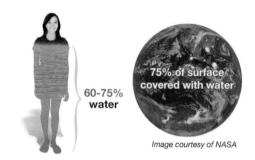

Figure 19.1: *The Earth's surface and our bodies are mostly water.*

The shape of a water molecule

How a water molecule is formed

neutral molecule
(2⁻) + (1⁺) + (1⁺) = 0

The chemical formula for water is H_2O. Why is that? From the formula, we know that for each water molecule there are two hydrogen atoms that are each attached to an oxygen atom by a chemical bond. Recall that oxygen has an oxidation number of 2^- and has six valence electrons. Hydrogen, with an oxidation number of 1^+, has only one valence electron. When two hydrogen atoms share their electrons with one oxygen atom, a neutral molecule is formed (shown at left). Note that the oxygen atom in the molecule (shown at right) now has eight valence electrons, the same number as a noble gas. Each hydrogen atom now has two valence electrons, giving them the same number of valence electrons as a helium atom.

The shape of a water molecule is a pyramid

A water molecule forms a pyramid shape called a *tetrahedron*. An oxygen atom is in the middle of the tetrahedron, and the electron pairs form the legs. Why does a water molecule form this shape? A water molecule has four pairs of electrons around the oxygen atom. Only two of these pairs are involved in forming the chemical bonds. These two pairs are called *bonding pairs*. The other two pairs of electrons are not involved in forming chemical bonds and are known as *lone pairs*.

Electron pairs repel each other

Because negative charges repel, the four electron pairs around the oxygen atom are located where they can be the farthest apart from each other, forming the tetrahedron shape (Figure 19.2). If you draw the molecule without the lone pairs, the oxygen and hydrogen atoms form a "V" (shown upside down in the diagram).

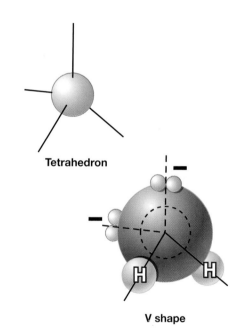

Figure 19.2: *The shape of a water molecule.*

Water is a polar molecule

What is a polar molecule? Water is a **polar molecule**, meaning it has a *negative* end (pole) and a *positive* end (pole). In a molecule of water, the oxygen atom attracts electrons so that they are shared *unequally* between the oxygen and hydrogen. The electrons are actually pulled toward the oxygen atom and away from the two hydrogen atoms. Therefore, the oxygen side of the molecule (the side with the lone pairs of electrons) has a partially negative charge and the hydrogen side of the molecule has a partially positive charge (Figure 19.3).

Ammonia is another polar molecule Ammonia (NH_3) is another example of a polar molecule. This molecule has one lone pair of electrons and three bonding pairs of electrons. This gives the ammonia molecule a pyramid shape. Figure 19.3 shows the shape of the molecule with the three hydrogens forming the base of the pyramid (the positive pole). The top of the pyramid is the negative pole.

Nonpolar molecules Methane (CH_4) is an example of a **nonpolar molecule**. Nonpolar molecules do not have distinct positive and negative poles. Figure 19.3 shows a methane molecule. This molecule does not contain any lone pairs of electrons. Because there are no lone pairs of electrons, the electrons are shared equally between the carbon atom and the four surrounding hydrogen atoms.

Comparing polar and nonpolar molecules It takes energy to melt and boil compounds. The fact that the melting and boiling points of a polar molecule (water) are much higher than those of a nonpolar molecule (methane) provides evidence that there are attractions *between* polar molecules. This is because it takes more energy to pull apart molecules that are polar compared to nonpolar molecules. The table below compares the melting and boiling points of water and methane. Notice that the melting and boiling points of water are much higher than those of methane.

Comparing water and methane		
Compound	**Melting point**	**Boiling point**
Water	0°C	100°C
Methane	−182°C	−164°C

VOCABULARY

polar molecule – a molecule that has a negative and a positive end or pole

nonpolar molecule – a molecule that does not have distinctly charged poles

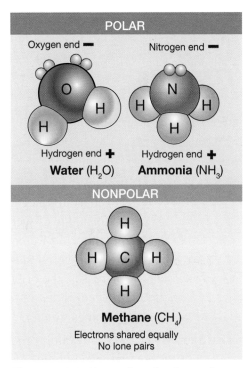

Figure 19.3: *Examples of polar and nonpolar molecules.*

Water molecules are connected by hydrogen bonds

VOCABULARY

hydrogen bond – an intermolecular force between the hydrogen on one molecule to an atom on another molecule

A water molecule is like a magnet   How can a water molecule be like a magnet? Think about what happens if you place a group of magnets together. Recall that a magnet has two sides or poles. This means that two side-by-side magnets are attracted to each other's opposite pole. A group of magnets will form an arrangement so that they alternate poles because similar poles repel each other. The same is true if you put a group of water molecules together. The positive pole of one water molecule is attracted to the negative pole of another. In a group of water molecules, the positive and negative poles align among the molecules in the group. These *polar* attractions create organization among water molecules.

Hydrogen bonds   Recall that a water molecule has two strong covalent bonds between the oxygen atom and the hydrogen atoms. The force that holds neighboring water molecules together is called a hydrogen bond. A **hydrogen bond** is an intermolecular force between a hydrogen atom on one molecule to an atom on another molecule. Hydrogen bonds are relatively weak. They constantly break and reform as water molecules collide.

A network of molecules   In Figure 19.4, you can see that the oxygen atom in a water molecule has two partially negative lone electron pairs. Each pair of electrons is available to form a hydrogen bond with the partially positive hydrogen atom of a neighboring water molecule. Many neighboring water molecules connected by hydrogen bonds form a network of water molecules. As temperature increases, the organized structure of the hydrogen bonds among water molecules decreases. As temperature decreases, the organized structure becomes greater.

Ice has a honeycomb structure   Frozen water, or ice, has a very organized structure that resembles a honeycomb because each water molecule forms hydrogen bonds with four other water molecules (Figure 19.5). This creates a six-sided arrangement of molecules that is evident if you examine snowflakes under a microscope. As water freezes, molecules of water separate slightly from each other as a result of hydrogen bonding. This causes the volume to increase slightly and the density to decrease. This explains why water expands when it is frozen and why ice floats. The density of ice is about 0.9 g/cm^3 whereas the density of water is about 1 g/cm^3.

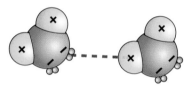

Figure 19.4: *A hydrogen bond between two water molecules.*

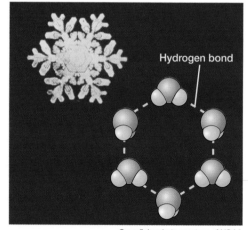

Snowflake photo courtesy of NOAA.

Figure 19.5: *The honeycomb structure of solid water (ice). Can you identify how each molecule forms four hydrogen bonds with other molecules?*

Properties of water related to hydrogen bonding

Water has a high specific heat You learned in Chapter 11 that water has a high *specific heat* compared to other substances. Recall that specific heat tells us how much heat is needed to raise the temperature of one kilogram of a substance by one degree Celsius. A high specific heat means that a lot of energy is needed for each degree of increase in temperature. For example, water's specific heat value is 4,184 J/kg°C whereas the specific heat for steel is 470 J/kg°C.

Water resists temperature changes Water has a high specific heat value because of hydrogen bonds. Between single molecules, hydrogen bonds are weak. However, at the group level, the polar attractions of the molecules (the hydrogen bonds) make more heat necessary to make water molecules move faster. The temperature finally rises once the water molecules begin to move faster. The same amount of energy that was used to heat a volume of water must be removed for it to cool. If a large amount of energy is needed to heat water, the same amount will have to be taken away to cool it back to the starting temperature. This also explains why water cools more slowly than other substances.

Figure 19.6: *In order for water to boil, enough energy must be added to separate the hydrogen bonds that hold the water molecules together.*

Water has a high boiling point Most of the water on Earth exists in liquid and solid states, rather than as a gas. This is because the hydrogen bonds hold the water molecules together strongly enough so that individual molecules cannot easily escape as a gas at ordinary temperatures. The hydrogen bonds in water explain why water has such a high boiling point (100°C). In order for water to boil and turn into a gas (water vapor), enough energy must be added to separate the hydrogen bonds that hold the molecules of water together. Once these molecules are separated, they are able to enter the gaseous state (Figure 19.6).

Hydrogen bonding and plants You may know that plants obtain water from their roots. How does water get from a plant root to its leaves? The stem of a plant has microscopically thin, straw-like structures that allow the water to rise up from the roots to the leaves. The water is able to make the entire journey from the roots to the leaves due to hydrogen bonding between water molecules (Figure 19.7). As water molecules evaporate from the leaves, other water molecules are pulled into place. It is as if water molecules hold hands. If one molecule moves, the ones behind follow because they are connected by hydrogen bonds!

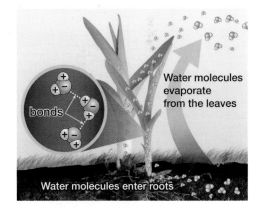

Figure 19.7: *Hydrogen bonds help water travel from roots to stem to leaves.*

Water is a universal solvent

Water dissolves many things
Water is often called the "universal solvent." While water doesn't dissolve *everything*, it does dissolve many different types of substances such as salt and sugar. Water is a good solvent because it is a polar molecule. This gives it the ability to dissolve ionic compounds and other polar substances.

How water dissolves salt
An example of an ionic compound is table salt (sodium chloride or NaCl). Recall that sodium chloride (NaCl) is an ionic compound that is made of sodium ions (Na+) and chloride ions (Cl−). Suppose a sodium chloride (table salt) crystal is mixed with water. The polar water molecules surround the sodium and chloride ions in the crystal. This causes the ions in the crystal to separate. Because opposites attract, the negative ends of the water molecules are attracted to the Na+ ions and the positive ends are attracted to the Cl− ions. Water molecules surround the Na+ and Cl− ions and make a solution. The process by which ionic compounds dissolve (become separated into positive and negative ions) is called *dissociation* (Figure 19.8).

How water dissolves sugar
Like water molecules, sugar (sucrose) molecules can form hydrogen bonds. In the case of sugar, these bonds hold the molecules together as solid crystals. When sucrose is mixed with water, the individual molecules of sucrose become separated from each other and are attracted to the opposite poles of the water molecules. Because sucrose is a covalent compound, the sucrose molecules do not dissociate into ions but remain as neutral molecules in the solution.

Like dissolves like. For example, polar solvents like water dissolve polar substances.

What doesn't water dissolve?

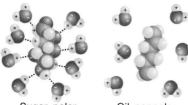

Sugar–polar dissolves in water

Oil–nonpolar insoluble in water

In general, "like dissolves like." This means water, a polar solvent, dissolves polar substances. Nonpolar solvents (like mineral oil) dissolve nonpolar substances. Mineral oil is insoluble in water because it lacks the ability to form hydrogen bonds. Figure 19.9 lists polar and nonpolar substances.

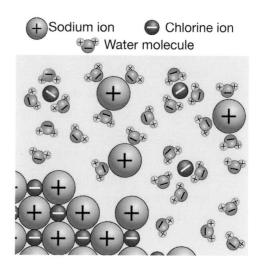

⊕ Sodium ion   ⊖ Chlorine ion
Water molecule

Figure 19.8: *Water dissolves sodium chloride to form a solution of ions.*

Polar substances	Nonpolar substances
Water	Vegetable oil
Vinegar	Mineral spirits
Alcohol	Turpentine
Sucrose	Wax

Figure 19.9: *Examples of polar and nonpolar substances.*

Section 19.1 *Review*

1. Why does an oxygen atom only form two covalent bonds with hydrogen atoms?

2. Draw the Lewis dot structure for water. Label lone pairs, bonding pairs, positive pole, and negative pole.

3. Why is a water molecule able to form four hydrogen bonds? Use a diagram to explain your answer.

4. Identify which of the molecules below are polar molecules and which are nonpolar molecules. Justify your answer.

A	B	C
H–C̈l:	H–N̈–H H	H H H–C–C–H H H

5. What is the difference between a bond in a polar molecule and a bond in a nonpolar molecule?

6. A single covalent bond is stronger than a single hydrogen bond, so why does a group of polar molecules tend to have a higher boiling point than a group of nonpolar molecules?

7. Compare and contrast a pair of magnets and a pair of water molecules.

8. Why is the density of ice less than the density of liquid water?

9. Water's specific heat value is 4,184 J/kg°C and the value for steel is 470 J/kg°C. Based on this information, compare water and steel with regard to the time it would take to heat up or cool down these substances.

10. List three properties of water that are related to hydrogen bonding.

11. How is the process of evaporation of water from a plant leaf involved in moving water up the stem of a plant? Use the term *hydrogen bond* in your answer.

19.2 Solutions

If you walk down the beverage aisle of your local grocery store, you might see mineral water, spring water, flavored water, and seltzer (carbonated water) for sale. While the labels on the bottles might call what's inside "water," each bottle contains more than just pure water. These varieties of water are actually *solutions* that also contain dissolved substances.

Solutions

Homogeneous at the molecular level A **solution** is a mixture of two or more substances that is homogeneous at the molecular level. The word *homogeneous* means the particles in the water are evenly distributed. For example, in mineral water, there are no clumps of hundreds of mineral ions. Figure 19.10 illustrates some examples of other solutions. The particles in a true solution exist as individual atoms, ions, or molecules. Each has a diameter of between 0.01 and 1.0 nanometer (nm). A nanometer is one one-billionth of a meter.

Heterogeneous mixtures

Muddy water is NOT a true solution

Muddy water is not homogeneous, and it is not a solution. Muddy water is *heterogeneous* because it contains larger particles of soil or plant debris. Of course, muddy water also contains individual atoms, ions, and molecules too.

An alloy is a solution of two or more solids We often think of solutions as liquid. However, solutions exist in every phase: solid, liquid, and gas. A solution of two or more solids is called an **alloy**. Steel is an alloy (solution) of iron and carbon. Fourteen-karat gold is an alloy of silver and gold. "Fourteen-karat" means that 14 out of every 24 atoms in the alloy are gold atoms and the rest are silver atoms. Carbonated water is a solution of a gas in a liquid. The sweet smell of perfume is a solution of perfume molecules in air. This is an example of a solution of gases.

Figure 19.10: *Examples of solutions.*

Mixtures that are not solutions

Colloids A **colloid** is a mixture with particles formed in clusters of atoms or molecules ranging in size from 1 to 1,000 nanometers. Colloids can look like solutions because the particles are too small to settle to the bottom of their containers. Instead, they stay evenly distributed throughout the mixture because they are constantly tossed about by the movement of the particles. Examples of colloids are mayonnaise, egg whites, and gelatin.

Suspensions You may notice that when you step into a pond or lake to go swimming, the water suddenly becomes cloudy. Your feet cause the mud and other particles on the bottom of the pond or lake to mix with the water. However, if you stand still, the water eventually becomes clear again because the individual particles sink. In a **suspension** like muddy water, the particles are greater than 1,000 nanometers in diameter and can range widely in size. A suspension will settle when it is left still for a period of time. Because a suspension is a heterogeneous mixture, the different-sized particles in a suspension can be separated by filtering.

The Tyndall effect It isn't easy to separate colloids by filtering. However, there is a way to visually distinguish colloids from true solutions. The **Tyndall effect** is the scattering of light by the 1- to 1,000-nanometer particles in a colloid. The Tyndall effect is occurring if you shine a flashlight through a jar of liquid and see the light beam. The Tyndall effect helps distinguish a translucent colloid from a true solution because the particles in a solution are too small to scatter light (Figure 19.11). An example of the Tyndall effect is when a car's headlights are seen cutting through fog. Fog is an example of a colloid.

VOCABULARY

colloid – a mixture that contains evenly distributed particles that are 1 to 1,000 nanometers in size

suspension – a mixture that contains particles that are greater than 1,000 nanometers

Tyndall effect – the scattering of light by the particles in a colloid

Colloid

Light beam is visible

Solution

Light beam is not visible

Figure 19.11: *The Tyndall effect helps you tell the difference between a translucent colloid and a solution.*

Table 19.1: Properties of solutions, colloids, and suspensions

	Approximate size of solute particles	Do solute particles settle?	Will filtering separate particles?	Do particles scatter light?
Solutions	0.01 to 1.0 nm	no	no	no
Colloids	1.0 to 1,000 nm	no	only with special equipment	yes, if translucent
Suspensions	> 1,000 nm	yes	yes	yes, if translucent

Solvents and solutes

What are solvents and solutes? A solution is a mixture of at least two substances: a *solvent* and a *solute*. The **solvent** is the substance that makes up the biggest percentage of the mixture. For example, the solvent in grape soda is water. Each of the remaining parts of a solution (other than the solvent) is called a **solute**. Sugar, coloring dyes, flavoring chemicals, and carbon dioxide gas are solutes in grape soda.

solvent – the component of a solution that is present in the greatest amount

solute – any component of a solution other than the solvent

dissolve – to separate and disperse a solid into individual particles in the presence of a solvent

Solutes dissolve in a solvent When the solute particles are evenly distributed throughout the solvent, we say that the solute has "dissolved." Solutes **dissolve** in a solvent to form a solution. For example, the illustration below shows the preparation of a sugar and water solution. The solute (sugar) is in the graduated cylinder on the left. Water (the solvent) is added and the mixture is carefully stirred. Once all the solid sugar has dissolved, the solution becomes clear.

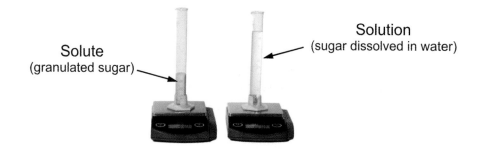

Solute
(granulated sugar)

Solution
(sugar dissolved in water)

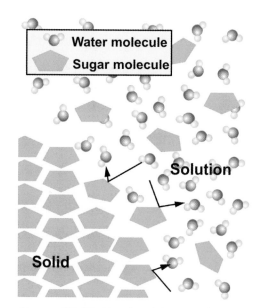

Water molecule
Sugar molecule

Solution

Solid

Dissolving and temperature Dissolving of a solid (like sugar) occurs when molecules of solvent interact with and separate molecules of solute (Figure 19.12). Most substances dissolve faster at higher temperatures. You may have noticed that sugar dissolves much faster in hot water than in cold water. This is because thermal energy is used to break the intermolecular forces between the solute molecules.

Figure 19.12: *For dissolving to take place, molecules of solvent interact with and carry away molecules of solute.*

The importance of surface area Dissolving only occurs where the solvent contacts the solute. A solute will dissolve faster if it has a large amount of exposed surface area. For this reason, most things that are meant to be dissolved, like salt and sugar, are sold as powders. A substance in powder form has a high amount of exposed surface area.

Solubility

What is solubility? **Solubility** describes the amount of solute (if any) that can be dissolved in a volume of solvent. Solubility is often listed in grams per 100 milliliters of solvent. Solubility is always given at a specific temperature because temperature strongly affects solubility. For example, Table 19.2 shows that 200 grams of sugar can be dissolved in 100 milliliters of water at 25°C.

Insoluble substances do not dissolve Notice that there are no solubility values for chalk and talc in Table 19.2. These substances are **insoluble** in water because they do not dissolve in water. You can mix chalk dust and water and stir them vigorously, but you will still just have a mixture of chalk dust and water. The water will not separate the chalk dust into individual molecules because chalk does not dissolve in water.

Saturation Suppose you add 300 grams of sugar to 100 milliliters of water at 25°C. What happens? According to Table 19.2, 200 grams will dissolve in the water. *The rest will remain solid.* That means you will be left with 100 grams of solid sugar at the bottom of your solution. Any solute added in excess of the solubility does not dissolve. A solution is **saturated** if it contains as much solute as the solvent can dissolve under certain conditions. Dissolving 200 grams of sugar in 100 milliliters of water at 25°C creates a saturated solution because no more sugar will dissolve under these conditions.

Table 19.2: Solubility values for common substances

Substance	Solubility (grams per 100 mL H_2O at 25°C)
Sugar ($C_{12}H_{22}O_{11}$)	200
Sodium nitrate ($NaNO_3$)	94
Calcium chloride ($CaCl_2$)	90
Table salt ($NaCl$)	38
Baking soda ($NaHCO_3$)	Approximately 10
Chalk ($CaCO_3$)	Insoluble
Talc (Mg silicates)	Insoluble

VOCABULARY

solubility – the amount of solute that can be dissolved in a specific volume of solvent under certain conditions

insoluble – when a solute is unable to dissolve in a particular solvent

saturated – describes a solution that has as much solute as the solvent can dissolve under the conditions

SCIENCE FACT

Dew point

Sometimes there is more water vapor dissolved in the air in your home than you might want. To prevent mildew, for example, a dehumidifier is used to remove water vapor dissolved in air. This device works by reducing the temperature of the air. The *dew point* is the temperature at which air is saturated with water vapor. If it becomes colder than the dew point, the air becomes *supersaturated* with water. This means there is more water than the air can hold. The excess water condenses out of the air as liquid water, which is then collected by the dehumidifier. This process is similar to the process in nature that causes dew to form on grass in the morning and causes raindrops to form.

Solving Problems: Solubility

Seawater is a solution of water, salt, and other minerals. How much salt can dissolve in 200 milliliters of water at 25°C?

1. **Looking for:**	Grams of solute (salt)	
2. **Given:**	Volume (200 mL) and temperature (25°C) of solvent	
3. **Relationships:**	38 grams of salt dissolves in 100 milliliters of water at 25°C (Table 19.2).	
4. **Solution:**	If 38 grams dissolves in 100 milliliters, then twice as much, or 76 grams, will dissolve in 200 milliliters.	

Your turn...

a. How much table salt can dissolve in 50 milliliters of water at 25°C?

b. How much sugar can dissolve in 300 milliliters of water at 25°C?

c. How much water would you need to dissolve about 30 grams of baking soda?

d. You want to make one liter of a sodium nitrate solution at 25°C. How much sodium nitrate will you need to make this solution?

e. In some laboratories, scientists only need to make a few milliliters or less of a solution at a time. For 25°C, how much of each solute would you need to make a 1-milliliter saturated solution of sugar? a 2-milliliter saturated solution of table salt? a 3-milliliter saturated solution of calcium chloride?

SOLVE FIRST/LOOK LATER

a. 19 grams

b. 600 grams

c. 300 milliliters

d. 940 grams

e. 2 grams of sugar; 0.76 gram of salt; 2.7 grams of $CaCl_2$

Temperature–solubility graphs

Solubility values on a graph   The solubility values for solutes are easily determined if you have a temperature–solubility graph like the one below. The *y*-axis on the graph represents how many grams of solute will dissolve in 100 milliliters of water. The *x*-axis represents temperature in degrees Celsius. You will notice that the solutes (NaCl, KNO_3, $NaNO_3$) dissolve differently as temperature increases. In order for something to dissolve in water, the water molecules need to break the intermolecular forces between the solute molecules. Water dissolves various substances differently because the chemical bond strengths between atoms found in different solutes are not the same.

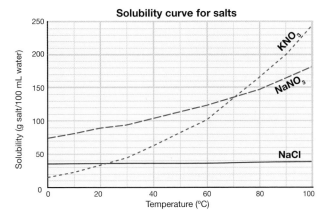

Question:
How many grams of potassium nitrate (KNO_3) will dissolve in 200 mL of water at 60°C?

Solution:
From the graph , you see that 106 grams of KNO_3 dissolves in100 mL of water at 60°C.

200 mL / 100 mL = 2
2 x 106 g = 212 g

212 grams of KNO_3 will dissolve in 200 mL of water at 60°C.

Figure 19.13: *Using a temperature–solubility graph helps you solve problems like the one above.*

Interpreting the graph   The solutes on the graph above are sodium chloride (NaCl), potassium nitrate (KNO_3), and sodium nitrate ($NaNO_3$). Notice that the solubility of NaCl does not change much as temperature increases. The effect of temperature on the solubility of KNO_3 and $NaNO_3$ is more noticeable. More KNO_3 and $NaNO_3$ will dissolve in 100 milliliters of water at higher temperatures than NaCl.

Using the graph   How many grams of potassium nitrate (KNO_3) will dissolve in 200 mL of water at 60°C? In this example, you are asked for the mass in grams of solute and given temperature and volume. From the graph above, you see that 106 grams of KNO_3 dissolve in 100 mL of water at 60°C. Therefore, 212 grams of KNO_3 will dissolve in 200 mL of water at this temperature (Figure 19.13).

████ SOLVE IT! ████

Answer the following using the temperature–solubility graph.

1. How much NaCl dissolves in 200 mL at 80°C?

2. Is a 100-mL solution saturated at 40°C if it has 40 grams of $NaNO_3$?

Concentration

How to express concentration   The **concentration** of a solution is the amount of solute dissolved in an amount of solvent. Two common ways of expressing the concentration of a solution are molarity and mass percent.

> **CONCENTRATION**
> $$\text{Concentration} = \frac{\text{Amount of solute}}{\text{Amount of solution}}$$

What is molarity?   The most common way of expressing concentration in chemistry is to use molarity (M). **Molarity** is equal to the moles of solute per liter of solution. Recall that one *mole* of a substance contains 6.02×10^{23} particles (atoms or molecules) and allows you to express the *formula mass* in grams. If you dissolve 5.00 moles of NaCl (292 g) in enough water to make 1.0 L, what is the molarity of the solution? It would be a 5.0 M solution. Do you see why?

> **MOLARITY**
> $$\text{Molarity (M)} = \frac{\text{Moles of solute}}{\text{Liter of solution}}$$

What is mass percent?   The **mass percent** of a solution is equal to the mass of the solute divided by the total mass of the solution multiplied by 100. Suppose you dissolve 10.0 grams of sugar in 90.0 grams of water (Figure 19.14). What is the mass percent of sugar in the solution?

> **MASS PERCENT**
> $$\text{Mass percent} = \frac{\text{Mass of solute}}{\text{Mass of solution}} \times (100\%) = \frac{10 \text{ g sugar}}{(10 \text{ g} + 90 \text{ g}) \text{ solution}} \times (100\%) = 10\%$$

Describing low concentrations   Parts per million (ppm), parts per billion (ppb), and parts per trillion (ppt) are used to describe small concentrations of substances. These terms are measures of the ratio (by mass) of one material in a much larger amount of another. For example, a pinch (gram) of salt in 10 tons of potato chips is about 1 gram of salt per billion grams of chips, or a concentration of 1 ppb.

VOCABULARY

concentration – the ratio of solute to solvent in a solution

molarity – the moles of solute per liter of solution

mass percent – the mass of the solute divided by the total mass of the solution multiplied by 100

10 g sugar   90 g water

+

Mix

100 g of solution that is 10% sugar

Figure 19.14: *Preparing a sugar solution with a concentration of 10%.*

STUDY SKILLS

You learned about the terms *mole* and *formula mass* in Chapter 17. Review these terms and study how they apply to molarity.

Solving Problems: Concentration

How many grams of salt (NaCl) do you need to make 500 grams of a solution with a mass percent of 5 percent salt? The formula mass of NaCl is 58.4 g/mol. What is the molarity of this solution?

1.	**Looking for:**	Mass of salt (solute) and molarity of the solution
2.	**Given:**	Mass percent (5%), total mass of solution (500 g), and formula mass
3.	**Relationships:**	Mass percent = (mass of solute ÷ total mass of solution) × 100
		Molarity = moles of solution ÷ liter of solution
4.	**Solution:**	0.05 = (mass of salt ÷ 500 g)
		mass of salt = 0.05 × 500 g = 25 g
		moles of solute = (25 g) ÷ (58.4 g/mol) = 0.4 moles NaCl
		Assume 1 L of solution. 0.4 moles/1 L = 0.4 M

Your turn...

a. What is the molarity of 2.0 L of a calcium chloride ($CaCl_2$) solution that contains 166.5 g of $CaCl_2$? The formula mass of $CaCl_2$ is 110.99 g/mol.

b. The formula mass for $NaNO_3$ is 69.00 g/mol. How many grams of $NaNO_3$ would you need to make 1 liter of a 2 M solution?

c. How many grams of solution would you have if you made a 20 percent salt solution with 10 grams of salt?

d. How many grams of sugar would you need to make 1,000 grams of solution with a mass percent of 20 percent? How many grams of water would you need?

e. You want to make a 30 percent solution of table salt. Describe how you would make 500 grams of this solution.

SOLVE FIRST LOOK LATER

a. 0.75 M

b. 138 g

c. 50 g of solution

d. I would need 200 g of sugar and 800 g of water.

e. I would take 150 g of salt and add it to 350 g of water.

Equilibrium and supersaturation

Dissolving and undissolving When a solute like sugar is mixed with a solvent like water, *two* processes are actually going on continuously: (1) molecules of solute dissolve and go into solution; and (2) molecules of solute come out of solution and become "undissolved."

When a solution is **unsaturated**, its concentration is lower than the maximum solubility. For an unsaturated solution, the dissolving process puts molecules into solution faster than they come out. In time, the concentration increases and the mass of undissolved solute decreases. However, the processes of dissolving and undissolving are still going on.

Equilibrium concentration The more molecules that are in solution (higher concentration), the faster molecules come out of solution. As the concentration increases, the undissolving process also gets faster until the dissolving and undissolving rates are exactly equal. When the rate of dissolving equals the rate of coming out of solution, we say **equilibrium** has been reached. At equilibrium, a solution is *saturated* because the concentration is as high as it can go.

Supersaturation According to the solubility table in Figure 19.15, at 80°C, 100 grams of water reaches equilibrium with 365.1 grams of dissolved sugar. At lower temperatures, less sugar can dissolve. What happens if we cool the saturated solution? As the temperature goes down, sugar's solubility also goes down and the solution becomes **supersaturated**. A supersaturated solution means there is more dissolved solute than the maximum solubility.

Growing crystals

Rock Candy

A supersaturated solution is unstable. The excess solute comes out of solution and returns to its undissolved state. This is how the large sugar crystals of rock candy are made. Sugar is added to boiling water until the solution is saturated. As the solution cools, it becomes supersaturated. Solid sugar crystals form as the sugar comes out of the supersaturated solution.

unsaturated – describes a solution with a concentration less than the maximum solubility.

equilibrium – the state of a solution in which the dissolving rate equals the rate at which molecules come out of solution

supersaturated – describes a solution with a concentration greater than the maximum solubility

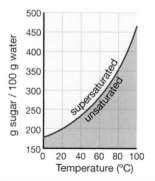

Temp (°C)	g sugar 100 g H_2O	Temp (°C)	g sugar 100 g H_2O
0	181.9	50	259.6
10	190.6	60	288.8
20	201.9	70	323.7
30	216.7	80	365.1
40	235.6	90	414.9

Figure 19.15: *A solubility graph and table for sugar in water.*

Solutions of gases and liquids

Gas dissolves in water Some solutions have a gas as the solute. For example, in carbonated soda, the fizz comes from dissolved carbon dioxide gas (CO_2). Table 19.3 lists the solubility of some gases in water.

Table 19.3: Solubility of gases in water at 21°C and 1 atm

Gas	Solubility (g/kg)
Oxygen (O_2)	0.04
Nitrogen (N_2)	0.02
Carbon dioxide (CO_2)	1.74

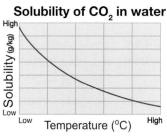

Solubility of CO_2 in water

Figure 19.16: *The solubility of gases in water decreases as temperature increases.*

Solubility of gas increases with pressure The solubility of gases in liquids increases with pressure. Soda contains a lot of carbon dioxide because this gas is dissolved in the liquid at high pressure. You release the pressure when you open a can of soda. Due to the decrease in pressure, the solution immediately becomes supersaturated. The result is that the CO_2 quickly bubbles out of the water and causes your drink to be fizzy.

Solubility of gas decreases with temperature When temperature goes up, the solubility of gases in liquid goes down. A graph of this relationship for carbon dioxide in water is shown in Figure 19.16. Because fish and other aquatic life depend on oxygen, it is important that oxygen gas dissolves in water (Figure 19.17). The amount of dissolved oxygen in water decreases when the water temperature rises. Oxygen enters a pond or river by being mixed in from the air, or it is produced as a by-product of photosynthesis from underwater plants. When the weather is warm, less oxygen is dissolved in the water near the surface, so fish stay near the bottom where there is cooler, more oxygenated water.

Figure 19.17: *Aquatic life is sustained by dissolved oxygen in water.*

Solubility of liquids Some liquids, such as alcohol, are soluble in water. Alcohol and water are polar substances. Other liquids, such as oil, are not soluble in water. Oil-and-vinegar salad dressing separates because oil is not soluble in water-based vinegar (Figure 19.18). Liquids that are insoluble in water may be soluble in other solvents. For example, vegetable oil is soluble in mineral spirits, a petroleum-based solvent used to thin paint. Both of these substances are nonpolar.

Oil

Vinegar (water solution)

Figure 19.18: *Oil, a nonpolar substance, does not dissolve in vinegar, a polar substance.*

Solubility rules

Ions versus water molecules In addition to nonpolar substances like oil, some ionic compounds do not dissolve in water. Why do you think this might be? Because water has charged poles, it is capable of attracting the positive or negative ions in an ionic compound. However, sometimes, the attraction of the ions for each other is stronger than their attraction to water. As a result, the ionic compound is insoluble in water.

What are solubility rules? A set of **solubility rules** helps predict when an ionic compound is soluble or insoluble (Table 19.4). Understanding the types of chemical reactions that can occur and knowing about solubility rules is useful for predicting the products of a chemical reaction. The Group 1 (alkali metals) and Group 2 (alkali earth metals) elements are included in the first two columns of the periodic table (Figure 19.19).

VOCABULARY

solubility rules – a set of rules used to predict whether an ionic compound will be soluble or insoluble in water

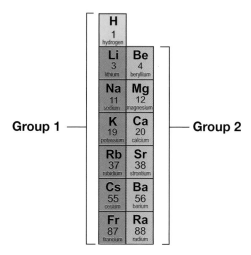

Figure 19.19: *Group 1 and 2 elements on the periodic table.*

Table 19.4: Solubility rules

Any ionic compound with ...	... is ...	Exceptions	Notes
Nitrate (NO_3^-)	Soluble	None	
Chloride (Cl^-)	Soluble	$AgCl$, Hg_2Cl_2, and $PbCl_2$	
Sulfate (SO_4^{2-})	Soluble	$BaSO_4$, $PbSO_4$, and $SrSO_4$	
Carbonate (CO_3^{2-})	Insoluble	NH^{4+} and those of the Group 1 elements	
Hydroxide (OH^-)	Insoluble	the Group 2 elements, $Ba(OH)_2$, and $Sr(OH)_2$	However, $Ca(OH)_2$ is slightly soluble
Sulfides (S^{2-})	Insoluble	Group 1 and Group 2 elements and NH^{4+}	

SOLVE IT!

Use the solubility rules in Table 19.4 to determine whether these compounds are soluble or insoluble.

a. NH_4OH

b. $Ca(NO_3)_2$

c. $PbCl_2$

d. CuS

Section 19.2 *Review*

Solution	Solvent	Solute(s)
Air	Nitrogen (gas)	Other gases
Carbonated water	Water (liquid)	CO_2 (gas)
Saline solution	Water (liquid)	Salt (solid)
Rubbing alcohol	Alcohol (liquid)	Water (liquid)
Sterling silver	Silver (solid)	Copper (solid)

Figure 19.20: *Question 9.*

1. Tell which of the following is *not* a solution and explain why.

    a. ocean water   **d.** orange soda

    b. water mixed with chalk powder   **e.** 24-karat gold

    c. steel   **f.** water with food coloring

2. For each of the following solutions, name the solvent and the solute.

    a. ocean water (salt water)

    b. carbonated water

    c. lemonade made from powdered drink mix

3. When can you say that a solute has completely dissolved?

4. Does sugar dissolve faster in cold or hot water? Explain your answer.

5. Is the solubility of oxygen higher in cold or hot water? Explain.

6. Jackie likes to put a lot of sugar in her hot tea. When she finishes drinking her tea, she notices that there are sugar crystals at the bottom of the teacup. Explain her observation in terms of saturation.

7. Describe exactly how you would make 100 grams of a saltwater solution that is 20 percent salt. In your description, tell how many grams of salt and how many grams of water you would need.

8. The table in Figure 19.20 lists the solvents and solutes for a variety of solutions. Make a list of three statements that you can make about solutions based on the information in this table.

Figure 19.21: *Question 11.*

9. The concentration of ocean water is 35 ppt salt. What is this concentration in grams of salt per grams of water?

10. Is the solution in Figure 19.21 in equilibrium? Why or why not?

11. Describe the solution in a can of soda before the can is opened and just after it is opened.

12. Use solubility rules to determine if these compounds will dissolve in water: (a) $FeSO_4$ and (b) K_2CO_3.

19.3 Acids, Bases, and pH

Acids and bases are among the most familiar of all chemical compounds. Some of the acids you may have encountered include acetic acid (found in vinegar), citric acid (found in orange juice), and malic acid (found in apples). You may also be familiar with some bases, including ammonia in cleaning solutions and magnesium hydroxide, found in some antacids. The pH scale is used to describe whether a substance is an acid or a base. This section is about the properties of acids and bases, and the pH scale.

What are acids?

Properties of acids   An **acid** is a compound that dissolves in water to make a particular kind of solution. Some properties of acids are listed below, and some common acids are shown in Figure 19.22. *Note*: You should *never* taste a laboratory chemical!

- Acids create the sour taste in foods like lemons.

- Acids react with metals to produce hydrogen gas (H_2).

- Acids change the color of blue litmus paper to red.

- Acids can be very corrosive, destroying metals and burning skin through chemical action.

Acids make hydronium ions   Chemically, an acid is any substance that produces *hydronium ions* (H_3O^+) when dissolved in water. When hydrochloric acid (HCl) dissolves in water, it ionizes, splitting up into hydrogen (H^+) and chlorine (Cl^-) ions. Hydrogen ions (H^+) are attracted to the negative oxygen end of a water molecule, combining to form hydronium ions.

What an acid does in water

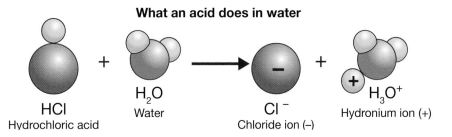

HCl	H_2O	Cl^-	H_3O^+
Hydrochloric acid	Water	Chloride ion (–)	Hydronium ion (+)

Some common acids
(relatively weak)

Oranges and citrus fruit

Vinegar   Lemon juice

Figure 19.22: *Some weak acids you may have around your home.*

Bases

Properties of bases A **base** is a compound that dissolves in water to make a different kind of solution, opposite in some ways to an acid. Some properties of bases are listed below, and some common bases are shown in Figure 19.23.

- Bases create a bitter taste.
- Bases have a slippery feel, like soap.
- Bases change the color of red litmus paper to blue.
- Bases can be very corrosive, destroying metals and burning skin through chemical action.

Bases produce hydroxide ions A base is any substance that dissolves in water and produces *hydroxide ions* (OH^-). A good example of a base is sodium hydroxide (NaOH), found in many commercial drain cleaners. This compound dissociates in water to form sodium (Na^+) and hydroxide (OH^-) ions.

What a base does in water

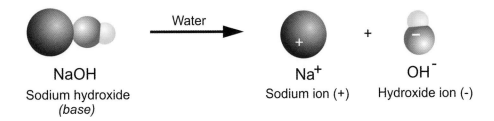

NaOH
Sodium hydroxide
(base)

Na$^+$
Sodium ion (+)

+

OH$^-$
Hydroxide ion (-)

Ammonia is a base Ammonia (NH_3), found in cleaning solutions, is a base because it increases the pH of water. It also is a base because it accepts a proton (H^+). This is another definition for a base—a proton acceptor. Notice that a hydroxide ion, from water, is formed in this reaction. How is this different from NaOH?

What ammonia (base) does in water

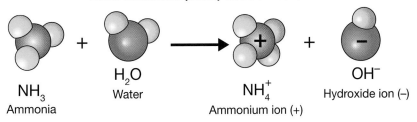

NH_3
Ammonia

+

H_2O
Water

NH_4^+
Ammonium ion (+)

+

OH^-
Hydroxide ion (–)

Some common bases

Baking soda

Ammonia

Soap

Drain cleaner

Figure 19.23: *Common bases include ammonia, baking soda, and soap.*

Strength of acids and bases

The strength of acids The strength of an acid depends on the concentration of the hydronium ions (H_3O^+) the acid produces when dissolved in water. Hydrochloric acid (HCl) is a *strong acid* because HCl completely dissolves into H^+ and Cl^- ions in water. This means that every molecule of HCl that dissolves produces one hydronium ion.

Acetic acid is a weak acid Acetic acid $(HC_2H_3O_2)$ is vinegar, a *weak acid*. When dissolved in water, only a small percentage of acetic acid molecules ionize (break apart) and become H^+ and $C_2H_3O_2^-$ ions. This means that only a small number of hydronium ions are produced compared to the number of acetic acid molecules dissolved (Figure 19.24).

The strength of bases The strength of a base depends on the relative amount of hydroxide ions (OH^-) produced when the base is mixed with water. Sodium hydroxide (NaOH) is considered a strong base because it dissociates completely in water to form Na^+ and OH^- ions. Every molecule of NaOH that dissolves creates one OH^- ion (Figure 19.25). Ammonia (NH_3), on the other hand, is a weak base because only a few molecules react with water to form NH_4^+ and OH^- ions.

Water can be a weak acid or a weak base One of the most important properties of water is its ability to act as both an acid and a base. In the presence of an acid, water acts as a base. In the presence of a base, water acts as an acid. In pure water, the H_2O molecule ionizes to produce both hydronium and hydroxide ions. This reaction is called the *dissociation of water*.

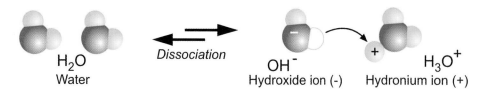

What does the double arrow mean? The double arrow in the illustration means that the dissociation of water can occur in *both* directions. This means that water molecules can ionize and ions can form water molecules. However, water ionizes so slightly that most water molecules exist whole, not as ions.

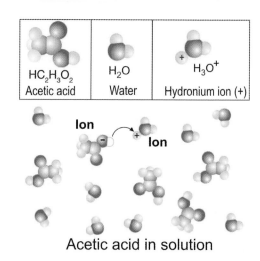

Acetic acid in solution

Figure 19.24: *Acetic acid dissolves in water, but only a few molecules ionize (break apart) to create hydronium ions.*

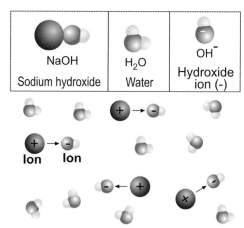

Sodium hydroxide in solution

Figure 19.25: *Sodium hydroxide (NaOH) is a strong base because every NaOH molecule contributes one hydroxide (OH^-) ion.*

The pH scale and pH

What is pH? The **pH scale** is a range of values from 0 to 14 that describe the acidity of a solution. The term **pH** is an abbreviation for "the power of hydrogen" and is a measure of the concentration of hydronium ions in a solution. The range of hydronium ion concentrations is 1 to 10^{-14} M. Notice that the pH for a solution equals the negative of the exponent of the hydronium ion concentration (Figure 19.26).

pH

$$pH = - (- exponent)$$

Negative exponent of H_3O^+ concentration

Determining pH Now let's use this equation to find pH. For example, a solution has a hydronium ion concentration of 10^{-9} M. What is its pH?

$$pH = - (-9)$$

$$pH = 9$$

What would the pH be for a solution with a hydronium ion concentration of 10^{-1}? It has a pH of 1! You may now be noticing a relationship—the higher the hydronium ion concentration, the lower the pH value. The reverse is true as well—the lower the hydronium ion concentration, the higher the pH value.

Why do we need a pH scale? The pH scale describes the hydronium concentration of different solutions using whole numbers. In reality, the range of possible hydronium ion concentrations in solutions is huge (from 1 M to 10^{-14} M), and the numbers can be small and require decimals. A number like 7 is always much easier to work with than 0.0000001 (10^{-7}).

pH scale – the pH scale goes from 1 to 14 with 1 being very acidic and 14 being very basic

pH – a measure of the concentration of hydronium ions in a solution

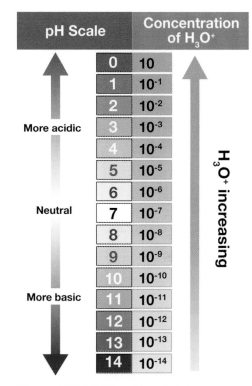

Figure 19.26: *The pH scale is based on the concentration of hydronium ions in a solution.*

Solving Problems: The pH Scale

A solution contains a hydronium ion concentration of $10^{-4.5}$ M. What is the pH value of the solution? Is this solution acidic or basic?

1. **Looking for:**	The pH value of a $10^{-4.5}$ M solution of hydronium ions.
2. **Given:**	The hydronium ion concentration is $10^{-4.5}$ M. The exponent is −4.5.
3. **Relationships:**	pH = − (exponent of the hydronium ion concentration)
	Acidic solutions have pH values that are less than 7 and basic solutions have pH values that are greater than 7.
4. **Solution:**	pH = − (−4.5)
	pH = 4.5
	Because the pH is less than 7, the solution is acidic.

Your turn...

a. A solution contains a hydronium ion concentration of 10^{-13} M. What is the pH value of the solution? Is the solution acidic or basic?

b. What is the pH of a solution with a hydronium ion concentration of 10^{-8} M?

c. What is the hydronium ion concentration for a solution with a pH value of 10? Is this solution acidic or basic?

d. A pH meter (a device that measures the pH of solutions) indicates that the pH value for your solution is 9.2, but you want it to be 8. Do you need more or less hydronium ions?

e. Write the following hydronium ion concentration as a decimal: 10^{-3} M.

SOLVE FIRST LOOK LATER

a. 13; basic

b. 8

c. 10^{-10}; basic

d. More

e. 0.001 M

More about pH

The numbers on the scale   A pH of 7 is *neutral*, neither acidic nor basic. Distilled water has a pH of 7. Acidic solutions have a pH less than 7. A concentrated solution of hydrochloric acid, a strong acid, has a pH of 1. Seltzer water is a weak acid at a pH of 4. Many foods we eat and many ingredients we use for cooking are acidic. Basic or alkaline solutions have a pH greater than 7. A concentrated solution of a *strong base* has the *highest* pH. For example, a strong sodium hydroxide solution can have a pH close to 14. Weak bases, such as baking soda, and weak acids have pH values that are close to 7. Many household cleaning products are basic (Figure 19.27).

Table 19.5: The pH of common substances

Common substances	pH	Acid or base?
Lemon juice	2	Acid
Vinegar	3	Acid
Soda water	4	Acid
Distilled water	7	Neutral
Baking soda	8.5	Base
Bar soap	10	Base
Ammonia	11	Base
Drain cleaner	14	Base

pH indicators   Certain chemicals turn different colors when pH changes. These chemicals are called pH indicators and they are used to determine pH. The juice of boiled red cabbage is a pH indicator that is easy to prepare. Red cabbage juice is deep purple and turns various shades ranging from purple to yellow at different values of pH. Litmus paper is another pH indicator that changes color. Red and blue litmus paper are pH indicators that test for acid or base.

Litmus paper strip

pH Scale

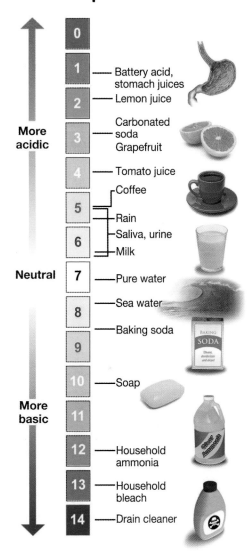

More acidic

0

1 — Battery acid, stomach juices

2 — Lemon juice

3 — Carbonated soda
Grapefruit

4 — Tomato juice

Coffee

5
Rain

6 — Saliva, urine
Milk

Neutral   7 — Pure water

8 — Sea water

Baking soda

9

10 — Soap

More basic   11

12 — Household ammonia

13 — Household bleach

14 — Drain cleaner

Figure 19.27: *The pH scale showing common substances.*

pH in the environment

The best pH for plants   The pH of soil directly affects nutrient availability for plants. Most plants prefer a slightly acidic soil with a pH between 6.5 and 7.0. Azaleas, blueberries, and conifers grow best in more acidic soils with a pH of 4.5 to 5.5 (Figure 19.28). Vegetables, grasses, and most shrubs do best in less acidic soils with a pH range of 6.5 to 7.0.

Effects of pH too high or low   In highly acidic soils (pH below 4.5), too much aluminum, manganese, and other elements may leach out of soil minerals and reach concentrations that are toxic to plants. Also, at these low pH values, calcium, phosphorus, and magnesium are less available to plant roots. At more basic pH values of 6.5 and above, iron and manganese become less available.

pH and fish   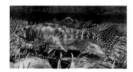 The pH of water directly affects aquatic life. Most freshwater lakes, streams, and ponds have a natural pH in the range of 6 to 8. Most freshwater fish can tolerate pH between 5 and 9, although some negative effects appear below a pH of 6. Trout (like the California Golden shown above) are among the most pH tolerant fish and can live in water with a pH from 4 to 9.5.

pH and amphibians    Frogs and other amphibians are even more sensitive to pH than fish. This California tree frog and other frogs prefer pH close to neutral and don't survive below a pH of 5. Frog eggs develop and hatch in water with no protection from environmental factors. Research shows that even pH values below 6 have a negative effect on frog hatching rates.

Figure 19.28: *Blueberries grow best in soils with a pH between 4.5 and 5.5.*

pH ranges in nature

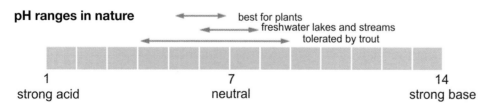

1						7						14
strong acid						neutral						strong base

best for plants
freshwater lakes and streams
tolerated by trout

CHALLENGE

Acid rain

Many environmental scientists are concerned about acid rain. Do research to answer the following questions.

1. What kinds of acids are in acid rain?

2. What is the typical pH of acid rain?

3. What is the cause of acid rain?

4. What are some environmental impacts of acid rain?

5. What can be done to reduce acid rain?

Acids and bases in your body

Acids and bases play a role in digestion
Many reactions, such as the ones that occur in your body, work best at specific pH values. For example, acids and bases are very important in the reactions involved in digesting food. The stomach secretes hydrochloric acid (HCl), a strong acid (pH 1.4). The level of acidity in your stomach is necessary to break down the protein molecules in food so they can be absorbed. A mucus lining in the stomach protects it from the acid produced (Figure 19.29).

Ulcers and heartburn
Deep fried foods, stress, or poor diet can cause the stomach to produce too much acid, or allow stomach acid to escape from the stomach. An *ulcer* may occur when the mucus lining of the stomach is damaged. Stomach acid can then attack the more sensitive tissues of the stomach itself. Infections by the bacteria *H. pylori* can also damage the mucus lining of the stomach, leading to ulcers. The uncomfortable condition called *heartburn* is caused by excessive stomach acid backing up into the esophagus. The *esophagus* is the tube that carries food from your mouth to your stomach. The esophagus lacks the mucus lining of the stomach and is sensitive to acid.

pH and your blood
Under normal conditions, the pH of your blood is within the range of 7.3–7.5, close to neutral but slightly basic. Blood is a watery solution that contains many solutes including the dissolved gases carbon dioxide (CO_2) and oxygen. Dissolved CO_2 in blood produces a weak acid. The higher the concentration of dissolved CO_2, the more acidic your blood becomes.

Blood pH is controlled through breathing
Your body regulates the dissolved CO_2 level by breathing. For example, if you hold your breath, more carbon dioxide enters your blood and the pH falls as your blood becomes more acidic. If you hyperventilate (breathe more quickly than usual), less carbon dioxide enters your blood and the opposite happens—blood pH starts to rise, becoming more basic. Your breathing rate regulates blood pH through these chemical reactions (Figure 19.30). You can learn more about pH and the human body in the Connection at the end of the chapter.

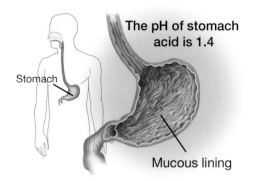

The pH of stomach acid is 1.4

Stomach

Mucous lining

Figure 19.29: *The stomach secretes a strong acid (HCl) to aid with food digestion. A mucus lining protects the stomach tissue from the acid.*

pH 6

pH 8

pH 7.4

Figure 19.30: *Under normal conditions, your blood pH ranges between 7.3 and 7.5. Holding your breath causes blood pH to drop. High blood pH can be caused by hyperventilating.*

Neutralization reactions

Mixing acid and base solutions
When acid and base solutions are mixed in the right proportions, their characteristic properties disappear. The positive ions from the base combine with the negative ions from the acid and a new ionic salt forms. Water is also a product of this type of reaction, called **neutralization**. The graphic below shows what happens when sodium bicarbonate is mixed with hydrochloric acid.

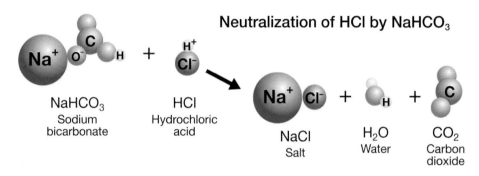

Neutralization of HCl by NaHCO$_3$

NaHCO$_3$
Sodium bicarbonate

HCl
Hydrochloric acid

NaCl
Salt

H$_2$O
Water

CO$_2$
Carbon dioxide

Neutralization in your body
Neutralization goes on in your body every day. As food and digestive fluids leave the stomach where the pH is very low, the pancreas and liver produce bicarbonate (a base) to neutralize the stomach acid. Antacids, which happen to be composed of sodium bicarbonate, have the same effect. The graphic above also illustrates what happens in your digestive system when you take an antacid. The antacid mixes with excess stomach acid to produce salt, water, and carbon dioxide!

Adjusting soil pH
Neutralization reactions are important in gardening and farming. For example, having soil that is too acidic (pH less than 5.5) is a common problem in the United States. Grass does not grow well in acidic soil. For this reason, many people add *lime* to their yard. A common form of lime is ground-up calcium carbonate (CaCO$_3$) made from natural crushed limestone. Lime is a weak base and undergoes a neutralization reaction with acids in the soil to raise the pH.

neutralization – the reaction of an acid and a base to produce a salt and water

Test your soil

Most garden centers carry soil test kits. These kits have pH test papers inside and are designed to help gardeners measure soil pH.

Get a soil test kit and test samples of soil from around your home or school. Repeat the test, taking new soil samples after a rainfall to see if the pH changes.

Answer the following questions.

1. What kinds of plants thrive in the pH of the soil samples you tested?

2. What kinds of treatments are available at your local garden center for changing soil pH?

Section 19.3 **Review**

1. What is a hydronium ion?

2. In this section, you learned about the properties of acids and bases. Make a table that organizes this information.

3. Both strong acids and strong bases are corrosive. Come up with a hypothesis for why this is so.

4. Answer the following questions about water.

   $$2\,H_2O \rightleftharpoons OH^- + H_3O^+$$

   a. Is pure water an acid, a base, neither, or both?
   b. What is the pH of pure water?
   c. What does the double-headed arrow mean in this reaction?

5. Nadine tests an unknown solution and discovers that it turns blue litmus paper red, and it has a pH of 3.0. Which of the following could be the unknown solution? Explain your choice.

   a. sodium hydroxide
   b. vinegar
   c. ammonia
   d. soap
   e. water

6. What is the concentration of hydronium ions for the solution in question 5? Is this solution acidic or basic?

7. Is tomato juice acidic or basic? Justify your answer.

8. Give two examples of a pH indicator.

9. Plants and animals live in environments that have conditions for which they are adapted. Is pH an important environmental factor for plants and animals? Why or why not?

10. Describe in your own words how the amount of carbon dioxide dissolved in your blood affects your blood pH.

11. What are the main reactants and products in a neutralization reaction?

12. Neutralization is an important part of digestion. Why?

STEM Are You Feeling
A Little Sour?

Have you ever heard someone describe a friend's personality or emotional state in terms usually used for food? "She's so sweet." "He's a bitter person." "Don't be sour." These expressions are figures of speech, but they reflect an underlying reality: Our body chemistry affects how we feel, and our body chemistry is influenced by what we eat. Knowing how food intake influences body chemistry can help you make choices that may keep you from feeling sour!

pH: A balancing act

You have learned that solutions can be described as having a certain pH. A solution can be an acid (acidic), a base (alkaline), or neutral. Acids and bases are everywhere. Many of our favorite foods are acidic: Oranges and lemons, for instance, contain citric acid. We depend on gastric acid in our stomachs to digest our food, and if we suffer from discomfort caused by that same gastric acid, we can help neutralize it with a base like baking soda. Your blood's normal pH range is between 7.35 and 7.45. You can compare this with many other ordinary solutions in the chart to the right. Our bodies are constantly adjusting to keep blood pH in a normal range.

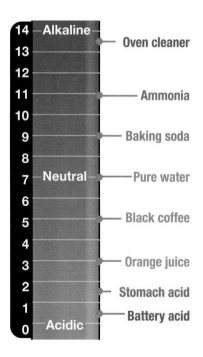

pH		
14	Alkaline	Oven cleaner
13		
12		
11		Ammonia
10		
9		Baking soda
8		
7	Neutral	Pure water
6		
5		Black coffee
4		
3		Orange juice
2		Stomach acid
1		Battery acid
0	Acidic	

Imbalances

The human body's many different processes produce a great deal of acid, which must then be removed. The lungs and kidneys handle most of this work. Lungs help dispose of excess acid when we breathe out carbon dioxide. Kidneys also remove excess acid from the blood and dispose of it in urine. Disease or extreme conditions can interfere with the body's self-adjusting system. There are two types of imbalance. We can have too much acid in our body fluids (acidosis), or those fluids can be too alkaline (alkalosis). When the lungs are not functioning properly, the imbalance is respiratory. When the body's physical and chemical processing of substances is not functioning properly, the imbalance is metabolic.

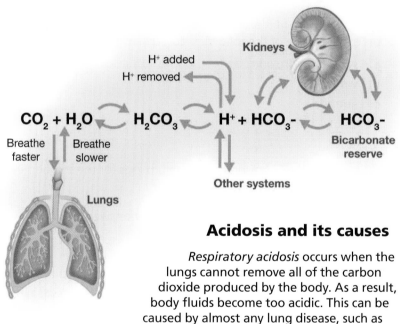

Kidneys

H⁺ added
H⁺ removed

$$CO_2 + H_2O \quad H_2CO_3 \quad H^+ + HCO_3^- \quad HCO_3^-$$

Breathe faster Breathe slower

Bicarbonate reserve

Other systems

Lungs

Acidosis and its causes

Respiratory acidosis occurs when the lungs cannot remove all of the carbon dioxide produced by the body. As a result, body fluids become too acidic. This can be caused by almost any lung disease, such as asthma. Treatment may include drugs that expand the air passages in the lungs.

Metabolic acidosis is a pH imbalance that occurs when the body does not have enough bicarbonate needed to neutralize the excess acid. This can be caused by a disease like diabetes, or by severe diarrhea, heart or liver failure, kidney disease, or even prolonged exercise. Prolonged exercise can result in a buildup of lactic acid, which causes the blood to become acidic. Sports drinks containing electrolytes can help restore the pH balance, which is why they are popular among athletes. Those drinks are specially formulated to help the body maintain its pH balance under stress.

Alkalosis and its causes

The opposite of acidosis, alkalosis is the result of too much base in the body's fluids. *Respiratory alkalosis* is caused by hyperventilation, that is, extremely rapid or deep breathing that makes the body lose too much carbon dioxide. It can be provoked by exertion at high altitudes, or even by anxiety. In such a case, the person may breathe (or be helped to breathe) into a paper bag. Why? Because the bag retains the exhaled carbon dioxide and it can be taken back in.

Metabolic alkalosis is a result of too much bicarbonate in the blood. Other types of alkalosis are caused by too little chloride or potassium. Alkalosis symptoms include: confusion, muscle twitching or spasms, hand tremors, nausea, and light-headedness.

Neutralizing acids

By nature, our slightly alkaline pH needs to remain balanced there. Yet what we eat and drink changes our pH. If you eat a lot of meat such as hamburgers, steak, and chicken, your body produces more acid than if you eat a lot of vegetables and fruits. If we don't balance what we eat, the body has to rely on reserves to neutralize the excess acid. For example, if you eat a lot of meat and no vegetables, your pH becomes acidic. Your kidneys can handle only so much acid and, if you have too much acid, the body must use reserve bicarbonate from your bones to help neutralize the acid. This is just one example of how the food we eat can affect our bodies. Maintaining a balanced diet is the first step toward good health and a normal pH level.

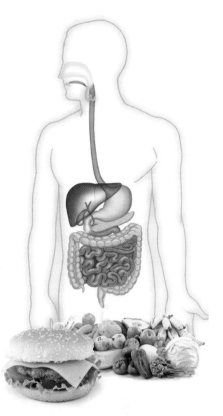

Questions:

1. What two organs regulate the acid–base balance?

2. What is a common cause of hyperventilation?

3. How is the alkalosis caused by hyperventilation treated?

4. Name a leading cause of respiratory acidosis.

Chapter 19 *Assessment*

Vocabulary

Select the correct term to complete each sentence.

acid	mass percent	solvent
alloy	molarity	solubility
base	nonpolar molecule	solubility rules
colloid	pH	supersaturated
concentration	pH scale	suspension
dissolve	polar molecule	Tyndall effect
equilibrium	saturated	unsaturated
hydrogen bond	solute	
insoluble	solution	

Section 19.1

1. A methane molecule is an example of a(n) _____ .

2. An attractive force between a hydrogen on one molecule and an atom on another molecule is called a(n) _____ .

3. A water molecule is an example of a(n) _____ .

Section 19.2

4. A solution with less than the amount of solute that can dissolve for a certain set of conditions is _____ .

5. _____ help you predict if a compound is soluble or not.

6. Muddy water is an example of a(n) _____ .

7. You can see the _____ if you shine a light through a(n) _____ but not if you shine a light through a solution.

8. The substance that dissolves particles in a solution is called the _____ .

9. The substance that is dissolved in a solution is called the _____ .

10. A mixture of two or more substances that is uniform at the molecular level is called a(n) _____ .

11. A solution of two or more metals is known as a(n) _____ .

12. A solvent is used to _____ a solute to make a solution.

13. When the dissolving rate equals the rate at which molecules come out of solution, the solution is in _____ .

14. The exact amount of solute dissolved in a given amount of solvent is the _____ of a solution.

15. A(n) _____ solution has a concentration greater than the maximum solubility.

16. If you make a solution using the solubility value for a substance, you will make a(n) _____ solution.

17. Talc is _____ in water.

18. Two ways to express a solution's concentration are _____ and _____ .

Section 19.3

19. A substance that produces hydronium ions (H_3O^+) in solution is called a(n) _____ .

20. A substance that produces hydroxide ions (OH^-) in solution is called a(n) _____ .

21. The neutral value on the _____ is 7.

22. The _____ of acids is less than 7.

Concepts

Section 19.1

1. The shape of a water molecule is a tetrahedron. Why?

2. Describe two main differences between a water molecule (H_2O) and a methane molecule (CH_4).

3. Describe a bond within a polar molecule and then describe how two polar molecules can "bond" with each other.

4. List two properties of water that are related to hydrogen bonding.

Section 19.2

5. Which of these substances is a colloid and which is a suspension?

6. Water is a solvent in which of the following solutions?

   a. air
   b. liquid sterling silver
   c. saline (salt) solution

7. What would happen to the solubility of potassium chloride in water as the water temperature increased from 25°C to 75°C? Why?

8. What are two ways to increase the dissolving rate of sugar in water?

9. What is the difference between mass percent and molarity?

10. Very small concentrations are often reported in ppm. What does "ppm" stand for? What does it mean?

11. What happens to a supersaturated solution when more solute is added? Use the word *equilibrium* in your answer.

12. How might the fish in a lake be affected if large amounts of hot water from a power plant or factory were released into the lake?

13. When you open a can of room-temperature soda, why is it more likely to fizz and spill over than a can that has been refrigerated?

14. In your own words, describe the solubility rule for hydroxide (OH^-).

Section 19.3

15. What determines the strength of an acid?

16. What determines the strength of a base?

17. What is the pH of a neutral solution?

18. Indicate whether the following properties belong to an acid (A), a base (B), or both (AB):

   a. _____ Creates a sour taste in food.
   b. _____ Creates a bitter taste in food.
   c. _____ Changes the color of red litmus paper to blue.
   d. _____ Changes the color of blue litmus paper to red.
   e. _____ Can be very corrosive.

19. Substance X has a pH of 6.5 and tastes sour. Is it an acid or a base?

20. Which of the following pH values is the most acidic?

   a. 1
   b. 3
   c. 7
   d. 8

21. When hydroxide ions are added to a solution, does the pH increase or decrease?

22. Are hydronium ions contributed to a solution by an acid or a base?

23. If you add water to a strong acid, how will the pH of the diluted acid compare to the pH of the original acid?

   a. lower

   b. higher

   c. the same

24. How can ammonia (NH_3) be a base if it doesn't contain any hydroxide ions?

25. What is the relationship between the values of the exponents for hydronium ion concentration and the corresponding pH values?

26. Describe what should happen and why when you mix vinegar and baking soda.

27. If you hold your breath for a while, how is your blood pH affected? Why?

Problems

Section 19.1

1. How much energy in joules would you need to raise the temperature of one kilogram of water by 2°C?

2. Why is this molecule a polar molecule?

Section 19.2

3. How much of the following materials will dissolve in 300 mL of water at 25°C?

   a. table salt

   b. sugar

   c. chalk

4. You add 20 grams of baking soda ($NaHCO_3$) to 100 mL of water at 25°C.

   a. Approximately how much of the baking soda will dissolve in the water?

   b. What happens to the rest of the baking soda?

   c. How could you increase the amount of baking soda that will dissolve in 100 mL of water?

5. How many grams of sugar do you need to make a 20 percent solution by mass in 500 g of water?

6. What is the mass percent of table salt in a solution of 25 grams of salt dissolved in 75 g of water?

7. The formula mass of KNO_3 is 101 g/mol. Describe how you would make 1 L of a 2 M solution of this salt.

8. Use the following graph to answer the questions below.

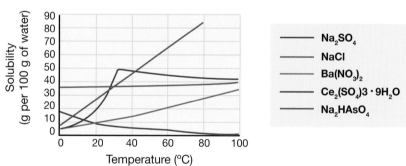

Solubility vs. temperature for a variety of salts

   a. What is the solubility of Na_2SO_4 at 80°C?

   b. Would a solution of $Ba(NO_3)_2$ be saturated with 20 grams dissolved in 100 grams of water at 80°C?

   c. How does solubility vary with temperature for Na_2HAsO_4? For Ce_2SO_4?

9. Figure 19.15 includes a temperature–solubility graph for table sugar. Where on this graph would you find saturated solution conditions?

Section 19.3

10. Recall that oxygen has an oxidation number of 2^- and hydrogen has an oxidation number of 1^+. Use this information to explain the charge on a hydroxide ion and a hydronium ion.

11. Solution A has a pH of 3 and solution B has a pH of 10.

   a. Which solution is a base?

   b. Which solution is an acid?

   c. What would happen if you combined both solutions?

12. Predict products of a chemical reaction between hydrochloric acid (HCl) and sodium hydroxide (NaOH). You may want to draw a diagram that illustrates what happens.

Applying Your Knowledge

Section 19.1

1. About 75 percent of Earth's surface is covered with water. Find out how much of this water is ice, how much is fresh water, and how much is part of the biggest solution of all—our ocean water.

2. Explain why ice forms on the top of ponds and lakes, not on the bottom. Use the following terms in your explanation: *water molecules, organized structure, hydrogen bonds,* and *density*. How does this property of water help support life in lakes and ponds?

Section 19.2

3. The concentration of ocean water is 35 ppt. Ocean water contains quite a bit of NaCl. Answer the following questions.

   a. What other solutes are dissolved in ocean water?

   b. Aside from it tasting bad, why is it not a good idea to drink ocean water? Answer this question in relation to your own body, which is mostly a watery solution.

4. Larry opens a new bottle of soda. He quickly stretches a balloon over the opening of the bottle. As he gently shakes the bottle, the balloon expands! Explain what is happening to cause the balloon to expand. Use at least three vocabulary words from this section.

Section 19.3

5. Just because an acid or a base is classified as weak does not mean that it is not important. Most of the acid–base chemistry that occurs inside your body occurs through reactions involving weak acids and bases. For example, the coiling of a DNA molecule into a "double-helix" is due to hydrogen bonding between weak bases. Find out more about the acid and base chemistry in your body. Possible topics include DNA, blood chemistry, digestion, and how your kidneys work.

6. Luke and Sian want to plant a vegetable garden in their yard. A soil testing kit measures the soil pH at 5.0, but the lettuce they want to plant in their garden does best at a pH of 6.5. Should they add an acid or a base to the soil to make it the optimum pH for growing lettuce?

7. Two years ago, you joined a project to study the water quality of a local pond. During the second spring, you noticed that there were not as many tadpoles (first stage in frog development) as there were the previous year. You want to know if the number of tadpoles in the pond is related to the pH of the pond. The records that document the water quality and wildlife started 10 years ago. Describe the steps you would take to determine whether a change in the pH of the pond water is affecting the population of frogs and their ability to reproduce.

Unit 7

Electricity and Magnetism

Try this at home

Find a strong magnet and some breakfast cereal that is advertised as "iron fortified." Place a small amount of cereal in a freezer-type zip top bag and gently crush the cereal into a fine powder. Spread some of the powder out into a thin layer inside the bag. Hold the magnet over the thin layer (outside the bag) and move it around. You will see pieces of the cereal moving with the magnet! Iron-fortified cereal actually has bits of iron metal in it, and the iron is attracted to the magnet.

Suppose you had a stationary bicycle that was connected to a light bulb, so that when you pedaled the bicycle, the energy from the turning wheels lit the bulb. How fast would you have to pedal to generate enough electrical energy to light the bulb? You would be surprised at how hard you would have to pedal to do something that seems so simple. Some science museums have interactive exhibits like this bicycle-powered light bulb to help people see how much energy is needed to accomplish everyday tasks.

What would your life be like without electricity? You can probably name at least a dozen aspects of your morning routine alone that would change if you didn't have electricity. Do you know how electrical circuits work? Do you know what *voltage* and *current* mean? This chapter will give you the opportunity to explore electricity, electrical circuits, and the nature of electrical energy. Electricity can be powerful and dangerous, but when you know some basic facts about how electricity works, you can use electricity safely and with confidence.

CHAPTER 20 INVESTIGATIONS

20A: Electricity
How do you measure voltage and current in electric circuits?

20B: Resistance and Ohm's Law
What is the relationship between current and voltage in a circuit?

20.1 **Charge**

Mass is one of the more obvious properties of matter. However, matter has other properties that are often hidden. *Charge* is a fundamental property of all matter that can be overlooked. All matter has electrical (and magnetic) properties because the atoms that make up matter are held together by electromagnetic forces.

Positive and negative charge

Two kinds of electric charge　Virtually all the matter around you has electric charge because all atoms contain electrons (-) and protons (+). Electrons have negative charge and protons have positive charge. However, unlike mass, electric charge is usually hidden inside atoms. Charge is hidden because atoms are made with equal amounts of positive and negative charges. The forces from **positive** charges are canceled by **negative** charges, the same way that +1 and -1 add up to 0. Because ordinary matter has zero *net* (total) charge, most matter acts as if there is no electric charge at all.

Like charges repel and unlike charges attract　Whether two charges attract or repel depends on whether they have the same or opposite sign. A positive and a negative charge will attract each other. Two positive charges will repel each other. Two negative charges will also repel each other. The force between charges is shown in Figure 20.1.

Charge is measured in coulombs　The unit of charge is the **coulomb (C)**. The name was chosen in honor of Charles Augustin de Coulomb (1736–1806), the French physicist who performed the first accurate measurements of the force between charges. One coulomb is a *huge* amount of charge, as you will see on the next page.

Fundamental property of matter　Electric charge, like mass, is a fundamental property of matter. An important difference between mass and charge is that there are two types of charge, which we call positive and negative. We know there are two kinds because electric charges can attract or repel each other. As far as we know, there is only one type of mass. All masses *attract* each other through gravity.

■■■■ VOCABULARY ■■■■

positive, negative – the two kinds of electric charge

coulomb (C) – the unit for electric charge

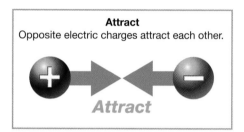

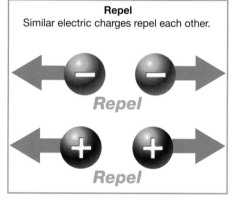

Figure 20.1: *The direction of forces on charges depends on whether they have the same or opposite charges.*

Static electricity

Neutral objects Matter contains trillions and trillions of charged electrons and protons because matter is made of atoms. Neutral atoms have the same number of electrons and protons. Therefore, the charge of an atom is *exactly zero*. Similarly, there is perfect cancellation between positive and negative in matter leaving a *net charge* of precisely zero. An object with a net charge of zero is described as being **electrically neutral**. Your pencil, your textbook, even your body are electrically neutral, at least most of the time.

Charged objects An object is **charged** when its net charge is *not* zero. If you have ever felt a shock when you touched a door knob or removed clothes from a dryer, you have experienced a charged object. An object with more negative than positive charge has a net negative charge overall. If it has more positive than negative charge, the object has a positive net charge. The net charge is also sometimes called *excess* charge because a charged object has an excess of either positive or negative charges.

Static electricity and charge A tiny imbalance in either positive or negative charge on an object is the cause of **static electricity**. If two neutral objects are rubbed together, the friction often pulls some electrons off one object and puts them temporarily on the other. This is what happens to clothes in the dryer and to your socks when you walk on a carpet. The static electricity you feel when taking clothes from a dryer or scuffing your socks on a carpet typically results from an excess charge of less than one one-millionth of a *coulomb*, the unit of charge.

What causes shocks You get a shock because excess charge of one sign strongly attracts charge of the other sign and repels charge of the same sign. When you walk across a carpet on a dry day, your body picks up excess negative charge. If you touch a neutral door knob, some of your excess negative charge moves to the door knob. Because the door knob is a conductor, the charge flows *quickly*. The moving charge makes a brief, intense electric current between you and the door knob. The shock you feel is the electric current as some of your excess negative charge transfers to the door knob (Figure 20.2).

VOCABULARY

electrically neutral – describes an object that has equal amounts of positive and negative charges

charged – describes an object whose net charge is not zero

static electricity – a tiny imbalance between positive and negative charge on an object

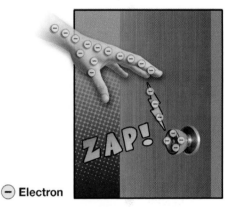

⊖ **Electron**

Figure 20.2: *The shock you get from touching a door knob on a dry day comes from moving charge.*

Electrical forces

The force between charges is very strong
Electric forces are incredibly strong. A millimeter cube of carbon the size of a pencil point contains about *77 coulombs* of positive and negative charge. If you could separate all the positive and negative charges by a distance of one meter, the attractive force between them would be about 50 thousand billion newtons! This is about the weight of *three thousand million cars*. This is all from the charge in a single pencil point (Figure 20.3). The huge force between charges is the reason objects are usually electrically neutral.

Lightning and charged particles
Lightning is caused by a giant buildup of static charge. Before a lightning strike, particles in a cloud collide and charges are transferred from one particle to another. Positive charges tend to build up on smaller particles and negative charges on bigger ones.

Storm clouds
The forces of gravity and wind cause the particles to separate. Positively charged particles accumulate near the top of the cloud, and negatively charged particles fall toward the bottom. Scientists from the National Aeronautics and Space Administration (NASA) have measured enormous buildups of negative charge in storm clouds. These negatively charged cloud particles repulse negative charges in the ground, causing the ground to become positively charged. This positive charge is why people who have been struck by lightning sometimes say they first felt their hair stand on end.

Lightning bolt
The negative charges in the cloud are attracted to the positively charged ground. The cloud, air, and ground can act like a giant circuit. All the accumulated negative charges flow from the cloud to the ground, heating the air along the path (to as much as 20,000°C) so that it glows like a bright streak of light. The air around a lightning bolt heats rapidly, and the expanding air creates sound waves that we hear as thunder. Thunder travels about 1 mile for every 5 seconds that you count between a flash of lightning and its thunder. If you see lightning and count 15 seconds before you hear the thunder, divide 15 by 5 and you know the lightning was about 3 miles away.

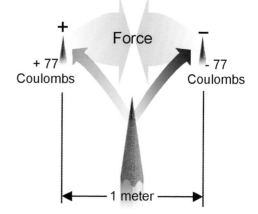

Figure 20.3: *If you could separate the charges in a pencil point by one meter, the force between the charges would be HUGE!*

Photo courtesy NOAA/ERL/NSSL.

Figure 20.4: *Lightning is created when negative charges in the cloud are attracted to the positively charged ground.*

Section 20.1 *Review*

1. Protons carry a _____ charge, and electrons carry a _____ charge.

2. Like charges _____, and opposite charges _____ .

3. What does it mean to say an object is electrically neutral?

4. Explain the difference between an electrically charged and a neutral object. Does a neutral object contain no electric charge at all?

5. Why is mass usually a more obvious property of matter than charge?

6. What is a coulomb (C)?

7. If you rub an air-filled balloon on your hair, you can make it stick to a wall. When the balloon and your hair are rubbed together, electrons are transferred from your hair to the balloon.

   a. What is the net charge on the balloon after it is rubbed on your hair? Is it positive, negative, or zero?

   b. What do you think happens to the atoms near the wall's surface when the balloon is brought near the wall? (*Hint*: The balloon will stick to the wall.)

   c. What happens when you try to stick a charged balloon to a metal object, like a door knob? Try it or do some research to find the answer and explain. Don't forget to include a Web site or book citation.

   d. The charged balloon experiments work better in dry weather than in damp weather. Why do you think this is so? Do some research to verify your answer. Don't forget to include a Web site or book citation.

8. What role do positive and negative charges play in the formation of lightning?

BIOGRAPHY

History of the terms positive and negative charge

Image courtesy of NOAA.

The terms *positive* and *negative* were first used by Benjamin Franklin (1706–1790). He and other scientists were seeking to describe their new observations about electricity. In 1733, French scientist Charles DuFay published a book describing how like charges repel and unlike charges attract. He theorized that two fluids caused electricity: vitreous (positive) fluid and resinous (negative) fluid.

Later that century, Franklin invented his own theory that argued that electricity is a result of the presence of a single fluid in different amounts. Although scientists no longer believe that electricity is caused by different kinds of fluids, the words *positive* and *negative* are still used to describe the two types of charge.

20.2 **Electric Circuits**

Think of how often you use TV, radio, computers, refrigerators, and light bulbs. All of these things are possible because of electricity. The use of electricity has become so routine that most of us never stop to think about what happens when we switch on a light or turn on a motor. This section is about electricity and electric circuits. Circuits are usually made of wires that carry electricity and devices that use electricity.

Electricity

What is electricity?   **Electricity** usually means the flow of **electric current** in wires, motors, light bulbs, and other inventions. Electric current is what makes an electric motor turn or an electric stove heat up. Electric current is almost always invisible and comes from the motion of electrons or other charged particles.

Electric current   Electric current is similar to a current of water, but electric current is not visible because it usually flows inside solid metal wires. Electric current can carry energy and do work just as a current of water can. For example, a waterwheel turns when a current of water exerts a force on it (Figure 20.5). A waterwheel can be connected to a machine such as a loom for making cloth, or to a millstone for grinding wheat into flour. Before electricity was available, waterwheels were used to supply energy to many machines. Today, the same tasks are done using energy from electric current. Look around you right now and you probably see wires carrying electric current into buildings.

Electricity can be powerful and dangerous   Electric current can carry a great deal of energy. For example, an electric saw can cut wood much faster than a hand saw. An electric motor the size of a basketball can do as much work as five big horses or 15 strong people. Electric current also can be dangerous. Touching a live electric wire can result in serious injury. The more you know about electricity, the easier it is to use it safely.

Figure 20.5: *A waterwheel uses the force of flowing water to run machines.*

Electric circuits

Electricity travels in circuits
An **electric circuit** is a complete path through which electricity travels. A good example of a circuit is the one in an electric toaster. Bread is toasted by heaters that convert electrical energy to heat. The circuit has a switch that turns on when the lever on the side of the toaster is pushed down. With the switch on, electric current enters through one side of the plug from the socket in the wall and goes through the toaster and out the other side of the plug.

Electric toaster   **Circuit inside toaster**

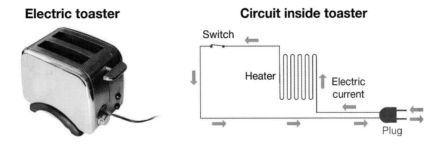

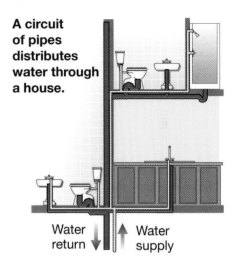

A circuit of pipes distributes water through a house.

Water return   Water supply

Wires are like pipes for electricity
Wires in electric circuits are similar in some ways to pipes and hoses that carry water (Figure 20.6). Wires act like pipes for electric current. Current enters the house on the supply wire and leaves on the return wire. The big difference between wires and water pipes is that you cannot get electricity to leave a wire the way water leaves a pipe. If you cut a water pipe, the water flows out. If you cut a wire, the electric current stops immediately.

Examples of circuits in nature
Circuits are not confined to appliances, wires, and devices built by people. The first experience humans had with electricity was in the natural world. These are some examples of natural circuits:

- The tail of an electric eel makes a circuit when it stuns a fish with a jolt of electricity.
- The Earth makes a gigantic circuit when lightning carries electric current between the clouds and the ground.
- The nerves in your body are an electrical circuit that carries messages from your brain to your muscles.

A circuit of wires distributes electric current through a house.

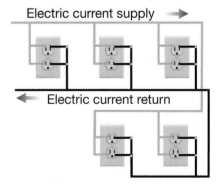

Electric current supply →

← Electric current return

Figure 20.6: *Comparing "circuits" for water and electricity.*

Circuit diagrams and electrical symbols

Circuit diagrams Circuits are made up of wires and electrical parts such as *batteries*, *light bulbs*, *motors*, and *switches*. When designing a circuit, people make drawings to show how the parts are connected. Electrical drawings are called *circuit diagrams*. In a circuit diagram, symbols are used to represent each part of the circuit. These *electrical symbols* make drawing circuits quicker and easier than drawing realistic pictures.

Electrical symbols A circuit diagram is a shorthand method of describing a working circuit. The electric symbols used in circuit diagrams are standard so that anyone familiar with electricity can build the circuit by looking at the diagram. Figure 20.7 shows some common parts of a circuit and their electrical symbols. The picture below shows an actual circuit on the left and its circuit diagram on the right. Can you identify the real parts with their symbols? Note that the switch is open in the circuit diagram, but closed in the photograph. Closing the switch completes the circuit so the light bulb lights.

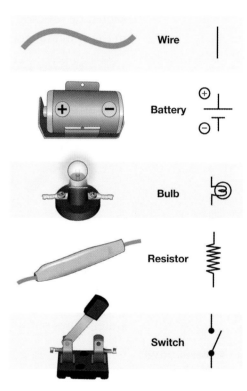

Figure 20.7: *These electrical symbols are used when drawing circuit diagrams.*

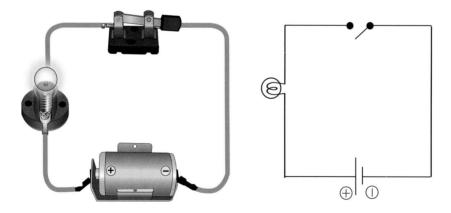

Resistors A **resistor** is an electrical device that uses the energy carried by electric current in a specific way. In many circuit diagrams, any electrical device that uses energy is shown with a resistor symbol. A light bulb, heating element, speaker, or motor can be drawn with a resistor symbol. When you analyze a circuit, many electrical devices may be treated as resistors when figuring out how much current is in the circuit.

Open and closed circuits

Batteries   All electric circuits must have a source of energy. Circuits in your home get their energy from power plants that generate electricity. Circuits in flashlights, cell phones, and portable radios get their energy from batteries. Some calculators have solar cells that convert energy from the sun or other lights into electrical energy. Of all the types of circuits, those with batteries are the easiest to understand. We will focus on battery circuits for now and will eventually learn how circuits in buildings work.

Open and closed circuits   It is necessary to be able to turn off light bulbs, radios, and most other devices in circuits. One way to turn off a device is to stop the current by "breaking" the circuit. Electric current can only flow when there is a complete and unbroken path from one end of the circuit to the other. A circuit with no breaks is called a **closed circuit** (Figure 20.8). A light bulb will light only when it is part of a closed circuit. Opening a switch or disconnecting a wire creates a break in the circuit and stops the current. A circuit with any break in it is called an **open circuit**.

Switches   **Switches** are used to turn electricity on and off. Flipping a switch to the *off* position creates an open circuit by making a break in the wire. The break stops the current because electricity cannot normally travel through air. Flipping a switch to the *on* position closes the break and allows the current to flow again, to supply energy to the bulb, radio, or other device.

Breaks in circuits   A switch is not the only way to make a break in a circuit. An incandescent light bulb burns out when the thin wire that glows inside it breaks. This creates an open circuit and explains why a burned-out bulb cannot light. You may also have seen compact fluorescent light bulbs (CFLs), which use less electrical energy than incandescent bulbs to put out the same amount of light. CFLs work differently than incandescent bulbs. Instead of heating a thin wire inside, a CFL is a coiled glass tube that contains a gas. When the circuit is closed, electricity passes through the gas-filled tube and causes the atoms in the gas to emit light. Just like incandescent bulbs, however, when a CFL bulb does finally quit working, the circuit will be broken and the CFL will need to be replaced.

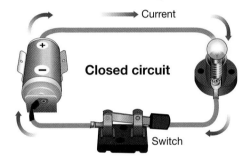
Current
Closed circuit
Switch

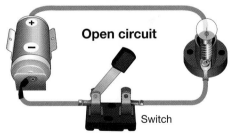
Open circuit
Switch

Figure 20.8: *There is current in a closed circuit but not in an open circuit.*

Section 20.2 *Review*

1. How are electric circuits and systems for carrying water in buildings similar?

2. Give one example of a circuit found in nature and one example of a circuit created by people.

3. Draw a circuit diagram for the circuit in Figure 20.9.

4. What is the difference between an open circuit and a closed circuit?

5. What does a resistor do in a circuit? Give an example.

6. Use the circuit diagram below to answer the following questions.

Figure 20.9: *Question 3.*

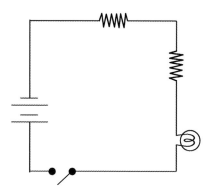

    a. How many bulbs are there in this circuit?
    b. How many batteries?
    c. How many resistors?
    d. How many switches?
    e. Is this circuit open or closed? Justify your answer.

7. When you turn "on" a light switch in a room, does this open or close the circuit? Explain.

20.3 **Current and Voltage**

Current is what carries energy in a circuit. Like water current, electric current only flows when there is a difference in energy between two locations that are connected. Water flows downhill from higher gravitational potential energy to lower energy. Electric current flows "downhill" from higher electrical potential energy to lower electrical potential energy.

VOCABULARY

ampere (A) – the unit of electric current

Current

Measuring electric current
Electric current is measured in units called **amperes (A)**, or amps for short. The unit is named in honor of Andre-Marie Ampere (1775–1836), a French physicist who studied electricity and magnetism. A small battery-powered flashlight bulb uses about 1/2 amp of electric current.

Current flows from positive to negative
Examine a battery and you will find a positive and a negative end. The positive end on a AA, C, or D battery has a raised bump, and the negative end is flat. A battery's electrical symbol uses a long line to show the positive end and a short line to show the negative end.

Current in equals current out
Electric current from a battery flows out of the positive end and returns back into the negative end. An arrow can be used to show the direction of current in a circuit (Figure 20.10). In most electric circuits, negative charge flows, and you would think the correct direction would be negative to positive. It is practical and conventional, however, to describe current as flowing from positive to negative, or from high voltage to low voltage. The amount of electric current coming out of the positive end of the battery must always be the same as the amount of current flowing into the negative end. You can picture this with steel balls flowing through a tube. When you push one in, one comes out. The rate at which the balls flow in equals the rate at which they flow out.

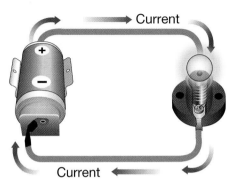

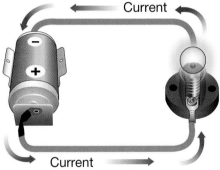

Figure 20.10: *Direction of electric current.*

SCIENCE FACT

Either positive or negative charges can make an electric current, depending on the circuit materials. In the human body, current is the movement of both positive and negative charges. In ordinary electric circuits, current is the movement of negative charge in metal conductors.

Voltage

Energy and voltage   **Voltage** is a measure of electric potential energy, just like height is a measure of gravitational potential energy. Voltage is measured in **volts (V)**. Like other forms of potential energy, a voltage difference means there is energy that can be used to do work. With electricity, the energy becomes useful when we let the voltage difference cause current to flow through a circuit. Current is what actually flows and does work. A difference in voltage provides the energy that causes current to flow (Figure 20.11).

What voltage means   A voltage difference of 1 volt means 1 amp of current does 1 joule of work in 1 second. Because 1 joule per second is a watt (power), *voltage is the power per amp of current that flows*. Every amp of current flowing out of a 1.5-V battery carries 1.5 watts of power. The voltage in your home electrical system is 120 volts, which means each amp of current carries 120 watts of power. The higher the voltage, the more power is carried by each amp of electric current.

Using a meter to measure voltage   A *voltmeter* measures voltage. A more useful meter is a **multimeter**, which can measure voltage or current, and sometimes also resistance. To measure voltage, the meter's probes are touched to two places in a circuit or across a battery. The meter shows the difference in voltage between the two places.

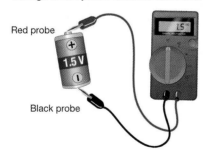

A multimeter can measure a battery's voltage if one probe touches each end.

Red probe

Black probe

The meter reads zero volts if both probes are connected at the same place.

Red probe

Black probe

Meters measure voltage difference   The meter reads *positive* voltage if the red (positive) probe is at a higher voltage than the black probe. The meter reads negative when the black probe is at the higher voltage. The meter reads voltage *differences* between its probes. If both probes are connected to the same place, the meter reads zero.

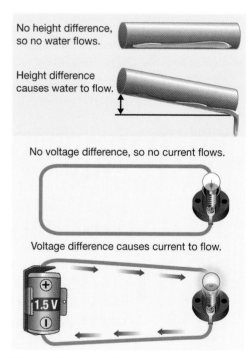

No height difference, so no water flows.

Height difference causes water to flow.

No voltage difference, so no current flows.

Voltage difference causes current to flow.

Figure 20.11: *A change in height causes water to flow in a pipe. Current flows in this circuit because a battery creates a voltage difference.*

Batteries

Batteries A **battery** uses chemical energy to create a voltage difference between its two terminals. When current leaves a battery, it carries energy. The current gives up its energy as it passes through an electrical device such as a light bulb. When a bulb is lit, the electrical energy is taken from the current and is transformed into light and heat energy. The current returns to the battery, where it gets more energy.

Batteries are like pumps Consider the water system shown below. The water pump raises the water level, increasing the potential energy of the water. As the water flows down, its potential energy is converted into kinetic energy at the waterwheel. How is a simple circuit similar? The pump keeps the water level different in the water system, and in the electrical circuit, the battery keeps the positive and negative charges separate. As long as the water level is different in the water system, the water can flow. As long as there is an area of charge separation in an electrical circuit, current can flow. This is why the battery is a sort of "pump." Chemical reactions in a battery give the energy to the current. The current then flows through the circuit, carrying the energy to any motors or bulbs (which are like the waterwheel in the water system). The current gets a "refill" of energy each time it passes through the battery, for as long as the battery's stored energy lasts.

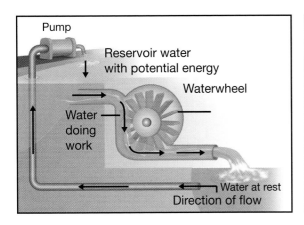

Pump
Reservoir water with potential energy
Waterwheel
Water doing work
Water at rest
Direction of flow

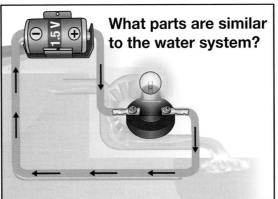

What parts are similar to the water system?

VOCABULARY

battery – a device that transforms chemical energy to electrical energy, and provides electrical force in a circuit

TECHNOLOGY

Batteries and cells

Battery voltage depends on how the battery is constructed and what chemicals it uses. A simple household zinc-carbon (alkaline) battery is 1.5 volts, and it is technically called a *cell*, not a battery. A, AA, AAA, C, and D cells all have 1.5 volts each. The D cell is the largest and carries the most energy, so a D cell can last longer than a smaller 1.5-volt cell.

If you have a device made up of more than one cell, you have a battery. A 9-volt battery is made up of three 1.5-volt alkaline cells. A car battery is usually 12 volts and is made up of 6 lead acid cells that are 2 volts each.

It is acceptable, although not entirely scientifically correct, to use the term *battery* when referring to A, AA, AAA, C, or D cells.

Measuring current in a circuit

Measuring current with a meter
Electric current can be measured with a multimeter. However, if you want to measure current, you must force the current to pass *through* the meter. That usually means you must break your circuit somewhere and rearrange wires so that the current must flow through the meter. For example, Figure 20.12 shows a circuit with a battery and bulb. The meter has been inserted into the circuit to measure current. If you trace the wires, the current comes out of the positive end of the battery, through the light bulb, *through the meter*, and back to the battery. The meter in the diagram measures 0.37 amps of current. Some electrical meters, called *ammeters*, are designed specifically to measure only current.

Setting up the meter
If you use a multimeter, you also must remember to set its dial to measure the type of current in your circuit. Multimeters can measure two types of electric current, called alternating current (AC) and direct current (DC). You will learn about the difference between alternating and direct current in the next chapter. For circuits with light bulbs and batteries, you must set your meter to read direct current, or DC. The symbols for AC and DC are shown in Figure 20.13.

Protect the meter
A meter can be damaged if too much current passes through it. Always be sure there is a light bulb or some other resistor in the circuit with the meter. This way, you are unlikely to overload the meter with too much current.

To protect its delicate electronics, most meters contain a *circuit breaker* or *fuse*. Circuit breakers and fuses are fast-acting, automatic switches that open a circuit if they sense too much current.

A circuit breaker can be reset the way a switch can be flipped. A broken fuse, however, is similar to a burned out light bulb and must be replaced for the meter to work again. The meter you use in your electric circuit investigations has a fuse inside. To replace the fuse, you will need a replacement fuse and a small screwdriver to open up the back of the meter. Your teacher can show you how this is done. To make your investigations easier, be careful when connecting current measurements, and you won't have to replace the fuse!

Measuring current

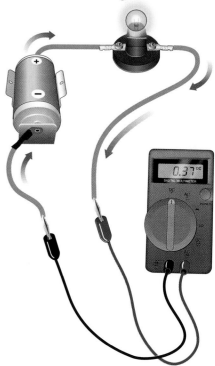

Figure 20.12: *Current must pass through the meter when it is being measured.*

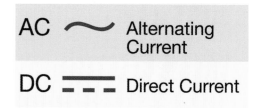

Figure 20.13: *A multimeter often uses these symbols for AC and DC settings.*

Section 20.3 *Review*

1. List the units for measuring current and voltage.

2. What is the difference between current and voltage, besides their units of measurement?

3. Why does a multimeter display a reading of zero when both of its probes are touched to the same end of a battery?

4. Study Figure 20.14 and answer the following questions. All batteries and bulbs are identical.
   a. Compare the voltage drop across the bulb in the one-bulb circuit with the voltage drops across each bulb in the four-bulb circuit.
   b. Which circuit will have more current, and why?
   c. Will there be a difference between the two circuits in bulb brightness? Why or why not?

5. The direction of electric current is away from the _____ end of the battery and toward the _____ end.

6. What voltage would the electrical meter show in each of the diagrams below?

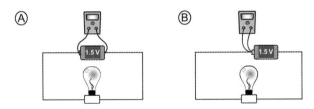

7. Which of the following diagrams shows the correct way to measure current in a circuit?

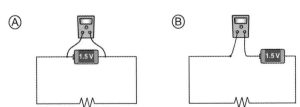

8. A flashlight needs three C batteries. How many volts of electricity does it need (Figure 20.15)?

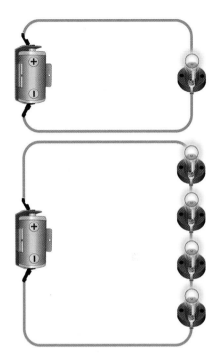

Figure 20.14: *Question 4.*

Figure 20.15: *Question 8.*

20.4 Resistance and Ohm's Law

You can apply the same voltage to different circuits and different amounts of current will flow. For example, when you plug in a desk lamp, the current through it is 1 amp. If a hair dryer is plugged into the same outlet (with the same voltage), the current is 10 amps. For a given voltage, the amount of current that flows depends on the *resistance* of the circuit.

VOCABULARY

resistance – determines how much current flows for a given voltage; higher resistance means less current

Electrical resistance

Current and resistance   **Resistance** is the measure of how strongly a wire or other object resists current flowing through it. A device with low resistance, such as a copper wire, can easily carry a large current. An object with a high resistance, such as a rubber band, can only carry a current so tiny it can hardly be measured.

A water analogy   The relationship between electric current and resistance can be compared with water flowing from the open end of a bottle (Figure 20.16). If the opening is large, the resistance is low and lots of water flows out quickly. If the opening is small, the resistance is greater and the water flow is slow.

Circuits   The total amount of resistance in a circuit determines the amount of current in the circuit for a given voltage. Every device that uses electrical energy adds resistance to a circuit. The more resistance the circuit has, the less the current. For example, if you string several light bulbs together, the resistance in the circuit increases and the current decreases, making each bulb dimmer than a single bulb would be.

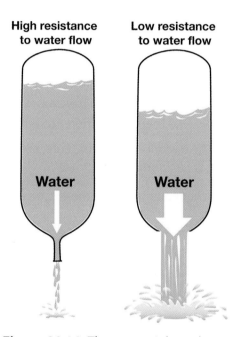

High resistance to water flow   Low resistance to water flow

Figure 20.16: *The current is less when the resistance is great.*

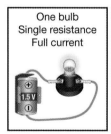
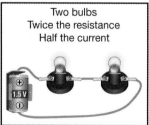

One bulb
Single resistance
Full current

Two bulbs
Twice the resistance
Half the current

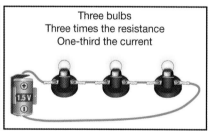

Three bulbs
Three times the resistance
One-third the current

Measuring resistance

The ohm   Electrical resistance is measured in units called **ohms**. This unit is abbreviated with the Greek letter *omega* (Ω). When you see Ω in a sentence, think or read "ohms." For a given voltage, the greater the resistance, the lesser the current. If a circuit has a resistance of one ohm, then a voltage of one volt causes a current of one ampere to flow.

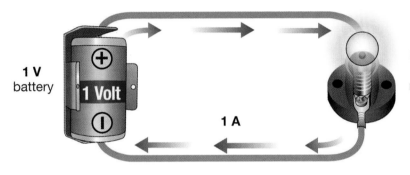

1 volt creates a current of 1 amp through a resistance of 1 ohm

1 V battery

1 Volt

1 A

Light bulb with a resistance of 1 Ω

Figure 20.17: *A multimeter can be used to measure resistance of a device.*

Resistance of wires   The wires used to connect circuits are made of metals such as copper or aluminum that have low resistance. The resistance of wires is usually so low compared with other devices in a circuit that you can ignore wire resistances when measuring or calculating the total resistance. The exception is when there are large currents. If the current is large, the resistance of wires may be important.

Measuring resistance   You can use a multimeter to measure the resistance of wires, light bulbs, and other devices (Figure 20.17). You must first remove the device from the circuit. Then set the dial on the multimeter to the resistance setting and touch the probes to each end of the device. The meter will display the resistance in ohms (Ω), kilo-ohms (× 1,000 Ω), or mega-ohms (× 1,000,000 Ω).

Ohm's law

Ohm's law
The current in a circuit depends on the battery's voltage and the circuit's resistance. Voltage and current are *directly* related. Doubling the voltage doubles the current. Resistance and current are *inversely* related. Doubling the resistance cuts the current in half. These two relationships form **Ohm's law** (Figure 20.18). The law relates current, voltage, and resistance with one formula. If you know two of the three quantities, you can use Ohm's law to find the third.

OHM'S LAW

$$\text{Current (A)} \quad I = \frac{V}{R} \quad \begin{array}{l} \text{Voltage (V)} \\[1em] \text{Resistance } (\Omega) \end{array}$$

Applying Ohm's law
Ohm's law shows how resistance is used to control the current. If the resistance is low, then a given voltage will result in a large amount of current. Devices that need a large amount of current typically have lower resistance. For example, a small electric motor might have a resistance of only 1 ohm. When connected in a circuit with a 1.5-volt battery, the motor draws 1.5 amps of current. By comparison, a small light bulb with a resistance of 2.5 ohms in the same type of circuit would draw only 0.6 amps.

Equation	Gives you...	If you know...
$I = V/R$	current (I)	voltage and resistance
$V = IR$	voltage (V)	current and resistance
$R = V/I$	resistance (R)	voltage and current

How much current flows when a 6 Ω bulb is connected to 3 V from batteries?

$$current = \frac{voltage}{resistance}$$

$$= \frac{3\ V}{6\ \Omega}$$

$$= 0.5\ A$$

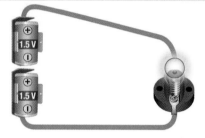

Figure 20.18: *An example of Ohm's law in action.*

Solving Problems: Using Ohm's law

A toaster oven has a resistance of 12 ohms and is plugged into a 120-volt outlet. How much current does it draw?

1. **Looking for:**　You are asked for the current in amperes.

2. **Given:**　　　You are given the resistance in ohms and voltage in volts.

3. **Relationships:** Ohm's law: $I = \dfrac{V}{R}$

4. **Solution:**　　Plug in the values for V and R: $I = \dfrac{120\ V}{12\ \Omega} = 10\ A$

Your turn...

a. A laptop computer runs on a 24-volt battery. If the resistance of the circuit inside is 16 ohms, how much current does it use?

b. A motor in a toy car needs 2 amps of current to work properly. If the car runs on four 1.5-volt batteries, what is the motor's resistance?

c. What is the current in the circuit below?

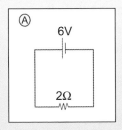

SOLVE FIRST LOOK LATER

a. 1.5 A

b. 3 Ω

c. 3 A

The resistance of common objects

Resistances match operating voltage   The resistance of electrical devices ranges from small (0.001 ohms) to large (10 × 10⁶ ohms). Every electrical device is designed with a resistance that causes the right amount of current to flow when the device is connected to the proper voltage. For example, a 100-watt light bulb has a resistance of 144 ohms. When connected to 120 volts from a wall socket, the current is 0.83 amps and the bulb lights (Figure 20.19). If you connect the same light bulb to a 1.5-volt battery, it will not light. According to Ohm's law, the current is only 0.01 amps when 1.5 volts are applied to a resistance of 144 ohms. This amount of current will not light the bulb. All electrical devices draw the right amount of power only when connected to voltage they were designed for.

The resistance of skin   Electrical outlets are dangerous because you can get a fatal shock by touching the wires inside. So why can you safely handle a 9-volt battery? The reason is Ohm's law. The typical resistance of dry skin is 100,000 ohms or more. According to Ohm's law, 9 V ÷ 100,000 Ω is only 0.00009 A. This is not enough current to be harmful. On average, nerves in the skin can feel a current of around 0.0005 amps. You can get a dangerous shock from 120 volts from a wall socket because that is enough voltage to force 0.0012 amps (120 V ÷ 100,000 Ω) through your skin, and you certainly can feel that!

Water lowers skin resistance   Wet skin has much lower resistance than dry skin. Because of the lower resistance, the same voltage will cause more current to pass through your body when your skin is wet. The combination of water and 120-volt electricity is especially dangerous because the high voltage and lower resistance make it possible for large (possibly fatal) currents to flow.

Changing resistance   The resistance of many electrical devices varies with temperature. For example, the amount of resistance a light bulb contributes to a circuit increases as its temperature increases (due to the current running through it). Devices that have a variable resistance like this are referred to as *non-ohmic*, because you can't use Ohm's law to predict the current when there is an ever-changing resistance (Figure 20.20). The small light bulbs in your circuit kit are non-ohmic, so you will use fixed resistors to apply Ohm's law to your simple circuits.

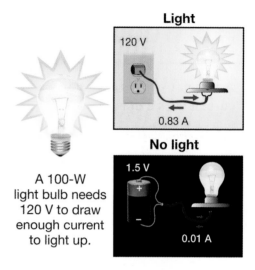

A 100-W light bulb needs 120 V to draw enough current to light up.

Figure 20.19: *A light bulb designed for use in a 120-volt household circuit does not light when connected to a 1.5-volt battery.*

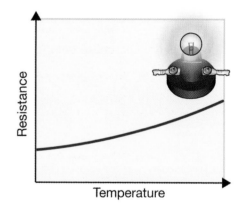

Figure 20.20: *The resistance of many materials, including those in light bulbs, increases as temperature increases. A light bulb is said to be "non-Ohmic" for this reason.*

Conductors and insulators

Conductors Current passes easily through some materials, such as copper, which are called conductors. A **conductor** can *conduct*, or carry, electric current. The electrical resistance of wires made from conductors is low. Most metals are good conductors.

Insulators Other materials, such as rubber, glass, and wood, do not allow current to easily pass through them. These materials are called **insulators,** because they insulate against, or block, the flow of current.

Semiconductors Some materials are in between conductors and insulators. These materials are called **semiconductors** because their ability to carry current is higher than an insulator but lower than a conductor. Computer chips, televisions, and portable radios are among the many devices that use semiconductors. You may have heard of a region in California called *Silicon Valley*. Silicon is a semiconductor commonly used in computer chips. An area south of San Francisco is called Silicon Valley because there are many semiconductor and computer companies located there.

Comparing materials No material is a perfect conductor or insulator. Some amount of current will always flow in any material if a voltage is applied. Even copper (a good conductor) has some resistance. Figure 20.21 shows how the resistances of various conductors, semiconductors, and insulators compare.

Applications of conductors and insulators Both conductors and insulators are necessary materials in human technology. For example, a wire has one or more conductors on the inside and an insulator on the outside. An electrical cable may have 20 or more conductors, each separated from the others by a thin layer of insulator. The insulating layer prevents the other wires or other objects from being exposed to the current and voltage carried by the conducting core of the wire.

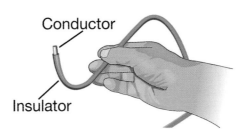

Conductor

Insulator

conductor – a material with low electrical resistance; metals such as copper and aluminum are conductors

insulator – a material with high electrical resistance; plastic and rubber are good insulators

semiconductor – a material between conductor and insulator in its ability to carry current

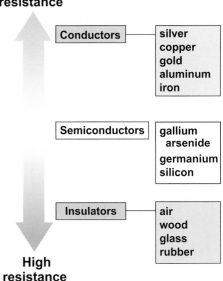

Low resistance

| Conductors | silver
copper
gold
aluminum
iron |

| Semiconductors | gallium
  arsenide
germanium
silicon |

| Insulators | air
wood
glass
rubber |

High resistance

Figure 20.21: *Comparing the resistance of materials.*

Resistors

Resistors are used to control current
Resistors are electrical components that are designed to have a specific resistance that remains the same over a wide range of currents. Resistors are used to control the current in circuits. They are found in many common electronic devices such as computers, televisions, telephones, and stereos.

Fixed resistors
There are two main types of resistors: fixed and variable. Fixed resistors have a resistance that cannot be changed. If you have ever looked at a circuit board inside a computer or other electrical device, you have seen fixed resistors. They can be small, skinny cylinders with colored stripes on them, or they can also be flat rectangles with a code written on top. They come in a range of sizes, depending on the resistance value and use.

Variable resistors
Variable resistors are used to vary the amount of current in a circuit. These resistors, which are a type of **potentiometer**, can be adjusted to have a resistance within a certain range. When would you use a variable resistor? Sometimes circuits need to be adjusted to keep them within tolerance. If something needs to be adjustable, like sound or light, a variable resistor  would be used. If you have ever turned a dimmer switch or volume control knob, you have used a potentiometer. When the resistance of a dimmer switch increases, the current decreases, and the bulb gets dimmer. Inside a potentiometer is a circular resistor and a little sliding contact called a wiper (Figure 20.22). If the circuit is connected at A and C, the resistance is always 100 Ω. But if the circuit is connected at A and B, the resistance can vary from 0 Ω to 100 Ω. Turning the dial changes the resistance between A and B and also changes the current (or voltage) in the circuit.

VOCABULARY

potentiometer – a type of variable resistor that can be adjusted to give resistance within a certain range

Potentiometer

The inside of a potentiometer

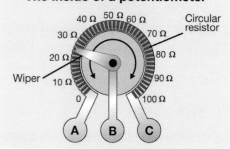

Circuit diagram

Figure 20.22: *The resistance of this potentiometer can vary from 0 Ω to 100 Ω.*

Section 20.4 *Review*

1. List the units and their abbreviations for resistance, voltage, and current.

2. What happens to the current if a circuit's resistance increases?

3. What happens to the current if a circuit's voltage increases?

4. A circuit contains one light bulb and a battery. What happens to the total resistance in the circuit if you replace the one light bulb with a string of four identical bulbs? Why?

5. Why can you safely handle a 1.5-V battery without being electrocuted?

6. A flashlight bulb has a resistance of about 6 Ω. It works in a flashlight with two AA alkaline batteries. About how much current does the bulb draw?

7. What voltage produces a 6-A current in a circuit that has a total resistance of 3 Ω?

8. What is a circuit's resistance if 12 V produces 2 A of current?

9. If you plug a device that has a resistance of 15 Ω into a 120-V outlet, how much current does it draw?

10. What is the difference between a conductor and an insulator? Give an example of each.

11. Do some research to find out why semiconductors are so important to computer technology. Don't forget to include Web site or book citations.

12. What is a fixed resistor, and where could you find fixed resistors in your home?

13. What is a variable resistor, and where could you find variable resistors in your home?

14. Look on the back or underside of different appliances and devices in your home. Find two that list the current and voltage each uses. Calculate the resistance of each.

TECHNOLOGY

Extension cord safety

The label on an extension cord will tell you how many amps of current it can safely carry. The length and wire thickness are both important. Always check to see if the extension cord can carry at least as much current as the device you plug in will require. Many fires have been caused by using the wrong extension cord!

Extension cords are made from 2 or 3 wires

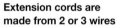

12-gauge wire

14-gauge wire

16-gauge wire

18-gauge wire

Wire Gauge	Current (amps)
12-gauge	20
14-gauge	15
16-gauge	10
18-gauge	7

The Shocking Truth:
You Are Wired!

Did you know that there are electric circuits in your body? Obviously, they aren't the kind made from batteries, bulbs, and wires—and there certainly isn't anything like lightning flashing around in there. However, there are electric circuits of a different type inside your body, and you couldn't survive without them.

Withdrawal reflexes

Have you ever accidentally touched a hot stove? The first thing you do is pull your hand back quickly—without even thinking about it. A withdrawal reflex like this happens because electrical signals are sent through the nerve network in your body. When you touch a hot stove, nerve endings in your fingers send a signal to nerves in your spinal cord. From the spinal cord, the signal is transferred to nerve fibers that control muscles in your hand and arm, causing them to contract, jerking your hand away from the stove. All of this happens in a split second!

Neurons

Your nervous system uses specialized cells called neurons to transfer electrical signals from one part of your body to another. A neuron has three basic parts: the cell body; a long, thin portion called the axon; and finger-like projections called dendrites.

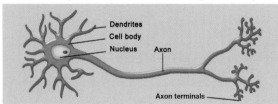

Battery circuits vs. body circuits

Unlike the components of the electric circuits you built in class, most neurons don't touch one another. Instead, as the electrical signal reaches the end of the axon, a chemical called a neurotransmitter is released. The chemical is picked up by receptors on the next neuron's dendrites. The dendrites then activate their own cell body to continue sending the signal along the axon.

Electricity in your home works because negatively charged electrons in the wires are free to carry the electrical current. This doesn't happen in the electric circuits of your body. Instead, the electrical current is carried by positively charged ions.

How does a nerve impulse work?

When a neuron is at rest, the inside of the cell membrane is electrically negative compared with the outside.

1. An outside stimulus, like touching a hot stove, causes the neuron's cell membrane to open tiny channels that let positively charged ions into the cell. One area of the neuron now has a positively charged inside relative to the outside.

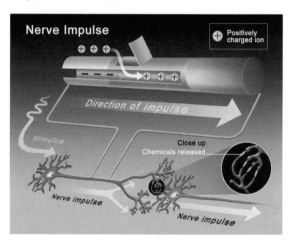

2. In a type of chain reaction, depolarization occurs along the entire neuron. As downstream channels open to let positive ions in, the previously depolarized areas let positive ions back out. As the ions leave, the membrane once again becomes negatively charged compared with the outside, as it was before the outside stimulus occurred.

3. The nerve impulse continues from neuron to neuron, across the gaps (synapses) between neurons, like a row of falling dominoes. The positive ions move in and out of one neuron, and at the gap between neurons, a chemical neurotransmitter is released to allow the depolarization to continue along the next neuron. In this way, nerve impulses or messages are conducted from one area of the body to another.

4. In a split second, your muscle receives the message to contract and pull your hand away from the source of heat. It all happens because of the flow of charged material. Your nervous system and your muscles are controlled by electrical impulses; some of them can move at upwards of 250 miles per hour!

Withdrawal reflexes are just one of many actions in your body that happen as a result of electrical signals. Your emotions, decisions, and physical actions all happen when nerve impulses transmit electrical signals through neurons in your brain, spinal cord, and body.

What makes your heart beat?

Did you know that electrical signals cause your heart to beat? There is specialized electrical tissue in your heart called the sinoatrial node. This specialized group of cells releases positive ions that carry an electrical message to the muscle cells all over the heart. This stimulates the heart to contract and pump blood throughout the body. People often refer to heart contractions as the "heartbeat."

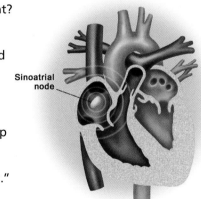

Sinoatrial node

For your heart to pump blood effectively, it must beat regularly, in a rhythmic pattern. The sinoatrial node is usually very good at sending rhythmic electrical impulses to the muscle cells, so the contractions happen steadily and regularly. However, if the sinoatrial node needs extra help, a surgeon can implant an artificial pacemaker to send the regular electrical impulses.

If you have ever seen doctors working in an emergency room on television or in a movie, you have probably seen a device called a defibrillator. A defibrillator uses an electric current to make a patient's heart start beating again after a heart attack or other trauma. Small, portable defibrillators are now being placed in schools, airports, and other public buildings. These devices have saved many lives by allowing trained people to help heart attack victims before paramedics arrive.

Electricity and living things

Many processes inside you (and other living things) depend on internal electric circuits. Most organisms keep this "shocking truth" to themselves, but not the electric eel! These South American river fish can stun unsuspecting prey with a 500-volt, 1 amp electric current generated through a flow of positive ions in specialized abdominal organs.

Questions:

1. Compare and contrast battery and bulb circuits with the circuits of your nervous system. How are they alike? How are they different?

2. Why would someone need to have a pacemaker? What do defibrillators do, and why are they made available in some public places?

3. There are hundreds of organisms that, like the electric eel, use electricity for more than just internal body processes. Do an Internet search. Choose two of the animals (other than the electric eel), name them, and write a brief description of how much electricity they produce and how they use it.

Neuron image courtesy of Wei-Chung Allen Lee, Hayden Huang, Guoping Feng, Joshua R. Sanes, Emery N. Brown, Peter T. So, and Elly Nedivi.

Chapter 20 *Assessment*

Vocabulary

Select the correct term to complete each sentence.

ampere	electricity	potentiometer
battery	insulator	resistance
charged	multimeter	resistor
closed circuit	negative	semiconductor
conductor	ohm	static electricity
coulomb	Ohm's law	switch
electric circuit	open circuit	volt
electric current	positive	voltage
electrically neutral		

Section 20.1

1. The unit in which charge is measured is the _____ .

2. An object is _____ when it has equal numbers of positive and negative charges.

3. All atoms have protons, which carry a(n) _____ charge.

4. All atoms have electrons, which carry a(n) _____ charge.

5. _____ is caused by a tiny imbalance of positive or negative charge.

6. A(n) _____ object is not electrically neutral.

Section 20.2

7. _____ is what flows and carries energy in a circuit.

8. A(n) _____ is used to create a break in a circuit.

9. A(n) _____ has a complete path for the current and contains no breaks.

10. A light bulb, motor, or speaker acts as a(n) _____ in a circuit.

11. A circuit diagram uses electrical symbols to represent a(n) _____ .

12. _____ is the science of electric current and charge.

13. When a light switch is in the "off" position, you have a(n) _____ .

Section 20.3

14. The unit for current is the _____.

15. A(n) _____ provides voltage for a circuit.

16. _____ is a measure of electric potential energy.

17. Use a(n) _____ to measure current or voltage in a circuit.

18. The _____ is the unit for measuring voltage.

Section 20.4

19. The _____ is the unit for measuring resistance.

20. _____ explains the relationship between current, voltage, and resistance in a circuit.

21. Wires in a circuit are made of a material that is a(n) _____ , such as copper.

22. _____ is the measure of how strongly a material resists current.

23. A(n) _____ like rubber or plastic has high electrical resistance.

24. Silicon is an example of a(n) _____ .

25. A(n) _____ is a type of variable resistor.

Concepts

Section 20.1

1. Like charges _____ and opposite charges _____ .

2. What does it mean to say an object is electrically neutral?

3. Is an object's net charge positive or negative if it loses electrons?

4. Why don't you usually notice electric forces between objects?

5. What unit is used for measuring charge, and where did the name come from?

6. Why do clothes sometimes stick together when you pull them out of the dryer?

Section 20.2

7. Use the illustrations below to answer the following questions.

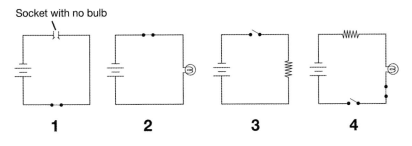

Socket with no bulb

1   **2**   **3**   **4**

    a. Which of the circuit(s) is/are closed?
    b. Which circuit(s) will *not* light a bulb?
    c. For any open circuits in the illustration, explain why the circuit is open.

8. Why are symbols used in circuit diagrams?

9. Draw the electrical symbol for each of the following devices.
    a. battery   **c.** switch
    b. resistor   **d.** wire

Section 20.3

10. How does voltage cause current to do work?

11. Explain how a battery in a circuit is similar to a water pump.

12. What are the differences between a multimeter, an ammeter, and a voltmeter?

13. Suppose you have a closed circuit containing a battery that is lighting a bulb.
    a. Explain how you would use a multimeter to measure the voltage across the bulb.
    b. Explain how you would use a multimeter to measure the current in the circuit.

14. What should you do to protect the multimeter when you measure current?

Section 20.4

15. What does it mean to say that current and resistance in a circuit are inversely related?

16. What does it mean to say that current and voltage in a circuit are directly related?

17. According to Ohm's law, the current in a circuit increases if the _____ increases. The current decreases if the _____ increases.

18. A battery is connected to a light bulb, creating a simple circuit. Explain what will happen to the *current* in the circuit if
    a. the bulb is replaced with a bulb having a higher resistance.
    b. the battery is replaced with a battery having a greater voltage.

19. Explain why electrical wires are made of copper covered in a layer of plastic. Use the terms *insulator* and *conductor* in your answer.

Problems

Section 20.1

1. Describe the forces between the positive and negative electric charges in each pair below.

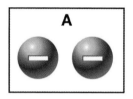

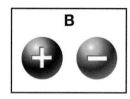

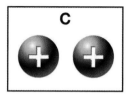

Section 20.2

2. Draw a circuit diagram of a circuit containing a battery, three wires, a light bulb, and a switch.

Section 20.3

3. What voltage would the multimeter show in each of the diagrams below?

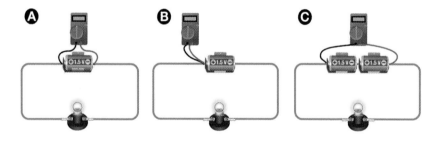

Section 20.4

4. What happens to the current in a circuit if the resistance triples? If the voltage triples?

5. A hair dryer draws a current of 10 A when plugged into a 120-V outlet. What is the resistance of the hair dryer?

6. A digital camera uses one 6-V battery. The circuit that runs the flash and takes the pictures has a resistance of 3 W. What is the current in the circuit?

Applying Your Knowledge

Section 20.1

1. On very dry days, when you use a comb or a brush, your hair sometimes stands on end and maybe even sticks to the comb or brush. Explain why this happens in terms of electric charge.

Section 20.2

2. A wire carrying an electric current is often likened to a pipe carrying water. What part of this analogy is incorrect?

Section 20.3

3. Design an experiment to determine whether more expensive household batteries last longer than cheaper ones. Don't forget to carefully select your controls! With your teacher's approval, try your experiment and use technology to report your findings. STEM

4. Standard voltage for electrical circuits in the United States is 120 volts. Is this the standard voltage in other countries? Do some research and if requested, cite your sources using a required format when you report your findings.

Section 20.4

5. Why can't you use an electric blender purchased in the United States in another country, like Spain or China?

Electrical Systems

You may recognize the abbreviations AC and DC. There is a classic rock band called AC/DC that helped make the acronyms famous. This chapter, however, is about the scientific meaning of AC and DC. Did you know that in the late 1800s, a major disagreement over the use of AC and DC methods for transmitting electricity erupted between two famous inventors? Thomas Edison favored the direct current (DC) method of moving electrical energy from electrical generation stations to homes and buildings. George Westinghouse argued that the alternating current (AC) method worked better. The feud became quite public, as each inventor tried to win support. The DC method works well over short distances, as between buildings in a densely populated city. AC works well over long distances but uses higher voltages than DC technology. Edison used some morbid methods for demonstrating his views of the danger involved with high-voltage electrical transmission through his opponent's AC method.

Which inventor won the AC/DC debate? Does the United States rely on AC or DC technology for transmitting electrical energy? In this chapter, you will find out how our country distributes electricity, and what the difference is between AC and DC. You will also learn how electricity is "purchased" and paid for, as well as how simple electrical circuits are constructed and how they work.

CHAPTER 21 INVESTIGATIONS

21A: Electrical Circuits
What are the different types of circuits?

21B: Electrical Energy and Power
How much energy is carried by electricity?

21.1 **Series Circuits**

We use electric circuits for thousands of things from flashlights to computers, cars, and satellites. There are two basic ways circuits can be built to connect different devices. These two types of circuits are called *series* and *parallel*. Series circuits have only one path for the current. Parallel circuits have branching points and multiple paths for the current. This section discusses series circuits. You will learn about parallel circuits in the next section.

What is a series circuit?

A series circuit has one path

A **series circuit** contains only one path for electric current to flow. That means the current is the same at all points in the circuit. All the circuits you have studied so far have been series circuits. For example, a battery, three bulbs, and a switch connected in a loop form a series circuit because there is only one path through the circuit (Figure 21.1). The current is the same in each bulb, so they are equally bright.

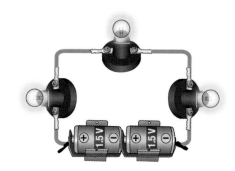

Figure 21.1: *A series circuit.*

A series circuit has only one path for the current so the current is the same at any point in the circuit.

Series circuit in holiday lights

If there is a break at any point in a series circuit, the current will stop everywhere in the circuit. Inexpensive strings of holiday lights are wired with the bulbs in series. The bulbs are only rated for 2.5 volts, but with 50 of them wired in series, the string runs well when plugged into a 120-volt outlet ($48 \times 2.5 = 120$, but manufacturers just add 2 bulbs to round out the number). If you remove one of the bulbs from its socket, the whole string of mini bulbs will go out. However, if a bulb's filament burns out, but the bulb is still in the socket, the string will stay lighted. How does this work? Modern 2.5-volt mini bulbs have a special backup wire to carry the current when a filament breaks (Figure 21.2). As long as the burned-out bulb is still in the socket, the series circuit will not be broken, because the current can travel through the backup wire (called a shunt).

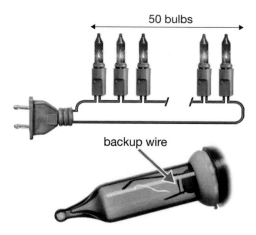

Figure 21.2: *A set of 50-strand mini bulbs have a backup wire inside, so if the filament burns out, current can flow through the backup wire and the rest of the bulbs in the strand can stay lit.*

Current and resistance in a series circuit

Use Ohm's law You can use Ohm's law to calculate the current in a circuit if you know the voltage and resistance. If you are using a battery, you know the voltage from the battery. If you know the resistance of each device, you can find the total resistance of the circuit by adding up the resistance of each device.

Adding resistances You can think of adding resistances like adding pinches to a hose (Figure 21.3). Each pinch adds some resistance. The total resistance is the sum of the resistances from each pinch. To find the total resistance in a series circuit, you add the individual resistances.

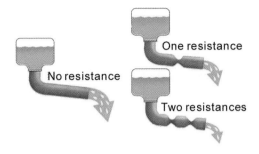

Figure 21.3: *Adding resistors in a circuit is like adding pinches in a hose.*

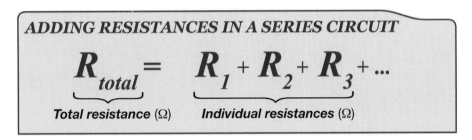

ADDING RESISTANCES IN A SERIES CIRCUIT

$$R_{total} = R_1 + R_2 + R_3 + ...$$

Total resistance (Ω)   Individual resistances (Ω)

Ignoring resistances in simple circuits Everything has some resistance, even wires. However, the resistance of a wire is usually so small compared with the resistance of light bulbs and other devices that we can ignore the resistance of the wire in the simple circuits we build and analyze.

Adding resistances in a series circuit

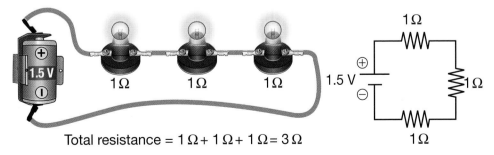

Total resistance = 1 Ω + 1 Ω + 1 Ω = 3 Ω

Solving Problems: Current in a Series Circuit

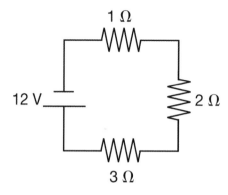

Figure 21.4: *What is the current in the circuit?*

A series circuit contains a 12-V battery and three bulbs with resistances of 1Ω, 2 Ω, and 3 Ω. What is the current in the circuit (Figure 21.4)?

1. **Looking for:**	You are asked for the current in amps.
2. **Given:**	You are given the voltage in volts and resistances in ohms.
3. **Relationships:**	$R_{tot} = R_1 + R_2 + R_3$
	Ohm's law: $I = V/R$
4. **Solution:**	$R_{tot} = 1\ \Omega + 2\ \Omega + 3\ \Omega = 6\ \Omega$
	$I = (12\ V)/(6\ \Omega) = 2\ A$

Your turn...

a. A string of 5 lights runs on a 9-V battery. If each bulb has a resistance of 2 Ω, what is the current?

b. A series circuit operates on a 6-V battery and has two 1 Ω resistors. What is the current?

c. A string of 50 mini-bulbs is wired in series. Each bulb has a resistance of 7 Ω. The string is plugged into a 120-V outlet. How much current does the string of lights draw?

SOLVE FIRST LOOK LATER

a. 0.9 A

b. 3 A

c. 0.3 A

Energy and voltage in a series circuit

Energy changes forms
Energy cannot be created or destroyed. The devices in a circuit convert electrical energy carried by the current into other forms of energy. As each device uses power, the power carried by the current is reduced. As a result, the *voltage is lower after each device that uses power*. This is known as the **voltage drop**. The voltage drop is the difference in voltage across an electrical device that has current flowing through it.

Charges lose their energy
Consider a circuit with three bulbs and two batteries (illustration C below). The voltage is 3 V, so each amp of current leaves the battery carrying 3 watts. As the current flows through the circuit, each bulb changes 1/3 of the power into light and heat. Because the first bulb uses 1 watt, the voltage drops from 3 V to 2 V as the current flows through the first bulb. Remember, the current in a series circuit is the same everywhere! As power gets used, voltage drops.

Voltage
If the three bulbs are identical, each gives off the same amount of light and heat. Each uses the same amount of power. A meter will show the voltage drop from 3 V, to 2 V, to 1 V, and finally down to 0 V after the last bulb. After passing through the last bulb, the current returns to the battery, where it is given more power, and the cycle starts over.

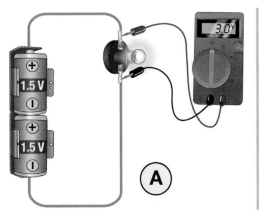

A

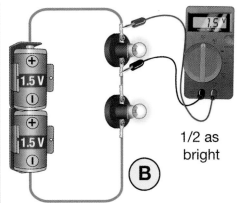

B

1/2 as bright

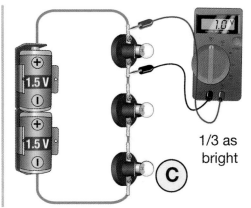

C

1/3 as bright

Voltage drops and Ohm's law

Voltage drops Each separate bulb or resistor creates a voltage drop. The voltage drop across a bulb is measured by connecting an electrical meter's leads at each side of the bulb (Figure 21.5). The greater the voltage drop, the greater the amount of power being used per amp of current flowing through the bulb.

Ohm's law The voltage drop across a resistance is determined by Ohm's law in the form $V = IR$. The voltage drop (V) equals the current (I) multiplied by the resistance (R) of the device. In a series circuit, the current is the same at all points, but devices may have different resistances. In the circuit below, each bulb has a resistance of 1 ohm, so each has a voltage drop of 1 volt when 1 amp flows through the circuit.

Applying Kirchhoff's law In the circuit below, three identical bulbs are connected in series to two 1.5-volt batteries. The total resistance of the circuit is 3 Ω. The current flowing in the circuit is 1 amp ($I = 3$ V ÷ 3 Ω). Each bulb creates a voltage drop of 1 V ($V = IR = 1$ A × 1 Ω). The total of all the voltage drops is 3 V, which is the same as the voltage of the battery.

VOCABULARY

Kirchhoff's voltage law – the total of all voltage drops in a series circuit must equal the voltage supplied by the battery

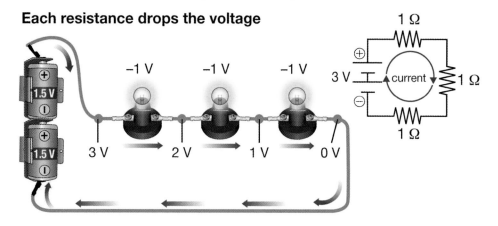

Each resistance drops the voltage

Figure 21.5: *Using a multimeter to measure the voltage drop across a bulb in a circuit.*

Energy conservation The law of conservation of energy also applies to a circuit. Over the entire circuit, the power used by all the bulbs must equal the power supplied by the battery. This means the total of all the voltage drops must add up to the battery's voltage. This rule is known as **Kirchhoff's voltage law**, after German physicist Gustav Robert Kirchhoff (1824–1887).

Finding voltage drops Ohm's law is especially useful in series circuits where the devices do *not* have the same resistance. A device with a larger resistance has a greater voltage drop. However, the sum of all the voltage drops must still add up to the battery's voltage. The example below shows how to find the voltage drops in a circuit with two different light bulbs.

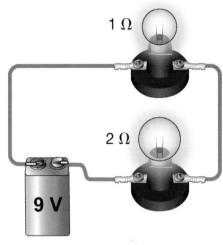

1 Ω

2 Ω

9 V

Figure 21.6: *What is the circuit's total resistance and current?*

 **Solving Problems: Voltage in Series Circuits**

The circuit shown at right (Figure 21.6) contains a 9-volt battery, a 1-ohm bulb, and a 2-ohm bulb. Calculate the circuit's total resistance and current. Then find each bulb's voltage drop.

1. **Looking for:**	You are asked for the total resistance, current, and voltage drops.
2. **Given:**	You are given the battery's voltage and the resistance of each bulb.
3. **Relationships:**	Total resistance in a series circuit: $R_{tot} = R_1 + R_2 + R_3$
	Ohm's law: $I = V/R$ or $V = IR$
4. **Solution:**	Calculate the total resistance: $R_{tot} = 1\ \Omega + 2\ \Omega = 3\ \Omega$
	Use Ohm's law to calculate the current: $I = (9\ V)/(3\ \Omega) = 3\ A$
	Use Ohm's law to find the voltage across the 1 Ω bulb:
	$V = (3\ A)(1\ \Omega) = 3\ V$
	Use Ohm's law to find the voltage across the 2 Ω bulb:
	$V = (3\ A)(2\ \Omega) = 6\ V$

SOLVE FIRST LOOK LATER

a. 4 A, 4 V across 1 W bulb, 8 V across 2 W bulb

b. 2 V, 10 V

Your turn...

a. The battery in the circuit above is replaced with a 12-volt battery. Calculate the new current and bulb voltages.

b. A 12-volt battery is connected in series to 1 Ω and 5 Ω bulbs. What is the voltage across each bulb?

Section 21.1 *Review*

1. What do you know about the current at different points in a series circuit?

2. Three bulbs are connected in series with a battery and a switch. Do all of the bulbs go out when the switch is opened? Explain.

3. What happens to a circuit's resistance as more resistors are added in series?

4. A series circuit contains a 9-volt battery and three identical bulbs. What is the voltage drop across each bulb?

5. A series circuit with three 1.5-volt batteries has two matching light bulbs (Figure 21.7).

   a. What is the voltage drop across each bulb?

   b. What would you have to know to find the value for the circuit's current?

6. A student builds a series circuit using three 1-ohm resistors. The current in the circuit is 1.5 amps.

   a. How many D-cells are in the circuit? (*Hint*: Use Ohm's law.)

   b. What is the voltage drop across each resistor?

7. A student builds a series circuit with four 1.5-volt batteries, a 5-ohm resistor, and two 1-ohm resistors.

   a. What is the total resistance in the circuit?

   b. Use Ohm's law to find the value of the current in the circuit.

8. How does Kirchoff's voltage law relate to the law of conservation of energy?

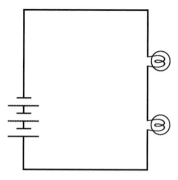

Figure 21.7: *Question 5.*

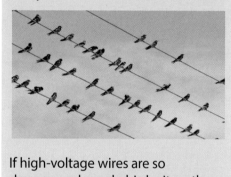

SCIENCE FACT

Why aren't birds electrocuted?

If high-voltage wires are so dangerous, how do birds sit on them without being instantly electrocuted? First, the bird's body has a higher resistance than the electrical wire. The current tends to stay in the wire because the wire is an easier path.

The most important reason, however, is that the bird has both feet on the same wire. That means the voltage is the same on both feet and no current flows through the bird.

21.2 Parallel Circuits

It would be a problem if your refrigerator went off when you switched off the overhead kitchen light! That is why houses are wired with parallel circuits instead of series circuits. Parallel circuits provide each device with a separate path back to the power source. This means each device can be turned on and off independently from the others. It also means that each device sees the full voltage of the power source without voltage drops from other devices.

What is a parallel circuit?

Parallel branches   A **parallel circuit** is a circuit with more than one path for the current. Each path in the circuit is sometimes called a *branch*. The current through a branch is also called the *branch current*. The current supplied by the battery in a parallel circuit splits at one or more branch points.

Example: three bulbs in parallel   All of the current entering a branch point must exit again. This rule is known as **Kirchhoff's current law** (Figure 21.8). For example, suppose you have three identical light bulbs connected in parallel as shown below. The circuit has two branch points where the current splits (green dots). There are also two branch points where the current comes back together. You measure the branch currents and find each to be 1 amp. The current supplied by the battery is the sum of the three branch currents, or 3 amps. At each branch point, the current entering is the same as the current leaving.

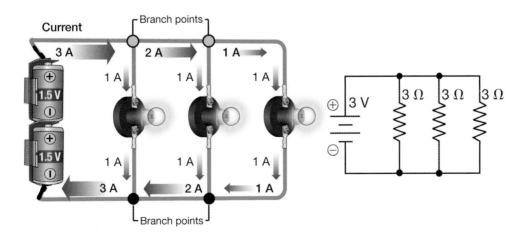

parallel circuit – an electric circuit with more than one path or branch

Kirchhoff's current law – states that all of the current entering a circuit branch must exit again

Kirchhoff's current law

All current flowing into a branch point must flow out again.

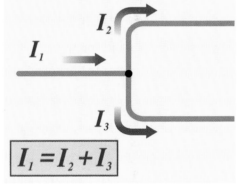

$$I_1 = I_2 + I_3$$

Figure 21.8: *All the current entering a branch point in a circuit must also exit the point.*

Voltage and current in a parallel circuit

Parallel circuit of two bulbs with different resistances

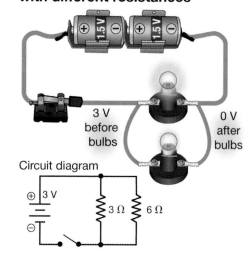

Each branch has the same voltage   The voltage is the same anywhere along the same wire. This is true as long as the resistance of the wire itself is very small compared to the rest of the circuit. If the voltage is the same along a wire, then the *same voltage appears across each branch of a parallel circuit*. This is true even when the branches have different resistances (Figure 21.9). Both bulbs in this circuit see 3 V from the battery because each is connected back to the battery by wires without any other electrical devices in the way.

The voltage is the same across each branch of a parallel circuit.

Parallel circuits have two big advantages over series circuits.

1. Each device in the circuit has a voltage drop equal to the full battery voltage.

2. Each device in the circuit may be turned off independently without stopping the current in the other devices in the circuit.

Figure 21.9: *The voltage across each branch of a parallel circuit is the same.*

Parallel circuits in homes   Parallel circuits need more wires to connect, but are used for most of the wiring in homes and other buildings. Parallel circuits allow you to turn off one lamp without all of the other lights in your home going out. They also allow you to use many appliances at once, each at full power.

Current in branches   Because each branch in a parallel circuit has the same voltage, the current in a branch is determined by the branch resistance and Ohm's law, $I = V/R$ (Figure 21.10). The greater the resistance of a branch, the smaller the current. Each branch works independently, so the current in one branch does not depend on what happens in other branches.

Total current   The total current in a parallel circuit is the sum of the currents in each branch. The only time branches have an effect on each other is when the total current is more than the battery or wall outlet can supply. A battery has a maximum amount of current it can supply at one time. If the branches in a circuit try to draw too much current, the battery voltage will drop and less current will flow.

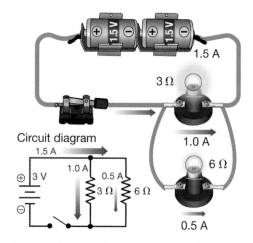

Figure 21.10: *The current in each branch may be different.*

Calculating current and resistance in a parallel circuit

More branches mean less resistance In series circuits, adding an extra resistor increases the total resistance of the circuit. The opposite is true in parallel circuits. Adding a resistor in a parallel circuit provides another independent path for current. More current flows for the same voltage, so the total resistance is *less*.

Example of a parallel circuit Compare the series and parallel circuits in Figure 21.11. In the series circuit, the current is 6 amps ($I = V/R = 12$ V $\div 2$ Ω). In the parallel circuit, the current is 6 amps *in each branch*. The total current is 12 amps. So what is the total resistance of the parallel circuit? Ohm's law solved for resistance is $R = V \div I$. The total resistance of the parallel circuit is the voltage (12 V) divided by the total current (12 A), which equals 1 ohm. The resistance of the parallel circuit is *half* that of the series circuit!

Parallel vs. series It can get confusing to keep track of what happens to three variables (current, voltage, and resistance) in different types of circuits. There is an easy way to remember the difference between series and parallel circuits. In a series circuit, current is the same everywhere, but voltage drops occur. In a parallel circuit, voltage is the same everywhere, but branch currents can be different. Voltage drops in series circuits and branch current differences in parallel circuits all depend on the resistance values for circuit resistors. So remember: series/same/current; parallel/same/voltage. You can use Ohm's law for both.

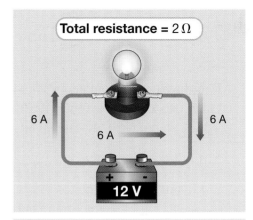

Total resistance = 2 Ω

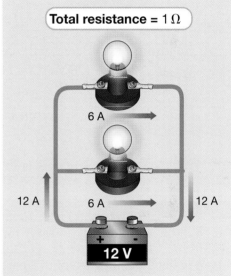

Total resistance = 1 Ω

Figure 21.11: *The parallel circuit has twice the current and half the total resistance of the series circuit.*

	Voltage	Current	Resistance
Series	Resistors must "share" total voltage available	Current is the same everywhere in the circuit	Resistance increases as you add resistors in series
Parallel	Each branch sees the full voltage available	Branch current can vary within the same circuit; add up branch currents to get total current	Resistance decreases as you add resistors in parallel, because current increases

Solving Problems: Current in Parallel Circuits

All of the electrical outlets in Jonah's living room are on one parallel circuit. The circuit breaker cuts off the current if it exceeds 15 amps. Will the breaker trip if he uses a light (240 Ω), stereo (150 Ω), and an air conditioner (10 Ω)?

1. **Looking for:**	You are asked whether the current will exceed 15 amps.
2. **Given:**	The resistance of each branch and the circuit breaker's maximum current
3. **Relationships:**	Ohm's law: $I = V/R$
4. **Solution:**	Because the devices are plugged into electrical outlets, the voltage is 120 volts for each.

$$I_{light} = (120 \text{ V})/(240 \text{ Ω}) = 0.5 \text{ A}$$

$$I_{stereo} = (120 \text{ V})/(150 \text{ Ω}) = 0.8 \text{ A}$$

$$I_{AC} = (120 \text{ V})/(10 \text{ Ω}) = 12 \text{ A}$$

The total is 13.3 A, so the circuit breaker will not trip.

Your turn...

a. Will the circuit breaker trip if Jonah also turns on a computer ($R = 60$ Ω)?

b. What is the total current in a parallel circuit containing a 12-V battery, a 2 Ω resistor, and a 4 Ω resistor?

Short circuits, circuit breakers, and fuses

Heat and wires When electric current flows through a resistor, some of the power carried by the current becomes heat. Toasters and electric stoves are designed to use electric current to make heat. Although the resistance of wires is low, it is not zero, so wires heat up when current flows through them. If too much current flows through too small a wire, the wire overheats and may melt or start a fire.

Short circuits

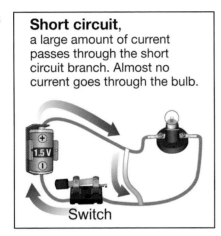

Short circuit, a large amount of current passes through the short circuit branch. Almost no current goes through the bulb.

Switch

A **short circuit** is a parallel path in a circuit with very low resistance. A short circuit can be created accidentally by making a parallel branch with a wire. A plain wire may have a resistance as low as 0.001 Ω. Ohm's law tells us that with a resistance this low, 1.5 V from a battery results in a (theoretical) current of 1,500 A! A short circuit is dangerous because currents this large can melt wires.

Circuit safety in homes Appliances and electrical outlets in homes are connected in many parallel circuits. Each circuit has its own fuse or circuit breaker that stops the current if it exceeds the safe amount, usually 15 or 20 amps (Figure 21.12). If you turn on too many appliances in one circuit at the same time, the circuit breaker or fuse cuts off the current. To restore the current, you must first disconnect some or all of the appliances. Then either flip the tripped circuit breaker (in newer homes) or replace the blown fuse (in older homes). Fuses are also used in car electrical systems and in electrical devices such as televisions.

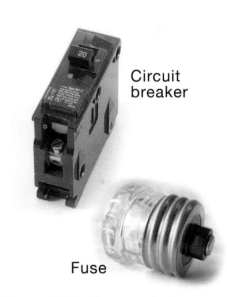

Circuit breaker

Fuse

Figure 21.12: *Houses and other buildings use either circuit breakers or fuses to cut off the current if it gets too high.*

Section 21.2 *Review*

1. Is the voltage across each branch of a parallel circuit the same? Is the current in each branch the same?

2. Give two reasons why parallel circuits are used for distributing electricity around homes and buildings instead of series circuits.

3. What happens to the total current in a parallel circuit as more branches are added? Why?

4. What happens to the total resistance of a parallel circuit if another resistor is added to the circuit?

5. For each diagram below, label the circuit *series*, *parallel*, or *short circuit*. The arrows show the flow of current. One circuit type is not shown.

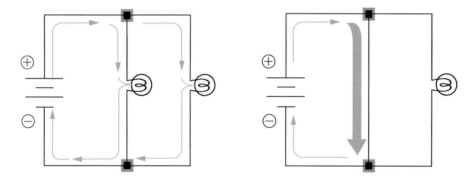

6. A circuit breaker in your house is set for 15 amps. You have plugged in a coffee maker that uses 10 amps. Plugging which of these four items into the same circuit will cause the circuit breaker to trip (because the current is too high)?

   a. a light that uses 1 amp

   b. a can opener that uses 2 amps

   c. a mixer that uses 6 amps

   d. an electric knife that uses 1.5 amps

21.3 **Electrical Power, AC, and DC Electricity**

If you look at a stereo, hair dryer, or other household appliance, you may find a label giving its power in watts. In this section, you will learn what the power ratings on appliances mean, and how to figure out the electricity costs of using various appliances.

Electric power

A watt is a unit of power
Electrical power is measured in watts, just like mechanical power. Electrical power is the rate at which electrical energy is changed into other forms of energy such as heat, sound, or light. Anything that "uses" electricity is actually converting electrical energy into some other type of energy. The **watt** is an abbreviation for one joule per second. A 100-watt light bulb uses 100 joules of energy *every second* compared to the 300 J/s used by the jogger (Figure 21.13).

The three electrical quantities
We have now learned three important electrical quantities:

Current (I)	Current is what carries power in a circuit. Current is measured in amperes (A).
Voltage (V)	Voltage measures the difference in electrical potential energy between two points in a circuit. Voltage is measured in volts (V). A difference in voltage causes current to flow. One volt is one watt per amp of current.
Resistance (R)	Resistance measures the ability to resist current. Resistance is measured in ohms (Ω). One amp of current flows if 1 V is applied across a resistance of 1 Ω.

Paying for electricity
Electric bills from utility companies don't charge by the volt, the amp, or the ohm. Electrical appliances in your home usually include another unit—the *watt*. Most appliances have a label that lists the number of watts or kilowatts. You may have purchased 60-watt light bulbs, a 1,000-watt hair dryer, or a 1,200-watt toaster oven. Electric companies charge for the energy you use, which depends on how many watts each appliance consumes and the amount of time each is used during the month.

100 watts
100 joules
each second

300 watts
300 joules
each second

Figure 21.13: *One watt equals one joule per second.*

Calculating power in a circuit

Calculating power   Power in a circuit can be measured using the tools we already have. Remember that one watt equals an energy flow of one joule per second.

Amps	One amp is a flow of one coulomb of charge per second.
Volts	One volt is an energy of one joule per coulomb of charge.

If these two quantities are multiplied together, you will find that the units of coulombs cancel out, leaving the equation we want for power.

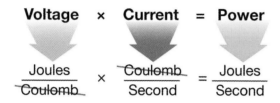

$$\text{Voltage} \times \text{Current} = \text{Power}$$

$$\frac{\text{Joules}}{\text{Coulomb}} \times \frac{\text{Coulomb}}{\text{Second}} = \frac{\text{Joules}}{\text{Second}}$$

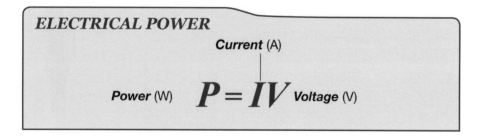

ELECTRICAL POWER

Current (A)

Power (W) $P = IV$ Voltage (V)

Watts and kilowatts   Most electrical appliances have a label that lists the power in watts (W) (Figure 21.14) or kilowatts (kW). The **kilowatt** is used for large amounts of power. One kilowatt (kW) equals 1,000 watts. Another common unit of power, especially on electric motors, is the horsepower. One horsepower is 746 watts. The range in power for common electric motors is from 1/25th of a horsepower (30 watts) for a small electric fan to 1 horsepower (746 watts) for a garbage disposal.

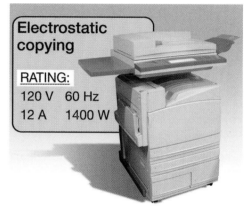

Electrostatic copying

RATING:
120 V   60 Hz
12 A   1400 W

Figure 21.14: *Most appliances have a label that lists the power in watts.*

Equation	...gives you ...	...if you know ...
$P = I \times V$	power (P)	current and voltage
$I = P \div V$	current (I)	power and voltage
$V = P \div I$	voltage (V)	power and current

Figure 21.15: *Different forms of the power equation.*

Solving Problems: Calculating Power

A 12-volt battery is connected in series to two identical light bulbs (Figure 21.16). The current in the circuit is 3 amps. Calculate the power output of the battery.

1. **Looking for:**   You are asked for the power in watts supplied by the battery.

2. **Given:**   You are given the battery voltage in volts and current in amps.

3. **Relationships:** Ohm's law: Power, $P = I \times V$

4. **Solution:**   Battery: $P = (3\ A)(12\ V) = 36\ W$

Your turn...

a. A 12-volt battery is connected in parallel to the same identical light bulbs as used in the example. The current through each bulb is now 6 amps. Calculate the power output of the battery.

b. The label on the back of a television states that it uses 300 watts of power. How much current does it draw when plugged into a 120-volt outlet?

 A compact fluorescent light bulb (CFL) uses a much different process than an incandescent bulb to produce light. As a result, the resistance of a CFL bulb is much less than that of an incandescent bulb that puts out the same amount of light.

c. What does this mean about the amount of current a CFL draws compared to the incandescent bulb?

d. What does this mean about the power a CFL uses compared to the incandescent bulb?

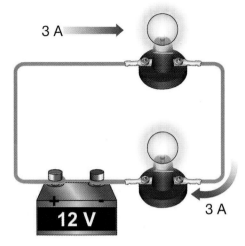

Figure 21.16: *What is the power output of the battery?*

SOLVE FIRST LOOK LATER

a. 144 W

b. 2.5 A

c. CFL draws less current than the incandescent bulb that puts out the same amount of light

d. CFL uses less power than the incandescent bulb that puts out the same amount of light

Buying electricity

Kilowatt-hours Utility companies charge customers for the number of **kilowatt-hours** (abbreviated kWh) used each month. One kilowatt-hour means that a kilowatt of power has been used for one hour. A kilowatt-hour is not a unit of power but a unit of *energy* like *joules*. A kilowatt-hour is a relatively large amount of energy, equal to 3.6 million joules. If you leave a 1,000-watt hair dryer on for one hour, you have used one kilowatt-hour of energy. You could also use 1 kilowatt-hour by using a 100-watt light bulb for 10 hours. The number of kilowatt-hours used equals the number of kilowatts multiplied by the number of hours the appliance was turned on. If you start with watts, be sure to move the decimal point to the left three places (divide by 1,000) before expressing your answer in kWh.

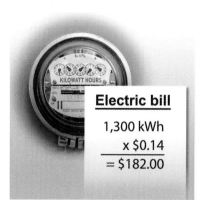

Electric bill

1,300 kWh
x $0.14
= $182.00

You pay for kilowatt-hours Electric companies charge for kilowatt-hours used during a period of time, often a month. Your home is connected to a meter that counts up total number of kilowatt-hours used, and the meter is read once a month. If you know the cost per kilowatt-hour the utility company charges, you can estimate the cost of operating any electrical appliance.

Save money on electricity How can you save money on your household's electric bill? Use less electricity, of course! There are many simple things you can do to use less electricity. When added up, these simple things can mean many dollars of savings each month, which adds up to a large amount of money over a one-year period. What can you do? Make sure your windows are locked so they seal properly. Turn off lights when you are not using them. Switch off electronic equipment that uses standby power. Electric utility companies will send an energy consultant to your home to give suggestions on how to conserve electricity. Conserving electricity means lower bills and a cleaner environment.

Appliance	Power (watts)
Electric stove	3,000
Electric heater	1,500
Toaster	1,200
Hair dryer	1,000
Iron	800
Washing machine	750
Television	300
Light	100
Small fan	50
Clock radio	10

Figure 21.17: *Typical power usage of some common appliances.*

Solving Problems: Estimating Electricity Costs

How much does it cost to run an electric stove for 2 hours? Use the power in Figure 21.17 and an electricity cost of $0.15 per kilowatt-hour.

1. **Looking for:** You are asked for the cost to run a stove for 2 hours.

2. **Given:** You are given the time, the power, and the price per kilowatt-hour.

3. **Relationships:** 1 kilowatt = 1,000 watts; number of kilowatt-hours = (# of kilowatts) × (hours appliance is used)

4. **Solution:** $3,000 \text{ W} = 3 \text{ kW}$ $3 \text{ kW} \times 2 \text{ hr} = 6 \text{ kWh}$ $6 \text{ kWh} \times \dfrac{\$0.15}{\text{kWh}} = \$0.90$

Your turn...

a. Suppose you run the electric stove in the example problem above for 2 hours a day, 5 days a week, for one 4-week month. What would the total cost of electricity be?

b. At $0.15 per kilowatt-hour, what is the cost of running an electric heater for 4 hours?

c. At $0.15 per kilowatt-hour, what is the cost of running a clock radio for 24 hours?

d. What is the cost of running the clock radio in the previous question for one year?

SOLVE FIRST LOOK LATER

a. $18.00

b. $0.90

c. $0.04 (rounded to nearest cent)

d. $13.14

Alternating (AC) and direct (DC) current

Direct current

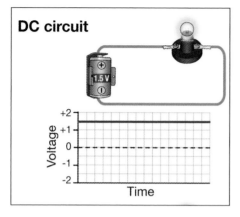

DC circuit

The current from a *battery* is always in the same direction, from the positive to the negative end of the battery. This type of current is called **direct current**, or **DC**. Although the letters "DC" stand for "direct current," the abbreviation "DC" is used to describe both voltage and current. A *DC voltage* is one that stays the same sign over time. The terminal that is positive stays positive, and the terminal that is negative stays negative. Your experiments in the lab use DC because they use batteries.

Alternating current   The electrical system in your *house* uses **alternating current**, or **AC**. Alternating current constantly switches direction. You can theoretically create alternating current with a battery if you keep reversing the way it is connected in a circuit. In the electrical system used in the United States, the current reverses direction 60 times per second. It would be hard to flip a battery this fast!

AC circuit

Flipping battery

Electricity in other countries   We use alternating current because it is easier to generate and to transmit over long distances. All the power lines you see overhead carry alternating current. Other countries also use alternating current. However, in many other countries, the current reverses itself 50 times per second rather than 60, and wall sockets are at a different voltage. When visiting Asia, Africa, Europe, and many other places, you need special adapters to use electrical appliances from the United States.

TECHNOLOGY

Peak and average voltages

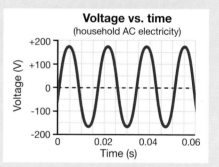

Voltage vs. time
(household AC electricity)

The 120-volt AC electricity used in homes and businesses in the United States alternates between peak values of +170 V and -170 V at a frequency of 60 Hz. This kind of electricity is called 120 VAC because +120 V is the average positive voltage and -120 V is the average negative voltage. AC electricity is usually defined by the average voltage, not the peak voltage.

Electricity in homes and buildings

Circuit breaker panel Electricity comes into most homes or buildings through a *circuit breaker service panel*. The circuit breakers prevent wires from overheating and causing fires. Each circuit breaker protects one parallel circuit that may connect many wall outlets, lights, switches, or other appliances.

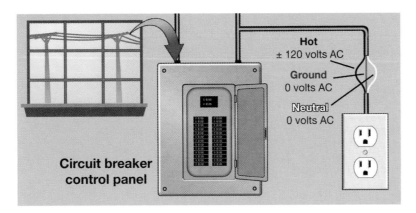

Ground fault interrupt (GFI) outlet

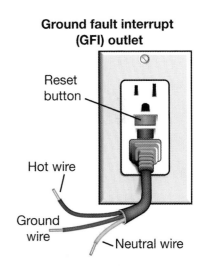

Figure 21.18: *A ground fault interrupt outlet can be found in bathrooms and kitchens where water may be near electricity.*

Hot, neutral, and ground wires Each wall socket has three wires feeding it. The hot wire carries 120 volts AC. The neutral wire stays at zero volts. When you plug something in, current flows in and out of the hot wire, through your appliance (doing work) and back through the neutral wire. The ground wire is for safety and is connected to the ground (0 V) near your house. If there is a short circuit in your appliance, the current flows through the ground wire rather than through you.

Ground fault interrupt (GFI) outlets Electrical outlets in bathrooms, kitchens, or outdoors are now required to have **ground fault interrupt (GFI) outlets** installed (Figure 21.18). A GFI outlet contains a circuit that compares the current flowing out on the hot wire and back on the neutral wire. If everything is working properly, the two currents should be exactly the same. If they are different, some current must be flowing to ground through another path, such as through your hand. The ground fault interrupter detects any difference in current and immediately breaks the circuit. GFI outlets are excellent protection against electric shocks, especially in wet locations.

Distributing electricity

Why electricity is valuable Electricity is a valuable form of energy because electrical power can be moved easily over large distances. You would not want a large power plant in your backyard! One large power plant converts millions of watts of chemical or nuclear energy into electricity. The transmission lines carry the electricity to homes and businesses, often hundreds of miles away.

Power transmission lines Overhead power lines use a much higher voltage than 120 V. That is because the losses due to the resistance of wires depend on the current. At 100,000 volts, each amp of current carries 100,000 watts of power, compared to the 120 watts per amp of household electricity. Big electrical transmission lines operate at very high voltages for this reason (Figure 21.19). The wires are supported high on towers because voltages this high are *very dangerous.* Air can become a conductor over distances of a meter at high voltages. *Never* go near a power line that has fallen on the ground in a storm or other accident.

Transformers A device called a **transformer** converts high-voltage electricity to lower voltage electricity. Within a few kilometers of your home or school, the voltage is lowered to 13,800 V or less. Right near your home or school, the voltage is lowered again to the 120 V or 240 V that actually comes into the circuits connecting your wall outlets and appliances.

Changing AC to DC Many electronic devices, like cell phones or laptop computers, use DC electricity inside, but also can be plugged into the AC electricity from a wall outlet with an *AC adapter* (Figure 21.20). An "AC adapter" is a device that changes the AC voltage from the wall outlet into DC voltage for the device. The adapter also steps the voltage down from 120 volts to the battery voltage, which is usually between 6 and 20 volts.

Why do some plugs have three prongs? The third round hole on an outlet is connected to grounding rods in the earth near your home or to a metal cold water pipe. When a plug with a third prong is plugged into the outlet, the ground prong allows an alternate pathway for electricity to flow. If a wire inside a metal-cased appliance should come loose and touch the metal case, the whole appliance becomes electrically charged. However, with the third safety prong plugged into a grounded circuit, the electricity follows the ground path and does not go through you!

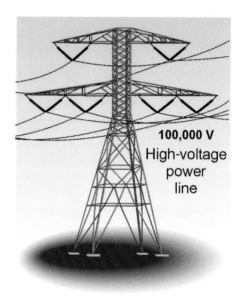

100,000 V
High-voltage power line

Figure 21.19: *Electrical power lines may operate at high voltages.*

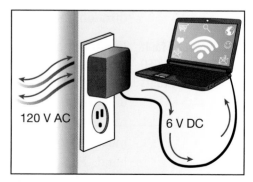

120 V AC 6 V DC

Figure 21.20: *Special adapters can change AC to DC and lower the voltage.*

Section 21.3 *Review*

1. How is an appliance's power related to the amount of energy it uses?

2. Which of the following does the electric utility company charge for each month?
   a. electrical power used
   b. electrical energy used
   c. electrical current used

3. What is the major difference between direct current and alternating current?

4. If a toaster oven draws 6 amps of current when plugged into a 120-volt outlet, what is the power rating of the appliance?

5. What is the current through a 60-watt light bulb if it is connected to a 120-volt circuit in your house?

6. A student used three appliances in her dormitory room: a 1,200-watt iron, which she uses 3.5 hours per month; a lamp with a 100-watt bulb, which she uses 125 hours per month; and a 700-watt coffee maker, which she uses 15 hours per month.
   a. How many kilowatt-hours of electrical energy are consumed in one month by each appliance?
   b. If the local utility company charges 15 cents per kilowatt-hour of electrical energy consumed, how much does it cost per month to operate each appliance?

7. Why are overhead power lines dangerous?

8. What is a transformer, and where would one be found?

9. List all the devices you own that have an AC adapter to convert AC to DC

⚙️ STEM **Bright** Ideas

The compact fluorescent lamp (CFL) was introduced in the early 1980s and was advertised as a highly efficient replacement for the standard incandescent light bulb. CFLs were slow to catch on, but by the year 2000, they began cutting into the market dominance of incandescent bulbs, which had been the preferred lighting source in homes across America for almost 100 years.

As public interest in conservation and efficiency grew, people began to see the value of this new technology. CFLs help the environment by reducing energy needs, and they help consumers save money on their electric bills. However, it did not take 100 years to come up with something even more efficient than the CFL. Watch out CFLs, LEDs are here!

LED stands for light emitting diode. A diode is an electronic device usually made out of layers of silicon that allow electric current to flow in one direction through a circuit while blocking current flowing in the opposite direction. Diodes are found in almost every electronic device in use today. The diode was invented in the 1870s, and by 1907, it was discovered that some diodes emit light when current passed through them. Since the 1960s, LEDs have been used as indicator lights on appliances and electronic equipment, and as number displays on alarm clocks, watches, televisions, and calculators. If you've ever pushed a button on a device and a little colored light lit up, it was probably an LED.

LED (light emitting diode)
Emitted light beams
Diode

One of the LED's best qualities is that it needs very little current to produce light. Recent technological breakthroughs have enabled LEDs to create more and more light. Designers have assembled arrays of LEDs into bulb shapes, and these new LED bulbs are even more efficient than CFLs.

Rating efficiency

So what makes one bulb more efficient than another? All three kinds of light bulbs convert electrical energy into light energy, and bulb efficiency is a measure of how much light can be produced using a particular amount of electricity. Scientists measure the perceived brightness of a light source with a unit called the lumen. The brighter an object appears, the more lumens it is rated to have. A 60-watt incandescent light bulb puts out about 1,000 lumens. This same amount of light can be produced by a CFL that only uses 18 watts. Lumens per watt (lumens/watt) are what really matter in light bulb efficiency; it's the measure of how much light is produced per watt used. After all, the electric company is selling you kilowatt-hours. The more watts (or kilowatts) you use, the more you will pay.

Producing the same amount of light using fewer watts means the light bulb is more efficient. The classic incandescent bulb produces 1,000 lumens/ 60 watts, or about 17 lumens/watt. The CFL produces 1,000 lumens/ 18 watts, or about 56 lumens/watt. That means you could get the same amount of light for less than one-third the amount of electricity—a huge savings. No wonder people used CFLs in great numbers to light their homes.

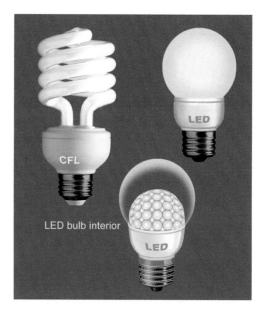

LED bulb interior

So, how much more efficient are the LEDs? The same 1,000 lumens of light can be generated with an LED bulb that only uses 6 watts. That's about 167 lumens/watt, more than three times the efficiency of the CFL!

Saving energy, saving money

What kind of savings does this mean in terms of electricity? Let's look at one day and one month of electricity usage at a cost of $0.15 per kilowatt-hour, and one light bulb used for five hours each day. How much does one light bulb cost to light?

	Incandescent	CFL	LED
Watts of equal light	60 W	18 W	6 W
Hours used	5	5	5
Watt hours used per day	300 Wh	90 Wh	30 Wh
Watt hours used per month (30 days per month)	9,000 Wh or 9 kWh	2,700 Wh or 2.7 kWh	900 Wh or 0.9 kWh
Cost of use per month at $0.15 per kWh	$1.35	$0.41	$0.14
Cost of use per year at $0.15 per kWh	**$16.20**	**$4.86**	**$1.62**

And that's just for one bulb! CFLs save a lot of energy compared to incandescent bulbs, but LEDs are by far the most efficient. Think about how many bulbs are in your house. Now consider there are approximately 100,000,000 households in the United States alone. Imagine how much energy we could save if every household swapped just one incandescent bulb for an LED bulb!

The efficiency of the LED is not the only benefit it offers. Because the LED is a silicon chip, it can't break like its glass bulb rivals the CFL and the incandescent bulb. It also has amazing powers of endurance. The typical lifespan of an incandescent bulb is 1,500 hours. That seems pretty long, but when compared to the CFL's lifespan of 10,000 hours, it's very short. But at the top is the LED at 60,000 to 100,000 hours of use.

Not only will you spend less money to power the LED bulb, it will last much longer, saving even more in replacement costs. Because LED bulb technology is relatively new, the cost for each bulb now is much greater than incandescent and CFL bulbs. However, the savings over the life of the LED are tremendous. As more bulbs are produced, the cost per bulb will undoubtedly decrease, much like the initial high cost of CFLs has decreased over the last few years. For these reasons, you are seeing the LED bulb pop up in new places all around you.

Traffic signals, scoreboards, and electronic billboards all use LED technology.

Questions:

1. How much more efficient is the 6 W LED bulb compared to the 60 W incandescent bulb?

2. How much energy could the United States save per day if all households switched one 60 W incandescent bulb that is lit for five hours a night to a 6 W LED bulb?

3. Assuming you pay $0.15/kWh for electricity, your household uses 500 kWh per month, and 50 percent of the electricity you use each month is from lighting, how much money would you save each month if you switched all the bulbs in use from incandescent to LED? How much would you save over one year?

Photos of traffic and billboard signs courtesy of David Pinsent

Chapter 21 *Assessment*

Vocabulary

Select the correct term to complete each sentence.

alternating current	kilowatt	series circuit
circuit breaker	kilowatt-hour	short circuit
direct current	Kirchhoff's current law	transformer
electrical power	Kirchhoff's voltage law	voltage drop
fuse	parallel circuit	watt
GFI outlet		

Section 21.1

1. In a(n) _____, there is one path for current and the value for current is the same everywhere.

2. _____ states that the sum of the voltage drops in a circuit must equal the battery voltage.

3. The _____ is the difference in voltage across an electrical device that has current flowing through it.

Section 21.2

4. _____ states that all current entering a branch point in a circuit must exit.

5. In a(n) _____ , there is more than one path or branch for current, and the voltage is the same everywhere.

6. A(n) _____ is an automatic device that trips like a switch to turn off an overloaded circuit.

7. A(n) _____ is used in car electrical systems, televisions, and electrical meters to protect the circuits from current overload.

8. A branch in a circuit with zero or very low resistance is a(n) _____ .

Section 21.3

9. Use a(n) _____ near water sources in the kitchen or bathroom for electrical safety.

10. A device that converts high voltages to lower voltages is a(n) _____ .

11. The unit of power that equals 1 joule per second is the _____ .

12. Electrical current supplied by a battery is called _____ .

13. Electric utility companies charge by the _____ .

14. _____ is the rate at which electrical energy is converted to other forms of energy.

15. Electrical appliances in your home use _____ .

16. A(n) _____ is equal to 1,000 watts.

Concepts

Section 21.1

1. Why is current the same everywhere in a series circuit?

2. Draw a circuit diagram for a circuit containing a battery and two bulbs in series.

3. As more bulbs are added to a series circuit, what happens to the resistance of the circuit? What happens to the current? What happens to the brightness of the bulbs?

4. Explain what is meant by a voltage drop.

5. How is Kirchhoff's voltage law useful for analyzing series circuits?

Section 21.2

6. A parallel circuit contains two bulbs in parallel. Why do the bulbs have the same voltage?

7. Draw the circuit diagram for a circuit containing two bulbs in parallel.

8. List two advantages of parallel circuits over series circuits.

9. What happens to the total resistance of a parallel circuit as more branches are added? Why?

10. What is a short circuit, and why can it be dangerous?

Section 21.3

11. Explain how to calculate the power of an appliance.

12. What is the difference between alternating current and direct current?

13. What is the definition of a kilowatt-hour?

14. What is the purpose of the AC adapter on the end of the cord used for mobile phones?

Problems

Section 21.1

1. A circuit contains a 5-ohm, a 3-ohm, and an 8-ohm resistor in series. What is the total resistance of the circuit?

2. A circuit contains a 9-volt battery and two identical bulbs. What is the voltage drop across each bulb?

3. A circuit contains a 12-volt battery and two 3-ohm bulbs in series. Draw a circuit diagram and use it to find the current in the circuit and the voltage drop across each bulb.

4. A circuit contains a 12-volt battery and three 1-ohm bulbs in series. Draw the circuit diagram and find the current in the circuit.

5. Calculate the total resistance of each circuit shown below. Then calculate the current in each.

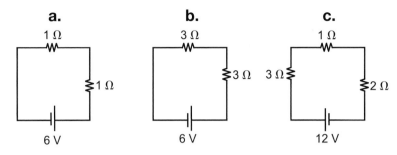

6. A circuit contains two 1-ohm bulbs in series. The current in the circuit is 1.5 amperes. What is the voltage provided by the batteries?

7. A circuit contains two identical resistors in series. The current is 3 amperes, and the batteries have a total voltage of 24 volts. What is the total resistance of the circuit? What is the resistance of each resistor?

Section 21.2

8. Find the amount and direction of the current through point P in each of the circuits shown below.

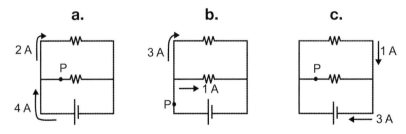

9. A parallel circuit contains a 24-volt battery, a 4-ohm bulb, and a 12-ohm bulb.

   a. Draw the circuit diagram for this circuit.
   b. Calculate the current through each branch.
   c. Calculate the total current in the circuit.
   d. Use Ohm's law to calculate the total resistance of the circuit.

10. Do the following for each of the three circuits shown.

a.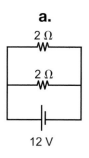

b.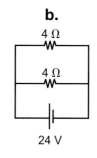

c.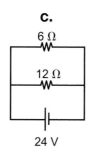

a. Find the voltage across each resistor.
b. Use Ohm's law to find the current through each resistor.
c. Find the total current in the circuit.
d. Find the total resistance of the circuit.

11. Find the unknown quantity in each of the circuits below.

a.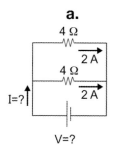

b.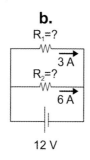

c.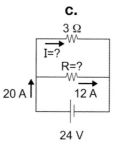

Section 21.3

12. Calculate the power of each of the following appliances when plugged into a 120-volt outlet.

a. an iron that draws 10 A of current
b. a stereo that draws 2 A of current
c. a light bulb that draws 0.5 A of current

13. Calculate the current each of the following appliances draws when plugged into a 120-volt outlet.

a. a 100-watt computer   **c.** a 30-watt radio
b. a 1,200-watt microwave

14. A smart phone requires 1.7 amps of current and has a power of 5 watts. What is the voltage of the battery it uses?

15. A flashlight contains a 6-watt bulb that draws 2 amps of current. How many 1.5-volt batteries does it use?

16. Alex uses a 1,000-watt heater to heat his room.

a. What is the heater's power in kilowatts?
b. How many kilowatt-hours of electricity does Alex use if he runs the heater for 8 hours?
c. If the utility company charges $0.15 per kilowatt-hour, how much does it cost to run the heater for 8 hours?

Applying Your Knowledge

Section 21.1

1. You can find series circuits throughout your home and classroom. A wall light switch that turns on a lamp is a series circuit. Look around and see how many examples of series circuits you can find.

Section 21.2

2. Many circuits combine resistors in both series and parallel, like in the circuit diagram below. Apply what you have learned about circuits to answer the questions below.

a. Which bulb(s) remain on when bulb A is removed?
b. Which bulb(s) remain on when bulb B is removed?
c. Which bulb(s) remain on when bulb C is removed?

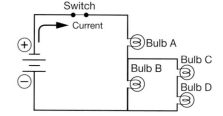

Section 21.3

3. Look at the backs or undersides of appliances in your home. Find the power of three appliances. Calculate the amount of current each draws when plugged into a 120-volt outlet.

Electricity and Magnetism

Electricity and magnetism may not seem very similar. You don't get a shock from picking up a magnet! However, you can create magnetism with electric current in an electromagnet. Why does electric current create magnetism?

In 1819, a teacher named Hans Christian Ørsted tried an experiment in front of his students for the first time. He passed electric current through a wire near a compass. To his surprise, the compass needle moved! A few years later, Michael Faraday built the first electric motor. Today we know electricity and magnetism are two faces of the same basic force: the force between charges. In this chapter, you will see how our knowledge of electricity and magnetism allows us to build both an electric motor and also an electric generator. It would be hard to imagine today's world without either of these important inventions.

As you read this chapter, you will see that our study of the atom, electricity, and magnetism has come full circle! This chapter will help you understand exactly how the electricity that we use in our homes, schools, and offices is generated. It is actually all about magnets! Isn't that amazing?

CHAPTER 22 INVESTIGATIONS

22A: Magnetism
How do magnets and compasses work?

22B: Electromagnets
How are electricity and magnetism related?

22.1 **Properties of Magnets**

Magnetism has fascinated people since the earliest times. We know that magnets stick to refrigerators and pick up paper clips and pins. They are also found in electric motors, computer disk drives, alarm systems, and many other common devices. This chapter explains some of the properties of magnets and magnetic materials. What is the source of Earth's magnetism? How does a compass work? Read on to find out.

██████ **VOCABULARY** ██████

magnetic – describes a material that can respond to forces from magnets

permanent magnet – a material that retains its magnetic properties, can attract or repel other magnets, and can attract magnetic materials

What is a magnet?

Magnets and magnetic materials   Magnets are usually made of the elements iron, cobalt, nickel, or some combination of these, such as steel (a mixture of iron and carbon). A magnet has an invisible force field that can attract or repel other magnets. A **magnetic** material, like a paper clip, can be attracted to a magnet, but is never repelled. Thus, magnetic materials are *affected* by magnets but do not actively create their own magnetic field.

Permanent magnets   A **permanent magnet** is a material that keeps its magnetic properties, even when it is not close to other magnets. Bar magnets, refrigerator magnets, and horseshoe magnets are good examples of permanent magnets.

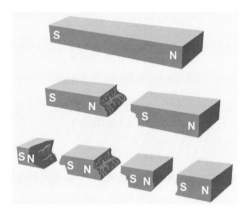

Figure 22.1: *If a magnet is cut in half, each half will have both a north pole and a south pole.*

Bar magnets   **Horseshoe magnet**   **Magnetic materials**

Poles   All magnets have two opposite *magnetic poles*, called the north pole and the south pole. If a magnet is cut in half, each half will have its own north and south poles (Figure 22.1). It is impossible to have only a north or south pole by itself. The north and south poles are like the two sides of a coin. You cannot have a one-sided coin, and you cannot have a north magnetic pole without a south pole.

The magnetic force

Attraction and repulsion When they are near each other, magnets exert forces. Two magnets can either attract or repel. Whether the force between two magnets is attractive or repulsive depends on which poles face each other. If two opposite poles face each other, the magnets attract. If two of the same poles face each other, the magnets repel.

The three interactions between two magnets

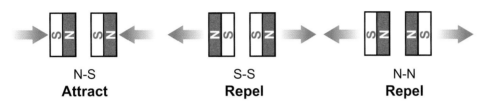

N-S	S-S	N-N
Attract	**Repel**	**Repel**

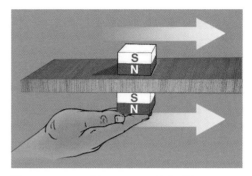

Figure 22.2: *Many materials, like this piece of wood, are transparent to magnetic forces.*

Most materials are transparent to magnetic forces Magnetic forces can pass through many materials with no apparent decrease in strength. For example, one magnet can drag another magnet even when there is a piece of wood between them (Figure 22.2). Plastics, wood, and most insulating materials are transparent to magnetic forces. Conducting metals, such as aluminum, also allow magnetic forces to pass through, but may change the forces. Iron and a few metals near it on the periodic table have strong magnetic properties. Iron and iron-like metals can either block or concentrate magnetic forces, and they are discussed later in this chapter.

Using magnetic forces

Magnetic forces are used for many applications because they are relatively easy to create and can be very strong. There are large magnets that create forces strong enough to lift a car or even a moving train (Figure 22.3). Small magnets are everywhere; for example, some doors are sealed with magnetic weather-stripping that blocks out drafts. There are several patents pending for magnetic zippers, and many handbags, briefcases, and cabinet doors close with magnetic latches. Magnetic repulsion is the principle behind how Magnetic Resonance Imaging (MRI) works. MRI is a process that uses magnetism and radio waves to scan the body for disease or injury.

Figure 22.3: *Powerful magnets are used to lift discarded cars in a junkyard.*

The magnetic field

Describing forces at a distance How does the force from one magnet get to another magnet? Does it happen instantly? How far does the force reach? These questions puzzled scientists for a long time. Eventually, they realized that the force between magnets acts in two steps. First, a magnet fills the space around itself with a kind of potential energy called a **magnetic field**. Then the magnetic field makes forces that act on other nearby magnets (and act on the original magnet, too).

The speed of magnetic forces When you move a magnet, the magnetic field spreads out around the magnet at the speed of light. The speed of light is nearly 300 million meters *per second*. That means the force from one magnet reaches a nearby magnet so fast it *seems* like it happens instantly. However, it actually takes a tiny fraction of a second.

Magnetic forces get weaker with distance The force from a magnet gets weaker as it gets farther away. You can feel this when you hold two magnets close together, then compare the force when you hold them far apart (Figure 22.4). Try this, and you will find that the force loses strength very rapidly with increasing distance. Separating a pair of magnets by twice the distance reduces the force by eight times or more.

The magnetic field 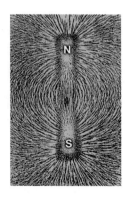 All magnets create a magnetic field in the space around them, and the magnetic field creates forces on other magnets. Imagine you have a small test magnet that you are moving around another magnet (Figure 22.5). The north pole of your test magnet feels a force everywhere in the space around the source magnet. To keep track of the force, imagine drawing an arrow in the direction in which the north pole of your test magnet is pulled or pushed as you move it around the source magnet. The arrows that you draw show you the magnetic field. If you connect all the arrows, you get lines called magnetic field lines. You can actually see the pattern of the magnetic field by sprinkling magnetic iron filings on cardboard with a magnet underneath (shown at left).

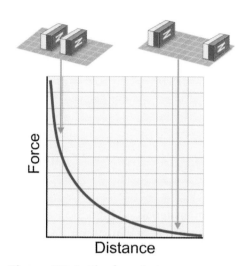

Figure 22.4: *The force between two magnets quickly gets weaker as the magnets are separated.*

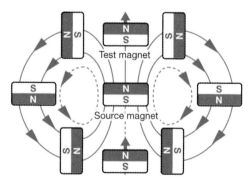

Figure 22.5: *The magnetic field lines show the force exerted by one magnet on the north pole of another magnet.*

Earth's magnetic field and compasses

A compass needle is a magnet
A compass needle is a freely spinning magnet. If you bring the south pole of a permanent magnet near the compass needle, the needle's north pole (identified by a red-painted tip) will spin around and point toward the south pole of the permanent magnet (Figure 22.6). This is because opposite poles attract.

North and south poles
The planet Earth *itself* has a magnetic field that comes from the core of the planet. A compass needle spins around until the north-seeking pole of the needle points toward Earth's North Pole. This action has been helpful to explorers for centuries. But doesn't this contradict Figure 22.6? Yes; it is contradictory to say that the north end of the compass needle points north, when you know that, scientifically, the north pole of the needle is always attracted (and points toward) a south magnetic pole. This is an example of an old naming convention that was decided long before people understood how a compass needle really worked. It is customary to say that the north pole of a compass needle points to Earth's North Pole, but technically, it does this because it is attracted to a south magnetic pole.

Geographic and magnetic poles

Geographic North Pole

Geographic South Pole

The true *geographic* North and South Poles are where the Earth's axis of rotation intersects its surface. The North Pole is the northernmost point on Earth's surface. However, as you can see on the illustration at left, Earth's internal magnetic field poles are actually the opposite of the geographic poles. Now, here is one more interesting point of confusion. Scientists still stick to the old naming convention, and refer to the magnetic pole that is near the geographic North Pole as the magnetic north pole (even though, technically, it's a south pole). Read on to find out *why* we make a distinction between Earth's geographic and magnetic poles.

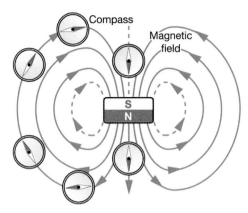

Figure 22.6: *How a compass needle interacts with a magnet. Remember, the compass needle is a magnet itself, and the red-painted end of the needle is a north pole.*

SCIENCE FACT

Some animals have biological compasses

Many organisms, including some species of birds, frogs, fish, turtles, and bacteria, can sense the planet's magnetic field. Migratory birds are the best known examples. Magnetite, a magnetic mineral made of iron oxide, has been found in bacteria and in the brains of birds. Tiny crystals of magnetite may act like compasses and allow these organisms to sense the small magnetic field of Earth. You can find more information about this topic by using your favorite search engine and the keywords "magnetite in birds."

Magnetic declination and "true north"

Magnetic declination   Because Earth's geographic North Pole (true north) and magnetic south pole are not located at the exact same place, a compass will not point *directly* to the geographic North Pole. Depending on where you are, a compass will point slightly east or west of true north. The difference between the direction a compass points and the direction of true north is called **magnetic declination**. Magnetic declination is measured in degrees and is indicated on topographical maps.

Finding true north with a compass   Most good compasses contain an adjustable ring with a degree scale and an arrow that can be turned to point toward the destination on a map (Figure 22.7). The ring is turned the appropriate number of degrees to compensate for the declination. Suppose you are using a compass and the map shown below and you want to travel directly north. You would not simply walk in the direction of the compass needle. To go north, you must walk in a direction 16 degrees west of the way the needle points.

magnetic declination – the difference between true north and the direction a compass points

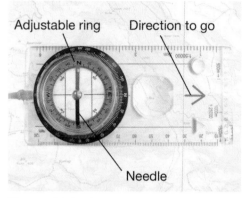

Figure 22.7: *An orienteering compass.*

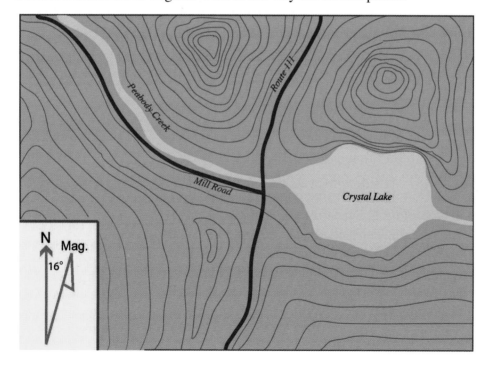

Position of magnetic North Pole

Year	Lat. (°N)	Long. (°W)
2007	83.9	120.7
2008	84.4	127.9
2009	84.7	130.8
2010	85.0	133.2
2011	85.4	138.6

Figure 22.8: *The National Oceanic and Atmospheric Administration reports that magnetic north is currently moving northwest at 55 km per year. The data above starts with the 2007 observed position of the magnetic north pole, and then shows the estimated positions for 2008–2011.*

The source of the Earth's magnetism

Earth's magnetic core While Earth's core is magnetic, we know it is not a solid permanent magnet. Studies of earthquake waves reveal that Earth's core is made of hot, dense, molten iron, nickel, and possibly other metals that slowly circulate around a solid metal core (Figure 22.9). Huge electric currents flowing in the molten iron produce Earth's magnetic field, much like a giant electromagnet.

The strength of Earth's magnetic field The magnetic field of Earth is weak compared to the magnetic field of the ceramic magnets you have in your classroom. For this reason, you cannot trust a compass to point north if any other magnets are close by. The *gauss* is a unit used to measure the strength of a magnetic field. A small ceramic permanent magnet has a field between 300 and 1,000 gauss at its surface. By contrast, Earth's magnetic field averages only about 0.5 gauss at the surface.

Reversing poles Historical data shows that both the strength of Earth's magnetic field and the location of the north and south magnetic poles change over time. Studies of magnetized rocks in Earth's crust provide evidence that the poles have reversed many times over the last tens of millions of years. The reversal has happened every 500,000 years on average. The last field reversal occurred roughly 750,000 years ago, so Earth is overdue for a pole reversal.

The next reversal Earth's magnetic field is currently losing approximately seven percent of its strength every 100 years. We do not know whether this trend will continue, but if it does, the magnetic poles could reverse sometime in the next 2,000 years. During a reversal, Earth's magnetic field would not completely disappear. However, the main magnetic field that we use for navigation would be replaced by several smaller fields with poles in different locations.

Movements of the magnetic poles The location of Earth's magnetic poles is always changing—slowly—even between full reversals. Currently, the magnetic north pole is located about 1,000 kilometers from the geographic North Pole. During the last century, the magnetic north pole has moved more than 1,000 km (Figure 22.10). Just remember—if you are using a handheld compass, and the red tip of the compass needle lines up with the north direction on the compass housing, you must adjust the compass to compensate for the fact that Earth's magnetic north and geographic north are in different places!

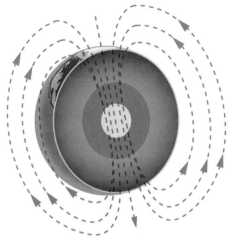

Figure 22.9: *Scientists believe moving charges in the molten core create Earth's magnetic field.*

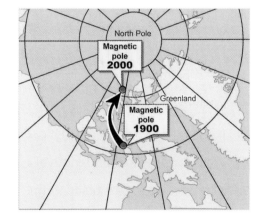

Figure 22.10: *Between 1900 and 2000, the location of the magnetic north pole moved approximately northwest at about 40 km per year, according to the Canadian Geological Survey.*

Using a compass Suppose you want to find a window sill in your house that faces east and would provide good light for growing African violets. Here's how to use a handheld compass to find an east-facing window.

- Turn the moveable ring so east is lined up with the direction arrow, as in Figure 22.11.

- Walk up to a window in your house.

- With the compass flat on your hand, turn until the red end of the compass needle is lined up with north on the compass housing.

- The direction arrow now points directly east. Is this pointing toward the window? If not, keep checking different windows in your house until you find one that faces east, in the direction of the arrow on the compass base.

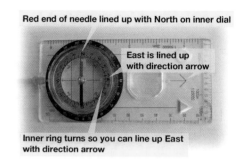

Red end of needle lined up with North on inner dial

East is lined up with direction arrow

Inner ring turns so you can line up East with direction arrow

Figure 22.11: *Using a compass.*

Section 22.1 *Review*

1. What is a magnetic material able to do?

2. Suppose you put a magnet on a refrigerator door. Is the magnet a magnetic material or a permanent magnet? Is the door a magnetic material, or is it a permanent magnet? Explain.

3. Describe three common uses of magnets.

4. What happens to a magnet if it is cut in half?

5. Is it possible to have a magnetic south pole without a north pole? Explain.

6. What happens to the strength of a magnetic field as you move away from a magnet?

7. Why does a compass point north?

8. How does the strength of Earth's magnetic field compare to the strength of the field of an average permanent magnet?

9. What is the cause of Earth's magnetism?

10. Is Earth's magnetic north pole at the same location as the geographic North Pole? Explain.

CHALLENGE

Antigravity magnets!

You can "float" a tethered magnet by attracting it to another magnet that has been glued to the bottom of a shelf or table. See if you can do it! How far apart can you get the two magnets before the lower one falls?

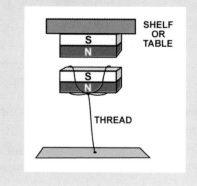

SHELF OR TABLE

THREAD

22.2 Electromagnets

In the last section, you learned about permanent magnets and magnetism. There is another type of magnet, one that is created by electric current. This type of magnet is called an electromagnet. What is an electromagnet? Why do magnets and electromagnets act the same way? In this section, you'll learn about electromagnets and how they helped scientists explain how magnetism works.

What is an electromagnet?

Searching for a connection For a long time, people thought about electricity and magnetism as different and unrelated topics. Around the beginning of the 19th century, scientists started to suspect that the two were related. As scientists began to understand electricity better, they searched for relationships between electricity and magnetism.

The principle of an electromagnet In 1819, Hans Christian Ørsted, a Danish physicist and chemist, noticed that a current in a wire caused a compass needle to deflect. He had discovered that moving electric charges create a magnetic field! A dedicated teacher, he made this discovery while teaching his students at the University of Copenhagen. He suspected that there might be an effect and did the experiment for the very first time in front of his class. With his discovery, Ørsted was the first to identify the principle of an electromagnet.

How to make an electromagnet **Electromagnets** are magnets that are created when there is electric current flowing in a wire. The simplest electromagnet uses a coil of wire, often wrapped around a piece of iron (Figure 22.12). Because iron is magnetic, it concentrates the magnetic field created by the current in the coil.

The north and south poles of an electromagnet The north and south poles of an electromagnet are located at the ends of the coil (Figure 22.12). Which end is the north pole depends on the direction of the electric current. If you curl the fingers of your right hand in the direction of the current, your thumb will point toward the magnet's north pole. This method of finding the magnetic poles is called the *right-hand rule*. You can switch the north and south poles of an electromagnet by reversing the direction of the current. This is a great advantage over permanent magnets. You can't easily change the poles of a permanent magnet.

VOCABULARY

electromagnet – a magnet created by a wire carrying electric current

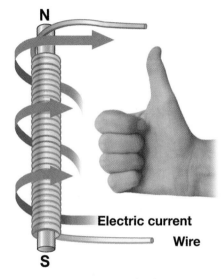

Figure 22.12: *A simple electromagnet uses a coil of wire, often wrapped around a piece of iron or steel. If you curl the fingers of your right hand in the direction of the current, your thumb will point toward the north pole of the electromagnet.*

Applications of electromagnets STEM

Current controls electromagnets By changing the amount of current, you can easily change the strength of an electromagnet or even turn its magnetism on and off. Electromagnets can also be much stronger than permanent magnets because the electric current can be large. For these reasons, electromagnets are preferable in many applications.

Magnetically levitated trains Magnetically levitated (maglev) train technology uses electromagnetic force to lift a train a few inches above its track (Figure 22.13). By "floating" the train on a powerful magnetic field, the friction between wheels and rails is eliminated. Maglev trains can achieve high speeds using less power than normal trains. In 2015, a manned seven-car maglev train in Japan reached a world-record speed of 603 kilometers (375 miles) per hour. Maglev trains are now being developed and tested in Germany, Japan, and the United States.

Electromagnets and toasters The sliding switch on a toaster does several things. First, it turns the heating circuit on. Second, it activates an electromagnet that then attracts a spring-loaded metal tray to the bottom of the toaster (Figure 22.14). When a timing device signals that the bread has been toasting long enough, current to the electromagnet is cut off. This releases the spring-loaded tray that then pushes up on the bread so that it pops out of the toaster.

Electromagnets and doorbells

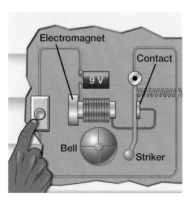

A doorbell contains an electromagnet. When the button of the doorbell is pushed, current is sent through the electromagnet. The electromagnet attracts a piece of metal called the striker. The striker moves toward the electromagnet but hits a bell that is in the way. The movement of the striker away from the contact breaks the circuit after it hits the bell. A spring pulls the striker back and reconnects the circuit. If a finger is still pressing on the button, the cycle starts over again and the bell keeps ringing.

Figure 22.13: *A maglev train track has electromagnets in it that both lift the train and pull it forward.*

Figure 22.14: *A toaster tray is pulled down by an electromagnet while bread is toasting. When the toast is done, current is cut off and the tray pops up. The cutaway shows the heating element— nichrome wires wrapped around a sheet of mica.*

Building an electromagnet

Making an electromagnet from wire and a nail
You can easily build an electromagnet from wire and a piece of iron, such as a nail. Wrap the wire in many turns around the nail and connect a battery as shown in Figure 22.15. When current flows in the wire, the nail becomes a magnet. Use the right-hand rule to figure out which end of the nail is the north pole and which is the south pole. To reverse north and south, reverse the connection to the battery, making the current flow the opposite way.

Increase the strength of an electromagnet
You might expect that more current would make an electromagnet stronger. You would be right, but there are, in fact, two ways to increase the current.

1. You can apply more voltage by adding a second battery.

2. You can add more turns of wire around the nail.

Why adding turns works
The second method works because the magnetism in your electromagnet comes from the *total* amount of current flowing *around* the nail. If there is 1 amp of current in the wire, each loop of wire adds 1 amp to the total amount that flows around the nail. Ten loops of 1 amp each make 10 total amps flowing around. By adding more turns, you use the same current over and over to get stronger magnetism (Figure 22.16).

More turns also means more resistance
Of course, nothing comes for free. By adding more turns, you also increase the resistance of your coil. Increasing the resistance makes the current a little lower and generates more heat. A good electromagnet has enough turns to get a strong enough magnet without too much resistance.

Factors affecting the force
The magnetic force exerted by a simple electromagnet depends on three factors:

1. the amount of electric current in the wire;

2. the amount of iron or steel in the electromagnet's core; and

3. the number of turns in the coil.

In more sophisticated electromagnets, the shape, size, material in the core, and winding pattern of the coil also have an effect on the strength of the magnetic field produced.

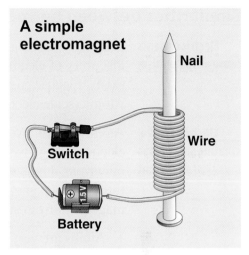

Figure 22.15: *Making an electromagnet from a nail and wire.*

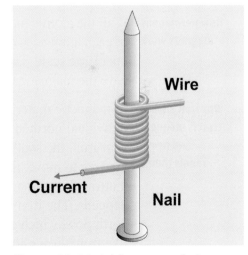

Figure 22.16: *Adding turns of wire increases the total current flowing around the electromagnet. The total current in all the turns is what determines the strength of the electromagnet.*

Similarities between permanent magnets and electromagnets

Atom

Electric currents cause all magnetism   Why do permanent magnets and electromagnets act the same way? The discovery of electromagnets helped scientists to determine why magnetism exists. Electric current through loops of wire creates an electromagnet. Atomic-scale electric currents create a permanent magnet.

Electrons move, creating small loops of current   Atoms contain two types of charged particles, protons (positive) and electrons (negative). The charged electrons in atoms behave like small loops of current. These small loops of current mean that atoms themselves act like tiny electromagnets with north and south poles! We don't usually see the magnetism from atoms for two reasons.

1. Atoms are very tiny and the magnetism from a single atom is far too small to detect without very sensitive instruments.

2. The alignment of the atomic north and south poles changes from one atom to the next. On average, the atomic magnets cancel each other out (Figure 22.17).

How permanent magnets work   If all the atomic magnets are lined up in a similar direction, the magnetism of each atom adds to that of its neighbors and we observe magnetic properties on a large scale. This is what makes a permanent magnet. Permanent magnets have the magnetic fields of individual atoms aligned in a similar direction.

Why iron always attracts magnets and never repels them?   In magnetic materials (like iron), the atoms are free to rotate and align their individual north and south poles. If you bring the north pole of a magnet near iron, the south poles of all the iron atoms are attracted. Because they are free to move, the iron near your magnet becomes a south pole and it attracts your magnet. If you bring a south pole near iron, the opposite happens. The iron atoms nearest your magnet align themselves to make a north pole, which also attracts your magnet. This is why magnetic materials like iron always attract your magnet, and never repel, regardless of whether your test magnet approaches with its north or south pole.

Nonmagnetic materials   The atoms in nonmagnetic materials, like plastic, are not free to move and change their magnetic orientation. This is why most objects are not affected by magnets.

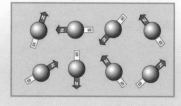

Nonmagnetic material

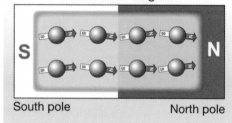

Permanent magnet

South pole   North pole

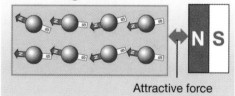

Magnetic iron

Attractive force

Figure 22.17: *Atoms act like tiny magnets. Permanent magnets have their atoms partially aligned, creating the magnetic forces we observe. The magnetic properties of iron occur because iron atoms can easily adjust their orientation in response to an outside magnetic field.*

Section 22.2 *Review*

1. Which of the following will *not* increase the strength of an electromagnet made by wrapping a wire around an iron nail?

   a. increasing the number of turns of the wire

   b. increasing the current in the electromagnet

   c. removing the nail from the center of the electromagnet

2. Explain why an electromagnet usually has a core of iron or steel.

3. Name two devices that use electromagnets. Explain the purpose of the electromagnet in each device.

4. In your own words, explain how atoms give rise to magnetic properties in certain materials.

5. Which picture shows the correct location of the north and south poles of the electromagnet? Choose A or B and explain how you arrived at your choice.

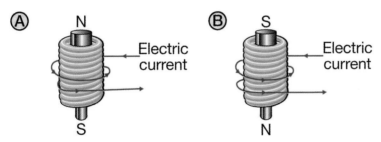

6. The north pole of a magnet is brought near a refrigerator door, and the magnet sticks. If the magnet is removed and the south pole is brought near the door instead, will it also stick? Explain.

7. What would happen if you placed a compass near an electromagnet when there is an electric current in the coil of the electromagnet? Why would this happen? What if you flipped the electromagnet around so the end that was closest to the compass is now farthest away?

22.3 **Electric Motors and Generators**

Permanent magnets and electromagnets work together to make electric motors and generators. In this section, you will learn about how an electric motor works. The secret is in the ability of an electromagnet to reverse its north and south poles. By changing the direction of electric current, the electromagnet attracts and repels other magnets in the motor, causing the motor to spin. **Electric motors** convert electrical energy into mechanical energy.

Using magnets to spin a disk

Imagine a spinning disk with magnets Imagine you have a disk that can spin on an axis at its center. Around the edge of the disk are several magnets. You have cleverly arranged the magnets so they have alternating north and south poles facing out. Figure 22.18 shows a picture of your disk.

Making the disk spin Imagine you also have another magnet that is not attached to the disk. You bring this loose magnet close to the disk's edge. The loose magnet attracts one of the magnets on the disk while at the same time repelling an adjacent magnet on the disk. These attract-and-repel forces make the disk spin a little way around (Figure 22.18).

Reversing the magnet is the key To keep the disk spinning, you need to reverse the magnet in your fingers as soon as the magnet that was attracted passes by. This way, you first attract the magnet on the disk, and then reverse the loose magnet to repel that magnet and attract the next one in line on the disk. You make the disk spin by using the loose magnet to alternately attract and repel the magnets on the disk.

Knowing when to reverse the magnet The disk is called a **rotor** because it can rotate. The key to making the rotor spin smoothly is to reverse your magnet when the disk is at just the right place. You want the reversal to happen just as each magnet in the rotor passes by. If you reverse too early, you will repel the magnet on the rotor backward before it reaches the loose magnet. If you reverse too late, you will attract the magnet backward after it has passed. For the best results, you need to change your magnet from north to south just as each magnet on the rotor passes by.

Using a magnet to spin a rotor

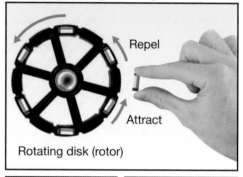

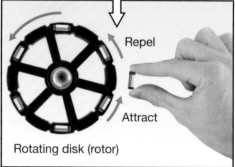

Figure 22.18: *Using a single magnet to spin a disk of magnets. Reversing the magnet in your fingers attracts and repels the magnets in the rotor, making it spin.*

How the electromagnets in a motor operate

How electromagnets are used in electric motors

In a working electric motor, an electromagnet replaces the magnet you reversed with your fingers. The switch from north to south is done by reversing the electric current in the electromagnet. The sketch below shows how an electromagnet switches its poles to make the rotor keep turning.

commutator – the device that switches the direction of electrical current in the electromagnet of an electric motor

First, the electromagnet repels magnet A and attracts magnet B.

Then, the electromagnet switches so it repels magnet B and attracts magnet C.

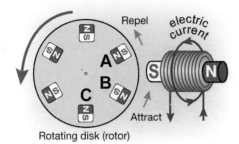

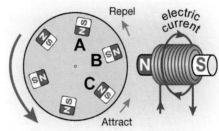

The main parts of an electric motor

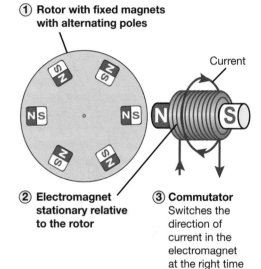

① **Rotor with fixed magnets with alternating poles**

Current

② **Electromagnet stationary relative to the rotor**

③ **Commutator** Switches the direction of current in the electromagnet at the right time

Figure 22.19: *An electric motor has three main parts.*

The commutator is a kind of switch

Just as with the magnet you flipped, the electromagnet must switch from north to south as each rotor magnet passes by to keep the rotor turning. The device that makes this happen is called a **commutator**. As the rotor spins, the commutator reverses the direction of the current in the electromagnet. This makes the electromagnet's side facing the disk change from north to south, and then back again. The electromagnet attracts and repels the magnets in the rotor, and the motor turns.

What do you need to make a motor?

All electric motors use the same idea: a rotor with magnets on it is driven by one or more magnets that are stationary relative to the motor. In the example on this page, the rotor has permanent magnets mounted on it with alternating poles. The electromagnet is stationary, and its magnetic field is reversed back and forth by the commutator switching the direction of its current. The electromagnet and the commutator work together to make the magnetic forces that drive the rotor. In any electric motor, electric energy is transformed into mechanical energy (or rotational energy) by magnetic fields.

How a battery-powered electric motor works

Inside a small electric motor   If you take apart an electric motor that runs on batteries, it doesn't look like the spinning disk motor illustrated on the previous page. However, the same three mechanisms are still there. The difference is in the arrangement of the electromagnets and permanent magnets. The illustration below shows a small battery-powered electric motor and what it looks like inside with one end of the motor case removed. The permanent magnets are on the outside, and they stay fixed in place.

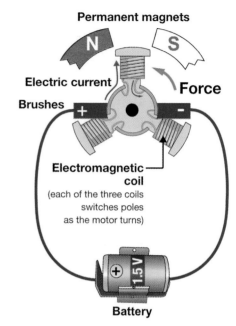

Figure 22.20: *A simple battery-powered motor has three electromagnets.*

Electromagnets and the armature   The electromagnets are in the rotor, and they turn. The rotating part of the motor, including the electromagnets, is called the *armature*. The armature in the illustration above has three electromagnets, corresponding to the three coils you see in Figure 22.20.

How the switching happens   The wires from each of the three coils are attached to three metal plates (the commutator) at the end of the armature. As the rotor spins, the three plates come into contact with positive and negative *brushes*. Electric current flows through the brushes into the coils. As the motor turns, the plates rotate past the brushes, reversing the positive and negative connections to the coils. As you know, when you change the direction of current through a coil, the electromagnet's magnetic poles switch positions. The turning electromagnets with alternating poles are thus attracted and repelled by the permanent magnets, and the motor turns.

AC motors   Motors that run on AC electricity are easier to make because the current switches direction all by itself. Almost all household, industrial, and power tool motors are AC motors. These motors use electromagnets for both the rotating and fixed magnets.

Electromagnetic induction

Motors and generators
Motors transform electrical energy into mechanical energy. Electric **generators** do the opposite. They transform mechanical energy into electrical energy. Generators are used to create the electricity that powers all of the appliances in your home.

Magnetism and electricity
An electric current in a wire creates a magnetic field. The reverse is also true. If you move a magnet near a coil of wire, an electric current (or voltage) is *induced* in the coil. The word *induce* means "to cause to happen." The process of using a moving magnet to create electric current is called **electromagnetic induction**. A moving magnet *induces* electric current to flow in a circuit.

Symmetry in physics
Many processes of physics display *symmetry*. In physics, symmetry means that a process works in both directions. Earlier in this chapter, you learned that moving electric charges create magnetism. The symmetry is that changing magnetic fields also cause electric charges to move. Many physical processes display symmetry in one form or another.

Making current flow
Figure 22.21 shows an experiment demonstrating electromagnetic induction. In the experiment, a magnet can move in and out of a coil of wire. The coil is attached to a meter that measures the electric current. When the magnet moves into the coil of wire, *as the magnet is moving,* electric current is induced in the coil and the meter swings to the left. The current stops if the magnet stops moving.

Reversing the current
When the magnet is pulled back out again, *as the magnet is moving,* current is induced in the opposite direction. The meter swings to the right as the magnet moves out. Again, if the magnet stops moving, the current also stops.

Current flows only when the magnet is moving
Current is produced only if the magnet is moving, because a *changing* magnetic field is what creates current. Moving magnets induce current because they create changing magnetic fields. If the magnetic field is not changing, such as when the magnet is stationary, the current is zero.

VOCABULARY

generator – a device that converts kinetic energy into electrical energy through induction

electromagnetic induction – the process of using a moving magnet to create a current

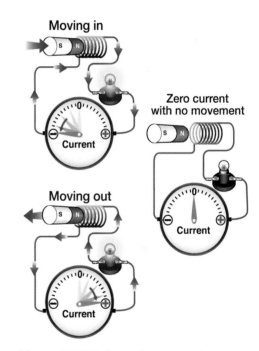

Figure 22.21: *A moving magnet produces a current in a coil of wire.*

Generating electricity

A simple generator A generator converts mechanical energy into electrical energy through magnetic induction. Most large generators use some form of rotating coil in a magnetic field (Figure 22.22). You can also make a generator by rotating magnets past a stationary coil (see the diagram below). As the disk rotates, first a north pole and then a south pole pass the coil. When a north pole is approaching, the current is in one direction. After the north pole passes and a south pole approaches, the current is in the other direction. As long as the disk is spinning, there is a changing magnetic field through the coil and electric current is created.

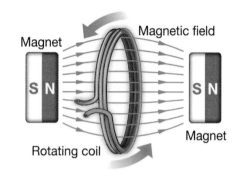

Figure 22.22: *Current is created when a coil rotates in a magnetic field.*

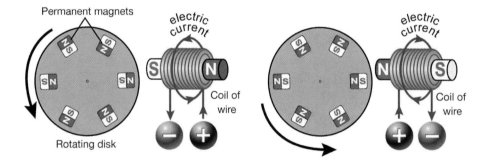

Alternating current The generator shown above makes AC electricity. The direction of current is one way when the magnetic field is becoming "more north" and the opposite way when the field is becoming "less north." It is impossible to make a situation where the magnetic field keeps increasing (becoming more north) forever. Eventually, the field must stop increasing and start decreasing. Therefore, the current always alternates. The electricity in your home is produced by AC generators.

Energy for generators The electrical energy produced by a generator must have a source. Energy must continually be supplied to keep the rotating coil (or magnetic disk) turning. In a hydroelectric generator, falling water turns a *turbine,* which spins the generator and generates electricity. Windmills can generate electricity in a similar way. Other power plants use gas, oil, or coal to heat steam to high pressures. The steam then spins turbines that convert the chemical energy stored in the fuels into electrical energy (Figure 22.23).

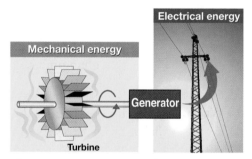

Figure 22.23: *A power plant generator contains a turbine that turns magnets inside loops of wire, generating electricity. Some other form of energy must be continually supplied to turn the turbine.*

Section 22.3 *Review*

1. A(n) _____ is used to convert mechanical energy into electrical energy.

2. A(n)_____ is used to convert electrical energy into mechanical energy.

3. Using a magnet to create electric current in a wire is called _____.

4. Why is it necessary to use at least one electromagnet in a motor instead of only permanent magnets?

5. At a certain instant, the electromagnet in the motor shown below has its north pole facing the rotor that holds the permanent magnets. In which direction is the rotor spinning?

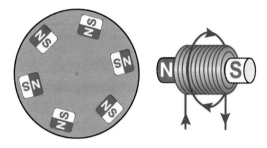

6. The rotor in the motor below is spinning clockwise. Is the direction of the current in the electromagnet from A to B or from B to A?

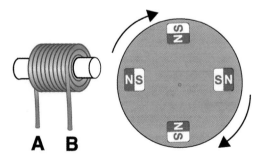

A B

7. In most electric power plants, the energy stored in gas, coal, oil, or nuclear energy is transformed into the movement of a turning turbine. Why is the turning turbine necessary in a power plant?

TECHNOLOGY

Electrical wiring

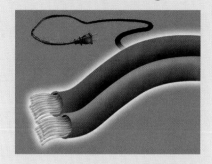

STEM — There is a magnetic field around all wires that carry current, but you don't notice magnetic fields created by electrical wiring in your house. Why not?

Well, the wires in your home are actually made of two parallel wires. If you look at an appliance wire, you will notice the two wires inside the plastic covering. At any instant, the current in one wire is opposite the other. Each creates a magnetic field, but the fields are in opposite directions so they mostly cancel each other out.

Because the wires are not at exactly the same location, and field strength depends on distance, the fields do not completely cancel each other right at the wire, but quickly fall off to nothing a short distance away.

A Walk on the Wild Side
Under the Sea!

Have you ever wondered what it would be like to take a "walk" on the sea floor? Engineer Jim Varnum gets to go "down under" regularly, because he is a pilot of an extremely complex robot called Jason II. *This robot dives deep into the ocean and is used to take pictures and collect data from the sea floor.* Jason II *is owned by Woods Hole Oceanographic Institution, in Woods Hole, Massachusetts.*

Although Jim resides in Seattle, Washington, he works wherever the unmanned submersible is needed for exploration. Scientists from all over the world use *Jason II* to explore the sea floor, but they need the expertise of pilots and engineers like Jim to "drive" the machine.

Just getting the submersible from the ship's deck and into the water can be a feat of its own! The 9,000-pound machine is lifted with a huge crane and settled into the water, and then it takes the plunge down to the sea floor.

Jim at the control station

Jim assists in monitoring the device as it sinks to the bottom of the ocean. This may take several hours and engineers take turns viewing the descent on a television screen. If a problem arises, they can attend to it immediately.

Jason II requires a lot of energy to power its lights, thrusters, hydraulics, and other systems. Power is produced by the shipboard generators (recall the power equation from your text).

However, there is a series of steps needed to make sure the correct flow of electricity is transferred to the submersible. First, the shipboard generator output is applied to a transformer, which increases the voltage to a level that is favorable to transmission along the 10-kilometer (6-mile) cable that connects *Jason II* to the ship. Another transformer drops the voltage to a level useful to *Jason II*, as the power reaches the submersible.

Why all this increase and decrease in voltage? It has to do with the fact that electricity is being transmitted over a very long distance, with a potential loss of energy (in the form of heat) in the cable. Copper conductors have low resistance to the flow of electricity, but with such a long cable, the resistance becomes significant. In fact, not only would excessive heating and destruction of the cable occur, but also a huge amount of energy would be lost. If the voltage is increased at the shipboard generator, then the required power could be obtained using a lower current, which results in less power and heat loss in the cable.

Jason II *takes the plunge*

The arrow is pointing to the fiber optics/insulated cable that sends immediate information to the scientists.

A typical fissure in the sea floor

Jason II's robotic arm taking a sample

Once *Jason II* is on the sea floor, Jim works diligently to maneuver the sub around pillars, over caverns, atop ridges, and through fissures. He sits in a room on the ship and uses a box with a joystick and levers to manipulate the robot in different directions. When he has to retrieve something from the sea floor, such as a rock or biology sample, he uses a robotic arm with a claw hand to accomplish the task. At times, he must take a temperature reading of a hydrothermal vent using a temperature probe.

Jason II collects a tremendous amount of data. Jim explains that getting the data from the sea floor to the ship is also interesting. As Jim says, "While we can do an incredible amount of physical work with *Jason II*, it's the system's ability to collect and transmit large amounts of sensor and video data that make it appealing to the scientific community as a tool for discovery. The physical work ends when the job ends, but data will be studied by scientists, in their labs, for years."

One half of the electromagnet on the wrist (red arrow)

As mentioned, the wrist is attached to a robotic arm that is used to perform data sample collection while on the sea floor. External instruments that utilize electrical and magnetic properties to provide measurements and information needed by scientists are abundant on the sub.

Jim enjoys his job as a *Jason II* pilot even though he could be away from home for as much as eight weeks at a time! It's all good though, because there are other times when he is home for several months in succession, spending time writing software or creating new instruments to attach to *Jason* for a special experiment.

On deck, Jason II picks up and manipulates a temperature probe.

Jim and his fellow engineers understand how each part of the *Jason II* robot works and are able to repair any electrical and mechanical problem as it arises. He and other engineers are responsible for fixing and maintaining all parts of the robot when the research cruise is complete. Some of the pieces of the robot, like the arms, include miniature gears and small electrical wires that can be broken when the moveable wrist is used. Each piece is carefully cleaned and analyzed by the engineering team.

Jim (right) and another pilot fix the wrist

Remember Ohm's law?

Questions:

1. If *Jason II* needs 15 kW of power at 300 volts to operate, what amperage of current must flow through the cable?

2. If the vehicle voltage is increased to 3,000 volts, and the power still needed is still 15 kW, what happens to the amperage? ($I = P/V$)

3. If the resistance of the cable is 60 ohms (Ω), at 300 volts, what amperage will flow through the cable? ($I = V/R$) What will the power be if you use these values? Is it enough power for *Jason II*?

Go to "www.WHOI.edu" for more information on Woods Hole Oceanographic Institution. Source: University of Washington. Control station photo courtesy of University of Washington. Jason II photos courtesy of Woods Hole Oceanographic Institution – Atlantis 15-17, and Laura Preston.

Chapter 22 *Assessment*

Vocabulary

Select the correct term to complete each sentence.

commutator	generator	magnetic field
electric motor	magnetic	permanent magnet
electromagnet	magnetic declination	rotor
electromagnetic induction		

Section 22.1

1. A(n) _____ material can create or respond to forces from magnets.

2. The region around a magnet is filled with a(n) _____ .

3. The difference between the direction a compass points and the direction of true north is called _____ .

4. A bar magnet, a refrigerator magnet, or a horseshoe magnet is a good example of a(n) _____ .

Section 22.2

5. A simple _____ uses a coil of wire, often wrapped around a piece of iron or steel.

Section 22.3

6. The process by which a moving magnet creates voltage and current in a loop of wire is called _____ .

7. A device that uses electromagnetic induction to make electricity is called a(n) _____ .

8. A(n) _____ is a device that converts electrical energy into mechanical energy.

9. A(n) _____ is the rotating disk of an electric motor.

10. A(n) _____ can switch the direction of electrical current in the electromagnet of an electric motor.

Concepts

Section 22.1

1. Name a metal that has strong magnetic properties.

2. Describe the types of forces that magnetic poles exert on each other.

3. Earth's magnetic north pole is:
   a. aligned with the north star.
   b. near Earth's geographic North Pole.
   c. near Earth's geographic South Pole.
   d. at the equator.

Section 22.2

4. If you reverse the direction of electrical current in an electromagnet, what happens to the electromagnet?

5. What are three ways you can increase the strength of an electromagnet?

6. Explain why an electromagnet usually has a core of iron or steel.

7. Only a few materials show magnetic properties because:
   a. their atomic magnets must line up with Earth's geographic South and North Poles, and this is rare.
   b. they contain a rare substance.
   c. their atomic magnets are much stronger than the atoms of other materials.
   d. we see magnetic properties only if atomic magnets line up in the same direction throughout a material.

8. Name two examples of machines that use electromagnets. Explain the purpose of the electromagnet in each machine.

9. Plastic and wood are not magnetic materials. Explain, in terms of their atoms, why they are not magnetic.

Section 22.3

10. What are the three key parts of any electric motor?

11. You can say that the battery used to power a DC motor is not directly responsible for making the rotor spin. What, then, is the battery directly responsible for? What actually causes the rotor to spin?

12. What is the purpose of a commutator in an electric motor?

13. A bar magnet is suspended so it is free to rotate. When you hold a second bar magnet near the suspended magnet, the suspended magnet begins to rotate. What do you have to do to keep the suspended magnet rotating?

Problems

Section 22.1

1. A student places two magnets with their north poles facing each other, about 50.0 cm apart. When she moves one magnet toward the other, the first magnet repels the second at a distance of 26.0 cm. She repeats the procedure, but now places the magnets so the south pole of one faces the north pole of the other (see below).

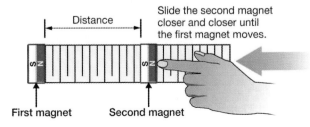

Distance

Slide the second magnet closer and closer until the first magnet moves.

First magnet   Second magnet

a. What is she likely to observe?

b. Next, she put one of the magnets on her wooden desk with the north pole down. If the desk's top is 2.5 cm thick, do you think she could move the magnet by placing another magnet under the desk? Explain.

2. The graph below shows the force between two magnets when they are at different distances from each other.

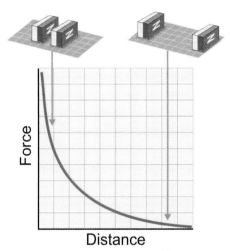

Force

Distance

a. What does this graph show about the force between magnets that are very close together?

b. What should you do to two magnets to decrease the force between them?

Section 22.2

3. The atoms of a permanent magnet can't move, and the electrons in the atoms are lined up so that a magnetic field is created around the magnet. The atoms in iron or steel can move. Describe what you think happens to the atoms of a steel paper clip when the paper clip is near a permanent magnet.

4. Draw an electromagnet. Label all parts including the magnetic poles.

Section 22.3

5. Which of the following things does a working electric motor need to have?

a. a device to switch the electromagnets at the right time

b. a moving element with magnets

c. an even number of magnets

d. an electromagnet

6. The picture shows a current-carrying coil. Toward which point is the magnetic field inside the coil directed?

a. A

b. B

c. C

d. D

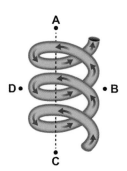

7. The diagram below represents the rotor of an electric motor. To cause the rotor to turn in a counterclockwise direction, the north pole of a magnet should be placed at which position (A, B, C, or D)?

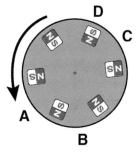

Applying Your Knowledge

Section 22.1

1. Would a magnetic screwdriver be useful? Why or why not?

2. Neodymium magnets are very strong. Do a keyword Internet search to find the answers to the following questions.

a. What materials does this type of magnet contain?

b. What are two uses for neodymium magnets?

c. Why should someone use extreme caution when using these magnets?

Section 22.2

3. Suppose you are walking in a wooded park and you want to use a handheld compass to walk directly west from your current position. Describe, using numbered steps, exactly how you would use the compass to direct you.

4. Why does Earth act like a giant magnet?

Section 22.3

5. In 1996, NASA scientists worked with Italian scientists to carry out an interesting experiment. They made a special satellite and connected it to the space shuttle with more than 20 kilometers of a special insulated copper cable. As the shuttle orbited Earth, scientists released the tethered satellite and conducted 12 different experiments while dragging the cable through Earth's magnetic field at speeds over 15,000 mph!

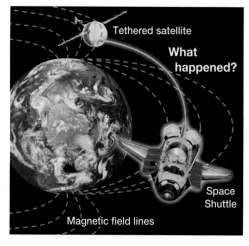

Tethered satellite

What happened?

Space Shuttle

Magnetic field lines

The satellite was equipped with many instruments to study the effects on the special copper cable. Based on your understanding of electromagnetic induction, what do you suppose happened to the copper cable?

6. Speakers and microphones use electromagnets to turn electric currents into sound, and vice versa. Research how electromagnets are used in sound systems. Draw a diagram that shows the location of permanent magnets and electromagnets in a speaker. How does the electromagnet produce vibrations that create sound?

7. A bicycle light generator is a device that you place on the wheel of your bike. When you turn the wheel, the generator powers a light. When you stop, the light goes out. Explain how you think the bike generator makes electricity.

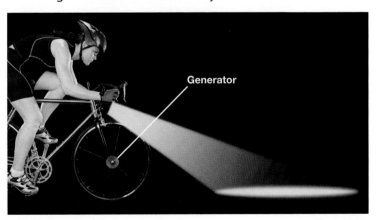

Generator

8. A clever inventor claims that he invented an electric car that makes its own electricity and never needs gas or recharging. The inventor claims that as the car moves, the wind created by its motion spins a propeller that turns a generator to make electricity and power the wheels. Do you believe the car can work? Why or why not? (*Hint*: Think about conservation of energy.)

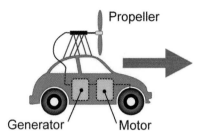

Propeller

Generator   Motor

9. Some electric toothbrushes contain rechargeable batteries that are charged by placing the toothbrush on a plastic charging base. Both the bottom of the toothbrush and the base are encased in plastic, so there is no connection between the circuits in the toothbrush and the base. How do you think the battery in the toothbrush gets charged?

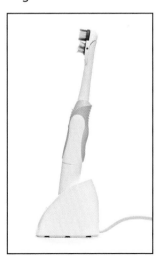

Unit 8

Waves, Sound, and Light

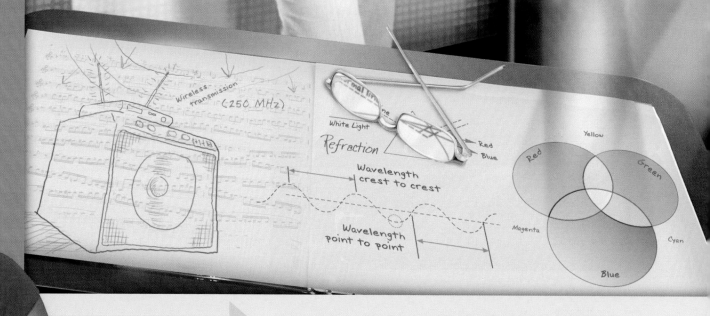

Try this at home ➡ What affects the pitch of a sound? Find a rectangular container, like a baking pan, plastic food storage container, or a shoebox. You will also need a few large rubber bands of different sizes and a pencil or long handled wooden spoon. Place the rubber bands around the container lengthwise. Pluck each rubber band and listen to the sound it makes. Insert the pencil or spoon so it runs widthwise under the rubber bands. Now pluck the rubber bands; do they make higher or lower pitched sounds? Why? Finally, try moving the stick up and down the length of the box. Does the position of the stick make a difference in the sound you get? Why do you think so?

Waves

The word **waves** might cause you to think of many things. Does going to a beach pop into your head? At the beach, you can enjoy the warm sunshine and swimming in water waves. Water waves come at regular intervals and seem to move up and down even as they move toward the shore. In this chapter, you will learn about water waves. You will also learn that you are surrounded by waves! Light and sound are waves. Our electronic devices depend on the transmission of waves. Whether a wave is occurring in water or in the air, it follows certain rules that you will learn about in this chapter. For example, wave-like motion is a type of harmonic, or repetitive, motion. A swing, a rocking chair, and all waves exhibit this kind of motion. Harmonic motion also includes motion that goes around and around, such as a Ferris wheel turning or Earth orbiting the Sun. In this sense, anywhere you go, it's possible to "catch" a wave!

CHAPTER 23 INVESTIGATIONS

23A: Harmonic Motion
How do we describe the back-and-forth motion of a pendulum?

23B: Natural Frequency and Resonance
What is resonance and why is it important?

23.1 **Harmonic Motion**

When you travel from one place to another—either on foot, by bicycle, or by car—you use linear motion. **Linear motion** goes from one place to another without repeating (Figure 23.1A). This chapter is about another kind of motion. **Harmonic motion** is motion that repeats over and over (Figure 23.1B). For example, our four seasons are caused by Earth's harmonic motion. Other types of harmonic motion cause your heartbeat and create sounds.

Motion in cycles

What is a cycle? To describe harmonic motion, we need to learn how to describe a repeating action or motion. A **cycle** is one unit of harmonic motion. This motion can be back-and-forth or a full revolution or rotation. One full swing of a child on a swing is one cycle. As the child continues to swing, the back-and-forth motion or cycle repeats over and over again.

Looking at one cycle A **pendulum** is a device that swings back and forth. We can use a pendulum to better understand a cycle. Each box in the diagram below is a snapshot of the motion at a different time in one cycle.

The cycle of a pendulum

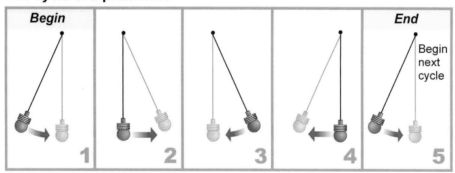

The cycle of a pendulum The cycle starts with (1) the swing from left to center. Next, the cycle continues with (2) center to right, and (3) back from right to center. The cycle ends when the pendulum moves (4) from center to left because this brings the pendulum back to (5) the beginning of the next cycle. Once a cycle is completed, the next cycle begins without any interruption in the motion.

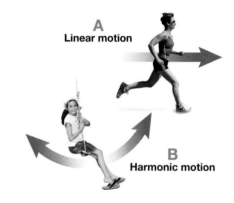

Figure 23.1: *(A) A sprinter is a good example of linear motion. (B) A person on a swing is a good example of harmonic motion.*

Oscillators

What is an oscillator? An **oscillator** is a physical system that has repeating cycles (harmonic motion). A child on a swing is an oscillator, as is a vibrating guitar string. A wagon rolling down a hill is not an oscillator. Which properties determine whether a system will oscillate or not?

Equilibrium Systems that oscillate move back and forth around a center or *equilibrium* position. You can think of equilibrium as the system at rest, undisturbed, with zero net force. A wagon rolling down a hill *is not* in equilibrium because the force of gravity that causes it to accelerate is not balanced by another force. A child sitting *motionless* on a swing *is* in equilibrium because the force of gravity is balanced by the tension in the ropes.

Restoring forces A **restoring force** is any force that always acts to pull a system back toward equilibrium. Restoring force is related to the force of gravity or weight and the lift force (or tension) of the string of a pendulum (Figure 23.2). If a pendulum is pulled forward or backward, gravity creates a restoring force that pulls it toward equilibrium. *Systems with restoring forces become oscillators.*

Inertia causes an oscillator to go past equilibrium The motion of an oscillator is the result of the interaction between a restoring force and inertia. For example, the restoring force pulls a pendulum toward equilibrium. But because of Newton's first law, the pendulum does not just stop at equilibrium. According to the first law, an object in motion tends to stay in motion. The pendulum has inertia that keeps it moving forward so it overshoots its equilibrium position every time.

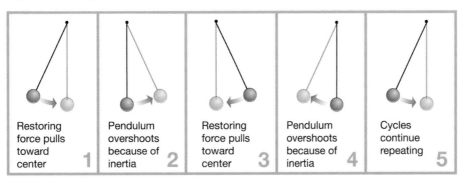

| 1 Restoring force pulls toward center | 2 Pendulum overshoots because of inertia | 3 Restoring force pulls toward center | 4 Pendulum overshoots because of inertia | 5 Cycles continue repeating |

oscillator – a physical system that has repeating cycles

restoring force – any force that always acts to pull a system back toward equilibrium

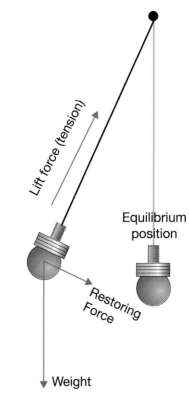

Figure 23.2: *Restoring force keeps a pendulum swinging. Restoring force is related to weight and the lift force (or tension) of the string of a pendulum.*

Frequency and period

A period is the time to complete one cycle

Harmonic motion can be fast or slow, but we don't use speed to tell the difference. This is because the speed of a pendulum constantly changes during its cycle. We use the terms *period* and *frequency* to describe how quickly cycles repeat themselves. The time it takes for one cycle to occur is called a **period**. A clock pendulum with a period of one second will complete one full back-and-forth swing each second.

Frequency is the number of cycles per second

The **frequency** is the number of complete cycles per second. The unit of one cycle per second is called a **hertz (Hz)**. Something that completes 10 cycles each second has a frequency of 10 Hz. A guitar string playing the note A vibrates back and forth at a frequency of 220 Hz (Figure 23.3). Your heartbeat has a frequency between one-half and two cycles per second (0.5 Hz–2 Hz).

Frequency is the inverse of period

Frequency and period are inversely related. The period is the number of seconds per cycle. The frequency is the number of cycles per second. For example, if the period of a pendulum is 2 seconds, its frequency is 0.5 cycles per second (0.5 Hz).

> **PERIOD AND FREQUENCY**
>
> $$\text{Period (s) } T = \frac{1}{f} \text{ Frequency (Hz)}$$
>
> $$\text{Frequency (Hz) } f = \frac{1}{T} \text{ Period (s)}$$

When to use period or frequency

While both period and frequency tell us the same information, we usually use period when cycles are slower than a few per second. A simple pendulum has a period between 0.9 and 2 seconds. We use frequency when cycles repeat faster. For example, the vibrations that make sound in musical instruments have frequencies between 20 and 20,000 Hz.

A guitar's "A" string vibrates at 220 Hz

Figure 23.3: *All musical instruments use harmonic motion to create sound.*

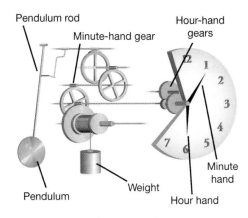 **Solving Problems: Frequency and Period**

The period of an oscillator is 2 minutes. What is the frequency of this oscillator in hertz?

1. **Looking for:** You are asked for the frequency in hertz.

2. **Given:** You are given the period in minutes.

3. **Relationships:** Convert minutes to seconds using the conversion factor 60 seconds/1 minute; use the formula:

$$f = \frac{1}{T}$$

4. **Solution:** 2 minutes × 60 seconds/1 minute = 120 seconds

The period (T) is 120 seconds.

$f = 1/120$ s = 0.008 Hz

Your turn...

a. Every 5 seconds, a pendulum completes one cycle. What are the period and frequency of this pendulum?

b. The period of an oscillator is 1 minute. What is the frequency of this oscillator in hertz?

c. How often would you push someone on a swing to create a frequency of 0.4 hertz?

d. Figure 23.4 shows the parts of a pendulum clock. The minute hand moves 1/60 of a turn after 30 cycles. What is the period and frequency of this pendulum?

e. A Ferris wheel spins 5 times in 10 minutes. Calculate the period and frequency of the Ferris wheel.

Figure 23.4: *The parts of a pendulum clock.*

SOLVE FIRST LOOK LATER

a. The period is 5 seconds, and the frequency is 0.2 Hz.

b. The frequency is 0.02 Hz.

c. You would need to push once every 2.5 seconds.

d. There are 30 cycles/second, so the frequency is 30 Hz. The period is 0.03 second.

e. The frequency is 0.008 Hz. The period is 125 seconds.

Amplitude

Amplitude describes the "size" of a cycle
The "size" of a cycle is called **amplitude**. Figure 23.5 shows a pendulum with a small amplitude and one with a large amplitude. With a moving object like a pendulum, the amplitude is often a distance or angle. With other kinds of oscillators, the amplitude might be voltage or pressure. The amplitude of an oscillator is measured in units appropriate to the kind of harmonic motion being described.

How do you measure amplitude?
The amplitude is measured as the maximum distance the oscillator moves away from its equilibrium position. For the pendulum in Figure 23.6, the amplitude is 20 degrees because the pendulum moves 20 degrees away from the equilibrium position in either direction. The amplitude can also be found by measuring the distance between the farthest points the motion reaches. The amplitude is half this distance. The amplitude of a water wave is often found this way.

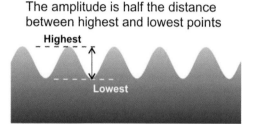

The amplitude is half the distance between highest and lowest points

Highest

Lowest

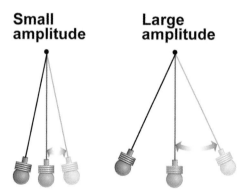

Figure 23.5: *Small amplitude versus large amplitude.*

Damping and friction
Look at the illustration below. Friction slows a pendulum down, just as it slows all motion. That means the amplitude gets reduced until the pendulum is hanging straight down, motionless. We use the word *damping* to describe the gradual loss of amplitude. If you wanted to make a clock with a pendulum, you would have to find a way to keep adding energy to counteract the damping of friction so the clock's pendulum would work continuously.

Damping is the gradual loss of amplitude

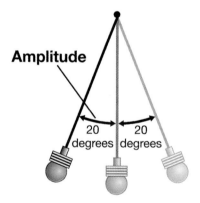

Figure 23.6: *A pendulum with an amplitude of 20 degrees swings 20 degrees away from the center in either direction.*

Graphs of harmonic motion

Graphing harmonic motion   It is easy to recognize cycles on a graph of harmonic motion. Figure 23.7 illustrates the difference between a graph of linear motion and a graph of harmonic motion. The most common type of harmonic motion graph places time on the horizontal (x) axis and position on the vertical (y) axis. The graph below shows how the position of a pendulum changes over time. The repeating "wave" on the graph represents the repeating cycles of motion of the pendulum.

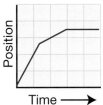

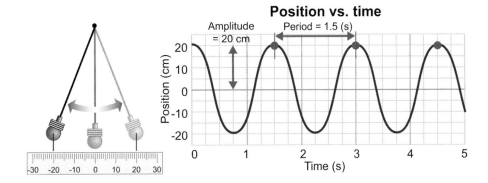

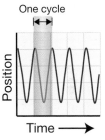

Figure 23.7: *A harmonic motion graph shows repeating cycles.*

Finding the period   In the graph above, the pattern repeats every 1.5 seconds. This repeating pattern represents the period of the pendulum, which is 1.5 seconds. If you were to cut out any piece of the graph and slide it left or right 1.5 seconds it would line up exactly.

Using positive and negative positions   Harmonic motion graphs often use positive and negative values to represent motion on either side of a center (equilibrium) position. Zero usually represents the equilibrium point. Notice that zero is placed halfway up the y-axis so there is room for both positive and negative values. This graph is in centimeters, but the motion of the pendulum could also have been graphed using the angle measured relative to the center (straight down) position.

Showing amplitude on a graph   The amplitude of harmonic motion can also be seen on a graph. The graph above shows that the pendulum swings back and forth from +20 centimeters to –20 centimeters. The equilibrium position is represented as the zero line. Therefore, the amplitude of the pendulum is 20 centimeters.

SOLVE IT!

Measuring amplitude

Use a protractor to find the amplitude (in degrees) of the pendulum in the graphic below.

23.1 Harmonic Motion   **559**

Natural frequency and resonance

Natural frequency An oscillator will have the same period and frequency each time you set it moving. This phenomenon is called **natural frequency**, the frequency at which a system naturally oscillates. Musical instruments use natural frequency. For example, guitar strings are tuned by adjusting their natural frequency to match musical notes (Figure 23.8).

Changing natural frequency The natural frequency of an oscillator changes according to its length. In the case of a vibrating guitar string, you can shorten the string to increase the force pulling the string back toward equilibrium. Higher force means higher acceleration so the natural frequency is higher and the period is shorter. Lengthening an oscillator results in a lower frequency and a longer period.

How mass affects oscillators For oscillators with side-to-side movement, increasing the mass means the oscillator moves slower and the period gets longer. This is because of Newton's second law of motion—as mass increases, the acceleration decreases proportionally. However, for a pendulum, changing the mass does *not* affect its period (also because of Newton's second law). The restoring force on a pendulum is created by gravity. Like in free fall, if you add mass to a pendulum, the added inertia is exactly equal to the added force from gravity. The acceleration is the same and therefore the period stays the same.

Periodic force and resonance 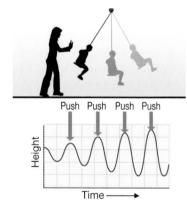 A force that is repeated over and over is called a **periodic force**. A periodic force supplies energy to an oscillator and has a cycle with an amplitude, frequency and period. **Resonance** happens when a periodic force has the same frequency as the natural frequency. For example, small pushes (a periodic force) to someone on a swing add together if they are applied at the right time (once each cycle). In time, the amplitude of the motion grows and can become very large compared to the strength of the force!

natural frequency – the frequency at which a system oscillates when disturbed

periodic force – a repetitive force

resonance – an exceptionally large amplitude that develops when a periodic force is applied at the natural frequency

Figure 23.8: *This guitarist is tuning his guitar by adjusting the natural frequency of the strings to match particular musical notes.*

Section 23.1 *Review*

1. Which is the best example of a cycle: a turn of a wheel or a slide down a ski slope?

2. Describe one example of an oscillating system you would find at an amusement park.

3. What is the relationship between period and frequency?

4. Every 10 seconds, a pendulum completes 2 cycles. What are the period and frequency of this pendulum?

5. What is the difference between a graph of linear motion and a graph of harmonic motion?

6. A graph of the motion of a pendulum shows that it swings from +5 centimeters to −5 centimeters for each cycle. What is the amplitude of the pendulum?

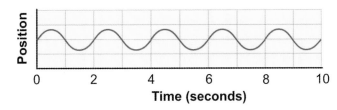

7. What is the period of the oscillation shown in the diagram above?

8. Figure 23.9 shows a sliding mass on a spring. Assume there is no friction. Will this system oscillate? Explain why or why not.

9. Which pendulum in Figure 23.10 will have the longer period? Justify your answer.

10. Why does mass *not* affect the period of a pendulum?

11. Resonance happens when:
    a. a periodic force is applied at the natural frequency.
    b. an oscillator has more than one natural frequency.
    c. a force is periodic and not constant.
    d. the amplitude of an oscillator grows large over time.

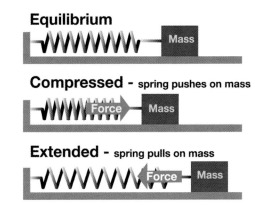

Figure 23.9: *Question 8.*

Figure 23.10: *Question 9.*

23.2 Properties of Waves

A **wave** is an oscillation that travels from one place to another. A musician's instrument creates waves of sound that move through the air to your ears. When you throw a stone into a pond, the energy of the falling stone creates waves in the water that carry energy to the edge of the pond. You are familiar with waves, but what are they exactly?

What is a wave?

Defining a wave If you poke a floating ball, it oscillates up and down. But something also happens to the water as the ball oscillates. The surface of the water oscillates in response and the oscillation spreads outward from where it started. An oscillation that travels is a wave.

Why do waves travel? When you drop a ball into water, some of the water is pushed aside and up by the ball (A). The higher water pushes the water next to it (B). The water that has been pushed then pushes on the water next to *it*, and so on. The waves spread or *propagate* through the connection between each drop of water and the water next to it (C).

 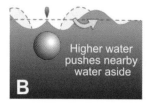 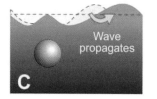

A — Water is displaced and pushed up

B — Higher water pushes nearby water aside

C — Wave propagates

Energy and information Waves are a traveling form of energy because they can cause changes in the objects they encounter. Waves also carry information, such as sound, pictures, or even numbers. Waves are used in many technologies because they quickly carry information over great distances. All the information you receive in your eyes and ears comes from waves. Figure 23.11 illustrates some of the many types of waves in our environment.

VOCABULARY

wave – a traveling oscillation that has properties of frequency, wavelength, and amplitude

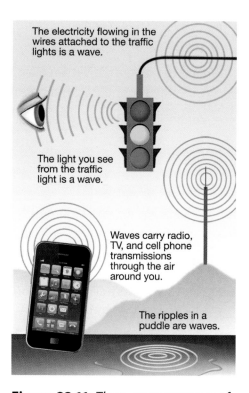

The electricity flowing in the wires attached to the traffic lights is a wave.

The light you see from the traffic light is a wave.

Waves carry radio, TV, and cell phone transmissions through the air around you.

The ripples in a puddle are waves.

Figure 23.11: *There are many types of waves in our environment.*

Frequency, amplitude, and wavelength

Waves are oscillators   Like all oscillators, waves have cycles, frequency, and amplitude. The frequency of a wave is a measure of how often it goes up and down at any one place (Figure 23.12). The frequency of one point on the wave is the frequency of the whole wave. Distant points on the wave oscillate up and down *with the same frequency*. A wave carries its frequency to every place it reaches. Like other frequencies, the frequency of a wave is measured in *hertz* (Hz). A wave with a frequency of one hertz (1 Hz) causes everything it touches to oscillate at one cycle per second.

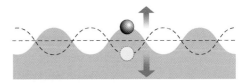

Figure 23.12: *The frequency of a wave is the rate at which every point on the wave moves up and down.*

Wavelength   You can think of a wave as a moving series of high points and low points. A *crest* is the high point of the wave, a *trough* is the low point. **Wavelength** is the distance from any point on a wave to the same point on the next cycle of the wave (Figure 23.13). The distance between one crest and the next crest is a wavelength. So is the distance between one trough and the next trough. We use the Greek letter "lambda" for wavelength. A lambda (λ) looks like an upside-down *y*.

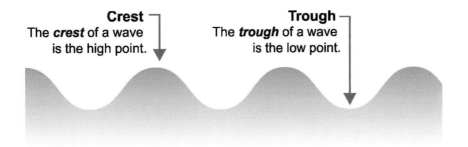

Crest
The **crest** of a wave is the high point.

Trough
The **trough** of a wave is the low point.

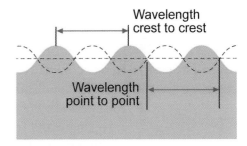

Wavelength crest to crest

Wavelength point to point

Figure 23.13: *The wavelength can be measured from crest to crest. This is the same as the distance from one point on a wave to the same point on the next cycle of the wave.*

Amplitude   You have learned that the amplitude of an oscillator—such as a wave—is measured as the maximum distance it moves away from its equilibrium position. For a wave, equilibrium is the average, or resting, position. You can measure amplitude as one-half the distance between the crest and the trough of a wave.

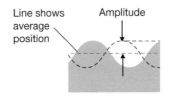

Line shows average position

Amplitude

The speed of waves

Waves spread Wave motion is due to the spreading of the wave from where it begins. For a water wave, the water itself stays in the same average place. Therefore, to gauge the speed of a wave, you measure how fast the wave spreads, *not* how fast the water surface moves up and down.

Measuring wave speed The graphic below shows what happens in water when you begin a wave in one location. You can measure the speed of this spreading wave by timing how long it takes the wave to affect a place some distance away. The speed of the wave is how fast the wave spreads, *not* how fast the water surface moves up and down. The speed of a typical water wave is about 1 m/s. Light waves are extremely fast—300,000 km/s (or 186,000 mi/s). Sound waves travel at about 1,000 km/hr (or 660 mph).

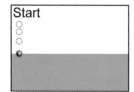 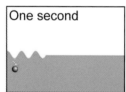 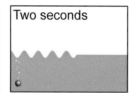 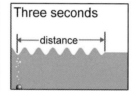

Speed is frequency times wavelength In one complete cycle, a wave moves one wavelength (Figure 23.14). The speed is the distance traveled (one wavelength) divided by the time it takes (one period). We can also calculate the speed of a wave by multiplying wavelength and frequency. This is mathematically the same because multiplying by frequency is the same as dividing by period. These formulas work for all kinds of waves, including water waves, sound waves, light waves, and even earthquake waves!

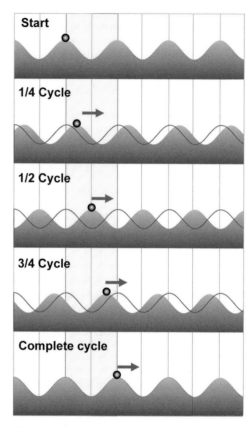

Figure 23.14: *A wave moves one wavelength in each cycle.*

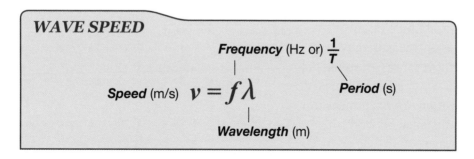

WAVE SPEED

Frequency (Hz or) $\frac{1}{T}$

Speed (m/s) $v = f\lambda$ Period (s)

Wavelength (m)

> **Remember these relationships...**
> period = T
> frequency = $^1/_T$
> Speed = wavelength ÷ period
> Speed = frequency × wavelength

Solving Problems: Wave Speed

The wavelength of a wave on a string is 1 meter, and its speed is 5 m/s. Calculate the frequency and the period of the wave.

1.	**Looking for:**	You are asked to find the frequency (f) and period (T) of a wave.
2.	**Given:**	You know the wavelength of the wave is 1 meter and its speed is 5 m/s.
3.	**Relationships:**	The formulas you know include:

$$speed = frequency \times wavelength$$

$$f = \frac{1}{T} \text{ and } T = \frac{1}{f}$$

4.	**Solution:**	Solve for frequency.

frequency = speed ÷ wavelength

frequency = 5 m/s ÷ 1 m = 5 Hz

Then solve for period.

$$period = \frac{1}{f} = \frac{1}{5 \text{ Hz}} = 0.20 \text{ s}$$

The frequency of the wave is 5 Hz, and the period is 0.20 second.

Your turn...

a. The wavelength of a wave is 0.5 meter, and its period is 2 seconds. What is the speed of this wave?

b. The wavelength of a wave is 100 meters, and its frequency is 25 hertz. What is the speed of this wave? What is its period?

c. If the period of a wave is 15 seconds, how many wavelengths pass a certain point in 2 minutes?

Section 23.2 *Review*

1. A wave and a pendulum are both oscillators. Why isn't a pendulum a wave?

2. Make a list of three types of waves that you encountered today.

3. The distance from the crest of a wave to the next crest is 10 centimeters. The distance from a crest of this wave to a trough is 4 centimeters.

   a. What is the amplitude of this wave?

   b. What is the wavelength of this wave?

4. The wavelength of the wave shown in this harmonic motion graph is about:

   a. 1.2 meters

   b. 2.5 meters

   c. 5.0 meters

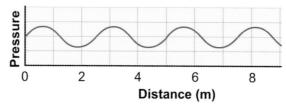

5. Which is the fastest way to send information, using sound waves, light waves, or water waves?

6. Is a wave that travels slower than 50 m/s most likely to be a sound wave, a light wave, or a water wave?

7. How far does a wave travel in three cycles?

8. How is the formula for finding the speed of a wave like the formula for finding the speed of a person running a race?

9. You are given the speed of a wave and its period. What kind of information can you also find out about this wave? Justify your answer.

10. You are watching a water wave in a long tank. Describe how you could determine the speed of the wave.

11. What is the speed of a wave that has a wavelength of 0.4 meter and a frequency of 10 hertz?

12. What is the period of a wave that has a wavelength of 1 meter and a speed of 20 m/s?

SCIENCE FACT

Waves and earthquakes

The outer layer of Earth is broken up into huge slabs called "plates." Sometimes a sudden slip happens between two plates and an earthquake occurs. The quake releases powerful seismic waves that travel along the surface and through Earth. Because these waves travel through the planet, they are used to investigate Earth's internal structure. For example, the way that seismic waves refract and reflect within the Earth provided scientists with the clues they needed to prove that Earth has a liquid core.

Investigate!

1. Find out more about seismic waves by doing research at your local library or on the Internet. Write up your findings in one or two paragraphs.

2. Is your city or town ever affected by seismic waves? If so, how do you know?

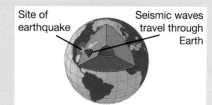

23.3 Wave Motion

Sometimes your car radio fades out. Why? It's because the radio waves are affected by objects. For example, if you drive into a tunnel, some or all of the radio waves get blocked. In this section, you will learn how waves move and discover what happens when they encounter objects or collide with other waves.

When a wave encounters objects

Wave fronts   A **wave front** is the leading edge of a moving wave and is often considered to be a wave crest rather than a trough. You can make waves in all shapes, but plane waves and circular waves are easiest to create and study (Figure 23.15). The crests of a **plane wave** look like parallel lines. The crests of a **circular wave** are circles. A plane wave can be started by disturbing water in a line. A circular wave can be started by disturbing water at a single point.

The direction a wave moves   The shape of the wave front determines the direction the wave moves. Circular waves have circular wave fronts that move outward from the center. Plane waves have straight wave fronts that move in a line perpendicular to the wave fronts.

The four wave interactions   Both circular and plane waves eventually hit surfaces. Four interactions are possible when a wave encounters a surface—reflection, refraction, diffraction, or absorption.

VOCABULARY

wave front – the leading edge of a moving wave

plane wave – moving waves that have crests in parallel straight lines

circular wave – moving waves that have crests that form circles around a single point where the wave began

Plane waves

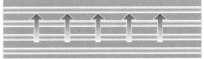

Circular waves

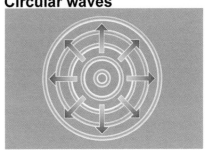

Figure 23.15: *Plane waves move perpendicular to the wave fronts. Circular waves radiate outward from a single point.*

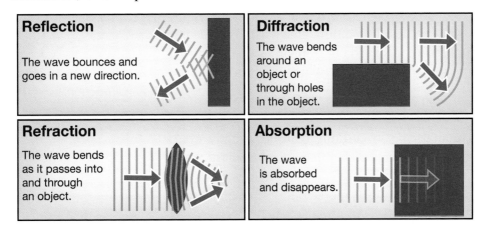

Reflection
The wave bounces and goes in a new direction.

Diffraction
The wave bends around an object or through holes in the object.

Refraction
The wave bends as it passes into and through an object.

Absorption
The wave is absorbed and disappears.

Wave interactions

Boundaries   A *boundary* is an edge or surface where one material meets a different material. The surface of a glass window is a boundary. A wave traveling in the air experiences a sudden change when it encounters the boundary between the air and the glass of a window. Reflection, refraction, and diffraction usually occur at boundaries. Absorption also occurs at a boundary, but happens to a greater extent within the body of a material.

Reflection   When a wave bounces off an object we call it **reflection**. A reflected wave is like the original wave but moving in a new direction. The wavelength and frequency are usually unchanged. An echo is an example of a sound wave reflecting from a distant object or wall. People who design concert halls pay careful attention to the reflection of sound from the walls and ceiling.

Refraction   **Refraction** occurs when a wave bends as it crosses a boundary. We say the wave is *refracted* as it passes through the boundary. The process of refraction of light through eyeglasses helps people see better. The lenses in a pair of glasses bend incoming light waves so that an image is correctly focused within the eye.

Diffraction   The process of a wave bending around a corner or passing through an opening is called **diffraction**. We say a wave is diffracted when it is changed by passing through a hole or around an edge. Diffraction usually changes the direction and shape of the wave. When a plane wave passes through a small hole, diffraction turns it into a circular wave (Figure 23.16). Diffraction explains why you can hear sound through a partially closed door. Diffraction causes the sound wave to spread out from any small opening.

Absorption   **Absorption** is what happens when the amplitude of a wave gets smaller and smaller as it passes through a material. The wave energy is transferred to the absorbing material. A sponge can absorb a water wave while letting the water pass. Theaters often use heavy curtains to absorb sound waves so the audience cannot hear backstage noise. The tinted glass or plastic in the lenses of your sunglasses absorbs some of the energy in light waves. Cutting down the energy of light makes your vision more comfortable on a bright, sunny day so you don't have to squint!

Diffraction through a small opening turns plane waves into circular waves.

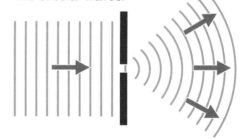

Figure 23.16: *An illustration of diffraction.*

Transverse and longitudinal waves

Wave pulses A wave *pulse* is a short "burst" of a traveling wave. A pulse can be produced with a single up-down movement. The illustrations below show wave pulses in springs. You can see the difference between the two basic kinds of waves—transverse and longitudinal—by observing the motion of a wave pulse.

Transverse waves The oscillations of a **transverse wave** are not in the direction the wave moves. For example, the wave pulse in the illustration below moves from left to right. The oscillation (caused by the boy's hand) is up and down. Water waves are an example of a transverse wave (Figure 23.17, top).

Making a transverse wave pulse

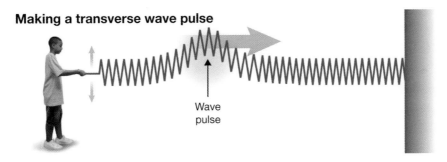

Wave pulse

Transverse waves

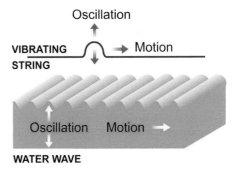

Oscillation

VIBRATING STRING → Motion

Oscillation   Motion →

WATER WAVE

Longitudinal waves The oscillations of a **longitudinal wave** are in the same direction that the wave moves (Figure 23.17, bottom). A sharp push-pull on the end of the spring makes a traveling wave pulse as portions of the spring compress then relax. The direction of the compressions are in the same direction that the wave moves. Sound waves are longitudinal waves.

Making a longitudinal wave pulse

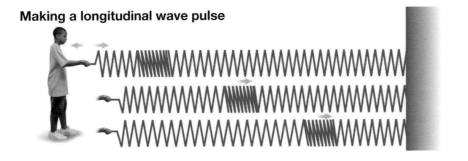

Longitudinal waves

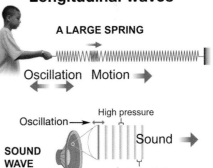

A LARGE SPRING

Oscillation   Motion →

High pressure

Oscillation →

Sound →

SOUND WAVE

Low pressure

Figure 23.17: *Transverse and longitudinal waves.*

Constructive and destructive interference

Wave pulses   If you have a long elastic string attached to a wall, you can make a wave pulse. First you place the free end of the string over the back of a chair. The string should be straight so that each part of it is in a neutral position. To make the pulse, you pull down a short length of the string behind the chair and let go. The pulse then races away from the chair all the way to the wall. You can see the wave pulse move *on* the string. Each section of string experiences the pulse and returns to the neutral position after the wave pulse has moved past it.

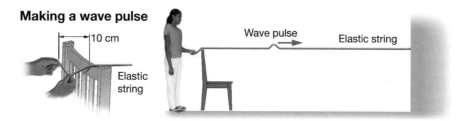

Constructive interference   Suppose you make two wave pulses on a stretched string. One comes from the left and the other comes from the right. When the waves meet, they combine to make a single large pulse. **Constructive interference** happens when waves combine to make a larger amplitude (Figure 23.18).

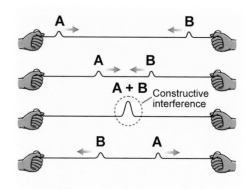

Figure 23.18: *This is an example of constructive interference.*

Destructive interference   There is another way to add two pulses. Sometimes one pulse is on top of the string and the other is on the bottom. When these pulses meet in the middle, they cancel each other out (Figure 23.19). One pulse pulls the string up and the other pulls it down. The result is that the string flattens and both pulses vanish for a moment. In **destructive interference**, waves add up to make a wave with smaller or zero amplitude. After interfering, both wave pulses separate again and travel on their own. This is surprising if you think about it. For a moment, the middle of the cord is flat, but a moment later, two wave pulses come out of the flat part and race away from each other. Waves still store energy, even during destructive interference. Noise-cancelling headphones are based on technology that uses destructive interference.

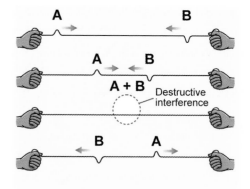

Figure 23.19: *This is an example of destructive interference.*

Section 23.3 *Review*

1. How does the motion of a plane wave differ from the motion of a circular wave?

2. For each of the examples below, identify whether reflection, refraction, diffraction, or absorption is happening.
   a. During a total solar eclipse, the Moon is in front of the Sun but you can still see some sunlight around the edges of the Moon.
   b. The black surface of a parking lot gets hot in the summer when exposed to sunlight.
   c. The image at the right of a straw in a glass looks funny.
   d. When you look in a mirror, you can see yourself.
   e. Sound seems muffled when it is occurring on the other side of a wall.
   f. Light waves bend when they move from water to air.
   g. A ball bounces back when you throw it at a wall.

3. When a wave is being absorbed, what happens to the amplitude of the wave? Use the term energy in your explanation.

4. Compare and contrast transverse waves and longitudinal waves.

5. Two waves combine to make a wave that is larger than either wave by itself. Is this constructive or destructive interference?

6. When constructive interference happens between two sound waves, the sound will get louder. What does this tell you about the relationship between amplitude and volume of sound?

7. One wave on a string is moving toward the right and another is moving toward the left. When, they meet in the middle, half of the cycle of the wave from the right overlaps with half of the cycle of the wave from the left. The result is that the string gets flat when the two waves meet. What happened? What will happen after the waves meet?

▮▮▮ TECHNOLOGY ▮▮▮

Noise-cancelling headphones

The graphic below illustrates how noise-cancelling headphones work. Study the graphic and write a description that explains why noise-cancelling technology is a good way to reduce noise. Verify your description by doing some research about these special headphones.

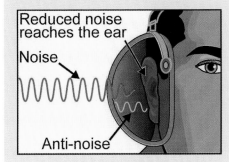

Cell Phones:
How They Work

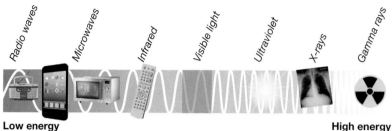

Radio waves | Microwaves | Infrared | Visible light | Ultraviolet | X-rays | Gamma rays

Low energy
Low frequency
Long wavelength

High energy
High frequency
Short wavelength

See if you can solve this puzzle: You dial a friend's number, and she answers on her cell phone. Name a three-word question that you can ask her only because she is on a cell phone. Thirty years ago, people never called someone and asked this question. Why? They already knew the answer!

Give up? The question is: "Where are you?"

Using electromagnetic waves

By now you know that waves of all kinds exist around us and that they are the result of the harmonic motion of an oscillator. You should also be familiar with the kinds of waves that make up the electromagnetic spectrum. Cell phones use electromagnetic waves in the low end of the microwave range of frequencies to send and receive signals. This does not mean that your cell phone can cook your food! The frequencies used by cell phones are lower in frequency and at a much lower power than the electromagnetic waves produced by microwave ovens.

The process that allows a cell phone to communicate is the same as for a radio or walkie-talkie. All of these devices use electromagnetic waves that are within a specific frequency range to send information. Walkie-talkies commonly use frequencies of 400–500 MHz (megahertz, or million hertz). FM radios use frequencies of 88–108 MHz, and cell phones are between 800 and 1,900 MHz.

Transmitting and receiving

Sound is translated into an electromagnetic wave at the desired frequency by a transmitter through a process called encoding. The electromagnetic wave is created by a rapidly changing electric current in a wire. Any device that creates a changing current creates electromagnetic waves. In theory, you can create radio waves by making a simple circuit with a battery, switch, and wire. Quickly flipping the switch would cause the current to flow and then stop in the wire. This process would create very low-energy electromagnetic waves that would sound like crackles of static if you could detect them on a radio!

When a cell phone, radio tower, or walkie-talkie sends out a signal, it travels at the speed of light (300,000 km/s) to the recipient. An antenna detects the wave because it causes electrons to move in the antenna. A tuner sorts through the thousands of electromagnetic waves coming into the antenna to find the correct one. Once the correct signal is detected, the information is taken from the signal, called decoding. An electric current is then sent to the speaker, where it is translated back into a sound wave.

A radio is only able to receive signals, while walkie-talkies and cell phones can both transmit and receive. A walkie-talkie uses only one frequency, so a transmitting and receiving must take place individually. You can't talk and listen at the same time. Cell phones use a more sophisticated type of technology called full-duplex radio. This process uses two separate frequencies at the same time, so a person can transmit on one frequency and receive on a different frequency simultaneously.

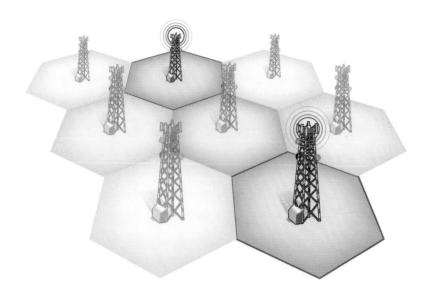

Catching the wave

The distance the electromagnetic wave can travel depends on the power of the signal, which is measured in watts. When taking a long trip in a car, you may have noticed that you can listen to your favorite radio station for an hour or so, and then the signal gradually fades out. A city in another state can use the same frequency as the radio station in your hometown without interference, as long as the cities are farther apart than the reach of their signals.

Many people can receive the same radio broadcast, so all of the radio stations in one city can transmit within a narrow range of frequencies and not interfere with each other. This is not true of cell phones, because each caller needs two frequencies to make his or her call–one to transmit and one to receive. When cell phones began to get popular, people quickly realized that there are not enough cell phone frequencies available for thousands of people in a city to be talking at the same time.

The solution to this problem was to divide regions into small areas called cells. The name "cell phone" comes from this idea. Each cell contains its own tower that sends and receives signals from

the phones located within that cell. Attached to each cell tower is a base station that connects the tower to the telephone system. The size of each cell depends on the population density. In a city, cell towers may be as close as one-half mile apart, while in rural areas with flat terrain, towers can be separated by up to 50 miles.

Dividing a city into small cells means that the same frequency can be reused in different locations, similar to the way a radio station frequency can be reused in a different state. Each cell phone company has a set of hundreds of frequencies used for their customers. The company divides their frequencies into several different groups. Each cell uses frequencies in only one of these groups. The diagram at left shows the arrangement of cells that use four groups of frequencies. A person in the top green cell can be using the same frequency as a person in the bottom green cell without the signals interfering.

But what happens if you make a call and then move from one cell to another? As you travel, your signal is handed off from one cell phone tower to the next. The frequencies your cell phone transmits and receives on can change many times changes without you ever noticing.

Questions:

1. How are cell phones, walkie-talkies, and radios similar? How are they different?

2. How are electromagnetic waves created?

3. Why are cities divided into regions called cells?

Chapter 23 Assessment

Vocabulary

Select the correct terms to complete each sentence.

wave	cycle	diffraction
period	reflection	absorption
refraction	pendulum	linear motion
frequency	constructive interference	harmonic motion
oscillator	amplitude	hertz
resonance	wavelength	wave front
circular wave	restoring force	plane wave
transverse wave	longitudinal wave	natural frequency
periodic force	destructive interference	

Section 23.1

1. This kind of force pulls a system back to equilibrium: _____ .

2. The harmonic motion of a boy on a swing is like the motion of a(n) _____ .

3. A pendulum is a kind of _____ in that it has repeating cycles of motion.

4. The note A in the musical scale has a(n) _____ of 220 Hz.

5. One unit of harmonic motion is called a(n) _____ .

6. The motion of a girl running is called _____ , and the motion of a girl riding a Ferris wheel is called _____ .

7. The formula for _____ is the inverse of the formula for frequency.

8. One _____ equals one cycle per second.

9. When the periodic force matches the natural frequency of an object, the object experiences _____ .

10. To have a high _____ on a swing, your friend needs to push you with a large _____ .

11. When I hit a drum, it will vibrate at its _____ .

Section 23.2

12. A(n) _____ is a travelling oscillation.

13. The distance from one crest to the next is a wave's _____ .

Section 23.3

14. The process of a wave bouncing off a surface is called _____ .

15. _____ is the process of the amplitude of a wave diminishing as it enters another material.

16. The _____ of a plane wave is perpendicular to the direction of motion of this wave.

17. The crests of _____ (s) look like parallel lines.

18. _____ is when waves bend when they enter another material and _____ is when waves bend around an object or outward after exiting a hole.

19. If you disturb water in a single point, _____ (s) will be created.

20. The amplitude of two waves will cancel when _____ occurs.

21. The amplitude of two waves gets larger when _____ occurs.

22. Sound waves are an example of this kind of wave: _____ .

23. Water waves are an example of this kind of wave: _____ .

Concepts

Section 23.1

1. State whether the following are linear or harmonic motions.
   - a. skiing downhill
   - b. riding on a merry-go-round
   - b. hiking uphill
   - c. jumping on a trampoline

2. How is the force of gravity involved in the motion of a pendulum? Use the words *equilibrium* and *restoring force* in your answer.

3. The motion of an oscillator is related to the interaction of what two factors? Describe each of these.

4. If the frequency of a heartbeat is 1 hertz, what is the period of this heartbeat?

5. Describe how you find the amplitude of a pendulum and of a water wave.

6. What information can you learn about the harmonic motion of an object by looking at a graph of its motion?

7. What will happen to the period of a pendulum if you:

   a. increase its mass?
   b. increase its length?
   c. Challenge: increase the amplitude?

Section 23.2

8. Identify how each of the following situations involves waves. Explain each of your answers.

   a. A person is talking to someone on a cell phone.
   b. An earthquake causes the floor of a house to shake.
   c. A person listens to her favorite radio station on the car stereo.
   d. A doctor takes an X-ray to check for broken bones.
   e. You turn on a lamp when you come home in the evening.

9. Arrange the equation relating wave speed, frequency, and wavelength for each of the following scenarios. Let v = wave speed, f = frequency, and l = wavelength.

   a. You know frequency and wavelength. Solve for v.
   b. You know frequency and wave speed. Solve for l.
   c. You know wave speed and wavelength. Solve for f.

10. Write a formula relating the speed of a wave to its period and wavelength.

11. How many wavelengths of a wave pass a point in one second if the frequency of the wave is 4 hertz?

12. For the wave in the diagram, which measurement shows the amplitude? Which measurement shows the wavelength?

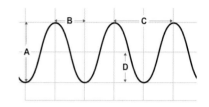

Section 23.3

13. Describe the shape of the light waves that would be created from a single, uncovered light bulb.

14. At the beach, describe where or when you would see wave fronts. How are wave fronts useful to surfers?

15. Below are diagrams representing interactions between waves and boundaries. Identify each interaction by name.

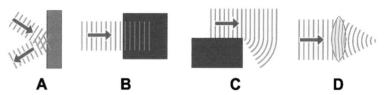

A   **B**   **C**   **D**

16. Read the descriptions below and indicate which of the four types of wave interactions (*absorption, reflection, refraction,* or *diffraction*) has occurred for each.

   a. The distortion of your partially submerged arm makes it look "broken" when viewed from the air.
   b. You hear the music even though you are seated behind an obstruction at a concert.
   c. You see yourself when you look at a highly polished car hood.
   d. Water ripples passing through a sponge become smaller.
   e. Heavy curtains are used to help keep a room quiet.

17. Can two waves interfere with each other so that the new wave formed by their combination has no amplitude? Explain your answer.

Problems

Section 23.1

1. The frequency of an oscillator is 20 hertz. What is its period? How long does it take this oscillator to complete one cycle?

2. A bicycle wheel spins 25 times in 5 seconds. Calculate the period and frequency of the wheel.

3. The piston in a gasoline engine goes up and down 3,000 times per minute. For this engine, calculate the frequency and period of the piston.

4. What is the period and frequency of the second hand on a clock? (*Hint*: How long does it take for the second hand to go around?)

5. The frequencies of musical instruments range between 20 and 20,000 Hz. Give this range in units of seconds per cycle.

6. Make a harmonic motion graph for a pendulum. Place time in seconds on the *x*-axis and position on the *y*-axis. The period of the pendulum is 0.5 second and the amplitude is 2 centimeters.

   a. What is the frequency of this pendulum?
   b. If you shortened the string of this pendulum, would the period get shorter or longer?

Section 23.2

7. A wave has a frequency of 10 hertz and a wavelength of 2 meters. What is the speed of the wave?

8. A sound wave has a speed of 400 m/s and a frequency of 200 Hz. What is its wavelength?

9. If the frequency of a wave is 30 hertz, how many wavelengths pass a certain point in 30 seconds?

10. Draw two cycles of a transverse wave with an amplitude of 4 cm and a wavelength of 8 cm. If the frequency of this wave is 10 Hz, what is its speed?

Section 23.3

11. A wave with a period of 1 second comes from the left. At the same time, a wave with a period of 2 seconds comes from the right. The amplitude of each wave is 5 centimeters. Draw a harmonic motion graph for each of these waves with time on the *x*-axis and position on the *y*-axis. Overlay two wavelengths of the 1-second wave on one wavelength of the 2-second wave. How do these two waves interfere—by constructive interference, destructive interference, or both?

Applying Your Knowledge

Section 23.1

1. Explain how Newton's laws of motion are helpful in understanding harmonic motion.

2. Does friction affect the amplitude of a pendulum as it is swinging? Does it affect the frequency? You may want to experiment to figure this out.

3. How might the period and frequency of the two rubber band oscillators in the figure at the right be different? Justify your answer.

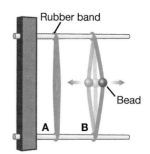

Section 23.2

4. When you watch fireworks, sometimes you see the explosion and then hear the sound. Why do you think this is?

Section 23.3

5. One of the four wave interactions is very important to how plants use light to grow. Guess which interaction this is, and write a couple of sentences justifying your answer.

Sound

Humans were making musical instruments to produce sounds 20,000 years before the wheel and axle were invented! Among instrument builders, Antonio Stradivari is one of the most famous. Between 1667 and 1730, Stradivari built violins in the small town of Cremona, Italy.

A violin's sound is rich and complex because vibrations of its wooden parts create a unique blend of frequencies. Stradivari worked tirelessly trying different woods and different varnishes, searching for the perfect sound. Over time, he developed a formula for varnish and special ways to carve and treat the all-important vibrating parts of the violin. In the 300 years since Stradivari, no one has completely figured out how he did it. Today, a Stradivarius violin is one of the most highly prized of all musical instruments. Its rich sound has never been duplicated.

CHAPTER 24 INVESTIGATIONS

24A: Properties of Sound
Does sound behave like other waves?

24B: Resonance in Other Systems
How can resonance be controlled to make the sounds we want?

24.1 Properties of Sound

Like other waves, sound has frequency, wavelength, and speed. Because sound is part of your daily experience, you already know its properties—but by different names. For example, the loudness of sound is related to its amplitude. Read on to find out more about sound's properties.

VOCABULARY

pitch – the perception of high or low that you hear at different frequencies of sound

The frequency of sound

Frequency and pitch
Your ears are very sensitive to the frequency of sound. The **pitch** of a sound is how you hear and interpret its frequency. A low-frequency sound has a low pitch, like the rumble of a big truck or a bass guitar. A high-frequency sound has a high pitch, like the scream of a whistle or siren. Humans can generally hear frequencies between 20 Hz and 20,000 Hz. Animals may hear a wider range of frequencies, or higher or lower frequencies than humans.

Most sound has more than one frequency
Almost all the sounds you hear contain *many frequencies at the same time*. In fact, the sound of the human voice contains thousands of different frequencies—all at once (Figure 24.1).

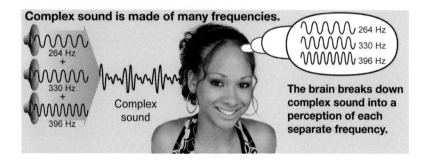

Complex sound is made of many frequencies.

264 Hz + 330 Hz + 396 Hz

Complex sound

264 Hz
330 Hz
396 Hz

The brain breaks down complex sound into a perception of each separate frequency.

The frequency spectrum
Why is it easy to recognize one person's voice from another's, even when both are saying the same word? The reason is that people have different mixtures of frequencies in their voices. A *frequency spectrum* shows loudness on the vertical axis and frequency on the horizontal axis. Figure 24.1 shows the frequency spectrum for three people saying *hello*. Can you see any difference between the graphs?

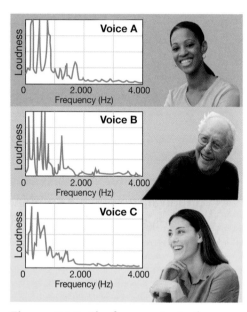

Figure 24.1: *The frequencies in three people's voices as they say the word hello. Each person's voice is made up of a mixture of frequencies.*

Intensity and loudness of sound

Decibels The unit for the intensity or strength of a sound is the **decibel (dB)**. We can measure sound intensity with scientific instruments just like we can measure mass with a balance. The decibel scale (shown below) is convenient to use because most sounds fall between 0 and 100. The amplitude of a sound increases 10 times for every 20-decibel increase (Figure 24.2).

0 dB	Threshold of human hearing; quietest sound we can hear
10–15 dB	A quiet whisper 1 meter away
30–40 dB	Background sound level in a house
45–55 dB	The noise level in an average restaurant
65 dB	Ordinary conversation 1 meter away
70 dB	City traffic
90 dB	A jackhammer cutting up the street 3 meters away
100 dB	Music through earbuds at maximum volume
110 dB	The front row of a rock concert
120 dB	The threshold of physical pain from loudness

Loudness When you experience a loud sound, you experience the effects of its intensity *and* frequency. An *equal loudness curve* compares how loud you hear sounds of different frequencies (Figure 24.3). As you can see, the human ear responds differently to high and low frequencies. This curve shows that low frequency sounds (below 100 Hz) need to have higher decibel values for you to hear them than the same as sounds between 100 and 1,000 Hz. Notice that the numbers are not evenly spread out on the *x*-axis of this graph. This type of spacing is called a *logarithmic scale*. You read the graph in the same way that you would read an evenly spaced graph.

Acoustics *Acoustics* is the science and technology of sound. Knowledge of acoustics is used to design facilities like libraries, recording studios, and concert halls. A design might address how to reduce sound intensity and/or whether sound needs to be absorbed, amplified, or even prevented from entering a room.

decibel (dB) – a unit of measure for the intensity or strength of a sound

Comparing decibels and amplitude

Decibels (dB)	Amplitude
0	1
20	10
40	100
60	1,000
80	10,000
100	100,000
120	1,000,000

Figure 24.2: *The decibel scale measures amplitude (loudness).*

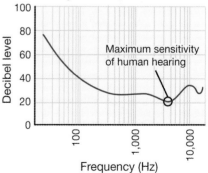

Figure 24.3: *All points on an equal loudness curve have the same loudness.*

The speed of sound

Sound is slower than light You have may have noticed that the sound of thunder often comes many seconds after you see lightning. Lightning is what creates thunder, so they really happen at the same time. You hear a delay because sound travels much slower than light. The speed of sound is about 1,000 km/h (660 mph). Light travels at 300,000 km/s (186,000 mi/s).

Subsonic and supersonic Objects that move faster than sound are called **supersonic**. If you were on the ground watching a supersonic plane fly toward you, there would be silence (Figure 24.4). The sound would be *behind* the plane, racing to catch up. Some military jets fly at supersonic speeds. Passenger jets are *subsonic* because they travel at speeds from 600 to 800 kilometers per hour.

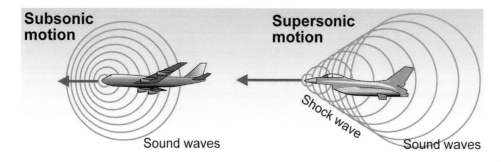

Sonic booms A supersonic jet compresses the sound waves that are created as its nose cuts through the air. A cone-shaped *shock wave* forms behind the point where the waves "pile up" at the nose of the plane. As a result, you only hear noise from a supersonic plane once it has passed overhead. At the boundary of hearing and not hearing the plane—the shock wave—the amplitude changes abruptly causing a very loud sound called a *sonic boom*.

Sound in liquids and solids Sound travels through most liquids and solids faster than through air (Figure 24.5). Sound travels about 5 times faster in water, and about 18 times faster in steel. This is because sound is a traveling oscillation. Like other oscillations, sound depends on restoring forces. The forces holding steel atoms together are much stronger than the forces between the molecules in air. Stronger restoring forces increase the speed of sound.

VOCABULARY

supersonic – a term to describe speeds faster than the speed of sound

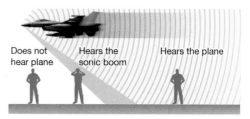

Figure 24.4: *The boundary between hearing and not hearing the plane is the "shock wave." The person in the middle hears a sonic boom as the shock wave passes over him.*

Material	Sound speed (m/s)
Air	330
Helium	965
Water	1,530
Wood (average)	2,000
Gold	3,240
Steel	5,940

Figure 24.5: *The speed of sound in various materials (helium and air at 0°C and 1 atmospheric pressure).*

The Doppler effect

The Doppler effect is caused by motion

The **Doppler effect** is a shift in the frequency of an oscillation caused by motion of the source of the oscillation. If a stationary object is producing sound, listeners on all sides will hear the same frequency. However, when the object is in motion, the frequency will *not* be the same to all listeners. People moving with the object or to the side hear the frequency as if the object were at rest. People in front hear a higher frequency. People behind hear a lower frequency. The Doppler effect occurs at speeds *below* the speed of sound.

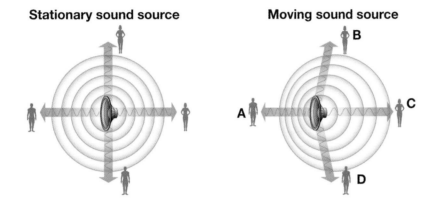

Stationary sound source

Moving sound source

The cause of the Doppler effect

The Doppler effect occurs because an observer hears the frequency at which wave crests arrive at his or her ears. For the moving sound source, observer (A) in the graphic above hears a higher frequency. This is because the object's motion causes the crests in front to be closer together. The opposite is true behind a moving object, where the wave crests are farther apart. Observer (C) in back hears a lower frequency because the motion of the object makes more space between successive wave crests. The greater the speed of the object, the larger the difference in frequency between the front and back positions.

Hearing the Doppler effect

You hear the Doppler effect when you hear a police or fire siren coming toward you, then going away from you. The frequency shifts up when the siren is moving toward you. The frequency shifts down when the siren is moving away from you.

VOCABULARY

Doppler effect – an increase or decrease in frequency caused by the motion of the source of an oscillation (such as sound)

TECHNOLOGY

Doppler radar

Doppler radar is a way to measure the speed of a moving object at a distance. A transmitter sends a pulse of microwaves. The waves reflect from a moving object, such as a car. The frequency of the reflected wave is increased if the car is moving toward the oncoming microwaves and decreased if the car is moving away. The difference in frequency between the reflected and transmitted wave is proportional to speed.

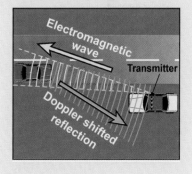

Recording sound

The microphone To record a sound, you must store the pattern of vibrations in a way that can be replayed and be true to the original sound. A common way to record sound starts with a microphone. A microphone transforms a sound wave into an electrical signal with the same pattern of vibration (Figure 24.6, top).

Analog to digital conversion In modern digital recording, a sensitive circuit called an analog to digital converter measures the electrical signal 44,100 times per second. Each measurement consists of a number between 0 and 65,536 corresponding to the amplitude of the signal. One second of compact-disc-quality sound is a list of 44,100 numbers. The numbers are recorded as data, and the information is digitally stored on the disc. The disc safely and reliably stores the digital information of the sound.

Playback of recorded sound To play the sound back, the string of numbers on the CD is read by a laser and converted into electrical signals again by a second circuit. This circuit is a digital to analog converter, and it reverses the process of the first circuit. The playback circuit converts the string of numbers back into an electrical signal. The electrical signal is amplified until it is powerful enough to move the coil in a speaker and reproduce the sound (Figure 24.6, bottom).

Stereo sound Most of the music you listen to has been recorded in stereo. A stereo recording is actually two recordings, one to be played from the left speaker, and the other from the right. Stereo sound seems almost "live" because it creates slight differences between when the sound reaches your left and right ears. Sound from all sources tends to reach you this way. The slight differences in how sound reaches your ears lets you know where sound is coming from. Another way to describe two sound waves that arrive at slightly different times is to say they are slightly *out of phase*.

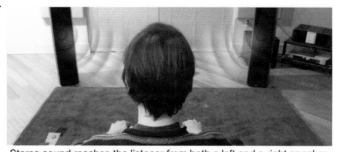

Stereo sound reaches the listener from both a left and a right speaker.

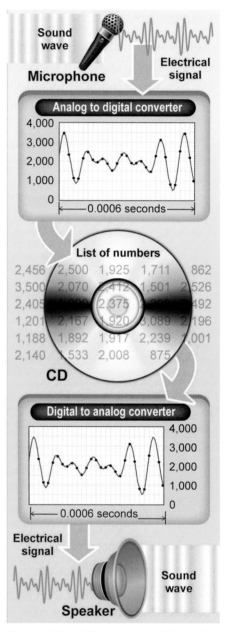

Figure 24.6: *The process of digital sound reproduction.*

Section 24.1 *Review*

1. What is the relationship between pitch and frequency?

2. If you looked at the frequency spectrums of two friends saying the word *dog*, would they look the same or different? Explain your answer.

3. Do two sound waves that seem equally loud always have the same amplitude? Explain.

4. What two variables affect how loud you hear sound?

5. How do the amplitudes of a 120-decibel sound and a 100-decibel sound compare?

6. Make a graph of the relationship between the amplitude (*x*-axis) and decibel level (*y*-axis) of sound. Describe this relationship.

7. Would an object moving at 750 km/h be supersonic or subsonic?

8. Would an object moving at 100 miles per hour be supersonic or subsonic? Use the conversion factor 1 mile = 1.6 kilometers.

9. Is it possible that a commercial passenger plane traveling at normal speeds could produce a shock wave or a sonic boom? Why or why not?

10. Why does sound travel faster through water than through air?

11. A paramedic in an ambulance does not experience the Doppler effect of the siren. Why?

12. You hear an ambulance in your neighborhood that is traveling a few blocks from where you are. The pitch of the siren seems to be getting lower and lower. Is the ambulance traveling toward you or away from you? How do you know?

13. *Research:* Find out how Doppler radar is used in weather forecasting.

14. What is the role of a microphone in recording sound?

15. The process of recording music involves converting between analog and digital information. Infer from the text what the terms *analog* and *digital* mean. Write a definition of these terms in your own words.

16. What about stereo sounds makes it seem like you are hearing the musicians play "live"?

CHALLENGE

Ultrasound

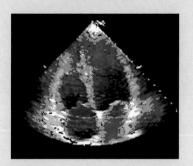

We cannot hear or see ultrasound waves, but they can pass through the human body. Doctors use ultrasound images to see "inside" patients, the same way they use X-rays. The ultrasound image pictured above is a heart.

Research the answers to the following questions.

1. What exactly is ultrasound?

2. How do the frequency and wavelength of ultrasound compare to sounds you can hear?

24.2 Sound Waves

How do we know that sound is a wave? For starters, it has both frequency and wavelength. We also know sound is a wave because it does all the things other waves do. Sound can be reflected, refracted, and absorbed. Sound also shows diffraction and interference. Resonance occurs with sound waves and is especially important for understanding how musical instruments work.

What is a sound wave?

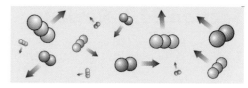

Figure 24.7: *Air is made of molecules in constant random motion, bumping off each other and the walls of their container.*

Sound in solids and liquids Sound is a traveling oscillation of atoms. If you push on one atom, it pushes on its neighbor. That atom pushes on the next atom, and so on. The push causes atoms to oscillate back and forth like tiny beads on springs. The oscillation spreads through the connections between atoms to make a sound wave. This is how sound moves through liquids and solids.

Sound in air and gases In air, the situation is different. Air molecules are spread far apart and interact by colliding with each other (Figure 24.7). The pressure is highest where atoms are closest together and lowest where they are farthest apart (Figure 24.8). Imagine pushing the molecules on the left side of the picture below. Your push squeezes atoms together creating a layer of higher pressure. That layer pushes on the next group of atoms and causes those atoms to squeeze together. This pattern repeats. The result is a traveling oscillation in pressure, which is a sound wave. Sound is a *longitudinal* wave because the oscillations are along the same direction that the wave travels.

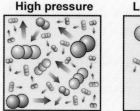 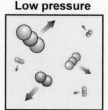

Figure 24.8: *If temperature is constant, high pressure means more molecules per unit volume. Low pressure means fewer molecules per unit volume.*

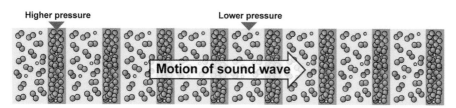

The frequency range of sound waves Anything that vibrates creates sound waves, as long as there is contact with other atoms. However, not all "sounds" can be heard. Humans can hear in the range between 20 and 20,000 Hz. Bats can hear high-frequency sounds from 2,000 to 110,000 Hz and elephants hear lower-frequency sounds from 16 to 12,000 Hz.

Sound and air pressure

Speakers If you touch the surface of a speaker, you can feel the vibration that creates a sound wave. Figure 24.9 shows an illustration of a speaker as well as an exaggerated sound wave and the oscillation of pressure. When music is playing, the surface of the speaker moves back and forth at the same frequencies as the sound waves. The back-and-forth motion of the speaker creates a traveling sound wave of alternating high and low pressure.

Air pressure The change in air pressure created by a sound wave is incredibly small. An 80-dB sound, equivalent to a loud stereo, changes the air pressure by only one part in a million. Our ears are very well structured to detect the small changes in pressure created by sound waves.

Frequency and pressure change The frequency of sound indicates how fast air pressure oscillates back and forth. The purr of a cat, for example, might have a frequency of 50 hertz. This means the air pressure alternates 50 times per second. The frequency of a fire truck siren may be 3,000 hertz. This corresponds to 3,000 vibrations per second in the pressure of the air.

Sound speed and temperature In air, the energy of a sound wave is carried by moving atoms and molecules bumping into each other. Anything that affects the motion of atoms affects the speed of sound. Molecules move more slowly in cold air and the speed of sound decreases. For example, at 0°C, the speed of sound is 330 meters per second, but at 21°C, the speed of sound is 344 meters per second.

Sound speed and molecular weight Lighter atoms and molecules move faster than heavier ones at the same temperature. The speed of sound is higher in helium gas because helium atoms are lighter (and faster) than either the oxygen (O_2) or nitrogen (N_2) molecules that make up air. Because water molecules are lighter than those of other gases that make up air, increasing the humidity of air also increases the speed of sound.

What does sound speed depend on? The speed of sound depends on what the sound wave is traveling through. The type of molecules, the temperature, and the phase of the material (solid, liquid, or gas) all affect the speed of a sound wave.

Speaker cone

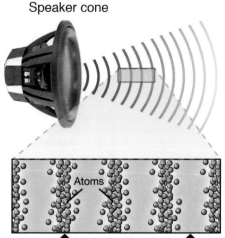

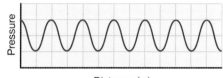

Higher pressure   Lower pressure

Pressure

Distance (m)

Figure 24.9: *This is what a sound wave might look like if you could see the atoms. The effect of sound on air molecules is exaggerated.*

■■■■■ **SOLVE IT!** ■■■■■

How many vibrations of air pressure occur per second when the note A (440 Hz) is played?

The wavelength of sound

Range of wavelengths of sound   The wavelengths of sound in air can be compared to the size of everyday objects (Table 24.1). As with other waves, the wavelength of a sound is inversely related to its frequency (Figure 24.10). A low-frequency, 20-hertz sound has a wavelength the size of a large classroom. At the upper range of hearing, a 20,000-hertz sound has a wavelength about the width of a finger.

Table 24.1: Frequency and wavelength for some typical sounds

Frequency (Hz)	Wavelength	Typical source
20	17 m	rumble of thunder
100	3.4 m	bass guitar
500	70 cm (27″)	average male voice
1,000	34 cm (13″)	female soprano voice
2,000	17 cm (6.7″)	fire truck siren
5,000	7 cm (2.7″)	highest note on a piano
10,000	3.4 cm (1.3″)	whine of a jet turbine
20,000	1.7 cm (0.67″)	highest-pitched sound you can hear

Wavelengths of sounds are important

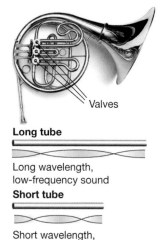

Long tube

Long wavelength, low-frequency sound

Short tube

Short wavelength, high-frequency sound

Differences in sound are due to differences in both frequency and wavelength. If you want to make a sound of a certain wavelength (or frequency), you need to have a vibrating object that is similar in size to the wavelength of that sound. So how is a French horn able to produce so many different sounds? A French horn makes sound by vibrating the air trapped in a long, coiled tube. Short tubes only fit short wavelengths and make higher-frequency sounds. Long tubes fit longer wavelengths and make lower-frequency sounds (Figure 24.10). Opening and closing the valves on a French horn allows the player to add and subtract different-length tubes, changing the frequency of the sound.

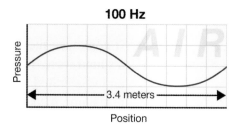

100 Hz

3.4 meters

Position

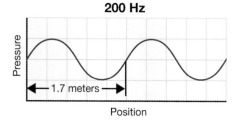

200 Hz

1.7 meters

Position

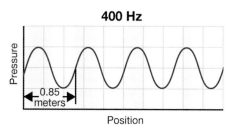

400 Hz

0.85 meters

Position

Figure 24.10: *The frequency and wavelength of sound are inversely related. When the frequency goes up, the wavelength goes down proportionally.*

Standing waves

What is a standing wave? You just learned that a French horn makes sounds by confining waves within tubes of different lengths. A wave that is confined in a space is called a **standing wave**. It is possible to make standing waves of almost any kind, including sound, water, and even light. You can experiment with standing waves using a vibrating string. Vibrating strings create sound on a guitar or piano.

VOCABULARY

standing wave – a wave that is confined in a space

fundamental – the lowest natural frequency of an oscillator

harmonic – one of many natural frequencies of an oscillator

Harmonics

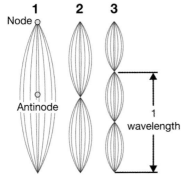

The first three harmonics of a vibrating string

A string with a standing wave is a kind of oscillator. Like all oscillators, a string has natural frequencies. The lowest natural frequency is called the **fundamental**. A vibrating string also has other natural frequencies called **harmonics**. The diagram at the left shows the first three harmonics. You can find the harmonic number by counting the number of "bumps" or places of greatest amplitude. The first harmonic has one bump, the second has two, the third has three, and so on. The place of highest amplitude on a string is the *antinode*. The place where the string does not move is called a *node*.

Resonance of sound Spaces enclosed by boundaries can create *resonance* with sound waves. Like a French horn, a panpipe makes music when sound resonates in tubes of different lengths (Figure 24.11). One end of each tube is closed and the other end is open. Blowing across the open end of a tube creates a standing wave inside the tube. The closed end of a pipe is a closed boundary, and it makes a *node* in the standing wave. The open end of a pipe is an open boundary to a standing wave and makes an *antinode*. The pipe resonates to a certain frequency when its length is one-fourth the wavelength of that frequency. If the pipe resonates at the fundamental frequency, then the wavelength of the fundamental is four times the length of the pipe.

Panpipes

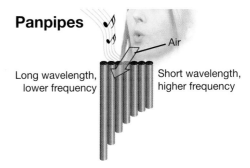

Standing Wave in a Panpipe

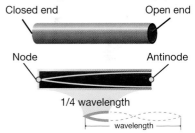

The wavelength of the fundamental is four times the length of the pipe.

Figure 24.11: *A panpipe is made from tubes of different lengths. The diagram shows the fundamental for a standing wave of sound in a panpipe.*

Interaction between sound waves and boundaries

VOCABULARY

reverberation – multiple echoes of sound caused by reflections of sound building up and blending together

Interactions of sound and materials
Like other waves, sound waves can be reflected by hard surfaces and refracted as they pass from one material to another. Diffraction causes sound waves to spread out through small openings. Carpet and soft materials can absorb sound waves. Figure 24.12 illustrates these four sound interactions.

Reverberation
In a good concert hall, the reflected sound and direct sound from the musicians, along with sound reflected from the walls, creates a multiple echo called **reverberation**. The right amount of reverberation makes the sound seem livelier and richer. Too much reverberation and the sound gets "muddy." Concert hall designers choose the shape and surface of the walls and ceiling to provide the best reverberation. Some concert halls have movable panels that can be raised or lowered from the ceiling to help shape the sound.

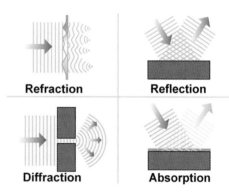

Refraction | Reflection

Diffraction | Absorption

Figure 24.12: *Sound displays all the properties of waves in its interactions with materials and boundaries.*

Constructing a good concert hall
Direct sound (**A**) reaches the listener along with reflected sound (**B**, **C**) from the walls. The shape of the room and the surfaces of its walls must be designed and constructed so that there is some reflected sound, but not too much.

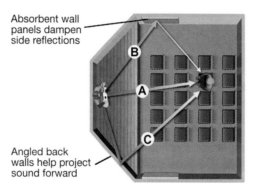

Absorbent wall panels dampen side reflections

Angled back walls help project sound forward

Interference can also affect sound quality
Reverberation also causes interference of sound waves. When two waves interfere, the total can be louder or softer than either wave alone. The diagram above shows a musician and an audience of one person. The sound reflected from each wall interferes as it reaches the listener. If the distances are just right, one reflected wave might be out of phase with the other. The result is that the sound is quieter at that spot. An acoustic engineer would call it a *dead spot* in the hall. Dead spots are areas where destructive interference causes some of the sound to cancel with its own reflections. It is also possible to make very loud spots where sound interferes constructively. The best concert halls are designed to minimize both dead spots and loud spots.

SOLVE IT!

Using your ears

Go to a concert hall, an auditorium, or even a smaller space. Make a map of the place. Play music from one location. While the music is playing, walk around and identify where you hear the music well and where you hear dead spots. Add these details to your sketch.

Section 24.2 *Review*

1. How could you increase the air pressure inside a bag containing a group of air molecules?

2. Is sound a longitudinal or transverse wave? Justify your answer.

3. A 200-Hz sound has a wavelength about equal to the height of an adult. Would a sound with a wavelength equal to the height of a two-year-old child have a higher or lower frequency than 200 Hz?

4. For each situation, identify when sound would travel faster and why.
   a. Outside on a winter day or outside on a summer day?
   b. Through water or air?
   c. When air pressure is high or low?
   d. Through a piece of wood that floats in water or through a piece of steel that sinks in water?
   e. Through a gas that is 90 percent nitrogen (N_2) and 10 percent helium (He) or through a gas that is 90 percent helium (He) and 10 percent nitrogen (N_2)?

5. The first five harmonics for a vibrating string are shown in Figure 24.13.
   a. For each harmonic, identify the number of wavelengths represented.
   b. For each harmonic, identify the number of nodes and antinodes that are present (include the ends of the string in your count).
   c. Which of the five harmonics has the highest natural frequency?
   d. Make a drawing that shows what the sixth harmonic would look like.

6. A panpipe is made of five pipes. The longest pipe is 25 centimeters long, and the shortest is 5 centimeters long. Which of these pipes produces the highest-frequency sound and why?

7. Would a full concert hall have different reverberation than an empty hall? Explain.

8. It is extremely difficult to play, record, and hear live music in a park or other open space. Explain why this is so. Use the word reverberation in your answer.

9. You and your band want to record a CD in your basement. What might you need to do to make your basement a good place for recording music?

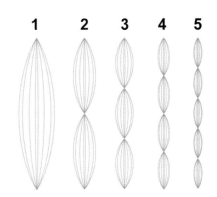

1   2   3   4   5

The first five harmonics of the vibrating string

Figure 24.13: *Question 5.*

████ CHALLENGE ████

How long does a pipe have to be to play the note E?

You wish to make a pipe that makes a sound with a frequency of 660 Hz (the note E). Use the relationship between wave speed and frequency to determine the wavelength of this note. The pipe length needs to be one-fourth the wavelength to make a resonance in the fundamental mode. Assume the speed of the sound is 340 m/s.

$$\frac{Wave\ speed}{Frequency} = Wavelength$$

24.3 Sound, Perception, and Music

Sound is everywhere in our environment. We use sound to communicate, and we listen to sound for information about what is going on around us. Our ears and brain are constantly receiving and processing sound. In this section, you will learn about how we *hear* a sound wave and how the ear and brain construct meaning from sound. This section will also introduce some of the science behind music. Musical sound is a rich language of rhythm and frequency, developed over thousands of years of human culture.

The perception and interpretation of sound

Constructing meaning from patterns   As you read this paragraph, you subconsciously recognize individual letters. However, the *meaning* of the paragraph is not in the letters themselves. The meaning is in the *patterns* of how the letters make words and the words make sentences. The brain does a similar thing with sound. A single frequency of sound is like one letter. It does not have much meaning. The meaning in sound comes from patterns of many frequencies changing together.

Ears hear many frequencies at once   When you hear a sound, the nerves in your ear respond to more than 15,000 different frequencies at once. This is like having an alphabet with 15,000 letters! The brain interprets all 15,000 different frequency signals from the ear and creates a "sonic image" of the sound.

Complex sound waves   Imagine listening to live music from a singer and a band. Your ears can easily distinguish the voice from the instruments. How does this occur? The microphone records a single "wave form" of how pressure varies with time. The recorded wave form is very complex, but it contains all the sound from the instruments and voice (Figure 24.14).

How the brain finds meaning   The brain makes sense of this sound because the ears separate the sound into different frequencies. Your brain, receiving signals from your ears, has learned to recognize certain patterns of how each frequency changes and gets louder and softer over time. One pattern might be a sung word. Another might be a musical note from a guitar. Inside your brain is a "dictionary" that associates a meaning with a pattern of frequency the same way an ordinary dictionary associates a meaning from a pattern of letters (a word).

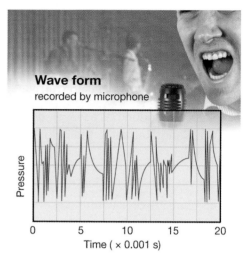

Figure 24.14: *The recorded wave form from 0.02 seconds of music.*

The frequency spectrum and the sonogram

Frequency spectrum A **frequency spectrum** is a graph that shows the amplitudes of different frequencies present in a sound. Amplitude, or loudness, is represented on the *y*-axis, and frequency is shown on the *x*-axis. Sound containing many frequencies has a wave form that is jagged and complicated. The wave form in Figure 24.15 is from an acoustic guitar playing the note E. The frequency spectrum shows that the complex sound of the guitar is made from many frequencies, ranging up to 10,000 Hz and beyond.

What is a sonogram? More information about a sound is available when a graph combines the variables—frequency, amplitude, and time. A **sonogram** shows frequency on the vertical axis and time on the horizontal axis. The loudness (amplitude) is shown by a color range.

Reading a sonogram The sonogram below (left) shows the word *hello* lasting from 1.4 to 2.2 seconds. A sonogram of your voice (or anyone else's) saying *hello* would look different because every voice is unique. In this example, you can see that there are many frequencies almost filling up the space between 0 and 5,000 Hz. The sonogram on the right is a simpler version of this type of graph. Which bar represents a loud sound of 100 Hz lasting from 1 to 3 seconds (A, B, C, or D)?

VOCABULARY

frequency spectrum – a graph that shows the amplitudes of different frequencies present in a sound

sonogram – a graph that shows the frequency, amplitude, and time length for a sound

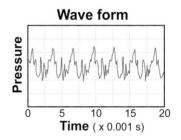

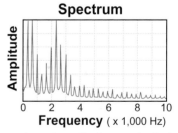

The *spectrum* shows the frequencies that make up a complex wave form.

Figure 24.15: *Each peak in the spectrum represents the frequency and amplitude of a wave that makes up the wave form.*

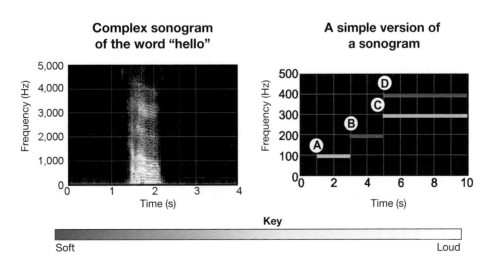

How we hear sound

The cochlea The *cochlea* provides us with our ability to interpret sound—in other words, our sense of hearing. However, the cochlea is in the inner ear (Figure 24.16). Sound has to reach the cochlea by first entering the ear canal, where it encounters the eardrum. Here, the sound waves cause the eardrum to vibrate. Then three delicate bones of the inner ear transmit these vibrations to the side of the cochlea. In turn, fluid in the spiral channel of the cochlea vibrates and creates waves. Nerves along the channel have tiny hairs that shake when the fluid vibrates. Near the entrance, the channel is relatively large so the nerves respond to longer-wavelength, lower-frequency sound. The nerves at the small end of the channel respond to shorter-wavelength, higher-frequency sound.

The semicircular canals As you know, the function of our ears is hearing. But did you know that your ears also provide you with your sense of balance? Near the cochlea in the inner ear are three semicircular canals. Like the cochlea, each canal contains fluid. The movement of this fluid in the canals indicates how the body is moving (left–right, up–down, or forward–backward).

Human hearing In general, the combination of the eardrum, bones, and the cochlea limit the range of human hearing to between 20 hertz and 20,000 hertz. However, hearing varies greatly among different people and changes with age. Some people can hear sounds above 15,000 Hz, and other people can't. On average, people gradually lose high-frequency hearing with age. Most adults cannot hear frequencies above 15,000 hertz, while children can often hear to 20,000 hertz.

Hearing can be damaged by loud noise Hearing is affected by exposure to loud or high-frequency noise. Listening to loud sounds for a long time can cause the hairs on the nerves in the cochlea to weaken or break off, causing permanent damage. Therefore, it is important to always protect your ears by keeping the volume of noise at a low or reasonable level. It is also important to wear ear protection if you have to stay in a loud place. In concerts, many musicians wear earplugs on stage to protect their hearing.

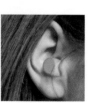
Earplug

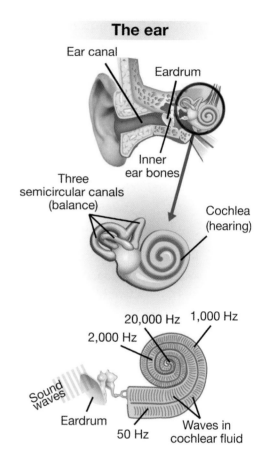

Figure 24.16: *The structure of the inner ear. When the eardrum vibrates, three small bones transmit the vibrations to the cochlea. The vibrations make waves inside the cochlea, which shake hairs attached to nerves in the spiral. Each part of the spiral is sensitive to a different frequency.*

Music

Pitch The *pitch* of a sound describes how high or low we hear its frequency. A higher-frequency sound is heard as a higher pitch. However, because pitch depends on the human ear and brain, the way we hear a sound can be affected by the sounds we heard before and after.

Rhythm **Rhythm** is a regular time pattern in a series of sounds. Here is a rhythm you can "play" on your desk: TAP-TAP-tap-tap-TAP-TAP-tap-tap. Play "TAP" louder than you play "tap." Rhythm can be made with sound and silence or with different pitches. People respond naturally to rhythm. Cultures are distinguished by their music and the special rhythms used in music.

The musical scale Music is a combination of sound and rhythm. Styles of music are vastly different, but all music is created from carefully chosen frequencies of sound. Most of the music you listen to is created from a pattern of frequencies called a **musical scale**. Each frequency in the scale is called a **note**. The C major musical scale that starts on the note C (262 Hz) is shown in the diagram below. The approximate frequencies of the notes in this scale are listed. Notice that this scale begins and ends with C and that the higher C is twice the frequency of the lower C. These two Cs are an octave apart. An **octave** is the range between any given frequency and twice that frequency. Notes that are an octave apart in frequency share the same name because they sound similar to the ear.

C major scale

Note	C	D	E	F	G	A	B	C
Approximate frequency (Hz)	262	294	330	349	392	440	494	524

VOCABULARY

rhythm – a regular time pattern in a series of sounds

musical scale – a pattern of frequencies

note – one frequency in a musical scale

octave – a range defined as being between a single frequency value and twice that frequency value; on a musical scale, these two notes would have the same name

SOLVE IT!

Getting to know octaves

1. What is the frequency and name of the note that is one octave *lower* than C-262 Hz?

2. What is the name and frequency of the note that is two octaves *higher* than A-440 Hz?

Consonance, dissonance, and beats

Harmony   *Harmony* is the study of how sounds work together to create effects desired by the composer. From experience, you know that music can have a profound effect on people's moods. For example, the tense, dramatic soundtrack of a horror movie is a vital part of the audience's experience. Harmony is based on the frequency relationships of the musical scale.

Beats   When two frequencies of sound are not exactly equal in value, the loudness of the total sound seems to oscillate or **beat.** The diagram below illustrates how beats occur for two waves occurring simultaneously. The *superposition principle* states that when sound waves occur at the same time, they combine to make a complex wave. The sound (amplitude) of this wave is louder than either wave separately when the waves are *in phase* due to constructive interference. When the waves are o*ut of phase*, the sound is quieter due to destructive interference. We hear the alternation in amplitude as beats.

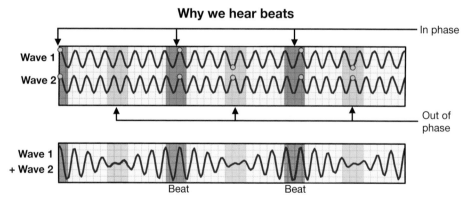

Why we hear beats

Consonance and dissonance   When we hear more than one frequency of sound and the combination sounds pleasant, we call it **consonance**. When the combination sounds unsettling, we call it **dissonance**. Consonance and dissonance are related to beats. When frequencies are far enough apart that there are no beats, we get consonance. When frequencies are too close together, we hear beats that are the cause of dissonance. In music, dissonance is often used to create tension or drama. Consonance can be used to create feelings of balance and comfort.

■■■■■ **VOCABULARY** ■■■■■

beat – the oscillation between two sounds that are close in frequency

consonance – a combination of frequencies that sounds pleasant

dissonance – a combination of frequencies that sounds unpleasant

■■■■■ **SCIENCE FACT** ■■■■■

Bats and beats

Bats use *echolocation* to navigate and find insects for food. Like a "sonic flashlight," the bat's voice "shines" ultrasound waves into the night. The sound occurs as "chirps," short bursts of sound that rise in frequency. When the sound reflects off an insect, the bat's ears receive the echo. Since the frequency of the chirp is always changing, the echo comes back with a slightly different frequency. The difference between the echo and the chirp makes beats that the bat can hear. The beat frequency is proportional to how far the insect is from the bat. A bat can even determine where the insect is by comparing the echo it hears in the left ear with what it hears in the right ear.

Making sounds

Voices The human voice is a complex sound that starts in the larynx, a small, hollow chamber at the top of your windpipe. The term *vocal cords* is a little misleading because the sound-producing structures are not really cords but folds of expandable tissue that extend across the larynx. The sound that starts in the larynx is changed by passing through openings in the throat and mouth (Figure 24.17). Different sounds are made by changing both the vibrations in the larynx and the shape of the openings.

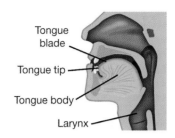

Figure 24.17: *Notice how the shape of the structures in the throat and mouth change as the human voice creates the sounds AH, EE, EH, and OH.*

The guitar

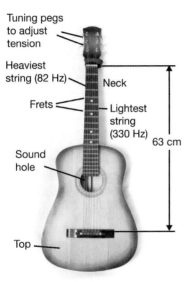

Tuning pegs to adjust tension

Heaviest string (82 Hz)

Neck

Frets

Lightest string (330 Hz)

63 cm

Sound hole

Top

The guitar has become a central instrument in popular music. Guitars come in many types but share the common feature of making sound from vibrating strings. A standard guitar has six strings that are stretched along the neck and body of the guitar. The strings have different weights and therefore different natural frequencies.

For a guitar in standard tuning, the heaviest string has a natural frequency of 82 hertz and the lightest a frequency of 330 hertz. Each string is stretched by a tension force of about 125 newtons (28 pounds). The combined force from six strings on a folk guitar is more than 750 newtons (170 pounds). The guitar is tuned by changing the tension in each string. Tightening a string raises its natural frequency and loosening lowers it.

Each string can make many notes A typical guitar string is 63 centimeters long. To make different notes, the vibrating length of a single string can be shortened by holding it down between one of many metal bars across the guitar's neck, called frets (Figure 24.18). The frequency goes up as the vibrating length of the string gets shorter.

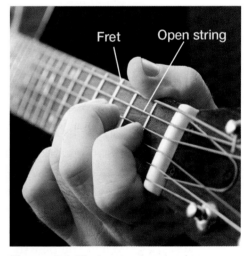

Fret   Open string

Figure 24.18: *A guitarist can play a note by playing an "open string", or he can shorten the length of a string by pressing down between frets.*

Harmonics and the sound of instruments

Same note, different sound
The same note sounds different when played on different instruments. As an example, suppose you listen to the note C (262 Hz) played on a guitar and the same C (262 Hz) played on a piano. A musician would recognize both notes as being C because they have the same frequency and pitch. However, as you know, a guitar sounds like a guitar and a piano sounds like a piano. If the frequency of the note is the same, what gives each instrument its characteristic sound?

Instruments make mixtures of frequencies
A guitar and a piano have recognizable sounds because each note played is not a single pure frequency. The most important frequency is still the fundamental note (C-262 Hz, for example). The variation comes from the *harmonics*. Remember, harmonics are frequencies that are multiples of the fundamental note. We have already learned that a string can vibrate at many harmonics. This is true for all instruments. A single C note from a grand piano might include 20 or more different harmonics.

Recipes for sound
Consider that every instrument has its own *recipe* for the frequency content of its sound. Another word for "recipe" in this context is *timbre*. In Figure 24.19, you can see how the mix of harmonics for a guitar compares to the mix for a piano when both instruments play the note C (262 Hz). Here, you can see that the timbre of a guitar is different from that of a piano.

Tuning and beats
A tuning fork is a useful tool for tuning an instrument because it produces a single frequency (Figure 24.20). Here's how a tuning fork is used. Let's say the A string on a guitar is out of tune and its natural frequency is 445 hertz. The correct frequency for A is 440 hertz. To tune the guitar, you need an A tuning fork, which will produce vibrations at 440 hertz when it is struck. When you play the guitar string and listen to the tuning fork, you will hear a beat frequency of 5 beats per second, or 5 hertz. The beat frequency becomes zero when the string is tuned to the tuning fork so that both it and the guitar string have a natural frequency of 440 hertz. The beats go away when the string is in tune.

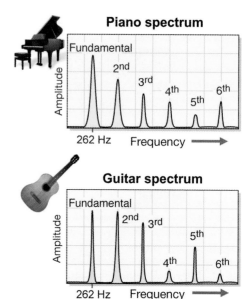

Figure 24.19: *The sound of the note C (262 Hz) played on a piano and on a guitar. Notice that the fundamental frequencies are the same but the harmonics have different amplitudes.*

Figure 24.20: *A tuning fork produces a single frequency.*

Section 24.3 *Review*

1. Do you hear sounds around you as one frequency at a time or as many frequencies at once?

2. Which of the frequencies in Figure 24.21 is a soft sound that lasts five seconds? What is the frequency of this sound?

3. What is the difference between a sonogram and a frequency spectrum?

4. If sound B has twice the amplitude of sound A, sound A is:
    a. louder
    b. softer
    c. higher pitched
    d. lower pitched

5. How does the cochlea allow us to hear both low-frequency and high-frequency sound?

6. What is the range of frequencies for human hearing?

7. If you were talking to an elderly person who was having trouble hearing you, would it be better to talk in a deeper voice (low-frequency sound) or a higher voice (high-frequency sound)?

8. What is one way that your body knows if it is upside-down or not?

9. If two sound waves have exactly the same frequency, will you hear beats? Why or why not?

10. A musician in a group plays a "wrong" note. Would this note disrupt the harmony or the rhythm of the song being played? Explain your answer.

11. The note G is 392 Hz. What is the frequency of this note one octave higher?

12. Explain the appearance of the complex wave in Figure 24.22. In particular, explain the areas of higher amplitude and lower amplitude.

13. Why does an A played on a violin sound different from the same note played on a guitar?

14. How is the length of a string on a stringed instrument related to the length of a pipe on panpipe? Use the words *frequency* and *wavelength* in your answer.

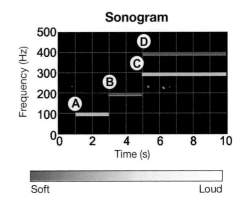

Figure 24.21: *Question 2.*

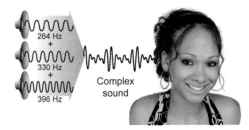

Figure 24.22: *Question 12.*

⚙ STEM Sound All Around

Acoustics is the science and technology of sound. Schools are crowded places with a variety of activities happening all at once. Engineers and architects who specialize in acoustics study how sound acts, so that they can design spaces that support clear communication.

Your school spaces

Think about the sounds you hear in the school gymnasium compared with the school library. From experience, you know that a gym can be a very loud place full of echoes, while libraries are quieter. Now, think about the designs of these spaces and the types of materials used in them. What comes to mind?

Most libraries have carpet on the floor, book shelves around the room, and acoustic tiles in the ceiling. All of these are sound-absorbing materials that reduce reverberation and noise levels. The floor of a gym, good for bouncing basketballs, is also good for bouncing sound. When sound is reflected rather than absorbed by a surface, the space tends to be noisy.

Of course, the sound in a library or gym is also controlled by rules and the types of activities happening there. In libraries, people study quietly; in a gym, they play sports and shout.

Is the space dead or alive?

Sound travels through a medium (solid, liquid, or gas) until the sound energy is absorbed by the medium. Sound waves move through a room and reflect off the surfaces they run into until the energy they carry is fully absorbed. The time for this varies depending upon many variables, including the strength of the initial sound, the absorption rate of the walls of the room, and the size of the room. A live room is one in which the materials have a low absorption rate. Live rooms include gymnasiums and cafeterias where sounds easily reflect off walls and last a long time. A library or auditorium is an example of a dead room. In these rooms, the absorption rate of the materials is much higher. Sounds are absorbed, and they don't last a long time, causing the space to be quieter.

Cancel the noise!

In a library, passive noise reduction (PNR) is used to help make this enclosed space quiet. Carpet, heavy curtains, or even ear plugs are examples of passive noise reduction items. These materials absorb sound.

Another way to reduce sound is by active noise reduction (ANR). ANR technology tries to cancel rather than absorb sound that is unwanted. Have you ever heard of noise reduction headphones? Specially-designed headphones are one part of the growing technology of ANR. There are three basic parts of ANR in these special headphones: a microphone, processing electronics, and a speaker. These parts, which must all fit into the ear pieces, work together to cancel unwanted sound waves.

How active noise reduction headsets work

PNR
Passive Noise Reduction

ANR
Anti-Noise Reduction

Noise

Loudness reduced by PNR. PNR can also remove high frequency sounds.

Anti-noise cancels noise, especially low frequency noise.

The ear covering provides passive noise reduction.

Noise is greatly reduced or eliminated.

Active noise reduction circuitry and power.

Headsets use a combination of passive and active noise reduction.

With carefully-designed ANR, the microphone very near the ear canal continually detects noise. The frequency and amplitude profile of the noise is detected. The processing electronics create another

noise that is just the opposite of the original. This new noise—or anti-noise—is sent into the ear canal by the speaker and cancels the offending sound.

ANR headsets for MP3 players are safe to wear because they only cancel the lower frequencies of sound, and not speech or warning sirens.

Which is better: ANR or PNR?

Scientists have learned that passive and active noise reduction are effective in different ways. ANR seems to work better with low frequency sounds. PNR is better at absorbing the higher frequencies. For example, studies have shown that the noise produced by propellers in airplanes is in the low frequency range. Therefore, specialized ANR headphones work well for airplane pilots. Extra soundproofing for passive noise reduction, although it would lessen high frequency noise, would add too much weight to a plane to be practical.

Presently, ANR technology is being tested to lower the noise from the cooling fans inside electronic devices like your computer, tailpipes of cars, or inside the cabin of the car. As ANR technology grows, new uses for ANR will be discovered. Can you think of a new use for ANR technology?

Questions:

1. What is the difference between a live room and a dead room in terms of sound?

2. If you wanted to create a recording studio for recording a new CD for your band, what would you do? You may want to do research on the Internet to find out the design features of recording studios.

3. Compare and contrast passive and active noise reduction.

4. Does active noise reduction work using constructive or destructive interference?

Chapter 24 *Assessment*

Vocabulary

Select the correct term to complete each sentence.

consonance	frequency spectrum	note
decibel	Doppler effect	beat
standing wave	musical scale	reverberation
octave	supersonic	pitch
fundamental	rhythm	harmonic
sonogram	dissonance	

Section 24.1

1. A moving object that makes a sound will sound differently if the object is moving toward or away from you due to the _____ .

2. The unit for measuring the loudness of a sound is the _____ .

3. How your ears hear and interpret a sound of a certain frequency is called the _____ .

4. _____ objects move faster than the speed of sound waves.

Section 24.2

5. You can tell which _____ a vibrating string is experiencing by counting the nodes and antinodes.

6. A(n) _____ is a wave confined or trapped in a certain space.

7. The _____ is the lowest natural frequency of an oscillator.

8. A multiple echo in a concert hall or other room is called a(n) _____ .

Section 24.3

9. A(n) _____ is a graph that shows the amplitudes of different frequencies that make up a sound.

10. As two sounds of slightly different frequencies go in and out of phase, _____ can be heard.

11. _____ is a regular time pattern in a series of sounds.

12. A(n) _____ is a pattern of frequencies used by musicians.

13. The range between a frequency on a musical scale and a frequency that is twice as great is called a(n) _____ .

14. A graph that shows the frequency, amplitude, and time of a sound such as a person saying a word is called a(n) _____ .

15. A combination of sounds of different frequencies that sounds pleasant is called _____ .

16. A combination of sounds of different frequencies that sounds unpleasant is called _____ .

17. Each frequency on a musical scale is called a(n) _____ .

Concepts

Section 24.1

1. Give an example of a sound with a high pitch and an example of a sound with a low pitch.

2. Explain how you can tell the difference between the voices of two people if they are saying the same word.

3. Approximately how many decibels is each of the following sounds?
   a. the cafeteria at your school at lunch
   b. an alarm clock
   c. a running sink faucet

4. Do all frequencies of sounds at 40 decibels seem equally loud to your ears? Explain.

5. How fast do sound waves travel in air? How does this compare to the speed of light waves?

6. What is a sonic boom?

7. Why do sound waves travel faster in steel than in air or water?

8. A car honking its horn moves toward you. Does the horn's pitch sound higher or lower than it would if the car were parked? Explain.

9. What does it mean to say a recording is in stereo?

Section 24.2

10. Draw a diagram that shows what air molecules look like when a sound wave is traveling through the air.

11. Does sound travel faster in warm or cold air? Why?

12. Does a person's voice sound higher or lower after inhaling helium gas? Why?

13. How is the wavelength of a sound wave related to its frequency?

14. Which would create sound waves with longer wavelengths, a cat meowing or a bear growling?

15. Why does a flute produce higher-pitched sounds than a tuba?

16. What is the difference between a node and an antinode on a standing wave?

17. Draw a standing wave on a string with six nodes and five antinodes. Which harmonic did you draw?

18. The diagram to the right shows a harmonic of a vibrating string.
    a. Which harmonic is shown?
    b. How many wavelengths does the standing wave contain?
    c. What is the wavelength of the standing wave?

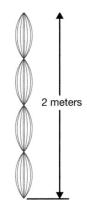
2 meters

19. List the four ways sound waves can interact with materials and boundaries.

Section 24.3

20. How many different frequencies do nerves in your ear sense at the same time when you hear a sound?

21. Which type of graph gives more information, a frequency spectrum or a sonogram? Explain.

22. What do your ears sense in addition to sounds?

23. Does the outer, larger part of the cochlea hear higher or lower frequencies?

24. What can happen if a person listens to loud sounds for a long time?

25. What causes the alternation of loud and soft sounds that occur when similar frequencies are played together?

26. Which of the following guitar strings would have the highest natural frequency?
    a. a thick string that is very loose
    b. a thick string that is tight
    c. a thin string that is very loose
    d. a thin string that is tight

27. What is the purpose of frets on a guitar?

28. How is the sound created by a tuning fork different from the sound created by plucking a guitar string?

Problems

Section 24.1

1. While you are at home, you hear the dishwasher with a loudness of 40 dB and a siren outside with a loudness of 60 dB. How much greater is the amplitude of the siren's sound than the amplitude of the dishwasher's sound?

2. According to an Equal Loudness Curve, a 100-Hz sound and a 10,000-Hz sound are heard at an equal loudness. If the 100-Hz sound is at 40 decibels, what is the intensity of the 10,000-Hz sound?

3. A sound wave takes 0.2 seconds to travel 306 meters. What is the speed of sound in this material? Through which of the materials in Figure 24.5 is the wave traveling?

Section 24.2

4. Suppose you stand in front of a tall rock wall that is 170 meters away. If you yell, how long does it take for the echo to get back to your ears if the speed of sound is 340 m/s?

5. A sound wave has a speed of 340 m/s and a wavelength of 10 meters. What is its frequency? Would you be able to hear this sound?

6. The range of human hearing is between 20 Hz and 20,000 Hz. If the speed of sound is 340 m/s, what is the longest wavelength you can hear? What is the shortest?

Section 24.3

7. The note E has a frequency of 330 Hz. What is the frequency of the E note one octave higher?

8. A note has a frequency of 988 Hz. What is the frequency of the note one octave lower? What note is this?

Applying Your Knowledge

Section 24.1

1. People can usually hear sounds with frequencies between 20 Hz and 20,000 Hz. Some animals can hear higher or lower frequencies than people can. Research to find out the hearing ranges of several different animals.

2. The Doppler effect is used to figure out whether stars are moving toward or away from Earth. Red light has a lower frequency than blue light. If the color of a star's light shifts to red, is it moving toward or away from Earth?

3. Light waves travel at 300,000 km/s. Sound waves in warm air travel at approximately 0.34 km/s. During a thunderstorm, a lightning bolt strikes 2 kilometers away from you. How long does it take you to see the lightning? How long does it take you to hear the thunder?

Section 24.2

4. Science fiction movies sometimes show explosions in outer space that make loud sounds. Explain why this is not scientifically correct.

Section 24.3

5. The *beat frequency* is the frequency of the loud and soft sounds heard when two sounds create beats. It is calculated by subtracting the frequencies of the two different sound waves. For example, playing 322-Hz and 324-Hz sounds will result in a beat frequency of 2 Hz. Suppose you strike two tuning forks and hear a beat frequency of 4 Hz. One tuning fork has a frequency of 440 Hz. Can you determine the frequency of the other fork? Explain.

6. How do noise-cancelling headphones work? Do they work equally well for all types of sounds? Review the chapters in this unit and do research to find the answers to these questions.

Light

Why do people catch colds? For thousands of years, people believed that colds and other illnesses came from evil spirits. The world changed in 1673, when Anton Leeuwenhoek peered through a primitive microscope he had made. To his astonishment, he saw tiny creatures swimming around! Leeuwenhoek's discoveries revealed a miniature universe no human had ever seen before. He was the first to see that a drop of pond water contains a tiny world of plants and animals.

Once the microscopic world was discovered, the causes of sickness could be investigated. Today, we know that small forms of life, bacteria and viruses, are usually what make you sick. Microscopes and telescopes are based on optics, the science and technology of light. By manipulating light, optical devices greatly enhance our eyesight so that we can see things that are miniscule or astronomically far away.

CHAPTER 25 INVESTIGATIONS

25A: Color
What happens when you mix different colors of light?

25B: Reflection and Refraction
How does light behave when its path is changed?

25.1 Properties of Light

Every time you "see," you are using *light*. You can't see *anything* in complete darkness! Whether you are looking at a light bulb, a car, or this book, light brings information to your eyes. In fact, *seeing* means receiving light and forming images in your mind from the light received by your eyes. This chapter is about light—where it comes from, its many useful properties, and how it is related to color.

What is light?

Light is a form of energy   **Light**, like sound and heat, is a form of energy. Our understanding of light starts with what light does and what its properties are (Figure 25.1). We know that light:

- travels extremely fast and over long distances;
- carries energy and information;
- travels in straight lines;
- bounces and bends when it comes in contact with objects;
- has color; and
- has different intensities and can be bright or dim.

Seeing with reflected light   What happens when you "see" this page? Light in the room reflects off the page and into your eyes. The reflected light carries information about the page that your brain uses to make a mental picture of the page. You see because light in the room *reflects* from the page into your eyes. If you were sitting in a perfectly dark room with no light, you would not be able to see this page at all because the page does not give off its own light. *We see most of the world by reflected light.*

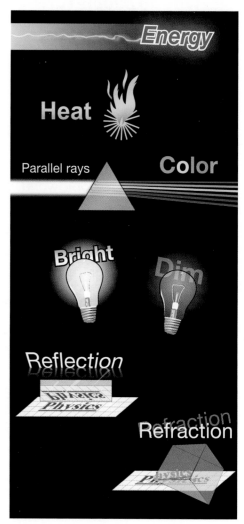

Figure 25.1: *Here are some words and properties that are associated with light. What words do you use to describe light?*

Most light comes from atoms

The electric light   For most of human history, people relied on the Sun, the Moon, and fire to provide light. Thomas Edison's electric light bulb (1879) changed our dependence on fire and daylight forever. The electric light is one of the most important inventions in the progress of human development.

Light is produced by atoms   Whether in an electric bulb or in the Sun, light is mostly produced by atoms. Here's an analogy. When you stretch a rubber band, you give the rubber band elastic energy. You can use that energy to launch a paper airplane. In this case, the energy is released as kinetic energy of the flying airplane. Unlike a rubber band, an atom releases the extra energy usually—but not always—as light!

Incandescent light bulbs   In order to get light out of an atom, you must first put some energy into the atom. One way to do this is with heat. Making light with heat is called **incandescence**. Incandescent bulbs use electric current to heat a thin metal wire called a *filament*. The atoms of the filament convert electrical energy to heat and then to light. Unfortunately, incandescent bulbs are not very efficient. Only a fraction of the energy of electricity is converted into light. Most of the energy becomes heat. In restaurants, this feature of incandescent bulbs is used to warm food.

Compact fluorescent light bulbs   Another common kind of electric light is a compact fluorescent light (CFL). CFLs are much more efficient than and provide four times as much light as incandescent bulbs. CFLs consists of a thin tube coiled into a bulb. High-voltage electricity is used to energize atoms of gas inside the tube. These atoms release the electrical energy as light (not heat) in a process called **fluorescence**. The atoms in the tube emit high-energy ultraviolet light—the same kind that gives you a sunburn. The ultraviolet light is absorbed by other atoms in a white coating on the inside surface of the tube. The atoms that make up this coating re-emit the energy as white light (Figure 25.2). Even with this two-step process, CFLs are still more efficient at producing light than incandescent bulbs.

LED bulbs   LED stands for light-emitting diode. A diode is an electronic device used in circuits to allow electricity to flow in one direction while blocking current from flowing in the other direction (Figure 25.3). Diodes are found in almost every electronic device in use today. The advantage of LEDs is that they need very little current to produce light. Arrays of LEDs are assembled into bulb shapes that are even more efficient than CFLs.

Compact fluorescent lamp

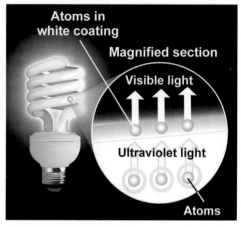

Figure 25.2: *Fluorescent lights generate light by exciting atoms with electricity in a two-step process.*

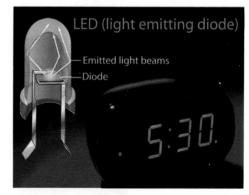

Figure 25.3: *The display on this alarm clock is made of LEDs.*

Color and energy

White light   When all the colors of the rainbow are combined, we see light without *any* color. We call the combination of all colors **white light**. The light that is all around us most of the time is white light. The light from the Sun and the light from most electric lights is white light.

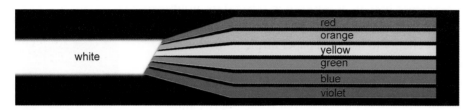

What is color?   Not all light has the same energy. **Color** is how we perceive the energy of light. This definition of color was proposed by Albert Einstein. All of the colors in the rainbow are light of different energies. Red light has the lowest energy we can see, and violet light has the highest energy. As we move through the rainbow from red to yellow to blue to violet, the energy of the light increases.

Blue flames = High energy

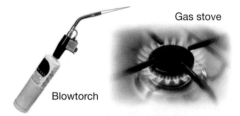

Gas stove

Blowtorch

Color and energy   What do we mean when we talk about the energy of light? Compare the blue flame from a gas stove to the orange flame of a match. The gas flame has more energy than the cooler flame of the match. The light from a gas flame is blue (high energy), and the light from a match is red-orange (low energy) (Figure 25.4).

Yellow-red flames = Low energy

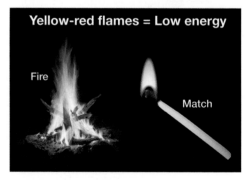

Fire

Match

Photons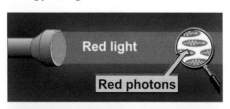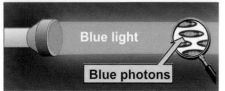

Just as matter is made of atoms, light energy comes in tiny wave-bundles called **photons**. In some ways, photons act like jellybeans of different colors. Each photon has its own color (energy), no matter how you mix them up. The lowest-energy photons we can see are dull red, and the highest-energy photons are blue-violet.

Figure 25.4: *High-energy flames such as the ones from a gas stove produce blue light. Fire flames are lower energy and produce yellow-red light.*

The speed of light

Comparing the speeds of sound and light
Think about what happens when you shine a flashlight on a wall that is far away. You don't see a time delay as the light leaves your flashlight, travels to the wall, bounces off, and comes back to your eyes. But that is exactly what happens. You don't notice it because it happens so *fast*. Suppose the wall is 170 meters away. The light travels to the wall and back in about one-millionth of a second (0.000001 s). Sound travels much slower than light. If you shout, you will hear an echo one full second later from the sound bouncing off the wall and back to your ears. Light travels almost a million times faster than sound!

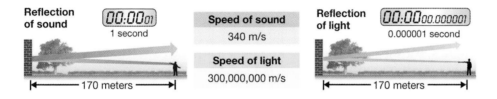

The speed of light, $c \approx 3 \times 10^8$ m/s
The speed at which light travels through air is about 300 million meters per second. Light is so fast that it can travel around the entire Earth 7.5 times in 1 second. The *speed of light* is so important in physics that it is given its own symbol, a lower case c. When you see this symbol in a formula, remember that it means the speed of light (c = 300,000,000 m/s).

Why you hear thunder after you see lightning
The speed of light is so fast that when lightning strikes a few miles away, we hear the thunder several seconds after we see the lightning. At the point of the lightning strike, the thunder and lightning are simultaneous. But just a mile away from the lightning strike, the sound of the thunder is already about 5 seconds behind the flash of the lightning. You can use this information to calculate how far you are away from a thunderstorm (see the sidebar at right).

The wavelength and frequency of light

Wavelength of light   The wavelength of visible light is very small. For example, waves of orange light have a length of only 0.0000006 meter. Because the wavelength of light is so small, scientists use **nanometers** to measure it. One nanometer (nm) is one-billionth of a meter (0.000000001 m). Figure 25.5 shows the size of a light wave relative to other small things. Thousands of wavelengths of red light would fit in the width of a single hair on your head!

Frequency of light   The frequency of light waves is very high. For example, red light has a frequency of 460 trillion, or 460,000,000,000,000 cycles per second. To manage these large numbers, scientists use units of terahertz (THz) to measure light waves. One THz is a trillion Hz (1,000,000,000,000 Hz).

Wavelength, frequency, color, and energy   As with other waves, the wavelength and frequency of light are inversely related. As frequency increases, wavelength decreases. Red light has a lower frequency and longer wavelength than blue light. Blue light has a higher frequency and shorter wavelength than red light.

Energy	Color	Wavelength (nanometers)	Frequency (THz)
Low	Red	650	462
	Orange	600	500
	Yellow	580	517
	Green	530	566
	Blue	470	638
High	Violet	400	750

As you can see from the table above, energy and frequency are directly related. The higher the frequency, the higher the energy. Because color is related to energy, the table also shows the relationships between color, frequency, and wavelength.

Size

Visible to the human eye
Bee 0.01 m
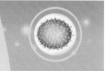

Microscopic
Pollen 0.00003 m

Appears as orange light to the human eye
Light wave 0.0000006 m
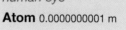

Invisible to the human eye
Atom 0.0000000001 m

Figure 25.5: *The sizes of some objects compared to the wavelength of a light wave.*

What kind of wave is light?

Light comes from electricity and magnetism

A sound wave is an oscillation of air. A water wave is an oscillation of the surface of water. What is oscillating in a light wave? The answer is electricity and magnetism. Imagine you have two magnets. One hangs from a string, and the other is in your hand. If you wave the magnet in your hand back and forth, you can make the magnet on the string sway back and forth, too (Figure 25.6). How does the oscillation of one magnet get to the other one? In Chapter 22, you learned that magnets create an invisible magnetic field around themselves. When you move a magnet in your hand back and forth, you make a change in the magnetic field. The changing magnetic field causes the other magnet to move. In a similar way, the force between two electric charges is carried by an *electric field*.

Electromagnetic waves

Any change in the electric or magnetic field travels at the speed of light. If you could shake your magnet (or electric charge) back and forth *100 million times per second,* you would make an electromagnetic wave. In fact, it would be an FM radio wave at 100 million Hz (100 MHz). An **electromagnetic wave** is a traveling oscillation in the electric and magnetic fields.

The hard way to make red light

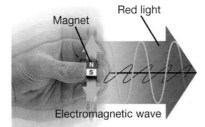

Magnet — Red light — Electromagnetic wave

If you could shake the magnet up and down 450 *trillion* times per second, you would make waves of red light. Red light is a traveling oscillation (wave) in the electric and magnetic fields with a frequency of about 450 THz.

Oscillations of electricity or magnetism create light waves

Anything that creates an oscillation of electricity or magnetism also creates electromagnetic waves. If you switch electricity on and off repeatedly in a wire, the oscillating electricity makes an electromagnetic wave. This is exactly how radio towers make radio waves. Electric currents oscillate up and down the metal towers and create electromagnetic waves of the right frequency to carry radio signals.

▬▬ VOCABULARY ▬▬

electromagnetic wave – a wave of electricity and magnetism that travels at the speed of light; light is an electromagnetic wave

Moving this magnet...

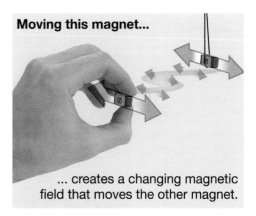

... creates a changing magnetic field that moves the other magnet.

Moving this charge...

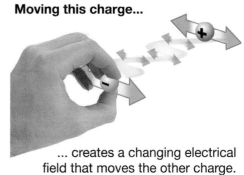

... creates a changing electrical field that moves the other charge.

Figure 25.6: *Magnets influence each other through the magnetic field. Charges influence each other through the electric field.*

The electromagnetic spectrum

Waves in the electromagnetic spectrum
The entire range of electromagnetic waves, including all possible frequencies, is called the **electromagnetic spectrum**. The electromagnetic spectrum includes radio waves, microwaves, infrared light, ultraviolet light, X-rays, and gamma rays. As you can see from the chart below, we use electromagnetic waves for all kinds of human technologies.

The electromagnetic spectrum

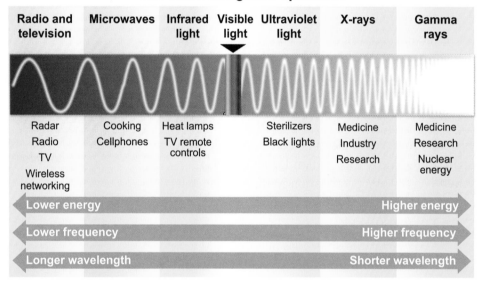

Radio and television	Microwaves	Infrared light	Visible light	Ultraviolet light	X-rays	Gamma rays
Radar	Cooking	Heat lamps		Sterilizers	Medicine	Medicine
Radio	Cellphones	TV remote controls		Black lights	Industry	Research
TV					Research	Nuclear energy
Wireless networking						

Lower energy → Higher energy

Lower frequency → Higher frequency

Longer wavelength → Shorter wavelength

Properties of electromagnetic waves
You can see that visible light is a small group of frequencies in the middle of the spectrum, between infrared and ultraviolet. The rest of the spectrum is invisible for the same reason you cannot see the magnetic field between two magnets. The energies are either too low or too high for the human eye to detect. Visible light includes only the electromagnetic waves with the range of energy that can be detected by the human eye. Some insects and animals can see other frequencies, including some infrared and some ultraviolet light.

TECHNOLOGY

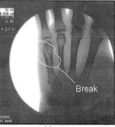

X-ray

X-rays

X-rays are high-energy electro-magnetic waves used in medicine and industry. The wavelength range is from about 10 nanometers to about 0.001 nm (or 10-trillionths of a meter). When you get a medical X-ray, the film darkens where bones are because calcium and other elements in your bones absorb the X-rays before they reach the film. X-rays show the extent of an injury such as a broken bone.

CHALLENGE

Who discovered that white light contains all colors? How was the discovery made? When was it made? This famous scientist is mentioned in this book, but not in connection with light!

Section 25.1 *Review*

1. Which of the following is *not* a property of light?
   a. light is a form of matter less dense than air
   b. light travels in straight lines
   c. light has different colors
   d. light has different intensities and can be bright or dim

2. If a room were completely dark, could you see your hand? Could you see a television screen? Explain the difference.

3. Most light comes from
   a. vibrating surfaces
   b. atoms
   c. conversion of frequency to wavelength

4. Compared to sound waves, the frequency of light waves is:
   a. much lower
   b. about the same
   c. much higher

5. Which electromagnetic wave has less energy than visible light and more energy than radio waves?
   a. microwaves
   b. ultraviolet light
   c. gamma rays
   d. X-rays

6. How are all electromagnetic waves similar? How are they different?

SOLVE IT!

The speed of light is frequency multiplied by wavelength, the same as for other waves. Suppose you make light with a frequency of 600 THz.

a. What is the wavelength of this light?

b. Describe what color the light would appear to your eye.

You will have to use scientific notation to solve this problem with your calculator. If necessary, ask your teacher or a friend for help.

25.2 Color and Vision

The energy of light explains how different colors are physically different. But it doesn't explain how we *see* colors. How does the human eye see color? The answer to this question also explains why computers and TVs can make virtually all colors by combining only three colors!

The human eye

Photoreceptors Light enters your eye through the lens then lands on the retina. On the surface of the retina are light-sensitive cells called *photoreceptors* (Figure 25.7). When light hits a photoreceptor cell, the cell releases a chemical signal that travels along the optic nerve to the brain. In the brain, the signal is translated into a perception of color.

Cone cells respond to color Our eyes have two kinds of photoreceptors, called *cones* and *rods*. Cones (or cone cells) respond to color (Figure 25.7). There are three types of cone cells. One type responds best to low-energy (red) light. Another type responds best to medium-energy (green) light. The third type responds best to higher-energy (blue) light.

Rod cells respond to light intensity The second kind of photoreceptors are called rods or rod cells. Rods respond to differences in light intensity, but not to color (Figure 25.7). Rod cells "see" black, white, and shades of gray. However, rod cells are much more sensitive than cone cells. At night, colors seem washed out because there is not enough light for cone cells to work. When the light level is very dim, you see "black-and-white" images from your rod cells.

Black-and-white vision is sharper than color vision A human eye has about 130 million rod cells and 7 million cone cells. Each cell contributes a "dot" to the image assembled by your brain. Because there are more rod cells, things look sharpest when there is a big difference between light and dark. That's why black letters on a white background are easier to read than colored letters. Each cone cell "colors" the signals from the surrounding rod cells. Because there are fewer cone cells, our color vision is less sharp than our black-and-white vision.

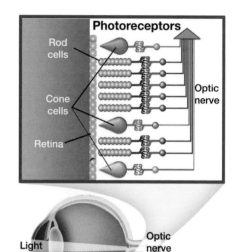

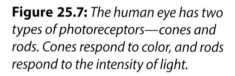

Figure 25.7: *The human eye has two types of photoreceptors—cones and rods. Cones respond to color, and rods respond to the intensity of light.*

How we see colors

The additive color process

Because there are three kinds of cone cells, our eyes work by adding three signals to "see" different colors. The color you "see" depends on how much energy is received by each of the three different types of cone cells. The brain thinks "green" when there is a strong signal from the green cone cells but no signal from the blue or red cone cells (Figure 25.8).

How we perceive color

What color would you see if light creates signals from both the green cones and the red cones? If you guessed *yellow*, you are right. We see yellow when the brain sees yellow light or when it gets an equally strong signal from both the red and the green cone cells at the same time. Whether the light is actually yellow, or a combination of red and green, the cones respond the same way and we perceive yellow. If the red signal is stronger than the green signal, we see orange (Figure 25.9). If all three cones send an equal signal to the brain, we see white.

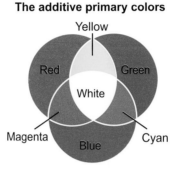

The additive primary colors

Two ways to see a color

The human eye can be "tricked" into seeing any color by adding different percentages of red, green, and blue. For example, an equal mix of red and green light looks yellow. However, *the light itself is still red and green!* The mix of red and green creates the same response in your cone cells as does true yellow light.

Do animals see colors?

To the best of our knowledge, primates (such as chimpanzees and gorillas) are the only animals with three-color vision similar to that of humans. Some birds and insects can see ultraviolet light, which humans cannot see. Dogs, cats, and some squirrels are thought to have only two color photoreceptors. Although both octopi and squid can change their body color better than any other animal, we believe they cannot detect color with their own eyes!

Color signals from only the green cones tell the brain the leaf is green.

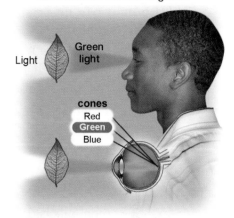

Figure 25.8: *If the brain gets a signal from only the green cone, we see green.*

A strong signal from the red cones and a weaker signal from the green cones tell the brain the fruit is orange.

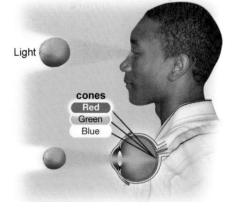

Figure 25.9: *If there is a strong red signal and a weak green signal, we see orange.*

25.2 Color and Vision   **613**

Making an RGB color image

The RGB color process Color images in TVs and computers are based on the **RGB color model**. RGB stands for "Red-Green-Blue." If you look at a TV screen with a magnifying glass, you see thousands of tiny red, green, and blue **pixels** (Figure 25.10). A television makes different colors by lighting red, green, and blue pixels to different percentages. For example, a light brown tone is 88 percent red, 85 percent green, and 70 percent blue. A computer monitor works the same way.

Pixels make up images TVs, digital cameras, and computers make images from thousands of pixels. An ordinary TV picture is 640 pixels wide × 480 pixels high, for a total of 307,200 pixels. A high-definition picture looks sharper because it contains more pixels. In the 720p format, HDTV images are 1,280 pixels wide × 720 pixels high, for a total of 921,600 pixels. This is four times as sharp as a standard TV image.

How video cameras create color images Like the rods and cones in your retinas, a video camcorder has tiny light sensors on a small chip called a CCD (Charge-Coupled Device). There are three sensors for each pixel of the recorded image, red, green, and blue. In HDTV, that means each recorded image contains 921,600 × 3 = 2,764,800 numbers. To create the illusion of motion, the camera records 30 images per second. In terms of data, the HDTV movie you watch represents 2,764,800 × 30, or about 83 million numbers every second.

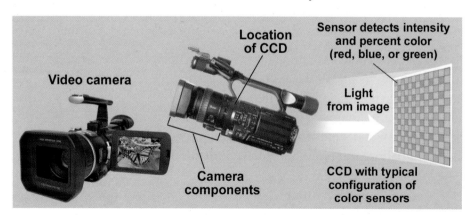

Video camera

Location of CCD

Sensor detects intensity and percent color (red, blue, or green)

Light from image

Camera components

CCD with typical configuration of color sensors

RGB color model – a model for tricking the eye into seeing almost any color by mixing proportions of red, green, and blue light

pixel – a single dot that forms part of an image made of many dots

Individual pixels

Figure 25.10: *A television makes colors using tiny glowing dots of red, green, and blue.*

How objects appear to be different colors

What gives objects their color? Your eye creates a sense of color by responding to red, green, and blue light. You don't see objects in their own light; you see them in reflected light! A blue shirt looks blue because it *reflects blue light into your eyes* (Figure 25.11). However, the shirt did not *make* the blue light. The color blue is not *in* the cloth. The blue light you see is blue light mixed into white light that shines on the cloth. You see blue because the other colors in white light have been subtracted out (Figure 25.12).

The subtractive color process Colored fabrics and paints get color from a *subtractive color process*. Chemicals known as *pigments* in dyes and paints absorb some colors and reflect other colors. Pigments work by taking away colors from white light, which is a mixture of all the colors.

The subtractive primary colors

Figure 25.11: *Why is a blue shirt blue?*

The subtractive primary colors To make all colors by subtraction, we need three primary pigments. We need one that absorbs blue (reflects red and green). This pigment is called *yellow*. We need another pigment that absorbs green (reflects red and blue). This is a pink-purple pigment called *magenta*. The third pigment is *cyan*, which absorbs red (reflects green and blue). Cyan is a greenish shade of light blue. Magenta, yellow, and cyan are the three *subtractive primary colors* (see illustration above). Different proportions of the three subtractive primary colors change the amount of reflected red, green, and blue light.

How white is white? A blue shirt won't look blue in red light! It will look *black*! The subtractive color model assumes that a painted or dyed surface is seen in white *sunlight* containing a precise mix of colors. If the "white" has a different mix than sunlight, colors don't look right. This is why home videos made under fluorescent lights often look greenish. The white from fluorescent lights has a slightly different mix of colors than the white from sunlight.

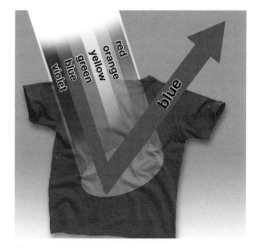

Figure 25.12: *The pigments in a blue cloth absorb all colors except blue. You see blue because blue light is reflected to your eyes.*

The CMYK color process

A subtractive color process The subtractive color process is often called **CMYK** for the four pigments it uses. CMYK stands for cyan, magenta, yellow, and *black*. The letter K stands for black because the letter B is used for the color blue in RGB. Color printers and photographs use CMYK.

■■■ VOCABULARY ■■■

CMYK – the subtractive color process using cyan, magenta, yellow, and black to create colors in reflected light

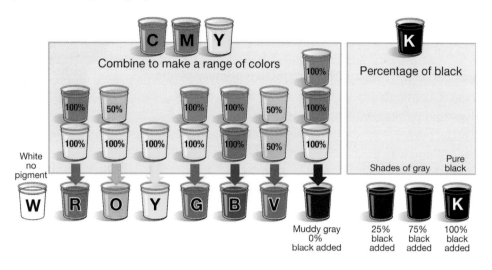

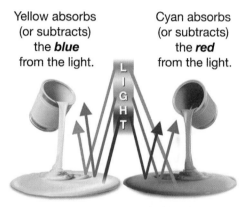

Yellow absorbs (or subtracts) the **blue** from the light.

Cyan absorbs (or subtracts) the **red** from the light.

The mix of cyan and yellow subtract both red and blue, so only **green** from the light remains.

CMYK are pigments The three pigments—cyan, magenta, and yellow—can combine in different proportions to make any color of reflected light. Figure 25.13 shows how CMYK pigments make green. Theoretically, mixing cyan, magenta, and yellow should make black, but in reality, the result is only a muddy gray. This is why a fourth color, pure black, is included in the CMYK process.

To make	Mix	Because	White light
Red	Magenta and yellow +	Magenta absorbs green Yellow absorbs blue Red gets reflected	
Blue	Magenta and cyan +	Magenta absorbs green Cyan absorbs red Blue gets reflected	
Green	Cyan and yellow +	Cyan absorbs red Yellow absorbs blue Green gets reflected	

Figure 25.13: *Creating the color green using cyan and yellow paints.*

Why plants are green

Light is necessary for photosynthesis   Plants absorb energy from light and convert it to chemical energy in the form of sugar. This process is called *photosynthesis*. The vertical (*y*) axis of the graph in Figure 25.14 shows the percentage of different colors of light that are absorbed by a plant. The *x*-axis on the graph shows the colors of light. The graph line shows how much and which colors of visible light are absorbed by plants. Based on this graph, can you explain why plants look green?

Why most plants are green

The important molecule that absorbs light in a plant is called *chlorophyll*. There are several forms of chlorophyll. They absorb mostly blue and red light, and reflect green light. This is why most plants look green. The graph in Figure 25.14 shows that plants absorb red and blue light to grow. A plant will die if placed under only green light!

Plants reflect some light to keep cool   Why don't plants absorb all colors of light? The reason is the same reason you wear light-colored clothes when it's hot outside. Like you, plants must reflect some light to avoid absorbing too much energy and overheating. Plants use visible light because the energy is just enough to change certain chemical bonds, but not enough to completely break them. Ultraviolet light has more energy but would break chemical bonds. Infrared light has too little energy to change chemical bonds.

Why leaves change color   The leaves of some plants, such as sugar maple trees, turn brilliant red or gold in the fall. Chlorophyll masks other plant pigments during the spring and summer. In the fall, when photosynthesis slows down, chlorophyll breaks down and red, orange, and yellow pigments in the leaves are revealed.

Absorption of light by plants

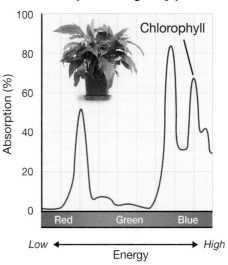

Figure 25.14: *Plants absorb energy from light. The plant pigment chlorophyll absorbs red and blue light, and reflects green light. This is why plants look green!*

CHALLENGE

What about red plants?

All plants that use sunlight to grow have chlorophyll, but some do not look green. Come up with a hypothesis to explain this observation.

Section 25.2 *Review*

1. If humans have only three kinds of color photoreceptors, how can we see so many different colors?

2. Why is it easier to read black text on a white background than to read green text or text of any light color?

3. Why might it be a good idea to put a light in your clothes closet? (*Hint*: What kind of vision do we have in dim light?)

4. Do you think this textbook was printed using the CMYK color process or the RGB color process? Explain your answer.

5. If you were going to design the lighting for a play, would you need to understand the CMYK color process, the RGB color process, or both? Explain your answer.

6. Suppose you have cyan, magenta, yellow, and black paint. Which colors would you mix to get blue?

7. Why does static on a television screen appear white?

8. How is the color black produced in the CMYK color process? How does this differ from the RGB color process?

9. A red shirt appears red because
   a. the shirt reflects red light
   b. the shirt absorbs red light
   c. the shirt emits green and blue light
   d. the shirt reflects magenta and blue light

10. What would happen if you tried to grow a green plant in pure green light? Would the plant live? Explain your answer.

11. Propose an explanation for how the top image in Figure 25.15 is related to the four images below it.

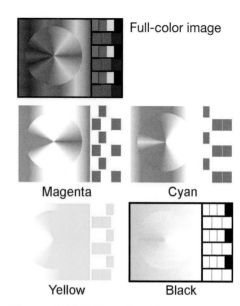

Full-color image

Magenta   Cyan

Yellow   Black

Figure 25.15: *Question 12.*

▰▰▰ CHALLENGE ▰▰▰

Pictures from dots

A color printer, such as an inkjet printer, makes color images by printing small dots. If there were only four dots per inch, your eye would see the individual dots instead of the picture the dots are supposed to make. How many dots must there be (per inch) to trick the eye into seeing a smooth image? How many dots per inch do printers around your home or office use?

25.3 Optics

Optics is the science and technology of light. Almost everyone has experience with optics. For example, trying on new glasses, checking your appearance in a mirror, or admiring the sparkle from a diamond ring all involve optics.

Basic optical devices

Lenses   A **lens** bends light in a specific way. A *converging lens* bends light so that the light rays come together in a point. This is why a magnifying glass makes a hot spot of concentrated light (Figure 25.16). A *diverging lens* bends light so it spreads light apart instead of bringing it together. An object viewed through a diverging lens appears smaller than it would look without the lens.

Mirrors   A **mirror** reflects light and allows you to see yourself. Flat mirrors show a true-size image. Curved mirrors distort images. The curved surface of a fun house mirror can make you look appear thinner, wider, or even upside-down!

Prisms   A **prism** is usually made of a solid piece of glass with flat polished surfaces. A common triangular prism is shown in the picture below. Prisms can both bend and/or reflect light. Telescopes, cameras, and supermarket laser scanners use prisms of different shapes to bend and reflect light in precise ways. A diamond is a prism with many flat, polished surfaces. The "sparkle" that makes diamonds so attractive comes from light being reflected many times as it bounces around the inside of a cut and polished diamond.

Figure 25.16: *A magnifying glass is a converging lens. This is why a magnifying glass can be used to make a hot spot of concentrated light. You should* not *try this yourself—the science is interesting, but can be unsafe.*

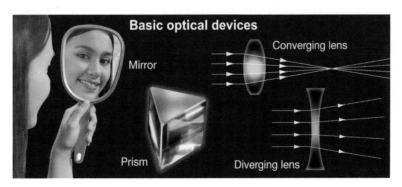

Basic optical devices

Mirror

Prism

Converging lens

Diverging lens

How light is affected by matter

transparent – allows light rays to pass through without scattering.

translucent – allows light rays through but scatters them in all directions

Interactions   Since light rays travel through oscillating electric and magnetic fields, they do not need matter to propagate. What happens when light does encounter matter? How light responds to an object or a surface is called an interaction. Light has interactions whenever it meets matter, whether the matter is a leaf, a glass, or a mirror.

Some materials bounce or scatter light   Like a mirror, some materials bounce, or reflect, light off in a new direction. Almost all surfaces reflect some light. A mirror is a very good reflector, but a sheet of white paper is also a good reflector. The difference is in how they reflect. Sometimes light is bounced back, and sometimes light is scattered in different directions.

Some materials allow light to pass through   Materials that allow light to pass through are called **transparent**. Glass is transparent, as are some kinds of plastic. Air is also transparent. You can see an image though a transparent material if the surfaces are smooth, like a glass window. An object is **translucent** if some light can pass through but the light is scattered in many directions. Tissue paper is translucent, and so is frosted glass. If you hold a sheet of tissue paper up to a light, some light comes through the paper, but you cannot clearly see an image through it.

Some materials can take energy from light   Sometimes, light's energy is absorbed by the material it is passing through. A black road surface gets hot on a sunny day because it absorbs energy from sunlight. A pair of sunglasses is also an example of a material that absorbs light. Tinted glass lenses absorb some of the bright light to keep it from reaching your eyes. Different color tints absorb different wavelengths of light.

More than one interaction at once   Most of the time, all four interactions happen together. A polished glass window absorbs about 10 percent of light, while still reflecting bright sunlight back outside and allowing light to pass through to the inside. The type of interactions light has with matter depends on the type of material and the frequency (color) of light. Materials that do not absorb light well look white; those that absorb all colors appear black. Most materials have a combination of interactions that appear as different colors. Look at the illustration (left). Green paper absorbs some light, and reflects some light. Can you tell which colors are absorbed and which colors are reflected?

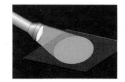

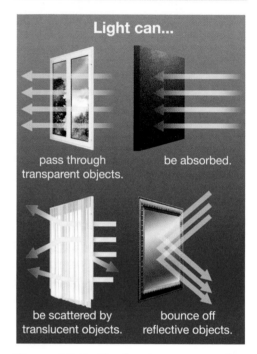

Light can...

pass through transparent objects.

be absorbed.

be scattered by translucent objects.

bounce off reflective objects.

Figure 25.17: *The four interactions of light with matter.*

Light rays

What are light rays? When light moves through a material, it travels in straight lines. Diagrams that show how light travels use straight lines and arrows to represent **light rays**. Think of a light ray as a thin beam of light, like a laser beam. The arrow shows the direction the light is moving.

Reflection and refraction When light rays move from one material to another, the rays may bounce or bend. **Reflection** occurs when light bounces off a surface. **Refraction** occurs when light bends while crossing a surface or moving through a material. Reflection and refraction cause many interesting changes in the images we see.

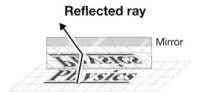
Reflected ray

Light rays are reflected in a mirror, causing an inverted image.

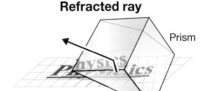
Refracted ray

Light rays are refracted (or bent) by a prism, causing the image to be distorted.

Reflection creates images in mirrors When you look in a mirror, objects that are in front of the mirror appear as if they are behind the mirror. Light from the object strikes the mirror and reflects to your eyes. The image reaching your eyes appears to your brain as if the object really *was* behind the mirror. This illusion happens because your brain "sees" the image where it would be if the light reaching your eyes had traveled in a straight line.

Refraction changes how objects look When light rays travel from air to water, they refract. This is why a straw in a glass of water looks broken or bent at the water's surface (Figure 25.18). Look at some objects through a glass of water; move the glass closer and farther away from the objects. What strange illusions do you see?

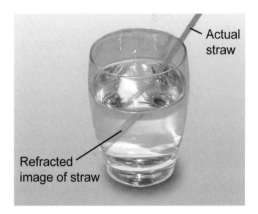

Figure 25.18: *Refraction bends light rays so the straw appears to be in a different place!*

Reflection

What is reflection? When you look directly into a mirror, your image appears to be the same distance from the other side of the mirror as you are on your side of the mirror. If you step back, so does your image. Reflected light forms images in mirrors.

The angle of incidence equals the angle of reflection Imagine a ray of light striking a mirror. The *incident ray* is the light ray that strikes the surface of the mirror. The *reflected ray* is the light ray that bounces off the surface of the mirror (Figure 25.19, top). The lower part of Figure 25.19 shows the reflection of a light ray. The angle of incidence is the angle between the incident ray and an imaginary line drawn perpendicular to the surface of the mirror called the *normal line. Perpendicular* means "at a 90-degree angle," also called a *right angle.* The angle of reflection is the angle between the reflected light ray and the normal line. The *Law of Reflection* states that the angle of incidence is equal to the angle of reflection.

Regular and scattered reflection When you look in a mirror, you can see your image because when parallel light rays hit the mirror at the same angle, they are all reflected at the same angle. This is called **specular reflection**. You can't see your image when you look at a white piece of paper because even though it seems smooth, its surface has tiny bumps on it. When parallel light rays hit a bumpy surface, the bumps reflect the light rays at different angles. Light rays reflected at different angles cause **scattered reflection**. Many surfaces, for example, polished wood, are in-between rough and smooth and create both types of reflection.

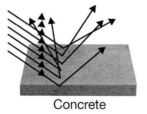

Regular reflection

Mirror

Scattered reflection

Concrete

Regular and scattered reflection

Polished wood

Reflection

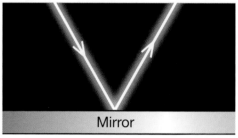

The angle of incidence is always equal to the angle of reflection.

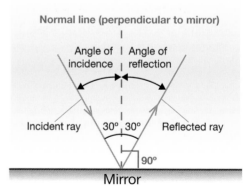

Figure 25.19: *The Law of Reflection states that the angle of incidence is always equal to the angle of reflection.*

Refraction

The index of refraction Eyeglasses, telescopes, binoculars, and fiber optics are a few inventions that use refraction to change the direction of light rays. Different materials have different abilities to bend light. Materials with a higher **index of refraction** bend light by a greater angle. The index of refraction for air is approximately 1.00. Water has an index of refraction of 1.33. A diamond has an index of refraction of 2.42. Diamonds sparkle because of their high index of refraction. Table 25.1 lists the index of refraction for some common materials.

Table 25.1: The index of refraction for some common materials

Material	Index of refraction
Air	1.00
Ice	1.31
Water	1.33
Glass	1.45–1.65
Diamond	2.42

The direction a light ray bends When light goes from air into glass (A), it bends toward the normal line because glass has a higher index of refraction than air. When the light goes from glass into air again (B), it bends away from the normal line. Coming out of the glass, the light ray is going into air with a lower index of refraction than glass.

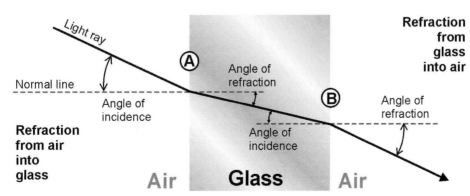

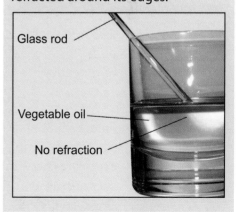

VOCABULARY

index of refraction – a number that measures how much a material is able to bend light

SCIENCE FACT

A trick of refraction

If two materials have the same index of refraction, light doesn't bend at all. Here's a neat trick you can do with a glass rod. You see the edges of a glass rod because of refraction. The edge appears dark because light is refracted away from your eyes.

Vegetable oil and glass have almost the same index of refraction. If you put a glass rod into a glass cup containing vegetable oil, the rod disappears because light is *not* refracted around its edges!

Lenses

A lens and its optical axis   An ordinary lens is a polished, transparent disc, usually made of glass. The surfaces are curved to refract light in a specific way. The exact shape of a lens's surface depends on how strongly and in what way the lens needs to bend light.

How light travels through a converging lens   The most common lenses, converging lenses, have surfaces shaped like part of a sphere. Any radius of a sphere is also a normal line to the surface. When light rays fall on a spherical surface from air, they bend *toward* the normal line (Figure 25.20). For a converging lens, the first surface (air to glass) bends light rays toward the normal line. At the second surface (glass to air), the rays bend *away* from the normal line. Because the second surface "tilts" the other way, it also bends rays toward the focal point.

Focal point and focal length   Light rays that enter a converging lens parallel to its axis bend to meet at a point called the *focal point* (see illustration below). Light can go through a lens in either direction, so there are always two focal points, one on either side of the lens. The distance from the center of the lens to the focal point is the *focal length*. The focal length is usually (but not always) the same for both focal points of a lens.

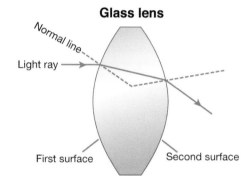

Figure 25.20: *Most lenses have spherically shaped surfaces.*

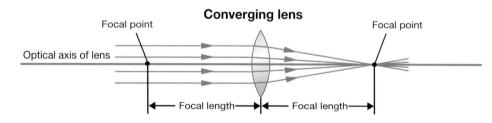

Converging and diverging lenses   Figure 25.21 shows how light rays enter and exit two types of lenses. The entering rays are parallel to the optical axis. A **converging lens** bends exiting rays toward the focal point. A **diverging lens** bends the rays outward, away from the focal point.

Figure 25.21: *Converging and diverging lenses.*

Section 25.3 *Review*

1. What process does a lens use to deflect light rays passing through it?
    a. reflection
    b. refraction
    c. absorption
    d. transparency

2. Can light be reflected and refracted at the same time? If so, give an example.

3. Make a list of all the optical devices you use on an average day.

4. Name an object that is mostly transparent, one that is translucent, one that is mostly absorbent, and one that is mostly reflective.

5. Windows that look into bathrooms are often translucent instead of transparent. Why?

6. Why can you see your own reflected image in a mirror but not on a dry, painted wall?

7. Why is the true surface of a perfect mirror invisible?

8. The index of refraction determines:
    a. the color of glass
    b. the ratio of thickness to focal length for a lens
    c. the amount a material bends light rays
    d. whether a material is transparent or translucent

9. A clear plastic rod seems to disappear when it is placed in water. Based on this observation and Table 25.1, predict the index of refraction for the plastic.

10. Fill in the blank. When light travels from water into the air, the refracted light ray bends _____ (away from or toward) the normal line.

11. What is the difference between a converging lens and a diverging lens?

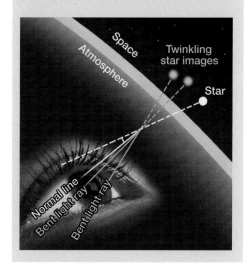

Searching the
Cosmos

The Hakeem Oluseyi Story

Astrophysicist Dr. Hakeem Oluseyi (Oh-lu-SHAY-ee) is fascinated by stars. A physics and space science professor at the Florida Institute of Technology, he has invented several new instruments to give astronomers a closer look at the cosmos.

Taking refuge in books

Dr. Hakeem Oluseyi remembers, "Growing up, I lived in tough neighborhoods all over the American south. My mom moved us every year, looking for better factory jobs, but things never really improved. In the ghettos where we lived, you always had to watch your back. I never felt entirely safe." As a result, he stayed inside a lot.

Hakeem's mother usually worked the evening shift, so he didn't see her a lot after school. But she loved to read, and often left library books for him on the kitchen table.

"By the time I was eleven, I had read our entire set of encyclopedias. My mom knew I loved science, and one time she brought home books on Albert Einstein and special relativity. That stuff just boggled my mind. Time slows down, lengths contract, masses increase? It shocked me!"

In high school, Hakeem taught himself computer programming, using a booklet that came with his girlfriend's father's home computer. He wrote a program demonstrating Einstein's theory of relativity. His science teacher encouraged him to enter it in a Mississippi state science fair competition. His was the only all-black high school to participate.

Eighth-grade graduation at Southside Middle School, Heidelberg, Mississippi

When Hakeem arrived at the fair, he found that his display board had been left behind at school. A friend told him, "Hakeem, you're our best chance to win this thing. Here—you take my board." They sat on the grass outside the exhibit hall and stapled his papers to the board. The wind was gusting hard and they had to chase after some pages, but managed to get it assembled just in time. Their perseverance paid off—Hakeem won first place in physics.

Hakeem went on to major in physics at Tougaloo College in Mississippi. He didn't want to ask his mother for money, so he worked as a hotel janitor to pay his living expenses. "There were hard times," he says, "where I ate the food people left on their room service trays—I was that hungry. But the worst part about those minimum wage jobs was that there was always somebody looking over my shoulder saying, 'Did you do what I told you to do?' and 'Why did you do that?' I never felt like I was trusted."

Freedom and respect in the research world

After his freshman year, Hakeem got a call from the University of Georgia, inviting him to participate in a summer research program.

"When I arrived, the professor assigned to be my mentor described what he wanted me to do. Then he said, 'Here are your keys to the building and the lab.' My jaw dropped to the floor."

"The first day I went into the lab at 8:00 a.m., and nobody was there. Finally, around 10:30, some graduate students showed up. So I asked them, 'What are the work hours around here?' A guy shrugged, 'You work when you want to work. It's all about getting the job done.'"

"So, the rest of the summer, I worked from about 6 p.m. to 3 a.m., and put my results on the advisor's desk. Afterward, they took me out to lunch and told me what a great job I had done. At that point I thought, this is where it's at. I'm going to have a career in research!"

Becoming an astrophysicist

After college, Oluseyi was accepted to a rigorous Ph.D. program at Stanford University. There, he learned to design space-borne astronomy tools under the guidance of the famous astrophysicist Dr. Arthur Walker II. Oluseyi remembers, "My first year at Stanford was really tough. I was used to making As in math and physics classes, but the level of expectation was so much higher here. At times I wanted to quit. But in my second year, I started to get the hang of things, and Dr. Walker encouraged me not to give up. By the time I was ready to defend my dissertation, I had published three papers and discovered a new class of structures on the Sun. The faculty recognized that I had made a significant contribution to the astrophysics field. That was a great moment in my life."

Hacking the stars

Dr. Oluseyi leads a research group at Florida Institute of Technology that "hacks" stars. To Dr. Oluseyi's group, hacking means studying stars not only to understand stars, but also to develop new technologies and investigate the structure and evolution of our galaxy and the universe. By studying how electromagnetic fields and plasmas interact at the Sun's surface, the group is already working on a new ion thruster in-space propulsion technology. They hope their work will help reveal secrets of the "dark energy" that may be responsible for the accelerating pace of the universe's expansion.

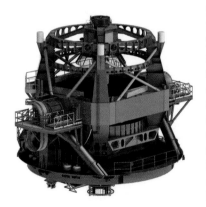

Render of the telescope model
Photo courtesy of LSST Project/J. Andrew

Dr. Oluseyi has been designing instruments for the Large Synoptic Survey Telescope (LSST). Once it is complete, the LSST will take high-quality images of an enormous area of the night sky, repeating every few nights, to help scientists study changes in the universe. The data gathered will include detailed catalogs of the night sky that will be thousands of times larger than anything done before. Dr. Oluseyi has worked on designing the LSST's camera and is developing programs for analyzing the enormous amounts of data it will collect.

Reaching out

In addition to his research, Dr. Oluseyi spends time sharing his passion for science. In 2002, he traveled to Africa with a group called Cosmos Education. They visited schools and orphanages,

teaching scientific principles and disease prevention through hands-on activities, skits, humor, and dance. Since then, Dr. Oluseyi has returned to Africa every year. He's currently working with an organization called Hands-on Universe to provide Kenyan high school students the opportunity to do real astronomy research by studying images downloaded from high-powered telescopes located at observatories in the United States and Australia.

Opportunities for astronomy research in Africa have been limited by a lack of Internet access. But, as Dr. Oluseyi put it, "That doesn't mean Africans aren't ready to do significant research. There are competent but under-utilized scientists across the continent and well-educated high school students eager for opportunities in science. So, we developed a new way to get data from observatories to areas with limited Internet access: We use a 'virtual observatory mode' that archives the data on DVDs, which can be delivered to researchers, teachers, and students without Internet access. A central server at a university provides periodic updates." What was once seen as a barrier to participation was just an obstacle to overcome.

Hakeem Oluseyi (right) with Kenyan high school students

"There's a thing that sometimes happens in the African community. It can be hard to see beyond your own little corner, to dream big dreams and believe they can happen. That's why I spend a lot of time in educational outreach. I want kids here and in Africa to know that they, too, can break out of that box."

Questions:

1. What is perseverance? Name three ways Dr. Oluseyi demonstrated this trait.

2. Give an example of perseverance from your own life.

3. What are two types of technology that Dr. Oluseyi has developed?

4. What does Dr. Oluseyi mean by "hacking stars"?

Chapter 25 *Assessment*

Vocabulary

Select the correct term to complete each sentence.

fluorescence	translucent	CMYK
visible light	RGB	incandescence
nanometer	index of refraction	photon
rod cell	white light	diffuse reflection
pixel	transparent	lens
cone cell	color	diverging lens
prism	converging lens	photoreceptor
electromagnetic spectrum	specular reflection	electromagnetic wave
	refraction	mirror

Section 25.1

1. _____ contains an equal mix of all colors.

2. You can use light produced by _____ to heat food.

3. Atoms produce light by _____ .

4. A(n) _____ travels at the speed of light.

5. A light wave at 500 THz is the _____ orange.

6. You see all the colors of _____ when you see a rainbow.

7. Light wavelengths are measured in _____(s).

8. Ultraviolet light and microwaves are part of the _____ .

9. A(n) _____ is the smallest possible amount of light.

Section 25.2

10. A(n) _____ specializes in detecting color.

11. A(n) _____ specializes in detecting light intensity.

12. The human eye has about 137 million _____(s).

13. Each tiny dot of color on your TV screen is called a(n) _____ .

14. Magenta is a pigment used in the _____ color model.

15. The _____ color model is used by video cameras to achieve a range of colors.

Section 25.3

16. A surface with _____ produces a single beam of reflected light rays.

17. _____ occurs when light enters a material and bends.

18. Three examples of optical devices are _____, _____, and _____ .

19. Glass is a(n) _____ material because light passes through it without scattering.

20. _____ materials allow light to pass but scatter it in all directions.

21. The _____ of water is 1.33.

22. Surfaces that scatter light when it reflects have _____ .

23. A(n) _____ bends light rays inward toward the focal point.

24. A(n) _____ bends light rays outward away from the focal point.

Concepts

Section 25.1

1. List four properties of light.

2. What is the main source of light?

3. Describe the difference between the light you would see from a flashlight and the light you see from a printed page.

4. Describe an electromagnetic wave. How is one made?

5. What is the relationship between the frequency of light and its wavelength?

6. Compare the speed, energy, wavelength, and frequency of red light and blue light.

7. A flame from a Bunsen burner is reddish at the top and blue near the opening of the burner. Where is the flame hottest? Explain your answer.

Section 25.2

8. Describe the kinds, number, and sensitivity of the photoreceptors in the human eye.

9. Your brain perceives color by an additive process. How would you see the following combinations of light colors?
   a. red + blue
   b. blue + green
   c. red + green
   d. red + blue + green

10. In the CMYK color process, why is black pigment used instead of mixing cyan, magenta, and yellow pigments?

11. Most objects do not make their own light, so how do we see the colors of these objects?

12. What colors of light are reflected by the color magenta?

13. For stage lighting for a play in a theater:
   a. a magenta spot of light is created along with a green spot of light. What happens when these two spots of light combine?
   b. light from a blue spotlight is combined with light from a green spotlight. What color light is produced?

14. What primary additive colors of light will be allowed to pass through a cyan filter?

15. Compare the way color is produced by a TV screen with how color is printed in an illustration in this book.

16. Why do the leaves of most plants look green?

Section 25.3

17. How do transparent and translucent materials differ?

18. Name the ways in which light may interact with matter. Give an example of a situation where more than one interaction happens at the same time.

19. What is a light ray?

20. Describe the difference between refraction and reflection.

21. Diamond has a higher index of refraction than water. What does this mean?

22. Explain the process lenses use to change the direction of light.

23. What is the difference between a converging lens and a diverging lens?

Problems

Section 25.1

1. Red light may have a wavelength of 0.00000078 meters. What is this wavelength in units of nanometers?

2. Frequencies of 462 THz, 517 THz, and 638 THz represent the frequencies of three colors: blue, red, and yellow. Match each frequency to its color.

3. Lightning strikes in the distance and six seconds later, thunder is heard. How far away was the lightning strike?

4. The Sun is about 150 million kilometers away from Earth. How long does it take for light to travel from the Sun to Earth?

Section 25.2

5. What color will a blue shirt appear in red light?

6. Which of the CMYK colors would you mix if you wanted to produce the following colors of ink?
   a. red
   b. green
   c. blue

7. Compare the quality of the images produced by your eyes to an HDTV screen in terms of pixels.

8. Identify the color process (RGB or CMYK) used in each step.
   a. taking a photograph with a digital camera
   b. the image appears on a computer monitor
   c. printing the image using a laser printer
   d. seeing the image on the paper with your eyes

9. Answer the following questions using the absorption graph shown.
   a. Which colors of light are absorbed the most by plants?
   b. Which colors of light are reflected the most by plants?
   c. Based on the information from the absorption graph, explain why a plant will grow more quickly if it is grown in white light rather than green light.

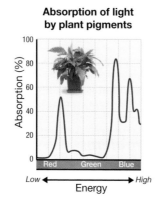

Absorption of light by plant pigments

Section 25.3

10. Glare from headlights can make it harder to see when driving at night. Glare is worse when the roads are wet than when roads are dry. Explain why, in terms of the two types of reflection.

11. Why do ambulances often have the letters for "AMBULANCE" reversed on the front of the vehicle?

12. A light ray crosses from a piece of glass into a liquid. You observe that the light ray bends away from the normal line when passing from the glass to the liquid. Based on this observation, how does the index of refraction for the liquid compare to the index of refraction for the glass?

13. A clear plastic ball seems to disappear when placed in a liquid. What does this tell you about the indices of refraction for the clear plastic and the liquid?

Applying Your Knowledge

Section 25.1

1. Thomas Edison is just one of many inventors who contributed to making electric light accessible to people. Research how one or more scientists contributed to the electric incandescent bulb. What is the general feeling about these types of light bulbs today?

2. Describe the health connection between ultraviolet light and vitamin D.

Section 25.2

3. Pick a common animal and find out about this animal's eyesight. Does it see colors? Is the animal nocturnal?

Section 25.3

4. Find out how white light is split into the colors of visible light by the following objects. Identify which light interaction is involved in splitting the light.
   a. a prism
   b. water droplets in the atmosphere
   c. a spectrometer

5. Your eye is an optical system that works together with your brain to help you see images. Research one of the following topics using these Internet search phrases: genetic diseases of the human eye, new technologies for vision correction, or preventing human eye problems. Use technology to present your findings.

Glossary

A

absolute zero – lowest possible temperature, at which thermal energy is as close to zero as it can be, approximately –273°C   **238**

absorption – what happens when the amplitude of a wave gets smaller and smaller as it passes through a material   **568**

acceleration due to gravity – the value of 9.8 m/s^2, which is the acceleration in free fall at Earth's surface, usually represented by the small letter g   **96**

acceleration – the rate at which velocity changes   **92**

accuracy – how close a measurement is to an accepted or true value   **20**

acid – a substance that produces hydronium ions (H_3O^+) when dissolved in water   **454**

activation energy – energy needed to break chemical bonds in the reactants to start a reaction   **411**

addition reaction – a chemical reaction in which two or more substances combine to form a new compound   **398**

alkali metals – elements in the first group of the periodic table   **338**

alloy – a solution of two or more solids   **442**

alternating current (AC) – electrical current that reverses direction at repeated intervals, as with household electricity   **518**

amino acids – a group of smaller molecules that are the building blocks of proteins   **373**

amorphous – a random arrangement of atoms or molecules in a solid   **272**

ampere (A) – the unit of electric current   **481**

amplitude – the amount that a cycle moves away from equilibrium   **558**

Archimedes' principle – states that the buoyant force is equal to the weight of the fluid displaced by an object   **285**

atmosphere – a layer of gases that surrounds a planet   **297**

atmospheric pressure – a measurement of the force of air molecules per unit of area in the atmosphere at a given altitude   **298**

atomic mass – the average mass of all the known isotopes of an element, expressed in amu   **337**

atomic mass unit – a unit of mass equal to 1.66×10^{-24} grams   **337**

atomic number – the number of protons in the nucleus of an atom; the atomic number determines what element the atom represents   **318**

atom – the smallest particle of an element that retains the chemical identity of the element   **229**

average speed – the total distance divided by the total time for a trip   **79**

Avogadro number – the number of atoms or molecules in a mole of any substance; the number equals 6.02×10^{23} **389**

axis – one of two (or more) number lines that form a graph   **83**

B

balanced forces – combined forces that result in a zero net force on an object   **124**

barometer – an instrument that measures atmospheric pressure   **300**

base – a substance that produces hydroxide ions (OH^-) when dissolved in water   **455**

battery – a device that transforms chemical energy to electrical energy, and provides electrical force in a circuit   **483**

beat – the oscillation between two sounds that are close in frequency **594**

Bernoulli's principle – a relationship that describes energy conservation in a fluid **279**

binary compound – a chemical compound that consists of two elements **365**

boiling point – the temperature at which a substance changes from liquid to gas (boiling) or from gas to liquid (condensation) **242**

Boyle's law – in a fixed quantity of a gas, the pressure and volume are inversely related if the mass and temperature are held constant **302**

brittleness – the tendency to crack or break; the opposite of elasticity **273**

buoyancy – the measure of the upward force that a fluid exerts on an object that is submerged **284**

C

carbohydrates – a group of energy-rich compounds that are made from carbon, hydrogen, and oxygen and that include sugars and starches **371**

catalyst – a molecule added to a chemical reaction that increases the reaction rate without getting used up in the process **419**

Celsius – a temperature scale in which water freezes at 0 degrees and boils at 100 degrees **234**

charged – describes an object whose net charge is not zero **473**

Charles's law – at constant pressure and mass, the volume of a gas increases with increasing temperature and decreases with decreasing temperature **305**

chemical bond – a bond that forms when atoms transfer or share electrons **354**

chemical change – a change that transforms one substance into a different substance **334**

chemical energy – a form of potential energy that is stored in molecules **165**

chemical equation – an expression of a chemical reaction using chemical formulas and symbols **391**

chemical equilibrium – the state at which the rate of the forward reaction equals the rate of the reverse reaction for a chemical reaction **420**

chemical formula – a representation of a compound that includes the symbols and ratios of atoms of each element in the compound **354**

chemical properties – characteristics that can only be observed when one substance changes into a different substance **270**

chemical reaction – the process of breaking chemical bonds in one or more substances and the reforming of new bonds to create new substances **385**

circuit breaker – an automatic device that trips like a switch to turn off an overloaded circuit **511**

circular wave – moving waves that have crests that form circles around a single point where the wave began **567**

closed circuit – a circuit with the switch in the *on* position, so there are *no* breaks and charge can flow **479**

CMYK – the subtractive color process using cyan, magenta, yellow, and black to create colors in reflected light **616**

coefficient – a whole number placed in front of a chemical formula in a chemical equation **392**

colloid – a mixture that contains evenly distributed particles that are 1 to 1,000 nanometers in size **443**

color – the sensation created by the different energies of light falling on your eye **606**

combustion reaction – a chemical reaction that results in a large amount of energy being released when a carbon compound combines with oxygen **401**

commutator – the device that switches the direction of electrical current in the electromagnet of an electric motor *541*

compound – a substance that contains two or more different elements chemically joined and has the same composition throughout *230*

compression – a "squeeze" or decrease in size *112*

concentration – the ratio of solute to solvent in a solution *448*

conductor – a material with low electrical resistance; metals such as copper and aluminum are conductors *491*

consonance – a combination of frequencies that sounds pleasant *594*

constant speed – speed that stays the same and does not change *87*

constructive interference – when waves add up to make a larger amplitude *570*

control variable – a variable that is kept constant (the same) in an experiment *64*

convection – the transfer of heat by the motion of matter, such as by moving air or water *260*

converging lens – a lens that bends exiting light rays toward the focal point *624*

conversion factor – a ratio that has a value of one, and is used when setting up a unit conversion problem *15*

coordinates – values that give a position relative to an origin *83*

coulomb (C) – the unit for electric charge *472*

covalent bond – a chemical bond formed by atoms that are sharing one or more electrons *354*

crystalline – an orderly, repeating arrangement of atoms or molecules in a solid *272*

cycle – a unit of motion that repeats *554*

D

decibel (dB) – a unit of measure for the intensity or strength of a sound *579*

decomposition reaction – a chemical reaction in which a compound is broken down into two or more smaller substances *399*

deduce – to figure something out from known facts using logical thinking *58*

density – the mass per unit volume of a given material; units for density are often expressed as g/mL, g/cm^3, or kg/m^3 *36*

dependent variable – the variable that you believe is influenced by the independent variable; the dependent variable can also be called the *responding variable* *43*

destructive interference – when waves add up to make a smaller, or zero, amplitude *570*

diffraction – the process of a wave bending around a corner or passing through an opening *568*

diffuse reflection – "dull" surface reflection, where each incident ray produces many scattered rays *622*

dimensional analysis – a method of using conversion factors and unit canceling to solve a unit conversion problem *15*

direct current (DC) – electrical current that flows in one direction, as with a battery *518*

direct relationship – a relationship in which one variable increases with an increase in another variable *45*

dissolution reaction – an endothermic reaction that occurs when an ionic compound dissolves in water to make an ionic solution *413*

dissolve – to separate and disperse a solid into individual particles in the presence of a solvent *444*

dissonance – a combination of frequencies that sounds unpleasant *594*

distance – the amount of space between two points *10*

diverging lens – a lens that bends exiting light rays outward, away from the focal point   *624*

DNA – a type of nucleic acid that contains the genetic code for an organism   *374*

Doppler effect – an increase or decrease in frequency caused by the motion of the source of an oscillation (such as sound)   *581*

double-displacement reaction – a chemical reaction in which ions from two compounds in solution exchange places to produce two new compounds   *400*

ductility – the ability to bend without breaking   *274*

E

efficiency – the ratio of usable output work divided by total input work, efficiency is often expressed as a percent, with a perfect machine having 100 percent efficiency   *194*

elasticity – the ability to be stretched or compressed and then return to original size   *273*

electrical conductor – a material that allows electricity to flow through easily   *343*

electrically neutral – describes an object that has equal amounts of positive and negative charges   *473*

electrical power – the rate at which electrical energy is changed into other forms of energy   *513*

electric charge – a fundamental property of matter that can be either positive or negative   *314*

electric circuit – a complete path through which electric charge can flow   *477*

electric current – the flow of electric charge   *476*

electricity – the science of electric charge and current   *476*

electric motor – a device that converts electrical energy into mechanical energy   *540*

electromagnet – a magnet created by a wire carrying electric current   *535*

electromagnetic induction – the process of using a moving magnet to create a current   *543*

electromagnetic spectrum – the entire range of electromagnetic waves, including all possible frequencies, such as radio waves, microwaves, X-rays, and gamma rays   *610*

electromagnetic wave – a wave of electricity and magnetism that travels at the speed of light; light is an electromagnetic wave   *609*

electron – a particle with an electric charge (–e) found inside atoms but outside the nucleus   *315*

element – a pure substance that cannot be broken down into simpler substances by physical or chemical means   *229*

elementary charge – the smallest unit of electric charge that is possible in ordinary matter; represented by the lowercase letter *e*   *314*

endothermic – describes a chemical reaction that uses more energy than it releases   *410*

energy – a quantity that describes the ability of an object to change or cause changes   *164*

energy level – one of the discrete allowed energies for electrons in an atom   *323*

engineer – a professional who uses scientific knowledge to create or improve inventions that solve problems and meet needs   *69*

engineering cycle – a process used to build and test devices that solve technical problems   *70*

English System – measurement system used for everyday measurements in the United States   *5*

enzyme – a type of protein used to speed up chemical reactions in living things   *373*

equilibrium – (1) the state in which the net force on an object is zero   *125*; (2) the state of a solution in which the dissolving rate equals the rate at which molecules come out of solution   *450*

excess reactant – a reactant that is not completely used up in a chemical reaction *417*

exothermic – describes a chemical reaction that releases more energy than it uses *410*

experimental technique – the exact procedure that is followed each time an experiment is repeated *65*

experimental variable – the variable you change in an experiment *64*

experiment – a situation specifically set up to investigate relationships between variables *64*

extension – a "stretch" or increase in size *112*

F

Fahrenheit – a temperature scale in which water freezes at 32 degrees and boils at 212 degrees *234*

fluid – any matter that flows when force is applied *276*

fluorescence – a process that makes light directly from electricity *605*

force – a push or pull, or any action that involves the interaction of objects and has the ability to change motion *108*

formula mass – the sum of the atomic mass values of the atoms in a chemical formula *389*

free-body diagram – a diagram showing all the forces acting on an object *127*

free fall – accelerated motion that happens when an object falls with only the force of gravity acting on it *96*

frequency – how often something repeats, expressed in hertz *556*

frequency spectrum – a graph that shows the amplitudes of different frequencies present in a sound *591*

friction – a force that resists motion *117*

fundamental – the lowest natural frequency of an oscillator *587*

fuse – a device with a thin wire that melts and breaks an overloaded circuit *511*

G

gas – a phase of matter that flows, does not hold its volume, and can expand or contract to fill a container *240*

gear – a rotating wheel with teeth that transfers motion and forces to other gears or objects *207*

generator – a device that converts kinetic energy into electrical energy through induction *543*

gram (g) – a unit of mass smaller than a kilogram; there are 1,000 grams in 1 kilogram *30*

graph – a visual representation of data *42*

ground fault interrupt (GFI) outlet – an outlet with an automatic device that protects you against electric shock *519*

group – a column of the periodic table *335*

H

half-life – a certain length of time after which half the amount of a radioactive element has undergone radioactive decay *427*

halogens – elements in the group containing fluorine, chlorine, and bromine, among others *338*

hardness – a measure of a solid's resistance to scratching *273*

harmonic motion – motion that repeats in cycles *554*

harmonic – one of many natural frequencies of an oscillator *587*

heat conduction – the transfer of heat by the direct contact of particles of matter *258*

heat – thermal energy that is moving or is capable of moving *252*

heat transfer – the flow of thermal energy from higher temperature to lower temperature   *258*

hertz (Hz) – the unit of frequency; one hertz is one cycle per second   *556*

heterogeneous mixture – a mixture in which different samples are not necessarily made up of the same proportions of matter   *231*

homogeneous mixture – a mixture that is the same throughout; all samples of a homogeneous mixture are the same   *231*

horsepower (hp) – a unit of power equal to 746 watts   *197*

hydrogen bond – an intermolecular force between the hydrogen on one molecule to an atom on another molecule   *438*

hypothesis – a possible explanation that can be tested by comparison with scientific evidence   *61*

I

incandescence – a process that makes light with heat   *605*

independent variable – a variable that you believe might influence another variable; the independent variable can also be called the *manipulated variable*   *43*

index of refraction – a number that measures how much a material is able to bend light   *623*

inertia – the property of an object that resists changes in its motion   *139*

inhibitor – a molecule that slows down a chemical reaction   *419*

input – forces, energy, or power supplied to make a machine accomplish a task   *206*

inquiry – a process of learning that starts with asking questions and proceeds by seeking the answers to the questions   *58*

insoluble – when a solute is unable to dissolve in a particular solvent   *445*

instantaneous speed – the actual speed of a moving object at any moment   *79*

insulator – a material that slows down or stops the flow of either heat or electricity   *343*

insulator (electrical) – a material with high electrical resistance; plastic and rubber are good insulators   *491*

intermolecular forces – forces between atoms or molecules in a substance that determine the phase of matter   *241*

inverse relationship – a relationship in which one variable decreases when another variable increases   *45*

ion – an atom (or group of atoms) that has an electric charge other than zero, created when an atom (or group of atoms) gains or loses electrons   *355*

ionic bond – a bond that transfers one or more electrons from one atom to another, resulting in attraction between oppositely charged ions   *355*

isotopes – atoms of the same element that have different numbers of neutrons in the nucleus   *319*

J

joule (J) – a unit of energy; 1 joule is enough energy to push with a force of 1 newton for a distance of 1 meter   *164*

K

Kelvin scale – a temperature scale that starts at absolute zero and has units the same as Celsius degrees   *238*

kilogram (kg) – the basic SI unit of mass   *30*

kilowatt-hours (kWh) – a unit of energy equal to one kilowatt of power used for one hour, equals 3.6 million joules   *516*

kilowatt (kW) – unit used to measure large amounts of power; 1 kilowatt equals 1,000 watts   *514*

kinetic energy – energy of motion   *170*

kinetic molecular theory – the concept that all atoms and molecules exhibit random motion   *419*

Kirchhoff's current law – states that all of the current entering a circuit branch must exit again   *507*

Kirchhoff's voltage law – the total of all voltage drops in a series circuit must equal the voltage supplied by the battery   *504*

L

law of conservation of energy – energy can never be created or destroyed, only transformed into another form; the total amount of energy in the universe is constant   *177*

law of conservation of mass – a principle that states that the total mass of the reactants equals the total mass of the products in a chemical reaction   *388*

law of conservation of momentum – a law that states that as long as interacting objects are not influenced by outside forces, the total amount of momentum is constant   *152*

length – a measured distance   *10*

lens – an optical device for bending light rays   *619*

lever – a stiff structure that rotates around a fixed point called a fulcrum   *207*

Lewis dot diagram – a method for representing an atom's valence electrons using dots around the element symbol   *359*

light – a form of electromagnetic energy that makes things visible   *604*

light ray – an imaginary line that represents a beam of light   *621*

limiting reactant – a reactant that is used up first in a chemical reaction   *417*

linear motion – motion that goes from one place to another without repeating   *554*

lipids – a group of energy-rich compounds that are made from carbon, hydrogen, and oxygen and that include fats, waxes, and oils   *372*

liquid – a phase of matter that holds its volume, does not hold its shape, and flows   *240*

longitudinal wave – a wave is longitudinal if its oscillations are in the direction it moves   *569*

M

machine – a device with moving parts that work together to accomplish a task   *206*

magnetic declination – the difference between true north and the direction a compass points   *532*

magnetic – describes a material that can respond to forces from magnets   *528*

magnetic field – the influence created by a magnet that exerts forces on other magnets and magnetic materials   *530*

malleability – the ability of a solid to be pounded into thin sheets   *274*

mass number – the number of protons plus the number of neutrons in the nucleus   *319*

mass percent – the mass of the solute divided by the total mass of the solution multiplied by 100   *448*

mass – the amount of matter in an object   *30*

matter – anything that has mass and takes up space   *30*

measurement – a determination of the amount of something; typically has two parts—a value and a unit   *4*

mechanical energy – a form of energy that is related to motion or position; potential and kinetic energy are examples   *165*

melting point – the temperature at which a substance changes from solid to liquid (melting) or liquid to solid (freezing)   *242*

metals – elements that are typically shiny and good conductors of heat and electricity   *336*

meter – a basic SI unit of length   *10*

mirror – a surface that reflects light rays   *619*

mixture – matter that contains a combination of different elements and/or compounds and can be separated by physical means   *231*

molarity – the moles of solute per liter of solution   *448*

molar mass – the mass, in grams, of one mole of a compound   *389*

mole – a unit of any substance that contains the Avogadro number of atoms or molecules   *389*

molecule – a group of two or more atoms joined together by chemical bonds   *230*

momentum – the mass of an object times its velocity   *152*

multimeter – a measuring instrument for current, voltage, and resistance   *482*

musical scale – a pattern of frequencies   *593*

N

nanometer – a unit of length equal to one-billionth of a meter (0.000000001 m)   *608*

natural frequency – the frequency at which a system oscillates when disturbed   *560*

natural law – a rule that describes an action or set of actions in the universe and that can sometimes be expressed by a mathematical statement   *58*

net force – the sum of all forces acting on an object   *124*

neutralization – the reaction of an acid and a base to produce a salt and water   *462*

neutron – a particle found in the nucleus with mass similar to the proton but with zero electric charge   *316*

Newton's first law – a law of motion that states that an object at rest will stay at rest and an object in motion will stay in motion with the same velocity unless acted on by an unbalanced force   *139*

Newton's second law – a law of motion that states that acceleration is force divided by mass   *144*

Newton's third law – a law of motion that states that for every action force, there is a reaction force equal in strength and opposite in direction   *149*

newton – the metric unit of force equal to the force needed to make a 1 kg object accelerate at 1 m/s^2 *109*

noble gases – elements in the group containing helium, neon, and argon, among others   *338*

nonmetals – elements that are poor conductors of heat and electricity   *336*

nonpolar molecule – a molecule that does not have distinctly charged poles   *437*

normal force – the perpendicular force that a surface exerts on an object that is pressing on it   *126*

note – one frequency in a musical scale   *593*

nuclear energy – a form of energy that is stored in the nuclei of atoms   *166*

nuclear fission – a nuclear reaction in which the nuclei of heavier atoms are split to make lighter atoms   *424*

nuclear fusion – a nuclear reaction in which the nuclei of lighter atoms are combined to make heavier atoms   *424*

nuclear reaction – a reaction in which the number of protons and/or neutrons is altered in one or more atoms   *422*

nucleic acids – compounds made of long, repeating chains of smaller molecules called nucleotides   *374*

nucleus – the tiny core at the center of an atom containing most of the atom's mass and all of its positive charge   *315*

O

objective – describes evidence that documents only what actually happened as exactly as possible **59**

octave – a range defined as being between a single frequency value and twice that frequency value; on a musical scale, these two notes would have the same name **593**

Ohm's law – states that the current is *directly* related to the voltage and *inversely* related to the resistance **488**

open circuit – a circuit with the switch in the *off* position, so there *is* a break and charge cannot flow **479**

organic chemistry – a branch of chemistry that specializes in the study of carbon compounds, also known as organic molecules **370**

origin – a place where the position has been given a value of zero **78**

oscillator – a physical system that has repeating cycles **555**

output – forces, energy, or power provided by the machine **206**

oxidation number – a quantity that indicates the charge on an atom when it gains, loses, or shares electrons during bond formation **361**

P

parallel circuit – an electric circuit with more than one path or branch **507**

Pascal's principle – the pressure applied to an incompressible fluid in a closed container is transmitted equally in all parts of the fluid **280**

pendulum – a device that swings back and forth due to the force of gravity **554**

percent yield – the actual yield of a product in a chemical reaction divided by the predicted yield and multiplied by 100 **417**

period – a row of the periodic table **335**

periodic force – a repetitive force **560**

periodicity – the repeating pattern of chemical and physical properties of the elements **342**

periodic table – a chart that organizes the elements by their chemical properties and increasing atomic number **335**

period – the time it takes for each complete cycle **556**

permanent magnet – a material that retains its magnetic properties, can attract or repel other magnets, and can attract magnetic materials **528**

pH – a measure of the concentration of hydronium ions in a solution **457**

photon – the smallest possible amount of light, in the form of a wave-bundle **606**

pH scale – the pH scale goes from 1 to 14 with 1 being very acidic and 14 being very basic **457**

physical change – a change that does not result in a new substance being formed **334**

physical properties – characteristics that you can observe directly **270**

pitch – the perception of high or low that you hear at different frequencies of sound **578**

pixel – a single dot that forms part of an image made of many dots **614**

plane wave – moving waves that have crests in parallel straight lines **567**

plasma – a phase of matter in which the matter is heated to such a high temperature that some of the atoms begin to break apart **243**

polar molecule – a molecule that has a negative and a positive end or pole **437**

polyatomic ion – an ion that contains more than one atom **366**

polymer – a compound that is composed of long chains of smaller molecules **370**

polymerization – the formation of polymers by a series of addition reactions   *398*

position – a variable that tells location relative to an origin   *78*

positive, negative – the two kinds of electric charge   *472*

potential energy – energy due to position   *169*

potentiometer – a type of variable resistor that can be adjusted to give resistance within a certain range   *492*

pound – the English unit of force equal to 4.448 newtons   *109*

power – the rate of doing work or moving energy; power is equal to energy (or work) divided by time   *197*

precipitate – a solid that forms and is insoluble in a reaction mixture   *385*

precision – how close together or reproducible repeated measurements are   *20*

pressure – the amount of force exerted per unit of area   *277*

prism – a glass shape with flat, polished surfaces that can both bend and reflect light   *619*

procedure – a collection of all the techniques you use to do an experiment   *65*

product – a new substance formed in a chemical reaction   *386*

projectile – an object moving through space and affected only by gravity   *98*

proteins – a group of very large molecules made of carbon, hydrogen, oxygen, nitrogen, and sometimes sulfur   *373*

proton – a particle found in the nucleus with a positive charge exactly equal and opposite to the electron   *316*

prototype – a working model of a design that can be tested to see if it works   *70*

pure substance – matter that cannot be separated into other types of matter by physical means; includes all elements and compounds   *231*

Q

quantum theory – the theory that describes matter and energy at very small (atomic) sizes   *324*

R

radiant energy – a form of energy that is represented by the electromagnetic spectrum   *166*

radioactive – a nucleus is radioactive if it spontaneously breaks up, emitting particles or energy in the process   *320*

reactant – a starting ingredient in a chemical reaction   *386*

reaction rate – the change in concentration of reactants or products in a chemical reaction over time   *419*

reflection – the process of a wave bouncing off an object   *568*

reflection – the process of light rays bouncing off a surface; light reflects from a mirror   *621*

refraction – the process of a wave bending as it crosses a boundary between two materials   *568*

refraction – the process of bending while crossing a surface; light refracts passing from air into water or back   *621*

repeatable – describes evidence that can be seen independently by others if they repeat the same experiment or observation in the same way   *59*

resistance – determines how much current flows for a given voltage; higher resistance means less current   *486*

resistor – a component that is used to control current in many circuits   *478*

resolution – the smallest interval that can be measured   *20*

resonance – an exceptionally large amplitude that develops when a periodic force is applied at the natural frequency   *560*

restoring force – any force that always acts to pull a system back toward equilibrium   **555**

reverberation – multiple echoes of sound caused by reflections of sound building up and blending together   **588**

RGB color model – a model for tricking the eye into seeing almost any color by mixing proportions of red, green, and blue light   **614**

rhythm – a regular time pattern in a series of sounds   **593**

rotor – the rotating disk of an electric motor or generator   **540**

S

saturated – describes a solution that has as much solute as the solvent can dissolve under the conditions   **445**

saturated fat – a fat in which the carbon atoms are surrounded by as many hydrogen atoms as possible   **372**

scatterplot (or XY graph) – a graph of two variables thought to be related   **42**

scientific method – a process of learning that begins with a hypothesis and proceeds to prove or change the hypothesis by comparing it with scientific evidence   **62**

semiconductor – a material between conductor and insulator in its ability to carry current   **491**

series circuit – an electric circuit that has only one path for current   **500**

short circuit – a branch in a circuit with zero or very low resistance   **511**

significant difference – two results are only significantly different if their difference is much larger than the estimated error   **21**

significant digits – meaningful digits in a measured quantity   **18**

SI – International System of Units used by most countries for everyday measurement and used by the scientific community worldwide   **5**

simple machine – an unpowered mechanical device that accomplishes a task with only one movement   **207**

single-displacement reaction – a chemical reaction in which one element replaces a similar element in a compound   **400**

sliding friction – the friction force that resists the motion of an object moving across a surface   **118**

slope – the ratio of the rise (vertical change) to the run (horizontal change) of a line on a graph   **89**

solid – a phase of matter that holds its shape and does not flow   **240**

solubility rules – a set of rules used to predict whether an ionic compound will be soluble or insoluble in water   **452**

solubility – the amount of solute that can be dissolved in a specific volume of solvent under certain conditions   **445**

solute – any component of a solution other than the solvent   **444**

solution – a mixture of two or more substances that is homogenous at the molecular level   **442**

solvent – the component of a solution that is present in the greatest amount   **444**

sonogram – a graph that shows the frequency, amplitude, and time length for a sound   **591**

specific heat – the amount of heat needed to raise the temperature of one kilogram of a material by 1 degree Celsius   **254**

spectral line – a bright, colored line in a spectroscope   **322**

spectroscope – an instrument that separates light into a spectrum   **322**

spectrum – the characteristic colors of light given off or absorbed by an element   **322**

specular reflection – "shiny" surface reflection, where each incident ray produces only on reflected ray   **622**

speed – describes how quickly an object moves, calculated by dividing the distance traveled by the time it takes   **79**

stable – a nucleus is stable if it stays together *320*

standing wave – a wave that is confined in a space *587*

static electricity – a tiny imbalance between positive and negative charge on an object *473*

static friction – the friction force that resists the motion between two surfaces that are not moving *118*

steel – an alloy of iron and carbon *344*

strength – the ability to maintain shape under the application of force *273*

supersaturated – describes a solution with a concentration greater than the maximum solubility *450*

supersonic – a term to describe speeds faster than the speed of sound *580*

suspension – a mixture that contains particles that are greater than 1,000 nanometers *443*

switch – a device for alternately allowing and not allowing charge to flow in a circuit *479*

system – a group of variables that are related *64*

T

technology – the application of science to meet human needs and solve problems *69*

temperature – a quantity that measures the kinetic energy per molecule due to random motion *236*

tensile strength – a measure of how much stress from pulling, or tension, a material can withstand before breaking *273*

tension – a pulling force that acts in a rope, string, or other object *112*

theory – a scientific explanation supported by a lot of evidence collected over a long period of time *60*

thermal conductor – a material that allows heat to flow easily *343*

thermal energy – energy due to temperature *236*

thermal equilibrium – when two objects are at the same temperature and no heat flows *258*

thermal expansion – the tendency of the atoms or molecules in a substance (solid, liquid, or gas) to take up more space as the temperature increases *274*

thermal radiation – electromagnetic waves produced by objects because of their temperature *261*

thermometer – an instrument that measures temperature *237*

transformer – converts high-voltage electricity to lower voltage electricity *520*

translucent – allows light rays through but scatters them in all directions *620*

transparent – allows light rays to pass through without scattering. *620*

transverse wave – a wave is transverse if its oscillations are not in the direction it moves *569*

trial – each time an experiment is tried *65*

Tyndall effect – the scattering of light by the particles in a colloid *443*

U

unbalanced forces – forces that result in a net force on an object and can cause changes in motion *139*

uncertainty principle – it is impossible to know variables precisely in the quantum world *324*

unit – a fixed amount of something, like a centimeter (cm) of distance *4*

unsaturated – describes a solution with a concentration less than the maximum solubility. *450*

unsaturated fat – a fat that has fewer hydrogen atoms because double bonds exist among some of the carbon atoms   *372*

V

valence electrons – the electrons in the highest, unfilled energy level of an atom   *357*

variable – a factor that affects how an experiment works   *64*

vector – a variable that gives direction information included in its value   *81*

velocity – a variable that tells you both speed and direction   *81*

viscosity – a measure of a fluid's resistance to flow   *282*

voltage – a measure of electric potential energy   *482*

voltage drop – the difference in voltage across an electrical device that has current flowing through it   *503*

volt (V) – the measurement unit for voltage   *482*

volume – the amount of space taken up by matter   *34*

W

watt (W) – a unit of power equal to 1 joule per second   *197*

wave – a traveling oscillation that has properties of frequency, wavelength, and amplitude   *562*

wave front – the leading edge of a moving wave   *567*

wavelength – the distance from any point on a wave to the same point on the next cycle of the wave   *563*

weight – a measure of the pulling force of gravity   *32*

white light – light containing an equal mix of all colors   *606*

work – a form of energy that comes from force applied over distance; a force of 1 newton does 1 joule of work when the force causes 1 meter of motion in the direction of the force   *188*

work input – the work that is done on an object   *193*

work output – the work that an object does as a result of work input   *193*

Index

644

Index

Index

Index

Periodic Table of the Elements

GROUP

Atomic Number → 80

Standard state of matter → (droplet)

Symbol → Hg

Name → mercury

Relative Atomic Mass → 200.59

PERIOD	1	2	3	4	5	6	7	8	9
1	1 **H** hydrogen 1.01								
2	3 **Li** lithium 6.94	4 **Be** beryllium 9.01							
3	11 **Na** sodium 22.99	12 **Mg** magnesium 24.31							
4	19 **K** potassium 39.10	20 **Ca** calcium 40.08	21 **Sc** scandium 44.96	22 **Ti** titanium 47.88	23 **V** vanadium 50.94	24 **Cr** chromium 52.00	25 **Mn** manganese 54.94	26 **Fe** iron 55.85	27 **Co** cobalt 58.93
5	37 **Rb** rubidium 85.47	38 **Sr** strontium 87.62	39 **Y** yttrium 88.91	40 **Zr** zirconium 91.22	41 **Nb** niobium 92.91	42 **Mo** molybdenum 94.94	43 **Tc** technetium (98)	44 **Ru** ruthenium 101.07	45 **Rh** rhodium 102.91
6	55 **Cs** cesium 132.91	56 **Ba** barium 137.33	72 **Hf** hafnium 178.49	73 **Ta** tantalum 180.95	74 **W** tungsten 183.85	75 **Re** rhenium 186.21	76 **Os** osmium 190.23	77 **Ir** iridium 192.22
7	87 **Fr** francium (223)	88 **Ra** radium (226)	104 **Rf** rutherfordium (267)	105 **Db** dubnium (268)	106 **Sg** seaborgium (271)	107 **Bh** bohrium (270)	108 **Hs** hassium (269)	109 **Mt** meitnerium (278)

The relative atomic mass is also known as atomic weight. The values listed in this table are current with the 2016 report from the IUPAC commission.

Atomic mass values shown in curved brackets indicate the mass number of the longest-lived isotope of an element with no stable isotopes.

For updates to this table, see iupac.org.

57 **La** lanthanum 138.91	58 **Ce** cerium 140.12	59 **Pr** praseodymium 140.91	60 **Nd** neodymium 144.24	61 **Pm** promethium (145)	62 **Sm** samarium 150.36
89 **Ac** actinium (227)	90 **Th** thorium 232.04	91 **Pa** protactinium 231.04	92 **U** uranium 238.03	93 **Np** neptunium (237)	94 **Pu** plutonium (244)